Lecture Notes in Computer Science   8885

Commenced Publication in 1973
Founding and Former Series Editors:
Gerhard Goos, Juris Hartmanis, and Jan van Leeuwen

T0212814

More information about this series at http://www.springer.com/series/7410

Willi Meier · Debdeep Mukhopadhyay (Eds.)

Progress in Cryptology – INDOCRYPT 2014

15th International Conference
on Cryptology in India
New Delhi, India, December 14–17, 2014
Proceedings

Springer

Editors
Willi Meier
Fachhochschule Nordwestschweiz
Hochschule für Technik
Windisch
Switzerland

Debdeep Mukhopadhyay
Computer Science and Engineering
Indian Institute of Technology
Kharagpur
India

ISSN 0302-9743
ISBN 978-3-319-13038-5
DOI 10.1007/978-3-319-13039-2

ISSN 1611-3349 (electronic)
ISBN 978-3-319-13039-2 (eBook)

Library of Congress Control Number: 2014953958

LNCS Sublibrary: SL4 – Security and Cryptology

Springer Cham Heidelberg New York Dordrecht London

Printed on acid-free paper

Springer is part of Springer Science+Business Media (www.springer.com)

Preface

We are glad to present the proceedings of INDOCRYPT 2014, held during 14–17 December in New Delhi, India. INDOCRYPT 2014 is the 15th edition of the INDOCRYPT series organized under the aegis of the Cryptology Research Society of India (CRSI). The conference has been organized by the Scientific Analysis Group (SAG), DRDO, New Delhi, India. The INDOCRYPT series of conferences began in 2000 under the leadership of Prof. Bimal Roy of Indian Statistical Institute.

In response to the call for papers, we received 101 submissions from around 30 countries around the globe. The submission deadline was July 28, 2014. The review process was conducted in two stages: In the first stage, most papers were reviewed by at least four committee members, while papers from Program Committee members received at least five reviews. This was followed by a week-long online discussion phase to decide on the acceptance of the submissions. The Program Committee was also suitably aided in this tedious task by 94 external reviewers to be able to complete this as per schedule, which was on September 7. Finally, 25 submissions were selected for presentation at the conference.

We would like to thank the Program Committee members and the external reviewers for giving every paper a fair assessment in such a short time. The refereeing process resulted in 367 reviews, along with several comments during the discussion phase. The authors had to revise their papers according to the suggestions of the referees and submit the camera-ready versions by September 22.

We were delighted that Phillip Rogaway, Marc Joye, and María Naya-Plasencia agreed to deliver invited talks on several interesting topics of relevance to INDOCRYPT. The program was also enriched to have Claude Carlet and Florian Mendel as Tutorial speakers on important areas of Cryptography, to make the conference program complete.

We would like to thank the General Chairs, Dr. G. Athithan and Dr. P.K. Saxena, for their advice and for being a prime motivator. We would also like to specially thank the Organizing Chair Saibal K. Pal and the Organizing Secretary Sucheta Chakrabarty for developing the layout of the program and in managing the financial support required for such a conference. Our job as Program Chairs was indeed made much easier by the software, easychair. We also say our thanks to Durga Prasad for maintaining the webpage for the conference. We would also acknowledge Springer for their active cooperation and timely production of the proceedings.

Last but certainly not least, our thanks go to all the authors, who submitted papers to INDOCRYPT 2014, and all the attendees. Without your support the conference would not be a success.

December 2014   Willi Meier
                                                                                      Debdeep Mukhopadhyay

Message from the General Chairs

Commencing from the year 2000, INDOCRYPT — the International Conference on Cryptology — is held every year in India. This event has been one of the regular activities of the Cryptology Research Society of India (CRSI) to promote R&D in the area of Cryptology in the country. The conference is hosted by different organizations including Academic as well as R&D organizations located across the country. The Scientific Analysis Group (SAG), one of the research laboratories of the Defence Research and Development Organization (DRDO), organized the conference in the years 2003 and 2009 in collaboration with the Indian Statistical Institute (Delhi Centre) and Delhi University, respectively. SAG was privileged to get an opportunity to organize INDOCRYPT 2014, the 15th conference in this series. Since its inception, the INDOCRYPT has proved to be a powerful platform for researchers to meet, share their ideas with their peers, and work toward the growth f cryptology, especially in India. For each edition of the conference in the past, the response from the cryptology research community has been overwhelming and the esponse for the current edition is no exception. As is evident from the quality of submissions and the a high rate of rejections due to a transparent and rigorous process of reviewing, the conference has been keeping its standards with proceedings published by LNCS. Even this year, the final set of selected papers amount to a net acceptance ratio of 25 percent.

On the first day of the conference, there were two Tutorials on the topics of S-Boxes and Hash Functions. They were delivered by Claude Carlet of University of Paris, France and Florian Mendel of Graz University of Technology, Austria. Both the Tutorials provided the participants with deep understanding of the chosen topics and stimulated discussions among others. Beginning from the second day, the main conference had three invited talks and 25 paper presentations for 3 days. Maria Naya-Plasencia of Inria (France), Marc Joye of Technicolor (USA), and Phillip Rogaway of University of California (USA) delivered the invited talks on Lightweight Block Ciphers and Their Security, Recent Advances in ID-Based Encryption, and Advances in Authenticated Encryption, respectively. We are grateful to all the Invited and Tutorial Speakers.

Organizing a conference having such wide ranging involvement and participation from international crypto community is not possible without the dedicated efforts of different committees drawn from the hosting and other support agencies. The Organizing Committee took care of all the logistic, coordination, and financial aspects concerning the conference under the guidance of the Organizing Chair Saibal K. Pal and the Organizing Secretary Sucheta Chakrabarty. We thank both of them and all the members of these committees for their stellar efforts.

Equally demanding is the task of the Program Committee in coordinating the submissions and in selecting the papers for presentation. The Program Co-Chairs Willi Meier and Debdeep Mukhopadhyay were the guiding forces behind the efforts of the Program Committee. Their love for the subject and the commitment to the cause of promoting Cryptology Research in India and elsewhere is deep and we thank them for putting together an excellent technical program. We also thank all the members of the Program Committee for their support to the Program Co-chairs. Special thanks are due to the Reviewers for their efforts and for sharing their comments with concerned persons, which led to completing the selection process in time.

We express our heartfelt thanks to DRDO and CRSI for being the mainstay in ensuring that the Conference received all the support that it needed. We also thank NBHM, DST, Deity, ISRO, CSIR, RBI, BEL, ITI, IDRBT, Microsoft, Google, TCS, and others for generously supporting/sponsoring the event. Finally, thanks are due to the authors who submitted their work, especially to those whose papers are included in the present Proceedings of INDOCRYPT 2014 and those who could make it to present their papers personally in the Conference.

December 2014   P.K. Saxena
                                                                G. Athithan

Organization

General Chairs

P.K. Saxena   SAG, DRDO, New Delhi, India
G. Athithan   SAG, DRDO, New Delhi, India

Program Chairs

Willi Meier   FHNW, Switzerland
Debdeep Mukhopadhyay   Indian Institute of Technology Kharagpur, India

Program Committee

Martin Albrecht   Technical University of Denmark, Denmark
Subidh Ali   NYU, Abu Dhabi
Elena Andreeva   KU Leuven, Belgium
Frederik Armknecht   Universität Mannheim, Germany
Daniel J. Bernstein   University of Illinois at Chicago, USA
Céline Blondeau   Aalto University School of Science, Finland
Christina Boura   Université de Versailles Saint-Quentin-en-
                          Yvelines, France
C. Pandurangan   Indian Institute of Technology Madras, India
Anne Canteaut   Inria, France
Nishanth Chandran   Microsoft Research, India
Sanjit Chatterjee   Indian Institute of Science Bangalore, India
Abhijit Das   Indian Institute of Technology Kharagpur, India
Sylvain Guilley   TELECOM-ParisTech and Secure-IC S.A.S.,
                          France
Abhishek Jain   MIT and BU, USA
Dmitry Khovratovich   University of Luxembourg, Luxembourg
Tanja Lange   Technische Universiteit Eindhoven,
                          The Netherlands
Willi Meier   FHNW, Switzerland
Debdeep Mukhopadhyay   Indian Institute of Technology Kharagpur, India
David Naccache   Université Paris II, Panthéon-Assas, France
Phuong Ha Nguyen   Indian Institute of Technology Kharagpur, India
Saibal K. Pal   SAG, DRDO, New Delhi, India
Goutam Paul   Indian Statistical Institute Kolkata, India
Christiane Peters   ENCS, The Netherlands

Thomas Peyrin	Nanyang Technological University, Singapore
Josef Pieprzyk	ACAC, Australia
Rajesh Pillai	SAG, DRDO, New Delhi, India
Axel Poschmann	NXP Semiconductors, Germany
Bart Preneel	KU Leuven, Belgium
Chester Rebeiro	Columbia University, USA
Vincent Rijmen	KU Leuven and iMinds, Belgium
Bimal Roy	Indian Statistical Institute, Kolkata, India
Dipanwita Roy Chowdhury	Indian Institute of Technology Kharagpur, India
S.S. Bedi	SAG, DRDO, New Delhi, India
Sourav Sen Gupta	Indian Statistical Institute, Kolkata, India
Francois-Xavier Standaert	UCL Crypto Group, Belgium
Ingrid Verbauwhede	KU Leuven, Belgium

External Reviewers

Tamaghna Acharya	Gregor Leander
Ansuman Banerjee	Wang Lei
Ayan Banerjee	Feng-Hao Liu
Harry Bartlett	Atul Luykx
Begül Bilgin	Subhamoy Maitra
Joppe Bos	Bodhisatwa Mazumdar
Seyit Camtepe	Florian Mendel
Sucheta Chakrabarti	Bart Mennink
Avik Chakraborti	Nele Mentens
Kaushik Chakraborty	Prasanna Mishra
Anupam Chattopadhyay	Paweł Morawiecki
Roopika Chaudhary	Imon Mukherjee
Chien-Ning Chen	Nicky Mouha
Kang Lang Chiew	Michael Naehrig
Dhananjoy Dey	Ivica Nikolić
Manish Kant Dubey	Ventzi Nikov
Pooya Farshim	Omkant Pandey
Aurélien Francillon	Sumit Pandey
Lubos Gaspar	Tapas Pandit
Benoît Gérard	Kenny Paterson
Hossein Ghodosi	Arpita Patra
Santosh Ghosh	Ludovic Perret
Shamit Ghosh	Léo Perrin
Vincent Grosso	Christophe Petit
Divya Gupta	Bertram Poettering
Indivar Gupta	Romain Poussier
Nupur Gupta	Michaël Quisquater
Jian Guo	Francesco Regazzoni
Sartaj Ul Hasan	Michał Ren

Yj Huang	Dhiman Saha
Andreas Hülsing	Abhrajit Sengupta
Hassan Jameel Asghar	Sujoy Sinha Roy
Kimmo Järvinen	Dale Sibborn
Jeremy Jean	Dave Singelee
Bhavana Kanukurthi	Ron Steinfeld
Sabyasachi Karati	Valentin Suder
Pierre Karpman	Aris Tentes
Oleksandr Kazymyrov	Tyge Tiessen
Manas Khatua	Meilof Veeningen
Dakshita Khurana	Muthuramakrishnan Venkitasubramaniam
Markulf Kohlweiss	Frederik Vercauteren
Sukhendu Kuila	Dhinakaran Vinayagamurthy
Manoj Kumar	Jo Vliegen
Vijay Kumar	Qingju Wang
Yogesh Kuma	Bohan Yang
Mario Lamberger	Wentao Zhang
Martin M. Lauridsen	Ralf Zimmermann

Local Organizing Committee

Saibal K. Pal (Organizing Chair)	Sucheta Chakrabarti (Organizing Secretary)
Kanika Bhagchandani	Vivek Devdhar
Dhananjoy Dey	Indivar Gupta
Sartaj Ul Hasan	Mohammad Javed
Gopal C. Kandpal	Sarvjeet Kaur
Ashok Kumar	Girish Mishra
P.R. Mishra	Bhartendu Nandan
Manoj Kumar Singh	Ajay Srivastava
Divya Anand Subba	

Invited Talks

S-boxes, Their Computation and Their Protection against Side Channel Attacks

Claude Carlet*

First Part of the Talk

After recalling the necessary background on S-boxes (see below), we shall study the criteria for substitution boxes (S-boxes) in block ciphers:

1. *bijectivity when used in SP networks, and if possible balancedness when used in Feistel ciphers,*
2. *high nonlinearity (for the resistance to linear attacks),*
3. *low differential uniformity (for the resistance to differential attacks),*
4. *not low algebraic degree (for resisting higher order differential attacks).*

We shall give the main properties of APN functions ((n, n)-functions having the best possible differential uniformity) and AB functions ((n, n)-functions having the best possible nonlinearity, which are APN).

Second Part of the Talk

We shall list the main known AB, APN, and differentially 4-uniform functions. These functions are defined within the structure of the finite field $\mathbb{F}_{2^n}$. We shall address the question of their implementation.

Satisfying the criteria 1-4 above is not sufficient for an S-box. It needs also to be fastly computable, for two reasons: (1) it is not always possible to use a look-up-table for implementing it, (2) the condition of being fastly computable more or less coincides with the constraint of allowing counter-measures to side-channel attacks (SCA) with minimized cost. The implementation of cryptographic algorithms in devices like smart cards, FPGA or ASIC leaks information on the secret data, leading to very powerful SCA if countermeasures are not included. Such counter-measures are costly in terms of running time and of memory when they need to resist higher order SCA. The most commonly used counter-measure is masking. We shall describe how an S-box can be protected with this counter-measure with minimized cost.

* LAGA, Universities of Paris 8 and Paris 13, CNRS; Address: Department of Mathematics, University of Paris 8, 2 rue de la liberté, 93526 Saint-Denis Cedex, France; e-mail: claude.carlet@univ-paris8.fr.

Background

Let n and m be two positive integers. The functions from $\mathbb{F}_2^n$ to $\mathbb{F}_2^m$ are called (n, m)-*functions*. Such function F being given, the Boolean functions $f_1, \ldots, f_m$ defined by $F(x) = (f_1(x), \ldots, f_m(x))$, are called the *coordinate functions* of F. The linear combinations of these coordinate functions, with non-all-zero coefficients, are called the *component functions* of F. When the numbers m and n are not specified, (n, m)-functions can be called *vectorial Boolean functions* and in cryptography we use the term of *S-boxes*.

The *Walsh transform* of an (n, m)-function F maps any ordered pair $(u, v) \in \mathbb{F}_2^n \times \mathbb{F}_2^m$ to the sum (calculated in $\mathbb{Z}$): $\sum_{x \in \mathbb{F}_2^n} (-1)^{v \cdot F(x) + u \cdot x}$, where the same symbol "$\cdot$" is used to denote inner products in $\mathbb{F}_2^n$ and $\mathbb{F}_2^m$. Note that the function $v \cdot F$ is a component function of F when $v \neq 0$. The *Walsh spectrum* of F is the multi-set of all the values of the Walsh transform of F, for $u \in \mathbb{F}_2^n, v \in \mathbb{F}_2^{m*}$ (where $\mathbb{F}_2^{m*} = \mathbb{F}_2^m \setminus \{0\}$). We call *extended Walsh spectrum* of F the multi-set of their absolute values.

The *algebraic normal form* (ANF) of any (n, m)-function F:

$$\sum_{I \subseteq \{1, \cdots, n\}} a_I \left(\prod_{i \in I} x_i \right) ; \ a_I \in \mathbb{F}_2^m \tag{1}$$

(this sum being calculated in $\mathbb{F}_2^m$) exists and is unique and satisfies the relation $a_I = \sum_{x \in \mathbb{F}_2^n / \ supp(x) \subseteq I} F(x)$; conversely, we have $F(x) = \sum_{I \subseteq supp(x)} a_I$.

The *algebraic degree* of the function is by definition the global degree of its ANF. It is a right and left *affine invariant* (that is, it does not change when we compose F by affine automorphisms). Vectorial functions for cryptography have better not too low algebraic degrees, to withstand higher order differential attacks.

A second representation of (n, m)-functions exists when $m = n$: we endow $\mathbb{F}_2^n$ with the structure of the field $\mathbb{F}_{2^n}$; any (n, n)-function F then admits a unique *univariate polynomial representation* over $\mathbb{F}_{2^n}$, of degree at most $2^n - 1$:

$$F(x) = \sum_{j=0}^{2^n - 1} b_j x^j , \ b_j \in \mathbb{F}_{2^n} . \tag{2}$$

We denote by $w_2(j)$ the number of nonzero coefficients j_s in the binary expansion $\sum_{s=0}^{n-1} j_s 2^s$ of j, i.e. $w_2(j) = \sum_{s=0}^{n-1} j_s$ and call it the *2-weight* of j. Then, the function F has algebraic degree $\max_{j=0,\ldots,2^n-1/ \ b_j \neq 0} w_2(j)$. If m is a divisor of n, then any (n, m)-function F can be viewed as a function from $\mathbb{F}_{2^n}$ to itself, since $\mathbb{F}_{2^m}$ is a sub-field of $\mathbb{F}_{2^n}$. Hence, the function admits a univariate polynomial representation, which can be represented in the form $tr_{n/m}(\sum_{j=0}^{2^n-1} b_j x^j)$, where $tr_{n/m}(x) = x + x^{2^m} + x^{2^{2m}} + x^{2^{3m}} + \cdots + x^{2^{n-m}}$ is the trace function from $\mathbb{F}_{2^n}$ to $\mathbb{F}_{2^m}$.

An (n, m)-function F is *balanced* (i.e. takes every value of $\mathbb{F}_2^m$ the same number 2^{n-m} of times) if and only if its component functions are balanced (i.e. have Hamming weight 2^{n-1}).

The *nonlinearity* $nl(F)$ of an (n, m)-function F is the minimum Hamming distance between all the component functions of F and all affine functions on n variables and quantifies the level of resistance of the S-box to the linear attack. We have:

$$nl(F) = 2^{n-1} - \frac{1}{2} \max_{v \in \mathbb{F}_2^{m*}; \, u \in \mathbb{F}_2^n} \left| \sum_{x \in \mathbb{F}_2^n} (-1)^{v \cdot F(x) + u \cdot x} \right|. \tag{3}$$

The two main known upper bounds on the nonlinearity are:
- the *covering radius bound*:

$$nl(F) \leq 2^{n-1} - 2^{n/2-1}$$

which is tight for n even and $m \leq n/2$ (the functions achieving it with equality are called *bent*);
- the *Sidelnikov-Chabaud-Vaudenay bound*, valid only for $m \geq n - 1$:

$$nl(F) \leq 2^{n-1} - \frac{1}{2}\sqrt{3 \times 2^n - 2 - 2\frac{(2^n - 1)(2^{n-1} - 1)}{2^m - 1}}$$

which equals the covering radius bound when $m = n - 1$ and is strictly better when $m \geq n$. It is tight only for $m = n$ (in which case it states that $nl(F) \leq 2^{n-1} - 2^{\frac{n-1}{2}}$), with n odd (the functions achieving it with equality are called *almost bent* AB).

An (n, m) function is bent if and only if all its *derivatives* $D_a F(x) = F(x) + F(x + a)$, $a \in \mathbb{F}_2^*$, are balanced. For this reason, bent functions are also called *perfect nonlinear* PN. According to Chabaud-Vaudenay's proof of the Sidelnikov-Chabaud-Vaudenay bound, any AB function is *almost perfect nonlinear* APN, that is, all its derivatives $D_a F$, $a \in \mathbb{F}_2^{n*}$, are 2-to-1 (every element of $\mathbb{F}_2^n$ has 0 or 2 pre-images by $D_a F$). Such functions, whose notion has been studied by Nyberg, contribute to an optimal resistance to the differential attack . More generally, F is called *differentially δ-uniform* if the equation $D_a F(x) = b$ has at most δ solutions, for every nonzero a and every b.

The nonlinearity and the δ-uniformity are invariant under affine, extended affine and CCZ equivalences (in increasing order of generality). Two functions are called *affine equivalent* if one is equal to the other, composed on the left and on the right by affine permutations. They are called *extended affine equivalent* (EA-equivalent) if one is affine equivalent to the other, added with an affine function. They are called *CCZ-equivalent* if their graphs $\{(x, y) \in \mathbb{F}_2^n \times \mathbb{F}_2^n \, | \, y = F(x)\}$ and $\{(x, y) \in \mathbb{F}_2^n \times \mathbb{F}_2^n \, | \, y = G(x)\}$ are affine equivalent, that is, if there exists an affine automorphism $L = (L_1, L_2)$ of $\mathbb{F}_2^n \times \mathbb{F}_2^n$ such that $y = F(x) \Leftrightarrow L_2(x, y) = G(L_1(x, y))$.

Cryptanalysis of Hash Functions

Florian Mendel

Graz University of Technology, Austria

Abstract. This extended abstract briefly summarizes a talk with the same title and gives literature pointers. In particular, we discuss recent advances in the cryptanalysis of ARX- and AES-based hash functions.

Overview

In the last few years, the cryptanalysis of hash functions has become an important topic within the cryptographic community. Especially the collision attacks on the MD4 family of hash functions (MD5, SHA-1) have weakened the security assumptions of these commonly used hash functions [17, 18]. As a consequence, NIST decided to organize a public competition in order to design a new hash function, which lead to the selection of Keccak as SHA-3 in 2012. In this talk, we discuss some recent advances in the cryptanalysis of hash functions. First, we will review the collision attacks of Wang et al. on the MD4 family and discuss the limitations of the techniques when applied to more complex functions such as the SHA-2 family. Due to the more complex structure of SHA-2 (compared to SHA-1 and MD5), several new challenges arise for the cryptanalyst. We show how to overcome these difficulties and present an automatic approach to construct complex differential characteristics and thus collisions for round-reduced SHA-2 with practical complexity [2, 10, 12]. The same techniques and tools also lead to new collision attacks on the Korean hash function standard HAS-160 [9] and the Chinese hash function standard SM3 [11], among others [6, 8, 13].

While the first part of the talk focuses on the analysis of the MD4 family and similar hash functions, the second part is dedicated to the analysis of AES-based hash functions. In the course of the SHA-3 competition, several advances have been made in the cryptanalysis of AES-based hash functions. In particular, several of the SHA-3 candidates turned out to be susceptible to the rebound attack [14], a new cryptanalytic technique that was introduced during the design of the SHA-3 finalist Grøstl. In the last years, the rebound attack and its extensions [3, 4, 7, 15] have become one of the most important tools for analyzing the security of AES-based hash functions. Even though the rebound attack was originally conceived to attack AES-based hash functions as well as their building blocks, it was later shown to also be applicable to other designs, including the SHA-3 finalists JH [16], Skein [5] and Keccak [1].

Finally, we will discuss directions of future work and open research problems at the end of this talk.

References

1. Duc, A., Guo, J., Peyrin, T., Wei, L.: Unaligned rebound attack: Application to keccak. In: Canteaut, A. (ed.) FSE 2012. LNCS, vol. 7549, pp. 402–421. Springer, Heidelberg (2012)
2. Eichlseder, M., Mendel, F., Schäffer, M.: Branching Heuristics in Di erential Collision Search with Applications to SHA-512. IACR Cryptology ePrint Archive 2014, 302 (2014)
3. Gilbert, H., Peyrin, T.: Super-sbox cryptanalysis: Improved attacks for AES-like permutations. In: Hong, S., Iwata, T. (eds.) FSE 2010. LNCS, vol. 6147, pp. 365–383. Springer, Heidelberg (2010)
4. Jean, J., Naya-Plasencia, M., Peyrin, T.: Improved Cryptanalysis of AES-like Permutations. J. Cryptology 27(4), 772–798 (2014)
5. Khovratovich, D., Nikolic, I., Rechberger, C.: Rotational Rebound Attacks on Reduced Skein. J. Cryptology 27(3), 452–479 (2014)
6. Kölbl, S., Mendel, F., Nad, T., Schläffer, M.: Differential cryptanalysis of keccak variants. In: Stam, M. (ed.) IMACC 2013. LNCS, vol. 8308, pp. 141–157. Springer, Heidelberg (2013)
7. Lamberger, M., Mendel, F., Rechberger, C., Rijmen, V.: Schäffer, M.: The Rebound Attack and Subspace Distinguishers: Application to Whirlpool. J. Cryptology (2013)
8. Mendel, F., Nad, T., Scherz, S., Schläffer, M.: Differential attacks on reduced RIPEMD-160. In: Gollmann, D., Freiling, F.C. (eds.) ISC 2012. LNCS, vol. 7483, pp. 23–38. Springer, Heidelberg (2012)
9. Mendel, F., Nad, T., Schläffer, M.: Cryptanalysis of round-reduced HAS-160. In: Kim, H. (ed.) ICISC 2011. LNCS, vol. 7259, pp. 33–47. Springer, Heidelberg (2012)
10. Mendel, F., Nad, T., Schläffer, M.: Finding SHA-2 characteristics: Searching through a minefield of contradictions. In: Lee, D.H., Wang, X. (eds.) ASIACRYPT 2011. LNCS, vol. 7073, pp. 288–307. Springer, Heidelberg (2011)
11. Mendel, F., Nad, T., Schläffer, M.: Finding collisions for round-reduced SM3. In: Dawson, E. (ed.) CT-RSA 2013. LNCS, vol. 7779, pp. 174–188. Springer, Heidelberg (2013)
12. Mendel, F., Nad, T., Schläffer, M.: Improving local collisions: New attacks on reduced SHA-256. In: Johansson, T., Nguyen, P.Q. (eds.) EUROCRYPT 2013. LNCS, vol. 7881, pp. 262–278. Springer, Heidelberg (2013)
13. Mendel, F., Peyrin, T., Schläffer, M., Wang, L., Wu, S.: Improved cryptanalysis of reduced RIPEMD-160. In: Sako, K., Sarkar, P. (eds.) ASIACRYPT 2013, Part II. LNCS, vol. 8270, pp. 484–503. Springer, Heidelberg (2013)
14. Mendel, F., Rechberger, C., Schläffer, M., Thomsen, S.S.: The rebound attack: Cryptanalysis of reduced whirlpool and grøstl. In: Dunkelman, O. (ed.) FSE 2009. LNCS, vol. 5665, pp. 260–276. Springer, Heidelberg (2009)
15. Naya-Plasencia, M.: How to improve rebound attacks. In: Rogaway, P. (ed.) CRYPTO 2011. LNCS, vol. 6841, pp. 188–205. Springer, Heidelberg (2011)
16. Naya-Plasencia, M., Toz, D., Varici, K.: Rebound attack on JH42. In: Lee, D.H., Wang, X. (eds.) ASIACRYPT 2011. LNCS, vol. 7073, pp. 252–269. Springer, Heidelberg (2011)
17. Wang, X., Yin, Y.L., Yu, H.: Finding collisions in the full SHA-1. In: Shoup, V. (ed.) CRYPTO 2005. LNCS, vol. 3621, pp. 17–36. Springer, Heidelberg (2005)
18. Wang, X., Yu, H.: How to break MD5 and other hash functions. In: Cramer, R. (ed.) EUROCRYPT 2005. LNCS, vol. 3494, pp. 19–35. Springer, Heidelberg (2005)

On Lightweight Block Ciphers and Their Security

María Naya-Plasencia

Inria, France
Maria.Naya_Plasencia@inria.fr

Abstract. In order to answer the requirements raised by a large number of applications, like RFID or sensor networks, the design of lightweight primitives has become a major interest of the cryptographic community. A (very) large number of lightweight block ciphers have been proposed. Correctly evaluating their security has become a primordial task requiring the attention of our community. In this talk we will make a survey of these proposed ciphers, some of the proposed cryptanalysis and their actual status. We will also try to provide links between some of these ciphers/attacks and the SHA-3 competition.

Keywords: lightweight block ciphers · cryptanalysis.

Recent Advances in ID-Based Encryption

Marc Joye

Technicolor, USA
marc.joye@technicolor.com

Abstract. Most ID-based cryptosystems make use of bilinear maps. A notable exception is a 2001 publication by Clifford Cocks describing an ID-based cryptosystem that works in standard RSA groups. Its semantic security relies on the quadratic residuosity assumption. Cocks's publication gave rise to several follow-up works aiming at improving the original scheme in multiple directions. This talk reviews Cocks' scheme and presents its known variants and extensions. It also discusses applications thereof. Finally it reports some recent developments the author made in the area.

Contents

Elliptic Curves

Side Channel Analysis - I

Side-Channel Analysis on Blinded Regular Scalar Multiplications

Benoit Feix[1], Mylène Roussellet[2], and Alexandre Venelli[3][(✉)]

[1] UL Security Transactions, UK Security Lab, Basingstoke, UK
benoit.feix@ul.com
[2] Gemalto, La Ciotat, France
mylene.roussellet@gemalto.com
[3] Thalès Communications and Security, Toulouse, France
alexandre.venelli@thalesgroup.com

Abstract. We present a new side-channel attack path threatening state-of-the-art protected implementations of elliptic curves embedded scalar multiplications. Regular algorithms such as the double-and-add-always and the Montgomery ladder are commonly used to protect the scalar multiplication from simple side-channel analysis. Combining such algorithms with scalar and/or point blinding countermeasures lead to scalar multiplications protected from all known attacks. Scalar randomization, which consists in adding a random multiple of the group order to the scalar value, is a popular countermeasure due to its efficiency. Amongst the several curves defined for usage in elliptic curves products, the most used are those standardized by the NIST. As observed in several previous publications, the modulus, hence the orders, of these curves are sparse, primarily for efficiency reasons. In this paper, we take advantage of this specificity to present new attack paths which combine vertical and horizontal side-channel attacks to recover the entire secret scalar in state-of-the-art protected elliptic curve implementations.

Keywords: Elliptic crves · Scalar multiplication · Side-channel analysis · Correlation analysis

1 Introduction

Elliptic Curve Cryptography (ECC) has become a very promising branch of cryptology. Since its introduction by Miller [25] and Koblitz [22] numerous studies have offered a rich variety of implementation methods to perform efficient and tamper resistant scalar multiplication algorithms in embedded products. Many standardized protocols like the *Elliptic Curve Digital Signature Algorithm* (ECDSA) [29] or the *Elliptic Curve Diffie-Hellman* (ECDH) are more and more used in payment and identity products. They have the strong advantage today to require significantly smaller parameters and key sizes than the well-known

---

Venelli: This work was carried out when the author was with INSIDE Secure.

© Springer International Publishing Switzerland 2014
W. Meier and D. Mukhopadhyay (Eds.): INDOCRYPT 2014, LNCS 8885, pp. 3–20, 2014.
DOI: 10.1007/978-3-319-13039-2_1

RSA [30] and Diffie-Hellman [15] cryptosystems. Most industrial ECC applications use elliptic curves defined in international standards [5,29,32]. These curves were generated with efficiency and security advantages for different classical security levels.

Besides these efficiency requirements in embedded environment, developers must also prevent their products from physical attacks. These techniques are split in two categories namely the *Side-Channel Analysis* (SCA) and the *Fault Analysis* (FA). In this paper, we use the full spectrum of *Side-Channel Analysis* namely classical *Vertical Correlation attacks* [7], *Horizontal Correlation attacks* [12], *Vertical Collision-Correlation* [27,38] and *Horizontal Collision-Correlation* [1,13].

A recent paper at Indocrypt 2013 from Bauer *et al.* [2] presented a new side-channel attack, combining vertical and horizontal techniques, on a standard RSA blinded exponentiation when the public exponent value is 3. Based on the same observation, we design new side-channel attack paths on regular scalar multiplication algorithms with blinded scalar implementations for most standardized curves. We present vertical and horizontal attacks with known and unknown input point values that successfully recover the whole secret scalar.

Our Proposed Attack Strategy. Our attack paths consist of three steps. First, the attacker uses the fact that the scalar blinding does not mask a large part of the secret. This side-channel vulnerability can be exploited vertically, *i.e.* using several execution traces. The attacker will recover the middle part of the secret. In a second step, he needs to recover the random value used for each scalar blinding. This part is performed horizontally, *i.e.* each random will be recovered using only one trace. The already recovered part of the secret in the first step can provide more side-channel information to exploit for the attacker. This step allows to recover the most significant part of the scalar. Finally, the third step consists in retrieving the least significant part of the scalar. Using the already recovered random values of each traces and the middle part of the secret, the attacker can perform a vertical attack.

Roadmap. The paper is organized as follows. Section 2 reminds basics on elliptic curve cryptography and embedded scalar multiplication. We also detail the classical side-channel countermeasures and explain the side-channel attack knowledge necessary for a good understanding of the rest of the paper. In Section 3, we describe our first attack that defeats a regular implementation when the secret scalar is blinded but not the input point. Section 4 extends our attack techniques to the unknown (or randomized) input point case. To illustrate our attacks efficiency, we present experimental results on simulated side-channel traces in Section 5. Discussion on countermeasures is done in Section 6. We finally conclude our paper in Section 7.

2 Preliminaries

2.1 Background on Elliptic Curves

Let $\mathbb{F}_p$ be a finite field of characteristic $\neq 2, 3$. Consider an elliptic curve E over $\mathbb{F}_p$ given by the short Weierstraß equation $y^2 = x^3 + ax + b$, where $a, b \in \mathbb{F}_p$ and with discriminant $\Delta = -16(4a^3 + 27b^2) \neq 0$. The set of points on an elliptic curve form a group under the chord-and-tangent law. The neutral element is the point at infinity O. Let $P = (x_1, y_1)$ and $Q = (x_2, y_2)$ be two affine points on $E(\mathbb{F}_p)$, their sum $R = P + Q = (x_3, y_3)$ belongs also to the curve. Generally on elliptic curves, the operation $P + P$, called doubling, has different complexity compared to the addition $P + Q$ with $Q \neq P$.

In practice, it is advantageous to use Jacobian coordinates in order to avoid inverses in $\mathbb{F}_p$. An affine point (x, y) is represented by a triplet $(X : Y : Z)$ such that $x = X/Z^2$ and $y = Y/Z^3$.

Let $n = \#E(\mathbb{F}_p)$ be the cardinality of the group of points $E(\mathbb{F}_p)$. Hasse's theorem states that n is close to p and bounded by: $(\sqrt{p} - 1)^2 \leq n \leq (\sqrt{p} + 1)^2$.

Given a point $P \in E(\mathbb{F}_p)$ and a scalar $d \in \mathbb{N}^*$, we note $[d]P$ the scalar multiplication of P by d. The scalar multiplication is the fundamental operation in most cryptographic algorithms that use elliptic curve arithmetic. In most protocols, the scalar is considered secret and the point public[1].

In the industry, elliptic curve cryptosystems are generally implemented using elliptic curves from standards such as the NIST FIPS186-2 [29], SEC2 [32] or recently generated curves by Bernstein and Lange [5]. All these curves are specified using both efficiency and security criteria. A classic efficiency criterion consists in choosing a special prime, *i.e.* Generalised Mersenne Numbers (GMN) [34], for the finite field $\mathbb{F}_p$. Those primes are sparse, *i.e.* they contain long patterns of zeros or ones, hence due to Hasse's theorem, the orders of the elliptic curves defined over those fields are also sparse.

2.2 Side-Channel Attacks Background

Side-channel analysis, also referred as *Passive Attacks*, was introduced by Kocher *et al.* in [23,24]. SCA regroups several different techniques. *Simple Side-Channel Analysis* (SSCA) exploits a single execution trace to recover the secret whereas *Differential Side-Channel Analysis* (DSCA) performs statistical treatment on several (possibly millions) traces.

Elliptic curves implementations have been subject to various side-channel attack paths. The simplest one uses SSCA. The attacker's objective is to distinguish a doubling from an addition operation using a single side-channel trace execution.

The principle of the classical DSCA on elliptic curve consists in guessing bit-per-bit (or w-bit per w-bit) the secret scalar and knowing the input point

---

[1] The problematic is different in pairing-based cryptography where the scalar is generally public and the point secret. We only consider here classic ECC protocols.

manipulated by the implementation. The attacker then recomputes an intermediate guessed value of the algorithm to validate the right guess with a statistical treatment applied to many side-channel execution traces [7,24]. A recent classification of attacks has categorized all these statistical attacks as *Vertical Analysis*. Indeed, these techniques combine a single time sample t on many side-channel traces to perform the analysis leading to the recovery of the secret data manipulated at this instant t.

Another class of side-channel attack, the *Horizontal Analysis*, has been presented by Clavier *et al.* [12], inspired by the Big Mac attack from Walter [37]. The technique has been later derived to present horizontal attacks on elliptic curves implementations by Hanley *et al.* [19] and Bauer *et al.* [1].

Correlation Analysis. Let $\mathcal{C}^{(i)}$ with $1 \leq i \leq N$ be a set of N side-channel traces captured from a device processing the targeted computations with input value $X^{(i)}$ whose processing occurs at time sample t with l the number of points acquired at time sample t. We consider $\Theta_0 = \{\mathcal{C}^{(1)}(t), \ldots, \mathcal{C}^{(N)}(t)\}$. We denote $S^{(i)}$ with $1 \leq i \leq N$ a set of N guessed intermediate sensible values based on a power model, which is generally linear in the Hamming weight of the data. Let $f(X^{(i)}, \hat{d})$ be a function of the input values $X^{(i)}$ and (a part of) the targeted guessed secret $\hat{d}$. All l points in the leakage trace are equal to this value $f(X^{(i)}, \hat{d})$ for the time sample t. We then consider $\Theta_1 = \{S^{(1)}, \ldots, S^{(N)}\}$. The objective is to evaluate the dependency between both sets Θ_0 and Θ_1 using the Bravais-Pearson correlation factor $\rho(\Theta_0, \Theta_1)$. The correlation value between both series is equal to 1 when the simulated model perfectly matches the measured power traces. It then indicates that the guess on the secret corresponds to the correct key value handled by the device in the computations.

Collision-Correlation Analysis. Correlation can also be used to determine the dependency between different time samples of the same side-channel trace. It will then allow the attacker to detect internal side-channel collisions at two different time samples t_0 and t_1. In this case, the term *collision-correlation* is used. The correlation is applied between the sets $\Theta_0 = \{\mathcal{C}^{(1)}(t_0), \ldots, \mathcal{C}^{(N)}(t_0)\}$ and $\Theta_1 = \{\mathcal{C}^{(1)}(t_1), \ldots, \mathcal{C}^{(N)}(t_1)\}$ where both sets correspond to points of the same side-channel trace taken at different time sample t_0 and t_1. We can expect a maximum correlation value when the same data is processed in the device at the time samples t_0 and t_1. If the attacker can then find a link between this information and the use of the secret, he can recover some information on the secret's value.

2.3   Side-Channel Resistant Scalar Multiplication

On embedded devices, a scalar multiplication needs to be protected against both *Simple Side-Channel Analysis* (SSCA) and *Differential Side-Channel Analysis* (DSCA). To resist SSCA, an attacker should not be able to distinguish an addition from a doubling operation. The main categories of countermeasures are:

- **Regular multiplication algorithms** – Specific scalar multiplication algorithms have been proposed such that they always compute a regular sequence of elliptic curve operations regardless of the value of the secret bits. The double-and-add-always [14] (see Alg. 1), the Montgomery ladder [21,26] or Joye's double-add [20] are the most well-known examples of regular algorithms. The recently proposed co-Z scalar algorithms [18] are one of the most efficient regular algorithms for ECC over $\mathbb{F}_p$.
- **Unified addition formulæ** – The same formula is used to compute both an addition and a doubling [35].
- **Atomic block** – The addition and doubling operations can be expressed such that the same sequence of field operations are performed. Propositions on the subject are numerous in the literature [10,17,31].

The resistance against DSCA can be achieved by using a combination of the following classic countermeasures:

- **Scalar blinding** [14] – We can add a random multiple of the order n of the group $E(\mathbb{F}_p)$ to the scalar d. This alters the representation of d without changing the output of the scalar multiplication. The blinded scalar d' is defined as $d' = d + r.n$ for a random r.
- **Scalar splitting** [9] – The scalar d can be split into several randomized scalars using different methods. The most efficient one consists in an Euclidean splitting [11] by writing $d' = \lfloor d/r \rfloor .r + (d \bmod r)$ for a random r. The scalar multiplication becomes $[d']\boldsymbol{P} = [d \bmod r]\boldsymbol{P} + [\lfloor d/r \rfloor].([r]\boldsymbol{P})$.
- **Randomized projective points** [14] – An affine point $\boldsymbol{P} = (x, y)$ can be represented in Jacobian coordinates as $(\lambda^2 X : \lambda^3 Y : \lambda Z)$ for any nonzero λ. The representation of a point can be randomized by choosing random values of λ.

Algorithm 1. Double-and-add-always

Input: $d = (d_{k-1}, \ldots, d_0)_2 \in \mathbb{N}$ and $\boldsymbol{P} \in E(\mathbb{F}_q)$
Output: $\boldsymbol{Q} = [d]\boldsymbol{P}$

1: $\boldsymbol{R_0} \leftarrow \boldsymbol{O}; \boldsymbol{R_1} \leftarrow \boldsymbol{O}$
2: **for** $j = k - 1$ to 0 **do**
3:      $\boldsymbol{R_0} \leftarrow [2]\boldsymbol{R_0}$
4:      $b \leftarrow d_j; \boldsymbol{R_{1-b}} \leftarrow \boldsymbol{R_0} + \boldsymbol{P}$
5: **end for**
6: **return** $\boldsymbol{R_0}$

The scalar blinding countermeasure [14] has been subject to several discussions in previous publications [11,33] as the order of the curves defined by the NIST is sparse. As we remind in next paragraphs, this property makes the blinding not fully efficient as several bits of the blinded scalar remain unmasked. Thanks to the combination of the recent collision correlation and new horizontal side-channel attack techniques, we define a new side-channel attack path which takes advantage of this sparse order to complete the full secret exponent recovery. The rest of the paper will consider an implementation using the double-and-add-always (see Alg. 1) in combination with first the scalar blinding technique and then the added randomized projective point countermeasure. Our attacks are applicable to other classical regular algorithm with minor changes as explained in the extended version of this paper [16].

3   Attack on a Blinded Regular Scalar Multiplication with Known Input Point

We first analyze a simple scenario where the input point of the scalar multiplication is known, *i.e.* no DSCA countermeasure on P is used. We consider that the scalar is protected against DSCA using the scalar blinding method. The targeted operation is then $[d']P$ where $d' = d + r.n$ for a random r and n the order of $E(\mathbb{F}_p)$.

Let $\{\mathcal{C}^{(1)}, \ldots, \mathcal{C}^{(N)}\}$ be the N side-channel leakage traces corresponding to the computations $[d'^{(i)}]P^{(i)}$ such that $d'^{(i)} = d + r^{(i)}.n$ are the blinded scalars using random values $r^{(i)}$ and known points $P^{(i)}$ with $1 \leq i \leq N$. We consider that the random factors $r^{(i)}$ are chosen relatively small such that $r^{(i)} \in [0, 2^m - 1]$ with $m \leq 32$ which is the case in many implementations for efficiency reasons.

We first detail the particular form of blinded scalars on standardized curves. Then, we present our attack which is composed of three steps. In a first step, we find the non-masked part of the secret d. Then, we recover each random value $r^{(i)}$ used for the scalar blinding. Finally, we look for the remaining least significant bits of d.

3.1   Representation of the Blinded Scalar using a Sparse Group Order

As noted before, most elliptic curve implementations use in practice curves from public standards [5,29,32]. Most standards consider the use of generalised Mersenne numbers to define the prime fields underlying the elliptic curves. These particular primes are very advantageous efficiency-wise as tricks can be applied to improve greatly the modular operations [8].

Classification of sparse group orders. The main standard that defines elliptic curves is the NIST FIPS186-2 [29]. It specifies curves defined over the following primes: $p_{192} = 2^{192} - 2^{64} - 1$, $p_{224} = 2^{224} - 2^{96} + 1$, $p_{256} = 2^{256} - 2^{224} + 2^{192} + 2^{96} - 1$, $p_{384} = 2^{384} - 2^{128} - 2^{96} + 2^{32} - 1$ and $p_{521} = 2^{521} - 1$. Due to Hasse's theorem, the orders of the curves defined over each of these fields have also a sparse representation in its upper half. We can categorize them in 3 sets:

- Type-1: the order has a large pattern of ones,
- Type-2: the order has a large pattern of zeros,
- Type-3: the order has a combination of large patterns of both ones and zeros.

Consider the notation $1^{[a,b]}$ with $a, b \in \mathbb{N}$ and $a > b$ a pattern of 1 bits from the bit positions a to b. Similarly, we note $0^{[a,b]}$ a pattern of 0 bits.

Let n, the order of the curve, be a k-bit integer. We can write it depending on its type:

- Type-1: $n = 1^{[k-1,a]} + x$ with $(k-1) > a$ and $0 \le x < 2^a$,
- Type-2: $n = 2^{k-1} + 0^{[k-2,a]} + x$ with $(k-2) > a$ and $0 \le x < 2^a$,
- Type-3: $n = 1^{[k-1,a]} + 0^{[a-1,b]} + 1^{[b-1,c]} + x$ with $(k-1) > a > b > c$ and $0 \le x < 2^c$,

where $a, b, c \in \mathbb{N}$.

Example 1. Here are some standard curves that belong to different types:

- Type-1: $n = 1^{[191,96]} + x$ (NIST P-192 [29]),
- Type-2: $n = 2^{225} + 0^{[224,114]} + x$ (SECP224k1 [32]),
- Type-3: $n = 1^{[255,224]} + 0^{[223,192]} + 1^{[191,128]} + x$ (NIST P-256 [29]).

Form of a random multiple of the order. Let $r \in [1, 2^m - 1]$ be an m-bit random used to mask the secret scalar d such as $d' = d + r.n$. Given the form of the orders of standard curves as seen previously, the mask $r.n$ also has a specific representation.

Let $\tilde{r} = r.(2^m - 1)$ be a $2m$-bit integer, we note $\tilde{r}_1$ and $\tilde{r}_0$ respectively the quotient and remainder of the Euclidean division of $\tilde{r}$ by 2^m. This product has a special form, $\forall r \in [1, 2^m - 1]$ we have:

$$\tilde{r}_1 = r - 1,$$
$$\tilde{r}_1 + \tilde{r}_0 = 2^m - 1.$$

This can be explained by noting the product: $r.(2^m - 1) = (r.2^m) - r$. Hence, $r.2^m$ equals to r followed by m zeros to which we subtract r. This subtraction is performed by computing $2^m - r$ and setting a carry for the most significant part. Hence the higher part equals to $r - 1$. If we add the two halves of $\tilde{r}$, we obtain $(2^m - r) + (r - 1) = 2^m - 1$.

Depending on the category of n, we have the following representations of the mask $r.n$:

- Type-1: $r.n = \tilde{r}_1.2^k + 1^{[k-1,a+m]} + x$, with $0 \le x < 2^{a+m}$,
- Type-2: $r.n = r.2^k + 0^{[k-1,a+m]} + x$, with $0 \le x < 2^{a+m}$,
- Type-3: $r.n = \tilde{r}_1.2^k + 1^{[k-1,a+m]} + \tilde{r}_0.2^{a+m} + 0^{[a-1+m,b+m]} + \tilde{r}_1.2^{b+m} + 1^{[b-1+m,c+m]} + x$, with $0 \le x < 2^{c+m}$.

The patterns of zeros and ones are reduced by m bits for the 3 categories of group orders. Note that these representations of $r.n$ are exact up to possible carries that can happen after each pattern. However, their effect is very limited and does not impact our results.

Adding the random mask r.n to the scalar. The last part of the scalar blinding consists in adding the secret scalar d to the mask $r.n$. First, we observe that an addition $x + (2^m - 1)$ with $x \in [1, 2^m - 1]$ equals to $x - 1$ on the least significant m bits of the results with the $(m+1)$-th bit set at 1.

The notation $d^{[a,b]}$ corresponds to the bits of the scalar d from the bit position a to b. The 3 types of masking representations have an important impact on the (non-)masking of the secret:

- Type-1: $d' = (\tilde{r}_1 + 1).2^k + d^{[k-1,a+m]} + x$, with $0 \le x < 2^{a+m}$,
- Type-2: $d' = r.2^k + d^{[k-1,a+m]} + x$, with $0 \le x < 2^{a+m}$,
- Type-3: $d' = (\tilde{r}_1 + 1).2^k + d^{[k-1,a+m]} + \tilde{r}_0.2^{a+m} + d^{[a-1+m,b+m]} + (\tilde{r}_1 + 1).2^{b+m} + d^{[b-1+m,c+m]} + x$, with $0 \le x < 2^{c+m}$.

Note that the addition of d to $r.n$ can add a carry to the least significant bit of the non masked part of d'.

3.2   First Step: Find the Non-Masked Part of d

From the previous observations on the representation of the blinded scalars $d'^{(i)}$, we can directly deduce chunks of the secret d. We note $\bar{d} = d^{[a,b]}$ the non-masked value of d, for some a, b. We note $\delta = (a-b)$ the bit size of $\bar{d} = (\bar{d}_{\delta-1}, \ldots, \bar{d}_1, \bar{d}_0)_2$. As we do not know the most significant part of the $d'^{(i)}$, we cannot compute an intermediate value based on a guess, we need to perform a *vertical collision-correlation attack*.

For each bit $\bar{d}_j$ of the scalar, a point doubling followed by a point addition are performed where the addition is dummy if $\bar{d}_j = 0$. If $\bar{d}_j = 1$, all the results of point doubling and point addition are used whereas, if $\bar{d}_j = 0$, the result of the point addition is discarded. This means that the next point doubling will take the same input as the previous point addition when $\bar{d}_j = 0$, resulting in a collision. We use the notations In, respectively Out, to indicate the input, respectively output, of a given operation.

1. To find the j-th bit $\bar{d}_j$ of $\bar{d}$ with $0 < j < \delta$, identify the two elliptic curve operations that possibly correspond to its processing. The processing of a bit $\bar{d}_j = 0$ generates a collision between the input of the point addition ECADD(j) and the input of the next point doubling ECDBL($j+1$) whereas there is no collision when $\bar{d}_j = 1$.
2. Construct a first vector $\Theta_0 = \{C^{(i)}(t_0)\}_{1 \le i \le N}$ that corresponds to the time sample t_0 of the N leakage traces $C^{(i)}$. The instant t_0 corresponds to the computation of In(ECADD(j)).
3. Construct similarly a second vector $\Theta_1 = \{C^{(i)}(t_1)\}_{1 \le i \le N}$ that corresponds to the time sample t_1 of the N leakage traces $C^{(i)}$. The instant t_1 corresponds to the computation of In(ECDBL($j+1$)).
4. Perform a collision-correlation analysis $\rho(\Theta_0, \Theta_1)$. We can expect that the correlation coefficient will be maximal when the operations ECADD(j) and ECDBL($j+1$) take the same input point, hence when $\bar{d}_j = 0$.

Remark 1. Note that, for the Type-3 orders, the attack has to be repeated on each interval of non-masked bits of d.

Remark 2. We remind that the success rate of collision-correlation attacks can heavily depend on the choice of the threshold value. A discussion of this point based on practical results is given in Section 5.1.

3.3 Second Step: Retrieve Random Masks with Horizontal Attacks

From Section 3.1, we know that the random r used in the scalar blinding directly appears in the most significant part of d'. The second part of our attack consists in retrieving the random values $r^{(i)} \in [1, 2^m - 1]$ from each blinded scalar $d''^{(i)}$ using an *horizontal correlation attack*. The following attack procedure is repeated for each trace $\mathcal{C}^{(i)}$, $1 \leq i \leq N$:

1. Try all possible m-bit values of $r^{(i)}$. In most implementations the random chosen for the scalar blinding is small, *i.e.* $r \leq 2^{32}$, hence this enumeration is generally feasible. A guess on $r^{(i)}$ directly gives a guess on the first m bits[2] of $d''^{(i)}$.
2. Let $\hat{r}$ be the guess on $r^{(i)}$. This guess gives the attacker a sequence of elliptic curve operations that appear at the beginning of the trace $\mathcal{C}^{(i)}$. Since the attacker knows the input point $\boldsymbol{P}^{(i)}$, he can compute the sequence of multiples of $\boldsymbol{P}^{(i)}$ that should be processed for a given $\hat{r}$. Note that from the previous section, we also know the following δ bits of the non-masked part of the blinded scalar. Then η intermediate points can be computed with[3] $\eta = 2(m + \delta)$.
3. Choose a leakage model function $\mathcal{L}$, *e.g.* the Hamming weight, and compute some predicted values derived from the η points T_j, $1 \leq j \leq \eta$. The attacker computes the values $l_j = \mathcal{L}(T_j)$ for $1 \leq j \leq \eta$ and creates the vector $\Theta_1 = (l_j)_{1 \leq j \leq \eta}$.
4. Construct η sub-traces from the trace $\mathcal{C}^{(i)}$ where the targeted values T_j, $1 \leq j \leq \eta$ are manipulated. The attacker constructs the vector $\Theta_0 = (o_j)_{1 \leq j \leq \eta}$ where o_j are the identified points of interest related to T_j.
5. Compute the correlation coefficient $\rho(\Theta_0, \Theta_1)$. If the guess $\hat{r}$ is correct, the sequence of T_j is also correct, hence we can expect a maximal coefficient of correlation.

Remark 3. The random r appears at the beginning of each pattern of ones in the order n. Hence, on curves of Type-3, the attacker could exploit this property to obtain more time samples per trace to recover the random values.

---

[2] Note that $(\tilde{r}_1^{(i)} + 1) = r^{(i)}$ for Type-1 and Type-3 orders.
[3] Depending on the point addition and point doubling formulæ used, an attacker could also include intermediate long-integer operations in order to work with even larger sets.

3.4 Third Step: Recover the Least Significant Part of d

From the previous parts of the attack, we know the most significant part of d as well as the random values $r^{(i)}$ of each blinded scalar $d'^{(i)}$. We need to recover the least significant part of the secret. By guessing the next w unknown bits of d, we can compute guessed blinded scalars $\hat{d}'^{(i)}$. We can then perform a classical *vertical correlation attack* to validate the guesses. The following steps need to be repeated until d is fully recovered (directly or with an easy brute-force):

1. Guess the following w unknown bits of d. From this guess and the known random $r^{(i)}$, compute the N guessed blinded scalars $\hat{d}'^{(i)}$ for $1 \leq i \leq N$.
2. Choose a leakage model function $\mathcal{L}$. For the i-th curve, the attacker can compute some predicted values derived from the η points $T_j^{(i)}$, $1 \leq j \leq \eta$ with $\eta = 2w$. He creates the vector $\Theta_1 = \left(l_j^{(i)} \right)_{i,j}$, with $1 \leq j \leq \eta, 1 \leq i \leq N$ and where $l_j^{(i)} = \mathcal{L}\left(T_j^{(i)} \right)$.
3. Construct a vector $\Theta_0 = \left(o_j^{(i)} \right)_{i,j}$ where $o_j^{(i)}$ is the point of interest of the trace $\mathcal{C}^{(i)}$ corresponding to the processing of $T_j^{(i)}$.
4. Compute the correlation coefficient $\rho(\Theta_0, \Theta_1)$. We can expect a maximal correlation coefficient when the w guessed bits are correct, hence the η intermediate points of the N traces are correct.

Remark 4. Note that there can be a carry on the least significant bit of the w guessed bits of $\hat{d}'^{(i)}$. If a wrong guess is recovered in first position due to the carry, the following attack on the next w bits will give low correlation values. The attacker then needs to correct the previous guess with a carry in order to continue his attack.

4   Attack on a Protected Scalar Multiplication

The main attack strategy proposed in the previous section can also be applied on an implementation with point blinding. The first step is identical even with unknown input points. However as the input is unknown, classical correlation attacks where a guessed intermediate variable is correlated to leakage observations are not applicable anymore. We present in this section modifications to the second and third steps of our previous attack to recover the full secret scalar on a fully protected scalar multiplication.

4.1 First Step: Vertical Collision-Correlation

The first attack is identical to the known input point scenario. The proposed vertical collision-correlation in Section 3.2 does not require the knowledge of the inputs. Hence the same steps can be applied in the unknown input case in order to recover the non-masked bits of the scalar d, *i.e.* $\bar{d}$ of bit length δ.

4.2 Second Step: Horizontal Collision-Correlation

The horizontal correlation attack presented in Section 3.3 is not applicable without a known input point. We need to perform an *horizontal collision-correlation* on each leakage trace $\mathcal{C}^{(i)}$, $1 \leq i \leq N$, simply noted $\mathcal{C}$ below for readability:

1. Try all possible m-bit values of $r^{(i)}$.
2. The guessed random $\hat{r}$ gives the attacker the supposed starting sequence of elliptic curve operations that appears in the scalar multiplication. The known part of d also provides the following δ bits of the blinded scalar. Hence, the attacker works with $(m + \delta)$ bits of the blinded scalar $\hat{d}'$. The processing of a bit at 0 or 1 generates different possible collisions between elliptic curve coordinates:
   - if $\hat{d}'_j = 1$, we have a collision between the coordinates of the output of ECADD(j) and the coordinates of the input point of ECDBL($j + 1$),
   - if $\hat{d}'_j = 0$, we have a collision between the coordinates of the input of ECADD(j) and the coordinates of the input of ECDBL($j + 1$).
3. Construct two vectors Θ_0 and Θ_1 corresponding to different time samples of the leakage trace $\mathcal{C}$. They are defined as:

$$\Theta_0 = \left\{ \mathcal{C}\left(t_0^X(j)\right), \mathcal{C}\left(t_0^Y(j)\right), \mathcal{C}\left(t_0^Z(j)\right) \right\}_{0 \leq j < (m+\delta)},$$
$$\Theta_1 = \left\{ \mathcal{C}\left(t_1^X(j)\right), \mathcal{C}\left(t_1^Y(j)\right), \mathcal{C}\left(t_1^Z(j)\right) \right\}_{0 \leq j < (m+\delta)},$$

where

$$t_0^X(j) = \begin{cases} \mathsf{Out}^X\left(\mathsf{ECADD}(j)\right) & \text{if } \hat{d}'_j = 1, \\ \mathsf{In}^X\left(\mathsf{ECADD}(j)\right) & \text{if } \hat{d}'_j = 0, \end{cases}$$
$$t_1^X(j) = \mathsf{In}^X\left(\mathsf{ECDBL}(j+1)\right),$$

respectively t_0^Y, t_1^Y and t_0^Z, t_1^Z for the Y and Z coordinates of the corresponding elliptic points. The notations In and Out represent the time samples of the processing of respectively the input point and output point coordinates of the parametrized elliptic curve operation.
4. Compute the correlation analysis $\rho(\Theta_0, \Theta_1)$. For the correct guess $\hat{r}$, the sequence of collisions is correct and should give the maximum coefficient of correlation.

4.3 Third Step: Vertical Collision-Correlation

We need to apply a *vertical collision-correlation* side-channel attack in this third step as the input is unknown. Instead of recomputing the intermediate points of the scalar multiplication corresponding to guesses on d and computing a correlation with the leakage observation, we build collision vectors, as previously, depending on the bit values of the guess:

1. Guess w unknown bits of d. From this guess and the known random $r^{(i)}$, we can compute guessed blinded scalars $\hat{d}'^{(i)}$ for $1 \leq i \leq N$.

2. Construct collision vectors Θ_0 and Θ_1 as defined in the previous attack depending on the values of the bits of $\hat{d}'^{(i)}$. If we consider that $u \leq \delta$ bits of d are already recovered, the collision vectors are of size $(m + u + w)N$.
3. Compute the correlation analysis $\rho(\Theta_0, \Theta_1)$. For the correct w guessed bits, we can expect the highest correlation coefficient.

Remark 5. In order to find the bit d_j, the collision should be evaluated on the operations of the next iteration $(j + 1)$ of the scalar multiplication. Hence, the final least significant bit cannot be recovered using the attack but has to be guessed.

5   Experimentations

In order to validate our different attack paths on the blinded scalar multiplication, we performed simulations on a double-and-add-always algorithm using the standardized elliptic curve P-192 from NIST. For our implementation, we chose the classical jacobian projective coordinates and used the most efficient generic addition and doubling algorithms[4]. The particular choice of coordinates or group operation algorithms has no impact on the feasibility of our attacks. Its only effect is on the selection of time samples on which to compute correlations or collisions. We performed our attacks using 8-bit and 16-bit random for the scalar blinding. As the use of larger random size impacts the computational time of the attacks, we chose small random sizes in order to repeat several hundred of times our attacks for consistency.

Our simulation traces consist of the leakage of the inputs and outputs of long integer operations (multiplication, squaring, addition) that are used for the elliptic curve group operations. The leakage is modeled with the classical Hamming weight function. As nowadays most arithmetic coprocessors and chip have 32-bit architectures, we consider Hamming weight leakage of words of 32-bits[5]. Hence, the leakage of the long-integer multiplication $c = a.b \bmod p$ is represented by the vector $(\mathsf{HW}_{32}(a_i), \mathsf{HW}_{32}(b_i), \mathsf{HW}_{32}(c_i))$ where $\mathsf{HW}_{32}(a_i)$, respectively $\mathsf{HW}_{32}(b_i)$ and $\mathsf{HW}_{32}(c_i)$, represents the Hamming weight of the i-th 32-bit word of a, respectively b and c. We performed our simulations with different level of noise having a Gaussian distribution with mean 0 and standard deviation σ. Finally, we use the Pearson correlation as side-channel distinguisher.

5.1   Simulated Attack Results on Known Input Points

We first present results on the attack path with a known (non-masked) input point from Section 3. Table 1 details the success rates obtained for the three

---

[4] We selected the addition algorithm *add-2007-bl* with complexity $11\mathsf{M} + 5\mathsf{S}$ and the doubling algorithm *dbl-2007-bl* with complexity $1\mathsf{M} + 8\mathsf{S}$ from [3].

[5] We expect the horizontal parts of our attacks to give better results on smaller architectures as more time samples will be available per long integer number.

attack steps with various parameters. We recall that the parameter N is the number of traces and m is the bit size of the random for the exponent blinding.

The first step of the attack is a vertical collision-correlation. We tested its success using 500 and 1000 leakage traces. The results show a great success rate even when the noise becomes quite high. We can expect even better success rate for high σ if the attacker has access to more traces. Figure 1 illustrates the spreading of the correlation coefficient around its mean value. We clearly see the variance of the coefficient increasing for high levels of noise when a collision happens, *i.e.* the bit equals 0. This figure also gives a good idea on the threshold value for the correlation coefficient in practice, in order to decide if a collision happened. Its selection needs to be more precise the higher the noise level to obtain a good success rate. In practice, we observe that the last bits found by the attack are sometimes different to the expected scalar d. This is due to a possible carry propagation because of the addition of the masking value $r.n$. In this case, a bit equal to 1 is found as the correlation coefficient becomes low. This possible error is then corrected during the third part of the attack where the attacker can start the analysis a few bits before the ones retrieved at this step.

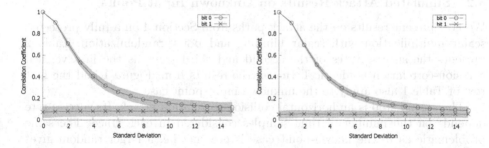

Fig. 1. First attack step: correlation coefficient spreading, left for 500 traces, right for 1000 traces

The second attack step is an horizontal correlation that needs to be repeated for each trace. As the horizontal attack uses only one trace, the parameters affecting its success rate are the size m of the random used for the exponent blinding as well as the noise level σ. A larger random gives more time samples per trace, hence better results for our attack. However, as we enumerate 2^m values, the computational times may be prohibitive for large bit sizes of random. The attack also uses the bits recovered in the first step to compute guessed intermediate variables and perform a correlation on even more time samples. The success rates are then very good even in the presence of high noise.

The last attack step is a vertical correlation. As the first part, we performed tests on 500 and 1000 traces to compare the evolution of the success rate. The results are very good until strong levels of noise ($\sigma > 10$).

Remark 6. As explained in Section 3.4, due to possible carry propagation instead of recovering the right guess we can obtain the correct guess ±1. However, we will be immediately informed as the correlation coefficients for the attack on the next w bits will be much lower. We consider the attack successful if the best guess is close to the right guess (±1).

Table 1. Success rate for known input points

Attack steps	N	m	Standard Deviation σ					
			0	1	2	5	10	15
Vertical	500	-	1.0	1.0	1.0	1.0	0.88	0.74
collision-correlation	1000	-	1.0	1.0	1.0	1.0	0.99	0.76
Horizontal	-	8	1.0	1.0	1.0	1.0	1.0	0.77
correlation	-	16	1.0	1.0	1.0	1.0	1.0	0.85
Vertical	500	-	1.0	1.0	1.0	1.0	0.64	0.42
correlation	1000	-	1.0	1.0	1.0	1.0	0.84	0.52

5.2 Simulated Attack Results on Unknown Input Points

We now present results on the attack paths from Section 4 on a fully protected scalar multiplication with scalar blinding and point randomization. Table 2 presents the success rates of the second and third steps as the first vertical collision-correlation is identical. Hence, the results from Figure 1 and the first row of Table 1 also apply to the unknown input point case.

The second step is an horizontal collision-correlation attack. Its success rate depends on the number of time samples considered in each trace. The same problematic as in the known-point case is present, *i.e.* a larger random gives better results for a higher computational cost. The success rate drops quicker than previous attacks for higher levels of noise. Indeed, the attack only uses time samples of computations on coordinates of intermediate elliptic curve points. Hence, contrary to vertical attacks the attacker is limited to a fixed number of time samples regardless of the noise level.

The third attack step is a vertical collision-correlation. As each vertical attack, we tested its success rate on 500 and 1000 traces. Its efficiency is very high even with a strong noise. The Remark 6 also applies here as possible carries can appear.

From our simulations, we observe that in the unknown input point case our attack retrieves the full scalar for noise levels up to $\sigma \approx 5$ whereas our attack works up to $\sigma \approx 10$ with a known input point.

6 Countermeasures

We propose here countermeasures that could be applied at different levels of the implementation.

Table 2. Success rate for unknown input points

Attack steps	N	m	Standard Deviation σ					
			0	1	2	5	10	15
Horizontal	-	8	1.0	1.0	0.9	0.1	0.02	0.01
collision-correlation	-	16	1.0	1.0	0.95	0.23	0.10	0.02
Vertical	500	-	1.0	1.0	1.0	1.0	1.0	0.97
collision-correlation	1000	-	1.0	1.0	1.0	1.0	1.0	0.99

Scalar splitting. The Euclidean splitting proposed in [11]: $[d]\boldsymbol{P} = [d \bmod r]\boldsymbol{P} + [\lfloor d/r \rfloor]([r]\boldsymbol{P})$ is generally preferred to the additive splitting [9] that could be vulnerable to advanced attacks [28] and to the multiplicative splitting [36] that requires a costly modular inversion. However the Euclidean splitting still remains less efficient than the scalar blinding and can be disregarded by developers. Note that exponent splitting with a mask of bit length m could be surmounted with $2^{m/2}$ traces due to the birthday paradox. The use of a scalar splitting method, with large enough random masks, thwarts the proposed attacks on standard curves.

Scalar blinding with larger random. As our attack path exploits the fact that, for small random values, the scalar blinding countermeasure does not mask part of the scalar, a possible solution could be to use larger random. A first selection parameter for the random size could be to have an implementation where all scalar bits are masked for the supported elliptic curves. Let $\mathcal{P}$ be the largest pattern size amongst all curves' order that are supported by an application[6]. Hence, in order to use the scalar blinding countermeasure, one would need to implement a random size m such that $m > \mathcal{P}$ to obtain a scalar fully masked.

A second selection parameter could be to select a random size large enough such that our proposed attack path is no more applicable. Indeed, in our second attack step the attacker needs to try all possible random values. Let $\mathcal{B}$ be the maximum brute force capability of an attacker, *i.e.* he can perform $2^{\mathcal{B}}$ operations in reasonable time. Hence, one would need to choose a random size m such that $m > \mathcal{B}$.

Generally, the more restrictive selection criterion is $m > \mathcal{B}$, as $\mathcal{B} << \mathcal{P}$ for most standardized curves. The overhead added to the scalar multiplication complexity by the larger m value then needs to be compared to other countermeasures. For example, the scalar splitting that has an overhead factor of 1.5 which can be more advantageous depending on the implementation requirements.

Atomic algorithm and unified formulæ. Our attack only targets regular scalar multiplication algorithms, hence an atomic algorithm could be considered. There are many atomic formulas for elliptic curves proposed in the literature [10,17,31].

---

[6] For example, amongst all NIST curves, the P-521 has the largest pattern of ones in its order with a pattern size of 262 bits, *i.e.* $\mathcal{P} = 262$.

This countermeasure generally offers an interesting time/memory trade-off for embedded devices. However a recent attack was presented by Bauer *et al.* [1] against the main atomic formulæ. Even if the practicality of their attack is subject to different parameters, it clearly demonstrates a vulnerability in many atomic schemes. As mentioned by the authors of [1], their technique can also be applied to unified formulas on Weierstraß curves [6] as well as Edward's curves [4].

7 Conclusion

We present in this paper a new side-channel attack combination targeting elliptic curves implementations of regular scalar multiplication on some standardized curves. We assume the scalar multiplication algorithm implements the classical scalar blinding and point randomization techniques, two of the most used countermeasures against differential side-channel attacks. Our attacks exploit the previously known weakness of the sparse order of the standardized curves in order to fully recover a blinded secret scalar. We combine techniques from collision correlation analysis as well as horizontal and vertical attack to design a new attack path.

Acknowledgments. The authors would like to thank Vincent Verneuil for his detailed and perceptive comments.

References

1. Bauer, A., Jaulmes, E., Prouff, E., Wild, J.: Horizontal collision correlation attack on elliptic curves. In: Selected Areas in Cryptography (2013)
2. Bauer, A., Jaulmes, É.: Correlation analysis against protected SFM implementations of RSA. In: Paul, G., Vaudenay, S. (eds.) INDOCRYPT 2013. LNCS, vol. 8250, pp. 98–115. Springer, Heidelberg (2013)
3. Bernstein, D.J., Lange, T.: Explicit-formulas database. http://hyperelliptic.org/EFD/g1p/auto-shortw.html
4. Bernstein, D.J., Lange, T.: Faster addition and doubling on elliptic curves. In: Kurosawa, K. (ed.) ASIACRYPT 2007. LNCS, vol. 4833, pp. 29–50. Springer, Heidelberg (2007)
5. Bernstein, D.J., Lange, T.: Safecurves: choosing safe curves for elliptic-curve cryptography. http://safecurves.cr.yp.to (accessed May 26, 2014)
6. Brier, E., Joye, M.: Weierstraß elliptic curves and side-channel attacks. In: Naccache, D., Paillier, P. (eds.) PKC 2002. LNCS, vol. 2274, pp. 335–345. Springer, Berlin Heidelberg (2002)
7. Brier, E., Clavier, C., Olivier, F.: Correlation power analysis with a leakage model. In: Joye, M., Quisquater, J.-J. (eds.) CHES 2004. LNCS, vol. 3156, pp. 16–29. Springer, Heidelberg (2004)
8. Brown, M., Hankerson, D., López, J., Menezes, A.: Software implementation of the NIST elliptic curves over prime fields. In: Naccache, D. (ed.) CT-RSA 2001. LNCS, vol. 2020, pp. 250–265. Springer, Heidelberg (2001)

9. Chari, S., Jutla, C.S., Rao, J.R., Rohatgi, P.: Towards sound approaches to counteract power-analysis attacks. In: Wiener, M. (ed.) CRYPTO 1999. LNCS, vol. 1666, pp. 398–412. Springer, Heidelberg (1999)
10. Chevallier-Mames, B., Ciet, M., Joye, M.: Low-cost solutions for preventing simple side-channel analysis: Side-channel atomicity. IEEE Transactions on Computers 53, 760–768 (2004)
11. Ciet, M., Joye, M.: (Virtually) free randomization techniques for elliptic curve cryptography. In: Qing, S., Gollmann, D., Zhou, J. (eds.) ICICS 2003. LNCS, vol. 2836, pp. 348–359. Springer, Heidelberg (2003)
12. Clavier, C., Feix, B., Gagnerot, G., Roussellet, M., Verneuil, V.: Horizontal correlation analysis on exponentiation. In: Soriano, M., Qing, S., López, J. (eds.) ICICS 2010. LNCS, vol. 6476, pp. 46–61. Springer, Heidelberg (2010)
13. Clavier, C., Feix, B., Gagnerot, G., Giraud, C., Roussellet, M., Verneuil, V.: ROSETTA for single trace analysis. In: Galbraith, S., Nandi, M. (eds.) INDOCRYPT 2012. LNCS, vol. 7668, pp. 140–155. Springer, Heidelberg (2012)
14. Coron, J.-S.: Resistance against differential power analysis for elliptic curve cryptosystems. In: Koç, Ç.K., Paar, C. (eds.) CHES 1999. LNCS, vol. 1717, pp. 292–302. Springer, Heidelberg (1999)
15. Diffie, W., Hellman, M.E.: New directions in cryptography. IEEE Transactions on Information Theory 22(6), 644–654 (1976)
16. Feix, B., Roussellet, M., Venelli, A.: Side-channel analysis on blinded regular scalar multiplications. IACR Cryptology ePrint Archive (2014)
17. Giraud, C., Verneuil, V.: Atomicity improvement for elliptic curve scalar multiplication. In: Gollmann, D., Lanet, J.-L., Iguchi-Cartigny, J. (eds.) CARDIS 2010. LNCS, vol. 6035, pp. 80–101. Springer, Heidelberg (2010)
18. Goundar, R., Joye, M., Miyaji, A., Rivain, M., Venelli, A.: Scalar multiplication on Weierstraß elliptic curves from co-z arithmetic. Journal of Cryptographic Engineering 1(2), 161–176 (2011)
19. Hanley, N., Kim, H., Tunstall, M.: Exploiting collisions in addition chain-based exponentiation algorithms. Cryptology ePrint Archive, Report 2012/485 (2012)
20. Joye, M.: Highly regular Right-to-left algorithms for scalar multiplication. In: Paillier, P., Verbauwhede, I. (eds.) CHES 2007. LNCS, vol. 4727, pp. 135–147. Springer, Heidelberg (2007)
21. Joye, M., Yen, S.M.: The Montgomery powering ladder. In: Kaliski, B., Koç, Ç.K., Paar, C. (eds.) CHES 2002. LNCS, vol. 2523, pp. 291–302. Springer, Heidelberg (2004)
22. Koblitz, N.: Elliptic curve cryptosystems. Mathematics of Computation 48, 203–209 (1987)
23. Kocher, P.C.: Timing attacks on implementations of Diffie-Hellman, RSA, DSS, and other systems. In: Koblitz, N. (ed.) CRYPTO 1996. LNCS, vol. 1109, pp. 104–113. Springer, Heidelberg (1996)
24. Kocher, P.C., Jaffe, J., Jun, B.: Differential power analysis. In: Wiener, M. (ed.) CRYPTO 1999. LNCS, vol. 1666, pp. 388–397. Springer, Heidelberg (1999)
25. Miller, V.S.: Use of elliptic curves in cryptography. In: Williams, H.C. (ed.) CRYPTO 1985. LNCS, vol. 218, pp. 417–426. Springer, Heidelberg (1986)
26. Montgomery, P.L.: Speeding the Pollard and elliptic curve methods of factorization. Mathematics of Computation 48(177), 243–264 (1987)
27. Moradi, A., Mischke, O., Eisenbarth, T.: Correlation-enhanced power analysis collision attack. In: Mangard, S., Standaert, F.-X. (eds.) CHES 2010. LNCS, vol. 6225, pp. 125–139. Springer, Heidelberg (2010)

28. Muller, F., Valette, F.: High-order attacks against the exponent splitting protection. In: Yung, M., Dodis, Y., Kiayias, A., Malkin, T. (eds.) PKC 2006. LNCS, vol. 3958, pp. 315–329. Springer, Heidelberg (2006)
29. National Institute Standards and Technology: Digital Signature Standard (DSS). Publication 186-2 (2000)
30. Rivest, R., Shamir, A., Adleman, L.: A method for obtaining digital signatures and public-key cryptosystems. Communications of the ACM **21**, 120–126 (1978)
31. Rondepierre, F.: Revisiting atomic patterns for scalar multiplications on elliptic curves. In: Francillon, A., Rohatgi, P. (eds.) CARDIS 2013. LNCS, vol. 8419, pp. 171–186. Springer, Heidelberg (2014)
32. SEC2: Standards for Efficient Cryptography Group/Certicom Research. Recommanded Elliptic Curve Cryptography Domain Parameters (2000)
33. Smart, N., Oswald, E., Page, D.: Randomised representations. IET Information Security **2**, 19–27(8) (2008)
34. Solinas, J.: Generalized Mersenne numbers. Technical report CORR-39, Dept. of C&O, University of Waterloo (1999)
35. Stebila, D., Thériault, N.: Unified point addition formulæ and side-channel attacks. In: Goubin, L., Matsui, M. (eds.) CHES 2006. LNCS, vol. 4249, pp. 354–368. Springer, Heidelberg (2006)
36. Trichina, E., Bellezza, A.: Implementation of elliptic curve cryptography with built-in counter measures against side channel attacks. In: Kaliski, B.S., Koç, Ç.K., Paar, C. (eds.) CHES 2002. LNCS, vol. 2523, pp. 98–113. Springer, Heidelberg (2003)
37. Walter, C.D.: Sliding windows succumbs to Big Mac attack. In: Koç, Ç.K., Naccache, D., Paar, C. (eds.) CHES 2001. LNCS, vol. 2162, pp. 286–299. Springer, Heidelberg (2001)
38. Witteman, M.F., van Woudenberg, J.G.J., Menarini, F.: Defeating RSA multiply-always and message blinding countermeasures. In: Kiayias, A. (ed.) CT-RSA 2011. LNCS, vol. 6558, pp. 77–88. Springer, Heidelberg (2011)

Online Template Attacks

Lejla Batina[1], Łukasz Chmielewski[2], Louiza Papachristodoulou[1](✉),
Peter Schwabe[1], and Michael Tunstall[3]

[1] Digital Security Group, Radboud University Nijmegen, P.O. Box 9010, 6500 GL
Nijmegen, The Netherlands
lejla@cs.ru.nl, louiza@cryptologio.org, peter@cryptojedi.org
[2] Riscure BV, Delft, The Netherlands
Chmielewski@riscure.com
[3] Cryptography Research Inc., 425 Market Street, San Francisco, CA 94105, USA
michael.tunstall@cryptography.com

Abstract. In the context of attacking elliptic-curve scalar multiplication with template attacks, one can interleave template generation and
template matching to reduce the amount of template traces. This paper
enhances the power of this technique by defining and applying the concept of *online template attacks* (OTA); a general attack technique with
minimal assumptions for an attacker, who has very limited control over
the target device. We show that OTA need only one power consumption trace of a scalar multiplication on the target device; they are thus
suitable not only against ECDSA and static Diffie-Hellman, but also
against elliptic-curve scalar multiplication in ephemeral Diffie-Hellman.
In addition, OTA need only one template trace per scalar bit and they
can be applied to almost all scalar-multiplication algorithms. To demonstrate the power of OTA we recover scalar bits of a scalar multiplication
using the double-and-add-always algorithm on a twisted Edwards curve
running on a smart card with an ATmega163 CPU.

Keywords: Side-channel analysis · Template attacks · Scalar
multiplication · Elliptic curves

1 Introduction

Side-channel attacks exploit various physical leakages of secret information or
instructions from cryptographic devices and they constitute a constant threat for
cryptographic implementations. We focus on power-analysis attacks that exploit

This work was supported in part by the Technology Foundation (STW) through
project 12624-SIDES, by the Netherlands Organization for Scientific Research
NWO through Veni 2013 project 13114 and project ProFIL-628.001.007, and
the ICT COST action IC1204 TRUDEVICE. Permanent ID of this document:
14c4b76aa264503f89f93abc9baf72c3. Date: 2014-07-16

W. Meier and D. Mukhopadhyay (Eds.): INDOCRYPT 2014, LNCS 8885, pp. 21–36, 2014.
DOI: 10.1007/978-3-319-13039-2_2

the power-consumption leakage from a device running some cryptographic algorithm. Attacking elliptic-curve cryptosystems (ECC) with natural protection against side-channel attacks, e.g. implementations using Edwards curves, is quite challenging. This form of elliptic curves, proposed by Edwards in 2007 [14] and promoted for cryptographic applications by Bernstein and Lange [3], has several advantages compared to elliptic curves in Weierstrass form. For instance, the fast and complete formulas for addition and doubling make these types of curves more appealing for memory-constrained devices and at the same time resistant to classical simple power analysis (SPA) techniques. Although considered a very serious threat against ECC implementations, differential power analysis (DPA), as proposed in [12,24], cannot be applied directly to ECDSA or ephemeral Diffie-Hellman because the secret scalar is used only once. This is incompatible with the requirement of DPA to see large number of power traces of computations on the same secret data. In order to attack various asymmetric cryptosystems, new techniques that reside between SPA and DPA were developed; most notably collision [1,15,18,33,34,37] and template attacks [28,30,32]. The efficiency of most of those collision-based attacks is shown only on simulated traces; no practical experiments on real ECC implementations have verified these results. To the best of our knowledge, only two practical collision-based attacks on exponentiation algorithms are published, each of which relies on very specific assumptions and deals with very special cases. Hanley et al. exploit collisions between input and output operations of the same trace [16]. Wenger et al. in [35] performed a hardware-specific attack on consecutive rounds of a Montgomery ladder implementation. However, both attacks are very restrictive in terms of applicability to various ECC implementations as they imply some special implementation options, such as the use of López-Dahab coordinates, where field multiplications use the same key-dependent coordinate as input to two consecutive rounds. In contrast, our attack is much more generic as it applies to arbitrary choices of curves and coordinates, and many scalar multiplication algorithms.

Previous Work. Collision attacks exploit leakages by comparing two portions of the same or different traces to discover when values are reused. The Big Mac attack [34] is the first theoretical attack on public key cryptosystems, in which only a single trace is required to observe key dependencies and collisions during an RSA exponentiation. Witteman et al. in [36] performed a similar attack on the RSA modular exponentiation in the presence of blinded messages. Clavier et al. introduced in [11] horizontal correlation analysis, as a type of attack where a single power trace is enough to recover the private key. They also extended the Big Mac attack by using different distinguishers. Horizontal correlation analysis was performed on RSA using the Pearson correlation coefficient in [11] and triangular trace analysis of the exponent in [10]. The first horizontal technique relevant to ECC is the doubling attack, presented by Fouque and Valette in [15]. Homma et al. in [18] proposed a generalization of this attack to binary right-to-left, m-ary, and sliding-window methods. The most recent attack, proposed by Bauer et al. in [1], is a type of horizontal collision correlation attack on ECC, which combines atomicity and randomization techniques.

Template attacks are a combination of statistical modeling and power-analysis attacks consisting of two phases, as follows. The first phase is the *profiling* or *template-building* phase, where the attacker builds templates to characterize the device by executing a sequence of instructions on fixed data. The second phase is the matching phase, in which the attacker matches the templates to actual traces of the device. The attacker is assumed to possess a device which behaves the same as the target device, in order to build template traces while running the same implementation as the target. Medwed and Oswald demonstrated in [28] a practical template attack on ECDSA. However, their attack required an offline DPA attack on the EC scalar-multiplication operation during the template-building phase, in order to select the points of interest. They also need 33 template traces per key-bit. Furthermore, attacks against ECDSA and other elliptic-curve signature algorithms only need to recover a few bits of the ephemeral scalar for multiple scalar multiplications with different ephemeral scalars and can then employ lattice techniques to recover the long-term secret key [2, 30, 32]. This is not possible in the context of ephemeral Diffie-Hellman: an attacker gets only a single trace and needs to recover sufficiently many bits of this ephemeral scalar from side-channel information to be able to compute the remaining bits through, for example, Kangaroo techniques.

Our Contribution. In this paper we introduce an adaptive template-attack technique, which we call *Online Template Attacks* (OTA). This technique is able to recover a complete scalar from only one power trace of a scalar multiplication using this scalar. The attack is characterized as *online*, because we create the templates *after* the acquisition of the target trace. While we use the same terminology, our attack is not a typical template attack; i.e. no preprocessing template-building phase is necessary. Our attack functions by acquiring one target trace from the device under attack and comparing patterns of certain operations from this trace with templates obtained from the attacker's device that runs the same implementation. Pattern matching is performed at suitable points in the algorithm, where key-bit related assignments take place by using an automated module based on the Pearson correlation coefficient.

The attacker needs only very limited control over the device used to generate the online template traces. The main assumption is that the attacker can choose the input point to a scalar multiplication, an assumption that trivially holds even without any modification to the template device in the context of ephemeral Diffie-Hellman. It also holds in the context of ECDSA, if the attacker can modify the implementation on the template device or can modify internal values of the computation. This is no different than for previous template attacks against ECDSA.

Our methodology offers a generic attack framework, which is applicable to various forms of curves (Weierstrass, Edwards and Montgomery curves) and implementations. As a proof of concept, we attack the doubling operation in the double-and-add-always algorithm. Contrary to the doubling attack [15], our attack can be launched against right-to-left algorithms and Montgomery ladder. We further note that Medwed and Oswald perform a very special template attack based on a set of assumptions: DPA performed in advance to find intermediate points for templates, implementation with Hamming-weight leakage and applicability only to ECDSA. Online template attacks do not have these restrictions, they need only a single target trace, and only a single template trace per key-bit. The advantages of our attack over previously proposed attacks are the following:

- It does not require any cumbersome preprocessing template-building phase, but a rather simple post-processing phase.
- It does not assume any previous knowledge of the leakage model.
- It does not require full control of the device under attack.
- It works against SPA-protected and some DPA-protected implementations with unified formulas for addition and doubling.
- Countermeasures such as scalar randomization and changing point representation from affine to (deterministic) projective representation inside the implementation do not prevent our attack.
- It is applicable to the Montgomery ladder and to constant-time (left-to-right and right-to-left) exponentiation algorithms.
- It is experimentally confirmed on an implementation of double-and-add-always scalar multiplication on the twisted Edwards curve used in the Ed25519 signature scheme.

Online template attacks require only one *target trace* and one *online template trace* per key-bit. We can, therefore, claim that our technique demonstrates the most efficient practical side-channel attack applicable to ephemeral-scalar ECC. When applied to ECDSA, the proposed attack can be used in combination with lattice techniques similar to [2,32], in order to derive the whole private key from a few bits of multiple ephemeral keys.

Organization of the Paper. This paper is organized as follows. We introduce and explain OTA in Section 2. Section 3 gives specific examples of how the attack applies to different scalar-multiplication algorithms. Section 4 presents our practical OTA on double-and-add-always scalar multiplication. A discussion of how the proposed attack can be applied to implementations that include countermeasures that randomize the algorithm or operands is given in Section 5. Finally, Section 6 summarizes our contribution and concludes the paper.

2   Online Template Attacks

We define an online template attack as a side-channel attack with the following conditions:

1. The attacker obtains only one power trace of the cryptographic algorithm involving the targeted secret data. This trace is called the *target trace*. We call the device from which the target trace is obtained the *target device*. This property makes it possible to attack scalar multiplication algorithms with ephemeral scalar and with randomized scalar.

2. The attacker is generating template traces *after* having obtained the target trace. These traces are called *(online) template traces*.
3. The attacker obtains the template traces on the target device or a similar device[1] *with very limited control over it*, i.e. access to the device to run several executions with chosen public inputs. The attacker does not rely on the assumption that the secret data is the same for all template traces.
4. At least one assignment in the exponentiation algorithm is made depending on the value of particular scalar bit(s), but there are no branches with key-dependent computations. Since we are attacking the doubling operation, this key-dependent assignment should be during doubling. As a counterexample, we note that the binary right-to-left add-always algorithm for Lucas recurrences [21] is resistant to the proposed attack, because the result of the doubling is stored in a non-key-dependent variable.

In the following we show that online template attacks are feasible and can be applied against implementations of various scalar-multiplication algorithms. In fact, we show that we need only a single template trace per scalar bit. Transfer of the approach to the corresponding exponentiation algorithms (for example in RSA or DSA) is straight-forward. Transfer to other cryptographic algorithms is clearly not trivial; we consider online template attacks as a specialized means to attack scalar multiplication and exponentiation algorithms.

2.1 Attack Description

Template attacks consist of two phases, *template building* for characterizing the device and *template matching*, where the characterization of the device together with a power trace from the device under attack are used to determine the secret [27]. Therefore, the first condition of our proposed attack is typically fulfilled by all attacks of this kind.

It is well known that template attacks against scalar multiplication can generate templates "on-the-fly", i.e., interleaving the template building and matching phases. See, for example, [28, Sec. 5.3]. We take this idea further by building templates after the target trace has been obtained (condition 2). The attacker, being able to do things in this order, needs only limited control over the target device. Moreover, the attacker is not affected by randomization of the secret data during different executions of the algorithm, since he always has to compare his template traces with the same target trace.

The basic idea consists of comparing the traces for inputs P (target trace) and $2P$ (online template trace) while executing scalar multiplication and then finding similar patterns between them, based on hypothesis on a bit for a given operation. The target trace is obtained only once. For every bit of the scalar, we need to obtain an online template trace with input $kP, k \in \mathbb{Z}$, where k is chosen as a function of our hypothesis on this bit. We hereby note that the

---

[1] By similar device we mean the same type of microcontroller running the same algorithm.

template trace is part of the target trace (for instance it corresponds to the first doubling) and it is compared bit-by-bit with the target trace. Therefore, alignment of traces is not necessary.

We performed pattern matching for our traces using an automated module based on the Pearson correlation coefficient, $\rho(X,Y)$, which measures the linear relationship between two variables X and Y. For power traces, the correlation coefficient shows the relationship between two points of the trace, which indicates the Hamming-weight leakage of key-dependent assignments during the execution of a cryptographic algorithm. Extensions to other leakage models and distinguishers are straightforward. Our pattern matching corresponds to a list of the correlation coefficients that show the relationship between all samples from the template trace to the same consecutive amount of samples in the target trace. If our hypothesis on the given key-bit is correct, then the pattern match between our traces at the targeted operation will be high (in our experiments it reached 99%).

In this way we can recover the first i bits of the key. Knowledge of the first i bits provides us with complete knowledge of the internal state of the algorithm just before the $(i+1)^{th}$ bit is processed. Since at least one operation in the loop depends on this bit, we can make a hypothesis about the $(i+1)^{th}$ bit, compute an online template trace based on this hypothesis, and correlate this trace with the target trace at the relevant predetermined point of the algorithm.

3   Applying the Attack to Scalar-Multiplication Algorithms

3.1   Attacking the Double-and-Add-Always Algorithm

The core idea and feasibility of the attack is demonstrated through an example to the double-and-add-always algorithm described in Algorithm 1. We note that the first execution of the loop always starts by doubling the input point $\boldsymbol{P}$, for all values of k. We assume that $k_{x-1}=1$. Depending on the second-most significant key bit k_{x-2}, the output of the first iteration of the algorithm will be either $2\boldsymbol{P}$ or $3\boldsymbol{P}$. For any point $\boldsymbol{P}$ we can, therefore, get a power trace for the operation $2\boldsymbol{P}$, i.e. we let the algorithm execute the first two double-and-add iterations. In our setup, we can zoom into the level of one doubling, which will be our target trace. Then we perform the same procedure with $2\boldsymbol{P}$ as the input point to obtain the online template trace that we want to compare with the target trace. If we assume that the second-most significant bit of k is 0, then we compare the $2\boldsymbol{P}$ template with the output of the doubling at first iteration. Otherwise, we compare it with the online template trace for $3\boldsymbol{P}$.

Assuming that the first $(i-1)$ bits of k are known, we can derive the i-th bit by computing the two possible states of $\boldsymbol{R}_0$ after this bit has been treated and in this way recover the key iteratively. Note that only the assignment in the ith iteration depends on the key-bit k_i, but none of the computations do, so we need to compare the trace of the doubling operation in the $(i+1)^{th}$ iteration with

---

Algorithm 1. The left-to-right double-and-add-always algorithm

---
Input: P, $k = (k_{x-1}, k_{x-2}, \ldots, k_0)_2$
Output: $Q = k \cdot P$

$R_0 \leftarrow P$;
for $i \leftarrow x - 2$ **down to** 0 **do**
    | $R_0 \leftarrow 2R_0$;
    | $R_1 \leftarrow R_0 + P$;
    | $R_0 \leftarrow R_{k_i}$;
end
return R_0

Algorithm 2. Binary right-to-left double-and-add-always algorithm

---
Input: P, $k = (k_{x-1}, k_{x-2}, \ldots, k_0)_2$
Output: $Q = k \cdot P$
$R_0 \leftarrow O$;
$R_1 \leftarrow P$;
For $i \leftarrow 0$ **up to** x-1 $b \leftarrow 1 - k_i$;
$R_b \leftarrow 2R_b$;
$R_b \leftarrow R_b + R_{k_i}$;
return R_0

---

our original target trace. To decide whether the ith bit of k is zero or one, we compare the trace that the doubling operation in the $(i + 1)^{th}$ iteration would give for $k_{i+1} = 0$ with the target trace. For completeness, we can compare the target trace with a trace obtained for $k_{i+1} = 1$ and verify that it has lower pattern match percentage; in this case, the performed attack needs two online template traces per key bit. However, if during the acquisition phase the noise level is low and the signal is of good quality, we can perform an efficient attack with only our target trace and a single trace for the hypothetical value of $R_{k_{i+1}}$.

Attacking the right-to-left double-and-add-always algorithm of [21] is a type of key-dependent assignment OTA. We target the doubling operation and note that the input point will be doubled either in the first (if $k_0 = 0$) or in the second iteration of the loop (if $k_0 = 1$). If k is fixed we can easily decide between the two by inputting different points, since if $k_0 = 1$ we will see the common operation $2O$. If the k is not fixed, we simply measure the first two iterations and again use the operation $2O$ if the template generator should use the first or second iteration. Once we are able to obtain clear traces, the attack itself follows the general description of Sect. 2. If we assume that the first i bits of k are known and we wish to derive the $(i + 1)^{th}$ bit, this means that we know the values of R_0 and R_1 at the start of the $(i + 1)^{th}$ iteration. By making a hypothesis on the value of the $(i+1)^{th}$ key bit, we can decide according to the matching percentage if R_0 or R_1 was used.

3.2   Attacking the Montgomery Ladder

The Montgomery Ladder, initially presented by Montgomery in [29] as a way to speed up scalar multiplication on elliptic curves, and later used as the primary secure and efficient choice for resource-constrained devices, is one of the most challenging algorithms for simple side-channel analysis due to its natural regularity of operations. A comprehensive security analysis of the Montgomery ladder given by Joye and Yen in [23] showed that the regularity of the algorithm makes it intrinsically protected against a large variety of implementation attacks (SPA, some fault attacks, etc.). For a specific choice of projective coordinates for the Montgomery ladder, as described in Algorithm 3, one can do computations with only X and Z coordinates, which makes this option more memory efficient than other algorithms.

Algorithm 3. The Montgomery Ladder

Input: P, $k = (k_{x-1}, k_{x-2}, \ldots, k_0)_2$
Output: $Q = k \cdot P$

$R_0 \leftarrow P$;
$R_1 \leftarrow 2P$;
for $i \leftarrow x - 2$ **down to** 0 **do**
$\quad b \leftarrow 1 - k_i$;
$\quad R_b \leftarrow R_0 + R_1$;
$\quad R_{k_i} \leftarrow 2 \cdot R_{k_i}$;
end
return R_0

The main observation that makes our attack applicable to the Montgomery ladder is that at least one of the computations, namely the doubling in the main loop, directly depends on the key-bit k_i. For example, if we assume that the first three bits of the key are 100, then the output of the first iteration will be $R_0 = 2P$. If we assume that the first bits are 110, then the output of the first iteration will be $R_0 = 3P$. Therefore, if we compare the pattern of the output of the first iteration of Algorithm 3 with scalar $k = 100$, we will observe higher correlation with the pattern of $R_0 = 2P$ than with the pattern of $R_0 = 3P$.

3.3   Attacking Side-Channel Atomicity

Side-channel atomicity is a countermeasure proposed by Chevallier-Mames et al. [9], in which individual operations are implemented in such a way that they have an identical side-channel profile (e.g. for any branch and any key-bit related subroutine). In short, it is suggested in [9] that the point doubling and addition operations are implemented such that the same code is executed for both operations. This renders the operations indistinguishable by simply inspecting a suitable side-channel. One could, therefore, implement an exponentiation as described in Algorithm 4.

Algorithm 4. Side-Channel Atomic double-and-add algorithm

Input: P, $k = (k_{x-1}, k_{x-2}, \ldots, k_0)_2$

Output: $Q = k \cdot P$

$R_0 \leftarrow O$; $R_1 \leftarrow P$; $i \leftarrow x - 1$;

$n \leftarrow 0$;

while $i \geq 0$ **do**

  $R_0 \leftarrow R_0 + R_n$;

  $n \leftarrow n \oplus k_i$;

  $i \leftarrow i - \neg n$;

end

return R_0

There are certain choices of coordinates and curves where this approach can be deployed by using unified or complete addition formulas for the group operations. For example, the Jacobi form [26] and Hessian [22] curves come with a unified group law. Edwards curves [6, 7] even have a complete group law. For Weierstrass curves, Brier and Joye suggest an approach for unified addition in [8].

Simple atomic algorithms do not offer any protection against online template attacks, because the regularity of point operations does not prevent mounting this sort of attack. The point $2P$, as output of the third iteration of Algorithm 4, will produce a power trace with very similar pattern to the trace that would have the point $2P$ as input. Therefore, the attack will be the similar as the one described in Sect. 3.1; the only difference is that instead of the output of the second iteration of the algorithm, we have to focus on the pattern of the third iteration. In general, when an attacker forms a hypothesis about a certain number of bits of k, the hypothesis will include the point in time where R_0 will contain the predicted value. This will mean that an attacker would have to acquire a larger target trace to allow all hypotheses to be tested.

4 Experimental Results

This section presents our experimental results. Firstly, in Sect. 4.1 we describe the attacked implementation and the experimental setup that we use to perform attacks. Secondly, we present experimental results of an OTA with projective input in Section 4.2; in particular, we present the results when we perform the attack bit-by-bit iteratively or in group of five bits. Finally, Sect. 4.3 presents an OTA with affine input.

4.1 Target Implementation and Experimental Setup

To validate feasibility and efficiency of our proposed method, we attack an elliptic-curve scalar multiplication implementation running on an "ATmega card", i.e., an ATmega163 microcontroller [13] in a smart card. To illustrate that our attack also works if the template device is not the same as the target device, we used two different smart cards: one to obtain the target trace and one to obtain the online template traces.

Our measurement setup uses a Picoscope 5203[2] with sampling rate of 125M samples per second for both target trace and online template traces.

This oscilloscope has limited acquisition memory buffer to 32M samples. Since 5 iterations of the scalar multiplication algorithm take around 235 ms, it means that with sampling rate of 125M samples per second we can record a trace of approximately 29.4M samples.

The scalar multiplication algorithm is based on the curve arithmetic of the Ed25519 implementation presented in [19], which is available online at http:// cryptojedi.org/crypto/#avrnacl. The elliptic curve used in Ed25519 is the twisted Edwards curve $E : -x^2 + y^2 = 1 + dx^2y^2$ with $d = -(121665/121666)$ and base point

$$P = {\scriptstyle (15112221349535400772501151409588533151145401269304185720604611328394984776220 2,}$$
$$\scriptstyle 46316835694926478169428394003475163141307993866256225615783033603165251855960).$$

For more details on Ed25519 and this specific curve, see [4,5].

We modified the software to perform a double-and-add-always scalar multiplication (see Algorithm 1). The whole underlying field and curve arithmetic is the same as in [19]. This means in particular that points are internally represented in extended coordinates as proposed in [17]. In this coordinate system a point $P = (x, y)$ is represented as $(X : Y : Z : T)$ with $x = X/Z, y = Y/Z, x \cdot y = T/Z$.

4.2   Online Template Attack with Projective Input

In this subsection we describe how to apply an OTA if the input supplied to the scalar multiplication is in projective (or extended) coordinates, i.e, if the attacker has full control over all coordinates of the starting point. This is a realistic assumption if a protocol avoids inversions entirely and protects against leakage of projective coordinates by randomization as proposed in [31, Sec. 6].

The attack targets the output of the doubling operation. We performed pattern matching for our traces as described in Section 2.1. In this way, we could determine the leakage of key-dependent assignments during the execution of the algorithm.

We first demonstrate how to attack a single bit and then we present our results from recovering the five most significant unknown bits of the scalar (recall that the highest bit is always set to one; see Algorithm 1). The remaining bits can be attacked iteratively in the same way as described in Section 2.1; as stated above we were not able to do so due to technical limitations of our measurement setup.

The first observation from our experiments is that when we execute the same algorithm with the same input point on two different cards, there is a constant vertical misalignment between the two obtained traces, but the patterns look

---

[2] http://www.picotech.com/discontinued/PicoScope5203.html

almost identical. This fact validates our choice of the correlation coefficient as our pattern-matching metric, since this metric does not depend on the difference in absolute values and therefore the constant misalignment does not affect the results.

For our target trace, we compute a multiple of a point P. We know that the most significant bit of the scalar is 1, so after the first iteration of the double-and-add-always loop the value of R_0 is either $2P$ (if the second bit of k is zero) or $3P$ (if the second bit of k is one).

To determine the second bit of the secret scalar k, we generate template traces by inputting exactly the projective representations of $2P$ and $3P$ and computing the correlation of the first iteration of the template trace with the second iteration of the target trace.

In fact, from our experiments we observe that the correlation between the correct template trace and the target trace is so much higher than between the wrong template trace and the target trace, that just one of the two template traces is sufficient to determine the second bit of k. [3]

For validation of our results, we conducted several experiments with different input points from the target card and the template card, and computed the correlation in the obtained power traces. We notice that the trace obtained from the point $2P$ is almost identical to the pattern obtained from the target trace; as expected the correlation is at least 97% for all our experiments. On the other hand, the percentage correlation of the target trace with the template trace for $3P$ is at most 83%. To determine the value of one bit, we can thus simply compute only one template trace, and decide the value of the targeted bit depending on whether the correlation is above or below a certain threshold set somewhere between 83% and 97%.

The results presented so far are obtained while attacking one single bit of the exponent. When we attack five bits with one acquisition, we observe lower numbers for pattern matching for both the correct and the wrong scalar guess. The correlation results for pattern matching are not so high, mainly due to the noise that is occurring in our setup during longer acquisitions. This follows from the fact that our power supply is not perfectly stable during acquisitions that are longer than 200 ms. However, the difference between correct and wrong assumptions is still remarkable. Correct bit assumptions have $84 - 88\%$ matching patterns, while the percentage for wrong assumptions drops to $50 - 72\%$. Therefore, we can set a threshold for recognizing a bit to be at 80%.

Note that the attack with projective inputs does not make any assumptions on formulas used for elliptic-curve addition and doubling. In fact, we carried out the attack for specialized doublings and for doublings that use the same unified addition formulas as addition and obtained similar results.

---

[3] Figures from experiments and measurements for different points and cards can be found in the full version of the paper in the IACR ePrint archive.

4.3   Online Template Attack with Affine Input

The attack as explained in the previous section makes the assumption that the attacker has full control over the input in projective coordinates. Most implementations of ECC use inputs in affine (or compressed affine) coordinates and internally convert to projective representation. We now explain how to adapt the attack to also handle those cases.

The input is now given as (x, y) and at the beginning of the computation converted to $(x : y : 1 : xy)$. However, already after the first iteration of the double-and-add-always loop, $Z = 1$ does not hold anymore. In the following we consider an attack on the second-most significant bit (which is again set to zero) and input point P of the target trace. After one iteration of the double-and-add-always loop, the value of R_0 is determined by the value of the second-most significant scalar bit. Choosing the affine versions of $2P$ and $3P$ to generate template traces does not help us now, because they do not have any coordinates in common with the projective representations used internally. To successfully perform the attack we need to modify our approach and take a closer look at the formulas used for point doubling. We illustrate the approach with the specialized doubling formulas from [17]. For details, see http://www.hyperelliptic.org/EFD/g1p/auto-twisted-extended-1.html#doubling-dbl-2008-hwcd. These doubling formulas begin with the following operations:

```
A = X^2,  B = Y^2,  C = 2*Z^2 ...
```

Our idea is not to attack a whole doubling operation but just a single squaring; in the following example we attack the squaring $B = Y^2$. The idea is to use input points Q and R for the template traces, such that the y-coordinate of Q is the same as the Y-coordinate in the internal projective representation of $2P$ and the y-coordinate of R is the same as Y-coordinate of the internal projective representation of $3P$. Unfortunately, such points do not always exist, but our experiments showed that it is sufficient to select points Q and R such that their y-coordinate is *almost the same* as the Y-coordinate of the respective internal projective representation. By almost the same we mean that the y-coordinate is allowed to differ in one bit. This flexibility in choosing the template input allows us to find suitable points with overwhelmingly large probability. When we compare the traces for P as input at the second iteration to the trace for Q at the first iteration during the second squaring operation (computing B) then we can observe that the two traces are almost identical; see Figure 1 for details. This figure is taken from an experiment where we have an exact match of the y-coordinate, i.e., we did not have to flip one bit in the expected internal value to find a suitable affine template point.

For validation of our result, we conducted several experiments with different input points using one card (for the sake of simplicity), and found the correlation in the obtained power traces. Let us assume that the scalar is $k = 10$ (let us recall the the most significant bit is always set to 1). Figure 2 shows the pattern match between a template trace during computation of B of input point Q (iteration 1) to the target trace for P' (iteration 2) and the pattern match

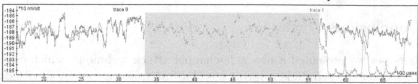

Fig. 1. Comparison between P' at the second iteration to Q at first iteration; the area of computing B is highlighted

between the template trace (iteration 1) for R to the target trace (iteration 2). We notice that the trace obtained from the point Q is almost identical to the pattern obtained from the target trace; as expected, the correlation is at least 96% for exactly matching y-coordinate of the template point and >91% for almost matching y-coordinate. For the non-matching template point the pattern match is at most 84%.

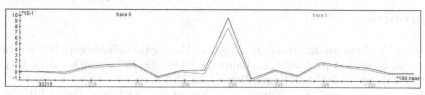

Fig. 2. Pattern Matching Q to P' and R to P' coming from the same card

5 Countermeasures and Future Work

Coron's first and second DPA countermeasures result in scalar or point being blinded to counteract the statistical analysis of DPA attacks [12]. Given that an attacker needs to predict the intermediate state of an algorithm at a given point in time, we can assume that the countermeasures that are used to prevent DPA will also have an effect on the OTA. All proposed countermeasures rely on some kind of randomization, which can be of either a scalar, a point or the algorithm itself. However, if we assume that the attacker has no technical limitations, i.e an oscilloscope with enough memory to acquire the power consumption during an entire scalar-multiplication, it would be possible to derive the entire scalar being used from just one acquisition. Therefore, if one depends on scalar blinding [12,25], this method provides no protection against our attack, as the attacker could derive a value equivalent to the exponent used.

There are methods for changing the representation of a point, which can prevent OTA and make the result unpredictable to the attacker. Most notably those countermeasures are randomizing the projective randomization and randomizing the coordinates through a random field isomorphism as described in [20]. However, inserting a point in affine coordinates and changing to (deterministic) projective coordinates during the execution of the scalar multiplication (compressing and decompressing of a point), does not affect our attack.

We aim exclusively at the doubling operation in the execution of each algorithm. Since most of the blinding techniques are based on the cyclic property of the elliptic curve groups, attacking the addition operation would be an interesting future research topic.

6   Conclusions

In this paper we presented a new side-channel attack technique, which can be used to recover the private key during a scalar-multiplication on ECC with only one target trace and one online template trace per bit. Our attack succeeds against a protected target implementation with unified formulas for doubling and adding and against implementations where the point is given in affine coordinates and changes to projective coordinates representation. By performing our attack on two physically different devices, we showed that key-dependent assignments leak, even when there are no branches in the cryptographic algorithm. This fact enhances the feasibility of OTA and validates our initial claim that one target trace is enough to recover the secret scalar.

References

1. Bauer, A., Jaulmes, E., Prouff, E., Wild, J.: Horizontal collision correlation attack on elliptic curves. In: Lange, T., Lauter, K., Lisoněk, P. (eds.) SAC 2013. LNCS, vol. 8282, pp. 553–570. Springer, Heidelberg (2014)
2. Benger, N., van de Pol, J., Smart, N.P., Yarom, Y.: "Ooh aah... just a little bit": A small amount of side channel can go a long way. Cryptology ePrint Archive, Report 2014/161 (2014)
3. Bernstein, D.J., Birkner, P., Joye, M., Lange, T., Peters, C.: Twisted Edwards curves. In: Vaudenay, S. (ed.) AFRICACRYPT 2008. LNCS, vol. 5023, pp. 389–405. Springer, Heidelberg (2008)
4. Bernstein, D.J., Duif, N., Lange, T., Schwabe, P., Yang, B.-Y.: High-speed high-security signatures. In: Preneel, B., Takagi, T. (eds.) CHES 2011. LNCS, vol. 6917, pp. 124–142. Springer, Heidelberg (2011)
5. Bernstein, D.J., Duif, N., Lange, T., Schwabe, P., Yang, B.Y.: High-speed high-security signatures. Journal of Cryptographic Engineering 2(2), 77–89 (2012)
6. Bernstein, D.J., Lange, T.: Faster addition and doubling on elliptic curves. In: Kurosawa, K. (ed.) ASIACRYPT 2007. LNCS, vol. 4833, pp. 29–50. Springer, Heidelberg (2007)
7. Bernstein, D.J., Lange, T., Rezaeian Farashahi, R.: Binary Edwards curves. In: Oswald, E., Rohatgi, P. (eds.) CHES 2008. LNCS, vol. 5154, pp. 244–265. Springer, Heidelberg (2008)
8. Brier, E., Joye, M.: Weierstraß elliptic curves and side-channel attacks. In: Naccache, D., Paillier, P. (eds.) PKC 2002. LNCS, vol. 2274, pp. 335–345. Springer, Heidelberg (2002)
9. Chevallier-Mames, B., Ciet, M., Joye, M.: Low-cost solutions for preventing simple side-channel analysis: Side-channel atomicity. IEEE Transactions on Computers 53(6), 760–768 (2004)
10. Clavier, C., Feix, B., Gagnerot, G., Giraud, C., Roussellet, M., Verneuil, V.: ROSETTA for single trace analysis. In: Galbraith, S., Nandi, M. (eds.) INDOCRYPT 2012. LNCS, vol. 7668, pp. 140–155. Springer, Heidelberg (2012)
11. Clavier, C., Feix, B., Gagnerot, G., Roussellet, M., Verneuil, V.: Horizontal correlation analysis on exponentiation. In: Soriano, M., Qing, S., López, J. (eds.) ICICS 2010. LNCS, vol. 6476, pp. 46–61. Springer, Heidelberg (2010)

12. Coron, J.-S.: Resistance against differential power analysis for elliptic curve cryptosystems. In: Koç, Ç.K., Paar, C. (eds.) CHES 1999. LNCS, vol. 1717, pp. 292–302. Springer, Heidelberg (1999)
13. Atmel Corporation. ATMEL AVR32UC technical reference manual. ARM Doc Rev. 32002F (2010)
14. Edwards, H.M.: A normal form for elliptic curves. In: Koç, Ç.K., Paar, C. (eds.) Bulletin of the American Mathematical Society, vol. 44, pp. 393–422 (2007)
15. Fouque, P.-A., Valette, F.: The doubling attack – *Why upwards is better than downwards*. In: Walter, C.D., Koç, Ç.K., Paar, C. (eds.) CHES 2003. LNCS, vol. 2779, pp. 269–280. Springer, Heidelberg (2003)
16. Hanley, N., Kim, H., Tunstall, M.: Exploiting collisions in addition chain-based exponentiation algorithms using a single trace. Cryptology ePrint Archive, Report 2012/485 (2012)
17. Hisil, H., Wong, K.K.-H., Carter, G., Dawson, E.: Twisted Edwards curves revisited. In: Pieprzyk, J. (ed.) ASIACRYPT 2008. LNCS, vol. 5350, pp. 326–343. Springer, Heidelberg (2008)
18. Homma, N., Miyamoto, A., Aoki, T., Satoh, A., Shamir, A.: Collision-based power analysis of modular exponentiation using chosen-message pairs. In: Oswald, E., Rohatgi, P. (eds.) CHES 2008. LNCS, vol. 5154, pp. 15–29. Springer, Heidelberg (2008)
19. Hutter, M., Schwabe, P.: NaCl on 8-Bit AVR microcontrollers. In: Youssef, A., Nitaj, A., Hassanien, A.E. (eds.) AFRICACRYPT 2013. LNCS, vol. 7918, pp. 156–172. Springer, Heidelberg (2013)
20. Joye, M.: Smart-card implementation of elliptic curve cryptography and DPA-type attacks. In: Quisquater, J.-J., Paradinas, P., Deswarte, Y., El Kalam, A.A. (eds.) Smart Card Research and Advanced Applications VI. IFIP, vol. 135, pp. 115–125. Springer, Heidelberg (2004)
21. Joye, M.: Highly regular right-to-left algorithms for scalar multiplication. In: Paillier, P., Verbauwhede, I. (eds.) CHES 2007. LNCS, vol. 4727, pp. 135–147. Springer, Heidelberg (2007)
22. Joye, M., Quisquater, J.-J.: Hessian elliptic curves and side-channel attacks. In: Koç, Ç.K., Naccache, D., Paar, C. (eds.) CHES 2001. LNCS, vol. 2162, pp. 402–410. Springer, Heidelberg (2001)
23. Joye, M., Yen, S.M.: The Montgomery powering ladder. In: Kaliski, B.S., Koç, Ç.K., Paar, C. (eds.) CHES 2002. LNCS, vol. 2523, pp. 291–302. Springer, Heidelberg (2002)
24. Kocher, P.C., Jaffe, J., Jun, B.: Differential power analysis. In: Wiener, M. (ed.) CRYPTO 1999. LNCS, vol. 1666, pp. 388–397. Springer, Heidelberg (1999)
25. Kocher, P.C.: Timing attacks on implementations of Diffie-Hellman, RSA, DSS, and other systems. In: Koblitz, N. (ed.) CRYPTO 1996. LNCS, vol. 1109, pp. 104–113. Springer, Heidelberg (1996)
26. Liardet, P., Smart, N.P.: Preventing SPA/DPA in ECC systems using the jacobi form. In: Koç, Ç.K., Naccache, D., Paar, C. (eds.) CHES 2001. LNCS, vol. 2162, pp. 391–401. Springer, Heidelberg (2001)
27. Mangard, S., Oswald, E., Popp, T.:Power Analysis Attacks: Revealing the Secrets of Smart Cards (Advances in Information Security). Springer New York Inc. (2007)
28. Medwed, M., Oswald, E.: Template attacks on ECDSA. In: Chung, K.-I., Sohn, K., Yung, M. (eds.) WISA 2008. LNCS, vol. 5379, pp. 14–27. Springer, Heidelberg (2009)
29. Montgomery, P.L.: Speeding the Pollard and elliptic curve methods of factorization. Mathematics of Computation **48**(177), 243–264 (1987)

30. De Mulder, E., Hutter, M., Marson, M.E., Pearson, P.: Using Bleichenbacher's solution to the hidden number problem to attack nonce leaks in 384-Bit ECDSA. In: Bertoni, G., Coron, J.-S. (eds.) CHES 2013. LNCS, vol. 8086, pp. 435–452. Springer, Heidelberg (2013)
31. Naccache, D., Smart, N.P., Stern, J.: Projective coordinates leak. In: Cachin, C., Camenisch, J.L. (eds.) EUROCRYPT 2004. LNCS, vol. 3027, pp. 257–267. Springer, Heidelberg (2004)
32. Römer, T., Seifert, J.-P.: Information leakage attacks against smart card implementations of the elliptic curve digital signature algorithm. In: Attali, S., Jensen, T. (eds.) E-smart 2001. LNCS, vol. 2140, pp. 211–219. Springer, Heidelberg (2001)
33. Schramm, K., Wollinger, T., Paar, C.: A new class of collision attacks and Its application to DES. In: Johansson, T. (ed.) FSE 2003. LNCS, vol. 2887, pp. 206–222. Springer, Heidelberg (2003)
34. Walter, C.D.: Sliding windows succumbs to Big Mac attack. In: Koç, Ç.K., Naccache, D., Paar, C. (eds.) CHES 2001. LNCS, vol. 2162, pp. 286–299. Springer, Heidelberg (2001)
35. Wenger, E., Korak, T., Kirschbaum, M.: Analyzing side-channel leakage of RFID-suitable lightweight ECC hardware. In: Hutter, M., Schmidt, J.-M. (eds.) RFIDsec 2013. LNCS, vol. 8262, pp. 128–144. Springer, Heidelberg (2013)
36. Witteman, M.F., van Woudenberg, J.G.J., Menarini, F.: Defeating RSA multiply-always and message blinding countermeasures. In: Kiayias, A. (ed.) CT-RSA 2011. LNCS, vol. 6558, pp. 77–88. Springer, Heidelberg (2011)
37. Yen, S.-M., Ko, L.-C., Moon, S.-J., Ha, J.C.: Relative doubling attack against montgomery ladder. In: Won, D.H., Kim, S. (eds.) ICISC 2005. LNCS, vol. 3935, pp. 117–128. Springer, Heidelberg (2006)

Improved Multi-Bit Differential Fault Analysis of Trivium

Prakash Dey and Avishek Adhikari[✉]

Department of Pure Mathematics, University of Calcutta, Kolkata 700019, India
{pdprakashdey,avishek.adh}@gmail.com

Abstract. Very few differential fault attacks (DFA) were reported on Trivium so far. In 2012, Yupu Hu et al. [4] relaxed adversarial power and allowed faults in random area within eight neighbouring bits at random time but with the major limitation that after each fault injection, the fault positions must not be from different registers. In this paper we present a generic attack strategy that allows the adversary to challenge the cipher under different multi-bit fault models with faults at any unknown random keystream generation round even if bit arrangement of the actual cipher device is unknown and thereby removing the limitation of Yupu Hu et al. To the best of our knowledge, this paper assumes the weakest adversarial power ever considered in the open literature for DFA on Trivium. In particular, if faults are allowed in random area within nine neighbouring bits at random time anywhere in the three registers and the fault injection (at keystream generation) rounds are uniformly distributed over $\{t, \ldots, t+49\}$, for any unknown $t \geq 1$, then 4 faults always break the cipher, which is a significant improvement over Yupu Hu et al.

Keywords: Stream Cipher · Differential Fault Attack · Multi-Bit Faults · SAT Solver

1 Introduction

In Differential Fault Attacks, faults are injected into the internal state of the cipher and from the difference of the normal and the faulty keystream, information about the internal state is partially or completely deduced. Faults can be injected in a register by under-powering and power spikes, clock glitches, temperature attacks, optical attacks, electromagnetic fault injection, etc.

State of the Art and Our Contribution. Very few differential fault attacks (such as [2–4,6]) were reported on Trivium. Papers [2,3,6] considered strong adversary capable of injecting single bit faults precisely, where as [4] relaxed

Research supported in part by National Board for Higher Mathematics, Department of Atomic Energy, Government of India (No 2/48(10)/2013/NBHM(R.P.)/R&D II/695)

© Springer International Publishing Switzerland 2014
W. Meier and D. Mukhopadhyay (Eds.): INDOCRYPT 2014, LNCS 8885, pp. 37–52, 2014.
DOI: 10.1007/978-3-319-13039-2_3

adversarial power and allowed faults in random area within eight neighbouring bits at random time but with the major limitation that after each fault injection, the fault positions must not be from different registers.

In this paper we present a generic attack strategy that allows the adversary to challenge the cipher under different multi-bit fault models with faults at any unknown random keystream generation (KSG) round even if bit arrangement of the actual cipher device is unknown and thereby, upto the best of our knowledge our work makes the adversarial power weakest ever considered in the open literature for DFA on Trivium.

In particular, this paper (in Appendix B) demonstrate and compare the attack strategy by choosing realistic fault models when a fault injection trial can only disturb a random area within k neighbouring bits, $k \in \{1, 2, 3, 4, 5\}$, at any known KSG round but the exact number of bits the injected fault has altered or their locations are unknown. Tapping statistical weakness of the cipher under DFA and using algebraic cryptanalysis method and SAT solver Cryptominisat-2.9.6 installed in SAGE, the internal state of the cipher can be completely recovered with as little as 3 (on average $2.021, 2.026, 2.054, 2.058$ respectively for $k = 2, 3, 4, 5$) distinct random faults. $k = 1$ results in the same fault model considered in [6]. For $k \in \{1, 2, 3, 4, 5\}$ fault locations are uniquely identified. Then we show that for $k = 5$, if faults are injected at unknown random KSG round, our attack strategy detects both the fault location and fault injection round with very high probability. Next we relax adversarial power and allow all possible faults. We show that with high probability bad-faults can be rejected (Section 7). Later we assume the fault model for $k = 9$ which is even weaker than [4] and show that 3 and 4 faults respectively always break the cipher in known and random fault injection KSG round.

2 Description of the Cipher

In this paper both the classical [1] as well as the floating representation [6] of Trivium are considered. Here we briefly describe the floating representation. The internal state IS_i of Trivium consists of three non-linear registers A, B and C with inner states $A_i = (\alpha_i, \ldots, \alpha_{i+92})$, $B_i = (\beta_i, \ldots, \beta_{i+83})$ and $C_i = (\gamma_i, \ldots, \gamma_{i+110})$ at the keystream generation round i (≥ 1). Thus,

$$IS_i = \underbrace{(\alpha_i, \ldots, \alpha_{i+92})}_{A_i} \Big| \underbrace{(\beta_i, \ldots, \beta_{i+83})}_{B_i} \Big| \underbrace{(\gamma_i, \ldots, \gamma_{i+110})}_{C_i}.$$

The secret key $(k_1, \ldots, k_{80})$ and IV $(IV_1, \ldots, IV_{80})$ are used to initialize the initial state (INS) as follows:

$$INS = \underbrace{(0, \ldots, 0, k_{80}, \ldots, k_1)}_{A} \Big| \underbrace{(0, 0, 0, 0, IV_{80}, \ldots, IV_1)}_{B} \Big| \underbrace{(1, 1, 1, 0, \ldots, 0)}_{C}.$$

The keystream bits z_i and the new inner state bits α_{i+93}, β_{i+84}, γ_{i+111} of Trivium registers A, B, C are generated respectively as follows:

$$z_i = \alpha_i + \alpha_{i+27} + \beta_i + \beta_{i+15} + \gamma_i + \gamma_{i+45}, \alpha_{i+93} = \alpha_{i+24} + \gamma_{i+45} + \gamma_i + \gamma_{i+1}\gamma_{i+2},$$

$$\beta_{i+84} = \beta_{i+6} + \alpha_{i+27} + \alpha_i + \alpha_{i+1}\alpha_{i+2}, \gamma_{i+111} = \gamma_{i+24} + \beta_{i+15} + \beta_i + \beta_{i+1}\beta_{i+2}.$$

3 Proposed Attack on *Trivium*: Attack Model, Tools and Definitions

Fault Location. Let after a fault injection trial, a multi-bit fault is injected at the KSG round t at positions given by $\phi = \{\phi_1, \phi_2, \ldots, \phi_p\}$ of the internal state, flipping exactly the bits at register locations $\phi_i \in [1, 288]$, $\forall i \in [1, p]$. The set ϕ will be called a *fault position* and the ordered pair (ϕ, t) will be called a *fault location* or simply a *fault* when no ambiguity arises.

Remark. A fault model is a set of properties of an injected fault that is used to characterize an attack. In this paper we consider faults at a single KSG round and the set of all possible fault positions of a fault model under consideration will be used to represent the corresponding fault model. An arbitrary fault model will be denoted by Γ.

The XOR Differential Keystream. Let $IS_i^{key,IV}$ be the internal state of the cipher at KSG round i ($i \geq 1$). Let us consider a fault (ϕ, t). Let $IS_i^{key,IV,\phi,t}$ be the faulty internal state and let $z_i^{key,IV,\phi,t}$ be the faulty output key bit at that KSG round i. Then $d_i^{key,IV,\phi,t} = z_i^{key,IV,\phi,t} + z_i^{key,IV}$ is the XOR difference of the normal (fault free) keystream bit $z_i^{key,IV}$ from the faulty one $z_i^{key,IV,\phi,t}$ at the KSG round i.

For given n we denote, $d^{key,IV,\phi,t,n} = (d_1^{key,IV,\phi,t}, \ldots, d_n^{key,IV,\phi,t})$. We shall analyse the XOR differential keystream $d^{key,IV,\phi,t,n}$ for DFA.

Remark. We shall drop the *key*, *IV* superscript when there is no need to emphasise them. Following the simplified notation, $d^{key,IV,\phi,t,n}$ becomes $d^{\phi,t,n} = (d_1^{\phi,t}, \ldots, d_n^{\phi,t})$. One should note that each $d_i^{\phi,t}$ may be thought of as a function of the *Key-IV* pair. Also at the fault injection KSG round t we have, $IS_t^{\phi,t}(e) = IS_t(e) + 1$, $\forall e \in \phi$ and $IS_t^{\phi,t}(e) = IS_t(e)$, $\forall e \in [1, 288] \setminus \phi$.

Since we are considering the XOR differential for finitely many KSG rounds (n), it may happen that XOR differential keystreams corresponding to two fault locations matches exactly with each other.

Signature of Fault Locations. Signature [8] for any fault location (ϕ, t) is nothing but a *key-IV* independent identified pattern in the XOR differential keystream expressed in a mathematical way. After a fault is injected in the KSG round t, *we shall study the l KSG rounds $t, t+1, \ldots, t+l-1$*. We now consider the XOR differential keystream $d^{\phi,t,t+l-1} = (d_t^{\phi,t}, \ldots, d_{t+l-1}^{\phi,t})$ and treat *key-IV* as variables and each $d_i^{\phi,t} : GF(2)^{80+80} \to GF(2)$ as a function of the *key (80 bit) - IV (80 bit)* pair. For simplicity we write $d_i^{\phi,t} \not\subseteq \{0, 1\}$, to mean that $d_i^{\phi,t}$ is not a constant function. It should be noted that $d_1^{\phi,t} = \cdots = d_{t-1}^{\phi,t} = 0$ as the fault is injected at the KSG round t. We denote $t + l - 1$ by n.

i	1	...	$t-1$	t	...	$t+l-1$
$d_i^{\phi,t}$	0	...	0	$d_t^{\phi,t}$	...	$d_{t+l-1}^{\phi,t}$

For each fault location (ϕ, t) we define its signature, $sig_{\phi,t}$ as the 5-tuple $sig_{\phi,t} = (sig_{\phi,t}^1, sig_{\phi,t}^0, sig_{\phi,t}^=, sig_{\phi,t}^{\neq}, sig_{\phi,t}^x)$ as explained below, where each of $sig_{\phi,t}^e$, $e \in \{1, 0, =, \neq, x\}$ will be called a component of $sig_{\phi,t}$.

1. For certain (ϕ, t) there may be some special values in $d^{\phi,t,n}$ such that $d_i^{\phi,t} = b$, $b \in \{0,1\}$ irrespective of *Key-IV*.

We define, $sig_{\phi,t}^b = \{i \in [t,n] : d_i^{\phi,t} = b\}$, $b \in \{0,1\}$.

If $d_i^{\phi,t} = b$, $b \in \{0,1\}$ holds irrespective of *Key-IV*, we shall say that: "all the XOR differential keystreams are fixed to b at the position i under the fault (ϕ, t)", i.e., $sig_{\phi,t}^b$ is the set of the positions (KSG rounds) where the XOR differential keystreams are fixed to b under the fault (ϕ, t).

2. For certain (ϕ, t) there may be some special values in $d^{\phi,t,n}$ such that $d_i^{\phi,t}, d_j^{\phi,t} \notin \{0,1\}$, $i \neq j$, but $d_i^{\phi,t} = d_j^{\phi,t}$ happens deterministically irrespective of *Key-IV*.

We define, $sig_{\phi,t}^{=} = \{\{i_1, \ldots, i_p\} : i_1, \ldots, i_p \in [t,n], d_{i_1}^{\phi,t} = \cdots = d_{i_p}^{\phi,t} \notin \{0,1\}$ and $\exists$ no $i_r \in [t,n]$ other than $i_1, \ldots, i_p$ such that $d_{i_r}^{\phi,t} = d_{i_1}^{\phi,t}\}$.

$sig_{\phi,t}^{=}$ gives sets of KSG rounds where the XOR differential keystreams are deterministically equal (but not fixed) irrespective of *Key-IV*.

3. For certain (ϕ, t) there may be some special values in $d^{\phi,t,n}$ such that $d_i^{\phi,t}, d_j^{\phi,t} \notin \{0,1\}$, $i \neq j$, but $d_i^{\phi,t} = d_j^{\phi,t} + 1$ happens deterministically irrespective of *Key-IV*.

We define, $sig_{\phi,t}^{\neq} = \{\{i,j\} : i,j \in [t,n], d_i^{\phi,t} + d_j^{\phi,t} = 1$ and $d_i^{\phi,t}, d_j^{\phi,t} \notin \{0,1\}\}$.

$sig_{\phi,t}^{\neq}$ gives pairs of KSG rounds where the XOR differential keystreams are deterministically different (but not fixed) irrespective of *Key-IV*.

4. The following signature component will be used to determine the fault injection KSG round. If $sig_{\phi,t}^1 \neq \emptyset$ and f_1 is the minimum of $sig_{\phi,t}^1$, then '1' surely occurs at the KSG round f_1 in the XOR differential keystream. We pickup those KSG rounds where the first '1' can occur in the XOR differential keystream.

We define, $S_{\phi,t}^x = \{i \in [t, f_1] : d_i^{\phi,t} \neq 0\}$, assuming $sig_{\phi,t}^1 \neq \emptyset$.

In case when $sig_{\phi,t}^1 = \emptyset$ we take $S_{\phi,t}^x = \{i \in [t,n] : d_i^{\phi,t} \neq 0\}$.

For each pair of possible indices $r_a, r_b \in S_{\phi,t}^x$ with $r_a < r_b$ if $d_{r_a}^{\phi,t} = d_{r_b}^{\phi,t}$ we remove the element r_b from $S_{\phi,t}^x$. Doing this iteratively we shall form $sig_{\phi,t}^x = \{i_1, i_2, \ldots, i_p\}$ from $S_{\phi,t}^x$ with the property that $i_a, i_b \in sig_{\phi,t}^x$ with $i_a \neq i_b \Rightarrow d_{i_a}^{\phi,t} \neq d_{i_b}^{\phi,t}$. One should note that, the elements of $sig_{\phi,t}^x$ are the only possible positions (KSG rounds) where the first 1 can occur in the XOR differential keystream, if fault is injected at the location (ϕ, t). This signature component is also *key-IV* independent.

Remark. The signatures are *Key-IV* independent and depend only on fault locations and the cipher design. Since signature of faults are constructed for finitely many KSG rounds it may happen that (1) for some fault locations some signature component becomes completely void and (2) signatures of two fault locations matches exactly with each other. In Appendix C, we illustrate signature of a fault location with an example.

The next theorem shows that *"for a fixed fault position ϕ we do not need to compute signatures for all fault locations (ϕ, t)"*. We use the following notations:

1. For any integer i, $\emptyset + i = \emptyset$.

2. For set S of integers and for any integer i, $S + i = \{s + i : s \in S\}$.

3. For any set S if $s \in S$ implies that s is a set of integers then for any integer i

$$i, S + i = \{s + i : s \in S\}.$$

4. $sig_{\phi,t} + i = (sig_{\phi,t}^1 + i,\ sig_{\phi,t}^0 + i,\ sig_{\phi,t}^{=} + i,\ sig_{\phi,t}^{\neq} + i,\ sig_{\phi,t}^x + i)$.

Theorem 1. *For any fault location (ϕ, t), $sig_{\phi,t} = sig_{\phi,1} + t - 1$.*

Proof. Let fault be injected in the KSG round 1 at position ϕ and $(s_1, s_2, \cdots, s_{288})$ be the corresponding internal state where each of s_j is a variable.

We now assume that the XOR differential keystream bit (DKB) $d_i^{\phi,1} = 1$. Then this happens (at the KSG round i) independent of the internal state (IS) at the KSG round 1.

Thus it is clear that if fault is injected at the KSG round t in the same position ϕ then $d_{t+i-1}^{\phi,t}$ must also be equal to 1, irrespective of the internal state.

Thus $d_i^{\phi,1} = 1 \Leftrightarrow d_{t+i-1}^{\phi,t} = 1$ and hence $i \in sig_{\phi,1}^1 \Leftrightarrow t + i - 1 \in sig_{\phi,t}^1$. This shows that $sig_{\phi,t}^1 = sig_{\phi,1}^1 + t - 1$. With similar arguments the theorem follows.

The Theorem 1 is pictorially illustrated in Appendix C.

Remark. In consequence of the above theorem it can be said that if fault is injected in the same position then a pattern is generated from the fault injection KSG round, in the XOR differential keystream, all previous keystream bits being 0's. For any fault (ϕ, t) if $t = 1$, we shall drop the subscript 't' from its signature and signature components. Thus with this simplified notation $sig_{\phi,1} = (sig_{\phi,1}^1, sig_{\phi,1}^0, sig_{\phi,1}^=, sig_{\phi,1}^{\neq}, sig_{\phi,1}^x)$ becomes $sig_\phi = (sig_\phi^1, sig_\phi^0, sig_\phi^=, sig_\phi^{\neq}, sig_\phi^x)$.

Attack Strategy in Brief. The attack is divided into three stages - Offline Processing Stage, Online Processing Stage and Internal State Recovery Stage. Offline Processing is just one time setup stage where signatures for all fault locations are pre-computed. In the Online Processing stage actual faults are injected to the cipher device and using the pre-computed signatures fault locations are identified or better to say possible fault locations are narrowed down. After Online Processing stage when different random fault locations are identified or narrowed down, we go to the Internal State Recovery stage. The actual attack strategy is described in the Sections 4 to 7.

4 Offline Processing

We now present two methods for computing the signature components of sig_ϕ, $\forall \phi \in \Gamma$ for any given computationally feasible fault model Γ.

4.1 Deterministic Signature Generation by Symbolic Computation

The algorithm *GenSigSym* can be used to compute all the signature components of sig_ϕ for any given computationally feasible fault model Γ.

Algorithm 1. *GenSigSym*(Γ, Number of KSG round $= L_1$)
output : sig_ϕ, $\forall \phi \in \Gamma$.
1. Define 288 symbolic variables over GF(2) and initialise the inner state with these symbolic variables. This will represent the inner state at the begining of the KSG round 1.
2. $\forall \phi \in \Gamma$ compute symbolically the XOR differential keystream $d^{\phi,1,L_1}$.
3. Observe the XOR differential keystream and compute sig_ϕ, $\forall \phi \in \Gamma$.

The deterministic algorithm *GenSigSym* has implementation limitations. However, the following probabilistic algorithm *GenSig*10 is more efficient.

4.2 Probabilistic Algorithm to Generate sig_ϕ^1 and sig_ϕ^0, $\forall \phi \in \Gamma$

Varying Ω and number of keystream generation round, say L_2, in the following randomised algorithm - *GenSig*10 (Generate Signature One-Zero), the least KSG round

n_2 for which sig_ϕ^1 and sig_ϕ^0 contains all possible fixed positions in the XOR differential keystream upto L_2 can be to determined together with sig_ϕ^1 and sig_ϕ^0 (for any computationally feasible fault model Γ).

Algorithm 2. $GenSig10(\Gamma, \Omega,$ Number of KSG round $= L_2)$

output : $sig_\phi^1, sig_\phi^0, \forall \phi \in \Gamma$ and n_2.

1. **for each** $\phi \in \Gamma$:
2.   **for** Ω number of distinct uniformly random independent *Key-IV* pair :
3.   Generate XOR differential keystream upto round L_2 under the fault $(\phi, 1)$
4.   Find positions (KSG rounds) at which all the generated XOR differential keystreams are fixed
5.   **if** $b \in \{0,1\}$ is at a fixed position : then add the position to sig_ϕ^b
6. **return** : $sig_\phi^1, sig_\phi^0 \ \forall \phi \in \Gamma$ and $n_2 = max(\bigcup_{\phi \in \Gamma}(sig_\phi^1 \bigcup sig_\phi^0))$.

Let i be an actual fixed position (KSG round) for $b \in \{0,1\}$ under $(\phi, 1)$ i.e., $i \in [1, L_2]$ be such that $d_i^{key, IV, \phi, 1} = b$ holds for all possible *key* and *IV*. Then the algorithm - $GenSig10$ will surely append i to sig_ϕ^b. Thus the algorithm will not miss any valid position in $[1, L_2]$ for sig_ϕ^1 and sig_ϕ^0 with probability 1.

Now we test the case when the above is not true i.e., $i \in [1, L_2]$ be such that $\exists$ some *key-IV* pair for which $d_i^{key, IV, \phi, 1} = 1$ and $\exists$ some *key-IV* pair for which $d_i^{key, IV, \phi, 1} = 0$. For that i, Pr(The algorithm appends i to sig_ϕ^b) $= 1/2^\Omega$ provided we assume (the idle case) that 0 and 1 are equally probable at the position i (since i is not a fixed position) and the XOR differential bit generated at the position i for each *Key-IV* pair are independent (*Key-IV* pairs are distinct uniformly random and independent).

Therefore taking large value of Ω it can be guaranteed that the algorithm generates correct signatures with very small failure probability. e.g., simply taking $\Omega = 2000$ we shall have $1/2^\Omega = 10^{-\Omega log_{10} 2} \approx 10^{-602}$ which is practically negligible.

Remark. The algorithm $GenSig10$ is generic in nature and is capable of coping with any computationally feasible fault model Γ. It taps statistical weakness of the cipher under DFA. We generate sig_ϕ^0, sig_ϕ^1 using the algorithm $GenSig10$ and $sig_\phi^=, sig_\phi^{\neq}$ and sig_ϕ^x using the algorithm $GenSigSym$. This increases the fault identification success probabilities.

Alternatively sig_ϕ^x can be computed upto KSG round L_2 using the algorithm $GenSig10$ by using the fact that $[1, L_2] \setminus sig_\phi^0 \bigcup sig_\phi^1$ gives the KSG rounds (upto L_2) where the XOR differential keystreams are not fixed.

We define, $n_1 = max(\bigcup_{\phi \in \Gamma}(\bigcup_{A \in sig_\phi^= \bigcup sig_\phi^{\neq}} A))$ and $end(\Gamma) = max(n_1, n_2)$.

5 Online Processing

In the online stage the adversary actually injects a fault into the cipher device and compares the XOR differential keystream with pre-computed signatures in Γ in order to identify the fault location.

5.1 Random Fault at Known KSG Round

1. Obtain the fault-free keystream. In this stage we need a XOR differential keystream of length $n = end(\Gamma) + T - 1$ in order to match it with all possible pre-computed signatures, if fault is injected at the known KSG round T.

2. Let a fault be injected at an unknown position ψ at the known KSG round T. Compute the faulty keystream and obtain $d^{\psi, T, n} = (d_1^{\psi, T}, \ldots, d_n^{\psi, T})$.

3. Define, $support^b = \{i \in [1,n] : d_i^{\psi,T} = b\}, \forall\, b \in \{0,1\}$.

4. $pf = allPossibleFaults\_known(\Gamma, T,\ d^{\psi,T,n})$ as described in Algorithm 3.

Algorithm 3. $allPossibleFaults\_known(\Gamma, t, d^{\psi,t,n})$

1. $pf = \emptyset$
2. **for all** $\phi \in \Gamma$:
3.   **if** $isaPossibleFault(\phi, t, d^{\psi,t,n}) == True$: // Algorithm 4
4. $pf = pf \bigcup \{(\phi, t)\}$
5. **return** pf

It should be noted that from the construction it immediately follows that the actual fault location $(\psi, T) \in pf$. Now if pf is singleton then, (ψ, T) is uniquely determined. When pf is not singleton we do not need to reject the case as a failure. We shall address the issue in section 6.4.

Algorithm 4. $isaPossibleFault(\phi, t, d^{\psi,T,n})$

output : "True" if (ϕ, t) is a possible fault location else "False".

1. $sig_{\phi,t} = sig_{\phi,1} + t - 1$
2. **if** $sig_{\phi,t}^1 \subseteq support^1$:
3.   **if** $sig_{\phi,t}^0 \subseteq support^0$:
4.   **if** $sig_{\phi,t}^= == \emptyset$:
5.   **if** $sig_{\phi,t}^{\neq} == \emptyset$:
6.   **return** *True*
7.   **else** :
8.   **if** $\forall \{i,j\} \in sig_{\phi,t}^{\neq} \Rightarrow d_i^{\psi,T} + d_j^{\psi,T} = 1$:
9.   **return** *True*
10.   **else** : **return** *False*
11.   **else** :
12.   **if** $\forall A \in sig_{\phi,t}^= $ and $\forall i,j \in A, i < j \Rightarrow d_i^{\psi,T} = d_j^{\psi,T}$:
13.   **if** $sig_{\phi,t}^{\neq} == \emptyset$:
14.   **return** *True*
15.   **else** :
16.   **if** $\forall \{i,j\} \in sig_{\phi,t}^{\neq} \Rightarrow d_i^{\psi,T} + d_j^{\psi,T} = 1$:
17.   **return** *True*
18.   **else** : **return** *False*
19.   **else** : **return** *False*
20.   **else** : **return** *False*
21. **else** : **return** *False*

5.2   Random Fault Injected at Random KSG Round

Equivalency of Two Fault Locations. Two fault locations (ϕ, t) and (ψ, T) are said to be equivalent [8] if they produce the same faulty keystream for all possible *Key-IV* pair.

Identifying Random Faults at Random KSG Rounds. Random fault identification at random KSG round is illustrated in Appendix C. We use the same technique as in [8] to identify equivalent faults.

1. Let a random fault (ψ, T) be injected where ψ and T are both unknown. Compute the XOR difference keystream $d^{\psi,T,n} = (d_1^{\psi,T}, \ldots, d_n^{\psi,T})$. The value of n will be discussed later.

2. $support^b = \{i \in [1,n] : d_i^{\psi,T} = b\}, \forall\, b \in \{0,1\}$.

3. Since the fault is injected at the KSG round T, we must have $d_1^{\psi,T} = \cdots = d_{T-1}^{\psi,T} = 0$. Let p be such that $d_1^{\psi,T} = \cdots = d_{p-1}^{\psi,T} = 0$ but $d_p^{\psi,T} = 1$. One should note that $p = min(support^1)$.

By testing all $r \in sig_{\phi,1}^x$ for which the first 1 can occur in $d^{\psi,T,n}$ we wish to recover the pair (ψ, T) as follows:

Algorithm 5. $allPossibleFaults\_unknown(\Gamma, d^{\psi,T,n})$
1. $p = min(support^1)$
2. $pf = \emptyset$
3. **for all** $\phi \in \Gamma$:
4.   **for all** $r \in sig_{\phi,1}^x$:
5.   if $p \geq r$:
6. $t = p - r + 1$
7.   **if** $isaPossibleFault(\phi, t, d^{\psi,T,n}) == True$:
8. $pf = pf \bigcup \{(\phi, t)\}$
9. **return** pf

It should be noted that from the construction it immediately follows that the actual fault location $(\psi, T) \in pf$. Now if pf is a singleton then, (ψ, T) is uniquely determined. When pf is not a singleton it may happen that pf contains equivalent fault locations. Equivalency of faults being an equivalence relation on pf induces a partition on pf, say $pf = pf_1 \bigcup \cdots \bigcup pf_c$. We sort each pf_i in increasing order with respect to the fault injection KSG round. If $pf_i = ((\phi_{i1}, t_{i1}), \ldots, (\phi_{id_i}, t_{id_i}))$ be one such sorting with $t_{i1} \leq \cdots \leq t_{id_i}$. We accept (ϕ_{i1}, t_{i1}) from pf_i and reject other faults and take $pf_i = \{(\phi_{i1}, t_{i1})\}$. Now we take $pf = pf_1 \bigcup \cdots \bigcup pf_c$. If pf is a singleton an equivalent fault of (ψ, T) is uniquely identified but if pf is not a singleton we do not need to reject the case as a failure. We shall address the issue in section 6.4.

Remark. For any $r \in sig_{\phi}^x$ if $p = min(support^1) \geq r$, then a keystream of length $n_r^{\phi} = p + end(\Gamma) - r$ will be enough for a comparison considering $r \in sig_{\phi}^x$. Thus a keystream of length $n^{\phi} = p + end(\Gamma) - min(sig_{\phi}^x)$ will be enough for comparisons considering all $r \in sig_{\phi}^x$. Hence $n = max(\{n^{\phi} : \phi \in \Gamma\})$ will be enough for comparisons considering all $r \in sig_{\phi}^x, \forall \phi \in \Gamma$.

6   Internal State Recovery Stage

To recover the internal state of the cipher using SAT solver we have adopted the procedure described in [6] and generalized it with modifications to cope with multi-bit faults at unknown KSG rounds.

In this section, given a fault location (ϕ_j, t_j) we denote the faulty keystream $z^{\phi_j, t_j, n} = (z_1^{\phi_j, t_j}, \ldots, z_n^{\phi_j, t_j})$ simply by $z^j = (z_1^j, \ldots, z_n^j)$.

We assume that the attacker has the following information:
1. m fault locations $(\phi_1, t_1), \ldots, (\phi_m, t_m)$ with $t_1 \leq \cdots \leq t_m < n$.
2. normal keystream $z = (z_1, \ldots, z_n)$ of length n.
3. m faulty keystreams $z^1, \ldots, z^m$ each of length n corresponding to the m fault locations $(\phi_1, t_1), \ldots, (\phi_m, t_m)$.

We shall recover the internal state of the cipher at the KSG round t_1.

6.1   Preliminaries on the DFA of Trivium

Each fault injection generates additional equations. Let us consider a fault (ϕ, t). At KSG round i, the difference values of the inner state is,

$$\Delta IS_i = \underbrace{(\Delta\alpha_i, \ldots, \Delta\alpha_{i+92})}_{\Delta A_i} \Big| \underbrace{(\Delta\beta_i, \ldots, \Delta\beta_{i+83})}_{\Delta B_i} \Big| \underbrace{(\Delta\gamma_i, \ldots, \Delta\gamma_{i+110})}_{\Delta C_i}.$$

As in [6], we use the difference of the keystream bit (Δz_i) and shift registers inputs ($\Delta\alpha_i, \Delta\beta_i, \Delta\gamma_i$) before and after performing the fault injection to generate additional polynomial equations as given by,

$$\Delta z_i = \Delta\alpha_i + \Delta\alpha_{i+27} + \Delta\beta_i + \Delta\beta_{i+15} + \Delta\gamma_i + \Delta\gamma_{i+45},$$
$$\Delta\alpha_{i+93} = \Delta\alpha_{i+24} + \Delta\gamma_{i+45} + \Delta\gamma_i + \Delta\gamma_{i+1}\gamma_{i+2} + \gamma_{i+1}\Delta\gamma_{i+2} + \Delta\gamma_{i+1}\Delta\gamma_{i+2},$$
$$\Delta\beta_{i+84} = \Delta\beta_{i+6} + \Delta\alpha_{i+27} + \Delta\alpha_i + \Delta\alpha_{i+1}\alpha_{i+2} + \alpha_{i+1}\Delta\alpha_{i+2} + \Delta\alpha_{i+1}\Delta\alpha_{i+2},$$
$$\Delta\gamma_{i+111} = \Delta\gamma_{i+24} + \Delta\beta_{i+15} + \Delta\beta_i + \Delta\beta_{i+1}\beta_{i+2} + \beta_{i+1}\Delta\beta_{i+2} + \Delta\beta_{i+1}\Delta\beta_{i+2}.$$

One should note that, for $e \in [1, 288]$,

$$\Delta IS_i(e) = 0, \forall i \in [1, t-1],$$
$$\Delta IS_t(e) = 1, \text{ if } e \in \phi \text{ else } \Delta IS_t(e) = 0.$$

6.2   Generating Low Degree Polynomials

The Algorithm $modEQgenerator$ is used to generate low degree polynomials. The Algorithm $modEQgenerator$ is pictorially illustrated in Appendix C. Let the internal state at the KSG round i ($i \geq t_1$) be $(a_{i-t_1+1}, \ldots, a_{i-t_1+93}, b_{i-t_1+1}, \ldots, b_{i-t_1+84}, c_{i-t_1+1}, \ldots, c_{i-t_1+111})$, the internal state at the KSG round t_1 being $(a_1, \ldots, a_{93}, b_1, \ldots, b_{84}, c_1, \ldots, c_{111})$. As in [6], we treat each a_i, b_i, c_i as variables and consider the $3N + 288$ variables $(a_1, \ldots, a_{N+93}, b_1, \ldots, b_{N+84}, c_1, \ldots, c_{N+111})$, $N = n - t_1 + 1$ and use similar procedure to obtain a system of low degree polynomials, changing only the *fault injection strategy*. The idea is to treat the KSG round t_1 as the base KSG round 1 and with this view the modified fault injection procedure now injects the actual fault (ϕ, t) to the internal state at position ϕ but at the KSG round $t - t_1 + 1$. One should note that N is the number of KSG rounds used by the algorithm $modEQgenerator$.

We assume that, in the Algorithm $modEQgenerator$, $da = (da_1, \ldots, da_{N+93})$, $db = (db_1, \ldots, db_{N+84})$, $dc = (dc_1, \ldots, dc_{N+111})$ and $dIS_i = (da_i, \ldots, da_{i+92}, db_i, \ldots, db_{i+83}, dc_1, \ldots, dc_{i+110})$. The procedure $InjectFault(da, db, dc, \phi, t)$ of $modEQgenerator$ injects fault to the internal state by setting $dIS_t(e) = 1$, for all $e \in \phi$.
The procedures $ExtractUnivariate$ and $Substitute$ are same as explained in [6].

Remark: One should note that, since we are considering $EQpoly$ as a set (of polynomials), $EQpoly$ has no duplicate polynomials.

6.3   Recovering *key-IV* with Known Fault Locations

The procedure $modEQgenerator$ (modified $EQgenerator$) returns a set of polynomials that are simply passed on to the SAT solver in SAGE (converting to a Boolean satisfiability problem) for extracting a solution for the inner state variables ($a_1, \ldots, a_{93}, b_1, \ldots, b_{84}, c_1, \ldots, c_{111}$) with the same cutting number as in [6]. The internal state obtained in this way corresponds to the KSG round t_1. Since t_1 is known, iterating backwards the *key-IV* can now be recovered. In this case we shall use the phrase "*The faults* (ϕ_1, t_1), \ldots, (ϕ_m, t_m) *are passed on to the SAT solver.*"

Algorithm 6. $modEQgenerator(n, \phi_1, \ldots, \phi_m, t_1, \ldots, t_m, z, z^1, \ldots, z^m)$

1. $t = t_1$, $N = n - t_1 + 1$
2. **for** $j = 1$ to m : $t_j = t_j - t + 1$
3. $P = \emptyset$
4. **for** $i = 1$ to N :
5. $\qquad P = P \bigcup \{a_i + a_{i+27} + b_i + b_{i+15} + c_i + c_{i+45} + z_{t+i-1}\}$
6. $\qquad P = P \bigcup \{a_{i+93} + a_{i+24} + c_{i+45} + c_i + c_{i+1}c_{i+2}\}$
7. $\qquad P = P \bigcup \{b_{i+84} + b_{i+6} + a_{i+27} + a_i + a_{i+1}a_{i+2}\}$
8. $\qquad P = P \bigcup \{c_{i+111} + c_{i+24} + b_{i+15} + b_i + b_{i+1}b_{i+2}\}$
9. $S = \emptyset$
10. **for** $j = 1$ to m :
11. $\qquad da = (0, \ldots, 0)$ // $\text{length}(da) = N + 93$
12. $\qquad db = (0, \ldots, 0)$ // $\text{length}(db) = N + 84$
13. $\qquad dc = (0, \ldots, 0)$ // $\text{length}(dc) = N + 111$
14. $\qquad InjectFault(da, db, dc, \phi_j, t_j)$
15. $\qquad$ **for** $i = 1$ to N :
16. $\qquad\qquad S_1 = \emptyset$
17. $\qquad\qquad dz = z_{t+i-1} + z^j_{t+i-1}$
18. $\qquad\qquad P = P \bigcup \{da_i + da_{i+27} + db_i + db_{i+15} + dc_i + dc_{i+45} + dz\}$
19. $\qquad\qquad da_{i+93} = da_{i+24} + dc_{i+45} + dc_i + dc_{i+1}c_{i+2} + c_{i+1}dc_{i+2} + dc_{i+1}dc_{i+2}$
20. $\qquad\qquad db_{i+84} = db_{i+6} + da_{i+27} + da_i + da_{i+1}a_{i+2} + a_{i+1}da_{i+2} + da_{i+1}da_{i+2}$
21. $\qquad\qquad dc_{i+111} = dc_{i+24} + db_{i+15} + db_i + db_{i+1}b_{i+2} + b_{i+1}db_{i+2} + db_{i+1}db_{i+2}$
22. $\qquad\qquad$ **do** :
23. $\qquad\qquad\qquad S_2 = ExtractUnivariate(P)$
24. $\qquad\qquad\qquad P = Substitute(P, S_2)$
25. $\qquad\qquad\qquad S_1 = S_1 \bigcup S_2$
26. $\qquad\qquad$ **while** $S_2 \neq \emptyset$
27. $\qquad\qquad da, db, dc = Substitute(da, db, dc, S_1)$
28. $\qquad\qquad S = S \bigcup S_1$
29. $EQpoly = P \bigcup S$
30. **return** $EQpoly$

Remark. Instead starting from the KSG round 1, we start from KSG round t_1. This gives us the advantage that the fault injection round t_1 could be very large.

6.4 Recovering the Internal State When Fault Locations Are Not Uniquely Identified

We now assume that P_i is the set of possible faults returned by the fault location determination algorithm at the i-th fault injection trial, $\forall i \in [1, m]$. We now consider the Cartesian product set $P = P_1 \times P_2 \times \cdots \times P_m$. It should be noted that one of the elements, say α of P corresponds to the actual m injected faults. Now we pass each element of P to SAT solver. Multiple solutions may be obtained. Solution considering the element α, if returned, will surely correspond to the actual Internal State of the cipher at some known KSG round. Assuming each returned solution as a possible internal state we simply use brute force "Guess and Determine Strategy" [7] to detect the correct internal state by matching fault free keystreams. If the cardinality of P and SAT solving time are low then the internal state can be recovered in reasonable time with 100% success. Otherwise we have to re-key the device for more fault injection trials.

7 Alien Fault Model

To the best of our knowledge, DFA described in all the existing open literature on Trivium, adversary must be 100% confident that the injected faults will be in the corresponding fault model and no work has been done if faults are from outside of the fault model. In this section we address this issue. Also the bit arrangement in the cipher device may be unknown to the adversary.

We now assume a weak adversary who is confident that any injected fault will be in the fault model Σ but Σ is large and the adversary is not capable of computing signatures for all the possible faults in Σ. In this case she chooses a proper subset Γ of Σ which she can process efficiently. She computes signature for all the possible faults in Γ. Later she matches the XOR differential keystream with the signatures to identify the fault location. Since she does not compute signatures of faults in $\Sigma \setminus \Gamma$, these faults, if occurs, cannot be identified. However the adversary wants to detect such a case and reject the faulty keystream.

1. Γ will be called a *"native fault model"*.
2. Faults in $\Sigma \setminus \Gamma$ will be called Σ-*relative alien faults of* Γ.
3. Σ will be called an *"alien fault model"* relative to the native fault model Γ.
4. Let Γ_∞ be the set of all possible faults in the first KSG round. Then faults in $\Gamma_\infty \setminus \Gamma$ will be called *"absolute alien faults with respect to Γ"*.
5. When no ambiguity arises we shall use the term *"alien fault"* to mean either of the faults which should be clear from the context.

The adversary may choose the fault model Γ depending on her knowledge or best guess on the cipher device. She may choose those fault positions which are more likely to occur by experimenting on similar cipher device if she has that privilege. If that is not possible she chooses more relaxed absolute alien fault model where any possible fault can occur but she will only use the *native faults* to recover the internal state of the cipher, rejecting *alien faults* with some probability. Success in a chosen native fault model is not guaranteed but is purely experiment based. One should note that different fault models might crucially affect the capabilities and the complexities of the attack strategy. The adversary is now very weak and will require more re-keying of the cipher device. Depending on the cipher device there may be some fault models that produces the best or the worst results.

Remark. One should note that, in this section all possible faults (i.e., both native and absolute alien faults) are allowed at a single KSG round. The adversary is very weak and simply could not guarantee that an injected fault is not an alien fault. Consequently if occurrence of alien faults are detected she must reject the keystream and re-key the cipher device for another fault injection trial. She must continue the process until required number of native faults are obtained since only the native faults are used for key-IV recovery.

Fault Location Determination in Alien Fault Model
The adversary will use the same fault location determination algorithms to identify a fault location. When considering an alien fault model, error will occur if the injected fault is alien but the fault location determination algorithm identifies it as a native fault. Thus error occurs if possible faults pf returned by the fault location determination algorithm is non-empty when injected fault is alien and $pf = \emptyset$ will imply that an alien fault is injected.

Recovering Internal State in This Case
Internal state recovery process in this case will be the same as described in Section 6. Depending on the chosen native and relative alien fault model, since the fault location determination algorithm is probabilistic, it may happen that the fault location determination algorithm returns an erroneous fault (Fault was alien but wrongfully

identified as native). In this case the adversary may have to use the "Guess and Determine" strategy to recover the internal state of the cipher device by matching fault free keystream with the original one and re-keying the cipher device again for more fault injection trials if necessary.

8   Conclusion

This paper removes the major restriction of [4] where the adversary must be certain that the fault positions must not be from different registers. Moreover we require very few faults to recover the internal state of the cipher even in a weaker setup (nbdMBF[9]) and thus the actual cipher device will be minimally stressed. Our DFA model works perfectly in a situation when the adversary is very weak. This paper allows alien fault rejection. Also adversary chosen custom fault models are allowed. During experimentation the fault detection algorithm never rejected an injected native fault which further justifies the correctness of the randomised algorithm *GenSig*10. To the best of our knowledge, this is the most practical, realistic and strongest DFA strategy reported on *Trivium* cipher so far. Our attack strategy is generic and can be adopted for other Trivium like stream ciphers.

Appendix A: Our Arsenal

1. One standalone desktop PC with AMD 4.0 GHz FX-8350 processor and 32 GB RAM, referred to as AMD.
2. A Beowulf Cluster of 20 desktop PC each with 2.60 GHz Intel Pentium E5300 Dual-core processor and 4 GB RAM connected via LAN and setupped for Distributed Parallel Computing, referred to as BEOWULF.
3. Ubuntu 12.04 LTS Operating System.
4. Mathematical software SAGE - 6.1.1.
5. SAT solver Cryptominisat - 2.9.6 installed in SAGE.

The BEOWULF cluster (all available cores) was used to (1) generate sig_ϕ^1 and sig_ϕ^0 (2) compute the success rates (probabilities) and (3) SAT solving where as the standalone AMD (only 2 cores was used) was used to generate $sig_\phi^=$, $sig_\phi^{\neq}$ and sig_ϕ^x.

Appendix B: Experimental Results

Attack strategy in this paper is generic. In particular we demonstrate the attack strategy for the scenario, denoted by the symbol nbdMBF^k, in which randomly chosen atmost k consecutive location (k neighbouring bits) can be disturbed by a single fault injection without knowing the locations or the exact number of bits the injected fault has altered. In this paper we consider the popular convention of treating $IS(e), IS(e + 1), IS(e + 2), \ldots, IS(e + p - 1)$ as p neighbouring bits (IS representing the internal state of the cipher) but in real life the arrangement may not follow this pattern. This does not affect our analysis.

Faults in the Model nbdMBF^k. Let Γ_k be set of all possible fault positions in this case. In the case when exactly r bits are flipped, the fault positions are of the form $i, i + j_1, i + j_2, \ldots, i + j_{r-1}$ where $i \in [1, 288 - j_{r-1}]$ and $(j_1, j_2, \ldots, j_{r-1})$ follows lexicographic ordering (increasing) without repetition of length $r - 1$ in the range $[1, k - 1]$. e.g., when $k = 4, r = 3$ possible (j_1, j_2) are $(1, 2), (1, 3), (2, 3)$ and hence the fault positions are of the form $(i, i + 1, i + 2)$, $(i, i + 1, i + 3)$, $(i, i + 2, i + 3)$.

i	$i+1$	$i+2$	$i+3$
*	*	*	
*	*		*
*		*	*

i	$i+1$	$i+2$	$i+3$
i	$i+1$	$i+2$	
i	$i+1$		$i+3$
i		$i+2$	$i+3$

j_1	j_2
1	2
1	3
2	3

We used the algorithm $GenSig10(\Gamma, \Omega, L_2)$ to generate sig_ϕ^1 and sig_ϕ^0 taking $L_2 = 1000$ and $\Omega = 2000$ and $GenSigSym(\Gamma, L_1)$ to generate $sig_\phi^=$, $sig_\phi^{\neq}$ and sig_ϕ^x taking $L_2 = 365$. One should also note that $\Gamma_1 \subset \Gamma_2 \subset \cdots \subset \Gamma_{288}$. Here we present experimental results for (comparing) $k = 1, 2, 3, 4, 5$.

We use the classical description of Trivium to generate the signatures and to identify fault locations and use the floating description for internal state recovery using SAT solvers.

1. Faults at Known KSG Round.
In 2^{20} experiments, for each $k = 1, 2, 3, 4, 5$ the fault location determination algorithm uniquely identified actual injected native fault locations with 100% success and rejected absolute alien faults with probability 1.0 for $k = 1$ and 0.999 for $k = 2, 3, 4, 5$. Thus alien faults are rejected with very high probability.

2. Faults at Random Unknown KSG Round.
Let t be the fault injection KSG round. We performed 2^{18} experiments, for $k = 5$ and $t = 1, 50, 100, 200, 300$ and also by varying $t \in [1, 300]$. In each case native faults were uniquely (equivalent faults) identified with probability 0.907 and average cardinality of pf was 1.15. Absolute alien faults were rejected with probability 0.999.

3. SAT Solving Results: With Faults at the Same KSG Round.
Abbreviations: (NOE, Number of Experiments), (NOT, Number of Timed out cases). Behaviour of SAT solvers are unpredictable. For each SAT solving trial (Cython compiled codes were used) we first generated a random inner state (as the fault injection KSG round) and $m(= 2)$ random faults uniformly and independently and allocated a time limit of 30 minutes. If the SAT solver does not self terminate within that time we agree to terminate it forcefully and mark the case as a TIMEOUT (which may have resulted in a success if enough time was given). Very few such cases occurred (for $m = 2$) during experimentation. For timed out cases we generated an additional random fault uniformly and independently and again pass them to the SAT solver. It turns out that $m = 3$ always resulted in a success. For $k = 2, 3, 4, 5$ on average $2.021, 2.026, 2.054, 2.058$ faults always broke the cipher (Table 1).

We also report that for $k = 9, m = 3, N = 450$ in 1200 random SAT solving experiments 100% success rate was obtained in recovering the internal state with MinTime, MaxTime and AvgTime respectively $3.06, 26.42$ and 7.92 (in seconds).

4. SAT Solving Results: With Faults at Random KSG Round.
For $k = 9, m = 4, N = 360$ in 1800 random SAT solving experiments 100% success rate was obtained in recovering the internal state with MinTime, MaxTime and AvgTime

respectively $0.11, 6782.98$ and 23.75 (in seconds) when fault injection KSG rounds are uniformly distributed over $\{t, \ldots, t+49\}$, for any t.

Table 1. SAT Solving Results with Faults at the Same KSG Round

Trivium					Time in Seconds if SUCCESS		
k	m	N	NOE	NOT	MinTime	MaxTime	AvgTime
2	2	500	1000	21	3.52	1377.86	47.06
3	2	500	1000	26	3.66	1308.87	69.71
4	2	500	1000	54	3.74	1491.47	88.58
5	2	500	1000	58	3.73	1592.26	98.81
2	3	500	21	0	4.62	35.88	10.89
3	3	500	26	0	4.58	19.07	10.61
4	3	500	54	0	3.57	54.21	17.24
5	3	500	58	0	4.37	151.68	26.62

Appendix C: Illustrations

Fault Location Determination when Fault Injection KSG Round Is Unknown.
Let us consider a fault (ϕ, t). As discussed earlier $d_1^{\phi,t} = \cdots = d_{t-1}^{\phi,t} = 0$. We consider the l (fixed) KSG rounds $t, t+1, \ldots, t+l-1$. We now assume a fake example for illustration. For the symbolic internal state $(s_1, \ldots, s_{288})$, taking $t = 1$ and $l = 17$, we assume that, $d^{\phi,1,17} = (0, x_1, 0, x_1, 0, x_2, 1, 0, x_3, 0, 1, 1, 0, x_3 + 1, x_4, x_5, x_6)$ where $x_i \notin \{0, 1\}$ (each x_i is treated as a function of $s_1, \ldots, s_{288}$ and is not a constant function) and are all distinct. Let $end(\Gamma) = 16$.

i	1	2	3	4	5	6	7	8	9	10	11	12	13	14	15	16	17
$d_i^{\phi,1}$	0	x_1	0	x_1	0	x_2	1	0	x_3	0	1	1	0	$x_3 + 1$	x_4	x_5	x_6

Then, $sig_{\phi,1}^1 = \{7, 11, 12\}$, $sig_{\phi,1}^0 = \{1, 3, 5, 8, 10, 13\}$, $sig_{\phi,1}^= = \{\{2, 4\}\}$, $sig_{\phi,1}^{\neq} = \{\{9, 14\}\}$, $sig_{\phi,1}^x = \{2, 6, 7\}$ ($4 \notin sig_{\phi,1}^x$).

We now assume that the same fault is injected at KSG round $t = 4$. For some particular *key-IV* let the first 1 occurs due to $r = 2 \in sig_{\phi,1}^x$ in the XOR differential keystream $d^{\phi,4,19}$.

i	1	2	3	$t=4$	$p=5$	6	7	8	9	10	11	12	13	14	15	16	17	18	$n_r^\phi = 19$
$d_i^{\phi,4}$	0	0	0	0	1	0	1	0	0	1	0	1	0	1	1	0	0	*	*
				0	x_1	0	x_1	0	x_2	1	0	x_3	0	1	1	0	x_3+1	*	*
			1	$r=2$	3	4	5	6	7	8	9	10	11	12	13	14	15	$end(\Gamma) = 16$

In this case $p = min(support^1) = 5$ and $t = p - r + 1$ and $n_r^\phi = p + end(\Gamma) - r$.
It should be noted that if $p < r$, then there is no need for a comparison. In the above example no comparison is needed for $r = 6, 7$.
We now assume that $t = 8$. For some particular *key-IV* let the first 1 occurs due to $r = 7 \in sig_{\phi,1}^x$ in the XOR differential keystream $d^{\phi,8,23}$.

i	1	2	3	4	5	6	7	$t=8$	9	10	11	12	13	$p=14$	15	16	17	18	19	20	21	22	$n_r^\phi=23$
$d_i^{\phi,8}$	0	0	0	0	0	0	0	0	0	0	0	0	0	1	0	0	0	1	1	0	1	*	*
								0	x_1	0	x_1	0	x_2	1	0	x_3	0	1	1	0	x_3+1	*	*
								1	2	3	4	5	6	$r=7$	8	9	10	11	12	13	14	15	$end(\Gamma)=16$

In this case $p = min(support^1) = 14$ and $t = p - r + 1$ and $n_r^\phi = p + end(\Gamma) - r$.
Pictorial Illustration of Theorem 1.

KSG round	Fault Free Internal State	XOR DKB	KSG round	Fault Free Internal State	XOR DKB
1	IS at KSG round 1	$d_1^{\phi,1}$	t	IS at KSG round t	$d_t^{\phi,t}$
2	IS at KSG round 2	$d_2^{\phi,1}$	$t+1$	IS at KSG round $t+1$	$d_{t+1}^{\phi,t}$
3	IS at KSG round 3	$d_3^{\phi,1}$	$t+2$	IS at KSG round $t+2$	$d_{t+2}^{\phi,t}$
i	IS at KSG round i	$d_i^{\phi,1}=1$	$t+i-1$	IS at KSG round $t+i-1$	$d_{t+i-1}^{\phi,t}=1$

modEQgenerator Pictorial Illustration.

ϕ	KSG round	i	Fault Free Internal State	Keystreams			
	1			z_1	z_1^1	...	z_1^m
	2			z_2	z_2^1	...	z_2^m
	3			z_3	z_3^1	...	z_3^m
	t_1-1			z_{t_1-1}	$z_{t_1-1}^1$	...	$z_{t_1-1}^m$
ϕ_1	t_1	$i=1$	$a_1,\ldots,a_{93},b_1,\ldots,b_{84},c_1,\ldots,c_{111}$	z_{t_1}	$z_{t_1}^1$	...	$z_{t_1}^m$
	t_1+1	$i=2$	$a_2,\ldots,a_{94},b_2,\ldots,b_{85},c_2,\ldots,c_{112}$	z_{t_1+1}	$z_{t_1+1}^1$	...	$z_{t_1+1}^m$
	t_1+2	$i=3$	$a_3,\ldots,a_{95},b_3,\ldots,b_{86},c_3,\ldots,c_{113}$	z_{t_1+2}	$z_{t_1+2}^1$	...	$z_{t_1+2}^m$
ϕ_j	t_j	$i=l_j-t_1+1$	$a_i,\ldots,a_{i+92},b_i,\ldots,b_{i+83},c_i,\ldots,c_{i+110}$	z_{t_j}	$z_{t_j}^1$	...	$z_{t_j}^m$
ϕ_m	t_m	$i=t_m-t_1+1$	$a_i,\ldots,a_{i+92},b_i,\ldots,b_{i+83},c_i,\ldots,c_{i+110}$	z_{t_m}	$z_{t_m}^1$	...	$z_{t_m}^m$
	n	$i=n-t_1+1=N$	$a_i,\ldots,a_{i+92},b_i,\ldots,b_{i+83},c_i,\ldots,c_{i+110}$	z_n	z_n^1	...	z_n^m

References

1. De Cannière, C., Preneel, B.: TRIVIUM Specifications. http://www.ecrypt.eu.org/stream/p3ciphers/trivium/trivium_p3.pdf
2. Hojsík, M., Rudolf, B.: Differential Fault Analysis of Trivium. In: Nyberg, K. (ed.) FSE 2008. LNCS, vol. 5086, pp. 158–172. Springer, Heidelberg (2008)
3. Hojsík, M., Rudolf, B.: Floating Fault Analysis of Trivium. In: Chowdhury, D.R., Rijmen, V., Das, A. (eds.) INDOCRYPT 2008. LNCS, vol. 5365, pp. 239–250. Springer, Heidelberg (2008)

4. Hu, Y., Gao, J., Liu, Q., Zhang, Y.: Fault analysis of Trivium. Designs, Codes and Cryptography **62**(3), 289–311 (2012)
5. Karmakar, S., Chowdhury, D.R.: Fault Analysis of Grain Family of Stream Ciphers. IACR Cryptology ePrint Archive 2014:261 (2014)
6. Mohamed, M.S.E., Bulygin, S., Buchmann, J.: Using SAT Solving to Improve Differential Fault Analysis of Trivium. In: Kim, T., Adeli, H., Robles, R.J., Balitanas, M. (eds.) ISA 2011. CCIS, vol. 200, pp. 62–71. Springer, Heidelberg (2011)
7. Rohani, N., Noferesti, Z., Mohajeri, J., Aref, M.R.: Guess and Determine Attack on Trivium Family. In: 2010 IEEE/IFIP 8th International Conference on Embedded and Ubiquitous Computing (EUC), pp. 785–790. IEEE (2010)
8. Sarkar, S., Banik, S., Maitra, S.: Differential Fault Attack against Grain family with very few faults and minimal assumptions. IACR Cryptology ePrint Archive 2013:494 (2013)

Recovering CRT-RSA Secret Keys from Message Reduced Values with Side-Channel Analysis

Benoit Feix[✉], Hugues Thiebeauld, and Lucille Tordella

UL Transaction Security - UK Evaluation Lab, Ashwood Park, Basingstoke, UK
{benoit.feix,hugues.thiebeauld,lucille.tordella}@ul.com

Abstract. Long integer modular reduction is an operation executed when processing public-key cryptographic algorithms such as a CRT-RSA signature. This operation is sensitive as it manipulates a part of the secret key. When computing a CRT-RSA signature or a decryption the input message is first reduced modulo the two secret prime values p and q. These two reductions are executed preliminarily before the exponentiations with d_p and d_q. Amongst the range of published side-channel attacks so far, few target these initial reductions whereas it represents a significant threat for the secret key confidentiality. One of them, the MRED attack from den Boer et al. makes use of chosen messages for attacking the reduced values. This attack is interesting as it does not require the knowledge of the algorithm used for the reduction. Besides it defeats the countermeasures aiming at randomizing the intermediate data during the reduction but not the final reduced value, as it is the case with the message additive blinding method. However this attack requires a large amount of traces to be successful. This paper introduces two efficient side-channel attacks considered more efficient than the MRED. Indeed it requires much less side-channel traces to expose the secret primes. The new techniques are exposed in this paper with practical results and discussion about their efficiency against the different existing countermeasures.

Keywords: Long integer arithmetic · Modular reduction · Exponentiation · Embedded devices · Side-Channel Analysis

1 Introduction

Public key cryptosystems are nowadays commonly used in a wide range of products for digital encryption or signature purposes. Indeed the use of these algorithms has increased together with the need of security in digital transactions, particularly in the embedded world from the smartcard technology to the smartphones, tablets or trusted platforms. These cryptographic operations typically protect users credentials and privacy against a number of intrusive attacks.

Among the public-key cryptosystems the RSA [22] and Diffie-Hellman (DH) [8] algorithms remain widely used in secure transaction, like in payment or for strong authentications. As a cost effective solution to perform a RSA computation, the

© Springer International Publishing Switzerland 2014
W. Meier and D. Mukhopadhyay (Eds.): INDOCRYPT 2014, LNCS 8885, pp. 53–67, 2014.
DOI: 10.1007/978-3-319-13039-2_4

CRT (Chinese Remainder Theorem) mode is commonly used. The CRT-RSA algorithm processes long integer modular operations involving the secret prime factors p and q of the public modulus all along the computations. It starts with the initial reductions of the input message to be signed or decrypted to the final recombination calculations. Although the security of the RSA has been considered as theoretically resistant providing the key is long enough, practical implementations in embedded processors must take into account physical attacks, such as side-channel analyses.

Indeed it has been known for years that naive implementations can be efficiently threatened by many existing *Side-Channel Analysis* (SCA) techniques. Since the initial publications by Kocher et al. [14,15] the side-channel threat has been deeply investigated by the academic and the industrial communities. This technique exploits the information contained in the physical traces, that are collected during the execution of the secret algorithm on a device. Indeed, the threat becomes apparent when a physical measurement holds information in line with the intermediate value manipulated during the processing of the cryptographic algorithms. To get such information, a side-channel trace typically makes use of the power fluctuations or the electromagnetic radiations measurement that are taken from the device.

Classically side-channel techniques can be split in 2 categories, the *Simple Side-Channel Analysis* (SSCA) making use of a single trace and the *Differential Side-Channel Analysis* (DSCA) exploiting a high number of traces to recover the secret value through observation and statistical processing. SSCA and DSCA are called *passive attacks* as they aim at observing only the activity of the device during the sensitive processing.

Among all side-channel publications on CRT-RSA, only few target the reductions occurring at the beginning of the exponentiations. In this paper, the reduction operation is particularly studied with the introduction of two new attacks threatening the secret primes p and q. More precisely these attacks focus on the initial reductions of the message modulo p or q. This area of interest in the CRT-RSA algorithm is not new as the MRED attack [4] by den Boer et al. took benefit of well-chosen messages to extract information about p or q with side-channels during the reduction and consequently exposing the whole secrets. Further to the MRED attack, new side-channel techniques are exposed in this paper taking benefit of another format of chosen messages. With a better efficiency than the MRED attack, the new attacks aim at recovering the prime factors p and q. The results were practically validated on simulated traces firstly and then on practical traces collected from a CRT-RSA implementation executed on a 32-bit architecture hardware device. One technique falls into the SSCA category while the second falls into the DSCA. For the latter, a comparative analysis is developed between the new technique and the MRED attack highlighting that the new attack turns out to be significantly more powerful than any existing attack targeting the reduction part of the CRT-RSA.

The paper is organized as follows. Section 2 reminds the reader of the necessary background on modular reductions and CRT-RSA algorithm in embedded products. A side-channel attack on CRT-RSA modular reduction operations will be presented. In Section 3, the principle of the chosen message construction used to recover the secret moduli involved in modular reductions (or division) with side-channel analysis is explained. Then the attack path is validated with experimental results on simulation traces as well as on traces collected from a 32-bit hardware device in Section 4. In Section 5 this new technique is compared with the MRED attack. In Section 6, side-channel countermeasures against the attacks presented are highlighted. Finally, Section 7 holds the conclusion of this paper.

2   Modular Reductions and Side-Channel Analysis

2.1   Reduction and RSA in Embedded Devices

The RSA cryptosystem is commonly used to generate signatures or is used for decryption. Signing with RSA a message m means computing the value $s = m^d \bmod n$. The signature s is then verified by computing $m = s^e \bmod n$. Integers e and d respectively denote the public and the private exponents when n is the public modulus. Embedded products generally take advantage of the Chinese remainder theorem [11,21] method to accelerate the RSA computations from three (practical ratio) to four (theoretical ratio) times compared to the straightforward method. Detail for this efficient computation is given in Algorithm 2.1.

Algorithm 2.1. CRT-RSA

Require: p, q, $d_p = d \bmod (p-1)$, $d_q = d \bmod (q-1)$, $i_q = q^{-1} \bmod p$
Ensure: $s = m^d \bmod n$

1: $m_p \leftarrow m \bmod p$ and $m_q \leftarrow m \bmod q$
2: $s_p \leftarrow m_p^{d_p} \bmod p$ and $s_q \leftarrow m_q^{d_q} \bmod q$
3: $s = s_q + ((s_p - s_q) \cdot i_q \bmod p) \cdot q$
4: **return** s

The input message m is firstly reduced modulo p and q leading to the corresponding values m_p and m_q. Interestingly this means that one of the first operation applied to the input message makes use of the secret primes p and q. Subsequently two exponentiations modulo p and q are computed resulting to S_p and S_q respectively. Finally the CRT recombination is performed leading to the final signature outcome modulo $n = p \cdot q$.

Notations. In this paper the following notations are being used:

- $n = ((n)_{2\ell-1} \ldots (n)_1(n)_0)_2$ corresponds to 2ℓ-bit integer n decomposition in base 2.
- $p = ((p)_{\ell-1} \ldots (p)_1(p)_0)_2$ corresponds to the decomposition of p in base 2.

- $m_p = ((m_p)_{\ell-1} \ldots (m_p)_1 (m_p)_0)_2$ corresponds to the decomposition of m_p in base 2.
- $n = ((n)_{\lambda-1} \ldots (x)_1 (x)_0)_b$ corresponds to the integer x decomposition in base b, i.e. the x decomposition in w-bit words with $b = 2^w$ and $\lambda = \lceil \log_b(x) \rceil$.
- $m_p = ((m_p)_{\nu-1} \ldots (m_p)_1 (m_p)_0)_b$ corresponds to the decomposition of m_p in base b and $\nu = \lambda/2$.
- $p = ((p)_{\nu-1} \ldots (p)_1 (p)_0)_b$ corresponds to the decomposition of p in base b.
- $\mathrm{HW}(x)$ is the Hamming weight of the value x.
- w is the number of bits to attack (divisor of ℓ)

Reduction Methods. Plenty of reduction and division algorithms have been published in the literature so far. Therefore operations like $m \bmod p$ and $m \bmod q$ are usually implemented by selecting one of these methods. As one of the main factor of choice for developers is the nature of the crypto accelerator implementing the arithmetic operations for processing long integers calculations. When available, the technologies used to perform such big computation can be very variable from a manufacturer to another.

As an example Montgomery [19] and Barrett [3] reduction methods are commonly used in a wide range of embedded hardware. The constant values can be precomputed once and kept in memory for any further computations like the numerous modular multiplications involved during both exponentiations modulo p and q. Division algorithms can also be selected to perform the reduction operations. More details on these different methods can be found in the literature [2,17]. Interestingly the attack introduced in this paper remains generic and turns out to be efficient regardless the algorithm selected to perform the modular operations.

2.2   Side-Channel Analysis on CRT-RSA

Since the original DPA (i.e. Difference of Means - DoM) published by Kocher et al. [15], some techniques were presented to improve the attack efficiency using the divide-and-conquer approach. Particularly statistical tools, also known as distinguishers, have been developed to improve the way the right key is being identified among all the other key values. We can for instance mention the Correlation Side-Channel Analysis (CSCA) introduced by Brier et al. [5], the Mutual Information Analysis from Gierlichs et al. [12] or the Linear Regression Side-Channel Analysis [9,23].

Applied to the straightforward RSA and to the CRT-RSA the literature is full of attack paths. However it is noticeable that most of them target the exponentiation [7,16,18,24,25] or the CRT recombination [1]. Only few of them are taking benefit of the initial reductions of the message m by the secret prime values p and q.

One of them was presented at CHES 2002 by den Boer et al.. It introduced a chosen message DSCA named MRED standing for *Modular Reduction on Equidistant Data* [4]. This technique relies on the assumption that *"the distance of the intermediate data equals the distance of the intermediate data after the modular reduction at least for a subgroup of single measurements"*. In principle the MRED targets the recovery of the value m_p (resp. m_q) resulting from the message reduction by the secret prime p (resp. q) of a message m defined randomly. A DSCA is performed byte per byte to disclose m_p (or m_q) from the least to the most significant byte. Once the pair of value m and m_p (or m_q) is known the secret values p (or q) can be easily retrieved by performing $\gcd(m - m_p, n)$ (or $\gcd(m - m_q, n)$).

To recover the jth byte of m_p, chosen messages are being submitted to the device within the set of values $\{m - b^j, m - 2 \cdot b^j, \ldots, m - 255 \cdot b^j\}$. As a result, internal value processed by the device after the reduction equal to respectively $m_p - b^j$, $m_p - 2 \cdot b^j$, etc. Hence MRED requires to run a high number k of transactions with the format of the message given above. The opportunity of doing this is subject to the cryptographic protocol in place. The decryption turns out to be a good candidate for being able to chose a message going through the CRT-RSA. For each execution it is necessary to collect the corresponding trace from a physical measurement. Thanks to the set of trace and taking advantage of the knowledge of the less significant bytes of m_p, it becomes possible estimate the jth byte of m_p by applying a DSCA.

Running the whole attack requires the collection of $\frac{\ell}{w}$ sets of traces. Moreover the success of each step relies on the ability to have obtained good results from the less significant bytes recovery, as the attack is recursive.

Among the different DSCA technique, the linear correlation factor from Bravais-Pearson is a good candidate to exploit the information following a linear leakage with the hamming weight of the estimated variable. Ideally the correlation factor between the estimated and the measured series converges towards 1. Practically this is never the case albeit exploitable when the correlation shows a peak.

Let's denote $\mathcal{C}^{(i)}$ with $1 \leq i \leq k$ a set of k side-channel traces captured from a device processing a CRT-RSA decryption giving $\Theta_0 = \{\mathcal{C}^{(1)}(t), \ldots, \mathcal{C}^{(k)}(t)\}$. Let's denote $X^{(i)}$ the corresponding input value that will be processed by the CRT-RSA algorithm and subsequently $S^{(i)} = f(X^{(i)}, \hat{p})$ the set of k guessed intermediate values. A typical function f matching a physical measurement with the corresponding value is the hamming weight of this value.

Having the physical traces $\Theta_0 = \{\mathcal{C}^{(1)}(t), \ldots, \mathcal{C}^{(k)}(t)\}$ on one hand and the corresponding set of estimated value $\Theta_1 = \{S^{(1)}, \ldots, S^{(k)}\}$ on the other hand, the dependency between Θ_0 and Θ_1 can be computed with the Pearson correlation factor:

$$\rho_{\Theta_0, \Theta_1} = \frac{\mathrm{Cov}(\Theta_0, \Theta_1)}{\sigma_{\Theta_0} \sigma_{\Theta_1}}$$

$$= \frac{N \sum (\mathcal{C}^{(i)}(t) \cdot S^{(i)}) - \sum \mathcal{C}^{(i)}(t) \sum S^{(i)}}{\sqrt{N \sum (\mathcal{C}^{(i)}(t))^2 - (\sum \mathcal{C}^{(i)}(t))^2} \sqrt{N \sum (S^{(i)})^2 - (\sum S^{(i)})^2}},$$

where sums are taken over $1 \leq i \leq k$ for each timing t.

3 Chosen Message Side-Channel Attack on Reduced Message Value

This section introduces two new side-channel techniques that could harm a CRT-RSA implementation when the initial reductions with the primes p and q are being processed. Both techniques use the same principle but their exploitation remains different as the first one is a SSCA and the second one is a DSCA.

3.1 Attack Underlying Principle

For both attacks, the principle is to take advantage of the following statement:

Property: for any given integer value $u < min(p,q)$, reducing the input message $m = n - u$ modulo p and q will respectively result to the values $m_p = p - u$ and $m_q = q - u$.

The previous property can be explained straightforwardly by highlighting the following observation:

$$m \bmod x = n \bmod x - u \bmod x = x - u$$

for x being equal to p or q and $u < min(p,q)$.

From this statement can be derived two very efficient side-channel attack paths. The first attack technique is a simple side-channel attack. Many executions are performed on the attacked device and each byte of secret is revovered in a single trace. The second attack is a differencial side-channel attack. Statistical processing is applied on the acquired traces to recover all the secret bytes. Both attacks are detailed in the sequel of this paper.

3.2 Chosen Message Simple Side-Channel Analysis

For the simple analysis, the idea is to exploit a similar principle as the power tag introduced by Clavier et al. in [6]. This technique exploits operands with word(s) equal to 0. The aim is to distinguish such operands from other operands, because 0 is being manipulated during the multiplication. When the hardware device shows a different level of leakage, this operand is said "tagged". In the following, the notation introduced in [6] will be used : $tag^i(m)$ denotes the event "the operand m has its ith word equal to zero".

The translation in base b of the value $A = (n - u) \bmod p = (p - u) \bmod p$ yields to: $A = (A_{\nu-1}A_{\nu-2}\ldots A_1 A_0)_b$. From this observation, it can be deduced that A_0 equals to $(p - u) \bmod b$ and then $(p \bmod b) - (u \bmod b) \bmod b$. In the context of the attack, it is assumed that $u \bmod b$ is chosen and known by the attacker. If the value $(u \bmod b)$ equals $(p \bmod b)$ then $A_0 = 0$ therefore $tag^0(A)$ could be distinguishable on the side-channel trace.

For this chosen message SSCA it is necessary to run several times the RSA operation with different messages. To find the value $(p \bmod b)$ the attacker has to try all the values possible for $(u \bmod b)$ until $tag^0(A)$ appears on the trace. When this event arises the value $(p \bmod b)$ is found. Therefore the w least significant bits of p are recovered. The same principle applied to A_1 and recursively the whole secret p can be disclosed.

Hence, an attack algorithm which takes advantage of this principle was designed to recover from a set of side-channel traces the whole secret p (and q). The full attack execution is given in algorithm 3.1.

Algorithm 3.1. Chosen message Simple Side-Channel Analysis on secret reduced value

---

Require: the 2ℓ-bit RSA public modulus n, number w of secret bit to guess per attack iteration (select a divisor of ℓ, used as integer decomposition base $b = 2^w$)
Ensure: bits or whole values p and q s.t. $n = p \cdot q$

1: $q \leftarrow 0$
2: **for** $i = 0$ **to** $\frac{\ell}{w} - 1$ **do**
3:      $guess \leftarrow 0$
4:      $A = n - (q + guess \cdot b^i)$
5:      Execute CRT-RSA with input message A
6:      **while** $(guess < b)$ **and** $(\text{not}(tag^i(A)))$ **do**
7:          $guess = guess + 1$
8:          $A = n - (q + guess \cdot b^i)$
9:          Execute CRT-RSA with input message A
10:     $q \leftarrow q + guess \cdot b^i$
11: **return** (q)

Attack Complexity. Using this algorithm, this SSCA requires at most $\frac{\ell}{w} \cdot b$ CRT-RSA executions to recover the secret RSA keys.

3.3   Chosen Message Correlation Side-Channel Analysis

In the same way as for the SSCA, this attack uses the fact that $A_0 = (p \bmod b) - (u \bmod b)$. As $u \bmod b$ is chosen and known by the attacker, it becomes possible to exploit the previous representation for estimating A_0 for each guess made on the secret value $p \bmod b$. The conditions are gathered to set up a differential side-channel analysis with the aim to estimate the value A_0. This can be achieved by executing a high number of transaction and by choosing a different value $u^{(i)}$ for each occurrence. Once the set of transaction collected, the aim is to apply the traditional divide and conquer approach for recovering the w least significant bits of the secret p on $(p)_0$.

With the knowledge of $(p)_0$, the same technique can subsequently be implemented to exhibit the secret value $(p)_1$. This does not require the collection of a new set of trace. In this second step, the statistical analysis targets the value A_1. Running the attack step by step in a recursive and similar manner, the values $(p)_2 \ldots (p)_{\nu-1}$ can be disclosed one after each other.

As previously, an efficient algorithm taking advantage of this principle was designed to recover the secret modulus. Full attack execution is given in Algorithm 3.2.

Algorithm 3.2. Chosen message correlation on secret reduced value

---

Require: 2ℓ-bit RSA public modulus n, number w of secret bit to guess per attack iteration (select a divisor of ℓ, used as integer decomposition base $b = 2^w$)
Ensure: p and q s.t. $n = p \cdot q$

1: Generate a set U of k random values: $U = \{u^{(0)}, \dots, u^{(k)}\}$ s.t $u^{(i)} < 2^{\nu-1}$ for $i = 0 \dots k - 1\}$
2: Execute k CRT-RSA with input messages $(n - u^{(0)}), \dots, (n - u^{(k-1)})$ and store the associated set of side-channel traces: $\theta_0 = \{T^{(0)}, \dots T^{(k-1)}\}$
3: $q \leftarrow 0$
4: **for** $i = 0$ **to** $\frac{\ell}{w} - 1$ **do**
5: $\quad c \leftarrow 0,\ g \leftarrow 0$
6: $\quad$**for** $Guess = 0$ **to** $b - 1$ **do**
7: $\quad\quad$**for** $v = 0$ **to** $k - 1$ **do**
8: $\quad\quad\quad qG = q + Guess \cdot b^i$
9: $\quad\quad\quad A = qG - (u \bmod b^{(i+1)}) = ((A)_{\nu-1} \dots (A)_i \dots (A)_1 (A)_0)_b$
10: $\quad\quad\quad S^{(v)} = \{\mathrm{HW}(A_i), \dots, \mathrm{HW}(A_i)\}$
$\quad\quad\quad \theta_1 = \{S^{(0)}, \dots S^{(k-1)}\}$
11: $\quad\quad \rho \leftarrow \rho_{\theta_0, \theta_1}$
12: $\quad\quad$**if** $(c < \rho)$ **then**
13: $\quad\quad\quad (c, g) \leftarrow (\rho, Guess)$
14: $\quad q \leftarrow q + g \cdot b^i$
15: **return** (q)

Remark. Algorithm 3.2 allows the full secret recovery. However, in some cases, some bits could not leak and render the algorithm less efficient. In that case recovering a byte (ie. $w = 8$) without the knowledge of the previous one would lead to recovering with certitude 7-bit of the targeted byte because of the carry uncertainty. Missing bits can then be recovered with exhaustive search.

4   Practical Results

During this study, practical experimentations were run to ascertain the validity of the attacks and analyse their efficiency and practicability. This experimental work was done on simulated traces and on several integrated circuits.

4.1   First Attack Technique with Simple Side-Channel Analysis

The first experiment made had the objective to validate the first attack presented. A CRT-RSA was implemented on a secure device embedding an arithmetic coprocessor and the attack presented in algorithm 3.1 was run.

In the traces, following the reduction(s) operations, the first steps of the CRT exponentiation(s) are performed. These calculations contain some multiplication operation which are performed with the reduced value. Focusing the analysis on this area of interest, it was possible to distinguish when zero bytes were being processed.

As visible on figure 1, the time required to process a multiplication varies depending on the number of word(s) to 0 in the multiplicand.

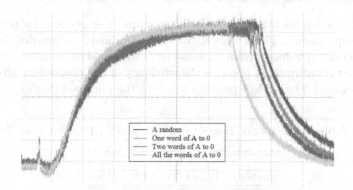

Fig. 1. Power traces of a multiplication between A and B

It was also observed that other components were not leaking in timing and had constant time multiplication whatever the value manipulated was. However in some cases the signal amplitude became lower when the zero value was manipulated by the coprocessor. It has also been previously discussed and experimented in the literature [6, 7].

4.2   Second Attack Technique with Differential Side-Channel Analysis

Here, the experimental work was executed in two steps. First the attack was tested on simulated side-channel traces. In a second phase a CRT-RSA was implemented on an ATMEGA chip with a 32-bit core. This IC is a standard chip that can be openly found on the market and does not include hardware countermeasures as some secure certified products could.

Simulated Traces Experiments. First of all the DSCA described in algorithm 3.2 was applied on simulated side-channel traces to validate its efficiency. Therefore traces for three standard core architectures were generated: $\lambda = 8$, $\lambda = 16$ and $\lambda = 32$ bits cores.

Classically the model chosen was a linear leakage model against the Hamming weight of the manipulated data — here $(m_p)_i$ and $(m_q)_i$ — and add a white Gaussian noise of mean $\mu = 0$ and standard deviation σ.

The process detailed hereafter has been followed :

- Generate CRT-RSA keys of 2ℓ bits for $\ell \in \{256, 384, 512\}$ with secret prime factors p and q of the public modulus n.
- Select random k ℓ-bit integer values $u^{(0)} \ldots u^{(k)}$ such that $u^{(i)} < 2^{\ell-1}$
- Consider CRT-RSA executions with input messages $n - u^{(i)}$
- Generate the side-channel traces $T^{(i)}$ of the reduced values $(m_p)^{(i)} \bmod p$ and $(m_q)^{(i)} \bmod q$. $T^{(i)}$ is made of the Hamming weights of values $(m_p^{(i)})_j$ for j from 0 to $\frac{\ell}{\nu}$.
- The attack from algorithm 3.2 was implemented and tested.

The success rate of this paper attack technique was assessed by performing 1 000 such experiments. These tests have been performed for three different noise standard deviation values[1]: from no noise ($\sigma = 0$) to a strong one ($\sigma = 7$).

These tests confirmed in practice the efficiency of the theoretical attack presented previously. The table 1 presents a few of the obtained results.

Table 1. Number of Traces to retrieve p in a CRT-RSA 1024

σ	0.1	2	7
Number of Traces	50	120	1500

Figures 2,3,4 and 5 show the attack success on some of the secret bytes. Here bytes 0, 5, 15 and 30 on CRT-RSA 1024 are given to illustrate the attack results. Figure 6 gives the attack success rate depending on the number of executions (i.e. messages).

Physical Experiments. In a second phase, the attack was experimented on a real hardware device. To do so a Barrett reduction on a 32-bit ATMEGA core running at a frequency of 20 MHz was implemented. Traces were collected by observing the electromagnetic radiations between the targeted reduction and the beginning of the exponentiation as it can be seen on figure 7.

The DSCA from the algorithm 3.2 was subsequently applied. The results showed that the secret prime was recovered using maximum 19000 traces.

The practical experimentations concluded the validity of the attack on real devices.

5   Comparison with MRED

As previously described, the attack introduced in this paper can be seen as comparable with the MRED attack from den Boer et al. [4]. As the MRED technique represents one of the most effective attack that targets the reduction

---

[1] Regarding the standard deviation of the noise, a unit corresponds to the side-channel difference related to a one bit difference in the Hamming weight.

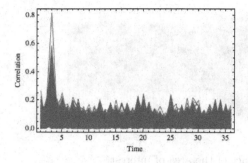

Fig. 2. 2^8 correlation traces attack result on (p_0) for $\lambda = 8$, $w = 8$, $k = 200$ and $\sigma = 1$

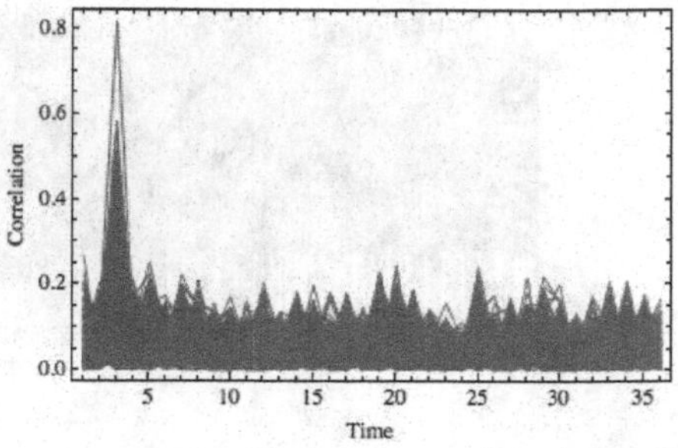

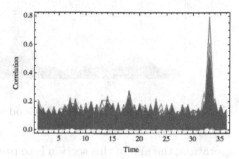

Fig. 3. 2^8 correlation traces attack result on (p_5) for $\lambda = 8$, $w = 8$, $k = 200$ and $\sigma = 1$

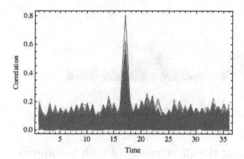

Fig. 4. 2^8 correlation traces attack result on (p_{15}) for $\lambda = 8$, $w = 8$, $k = 200$ and $\sigma = 1$

Fig. 5. 2^8 correlation traces attack result on (p_{30}) for $\lambda = 8$, $w = 8$, $k = 200$ and $\sigma = 1$

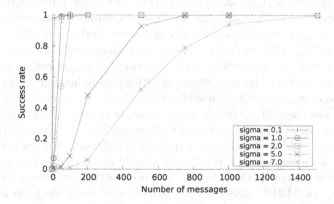

Fig. 6. Attack success rate for different noise values

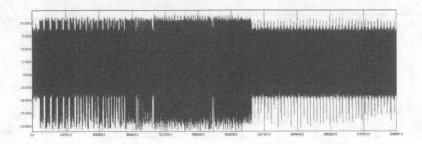

Fig. 7. Electromagnetic trace of the area of interest

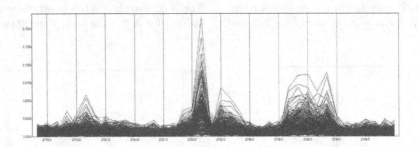

Fig. 8. Correlation traces. Good value in red, other values in blue.

operation, the aim of this section is to present the effectiveness of both techniques as well as their respective restrictions.

The attack presented in this paper, as well as the MRED attack, do not target the intermediate value of the reduction. Indeed, these attacks target the outcome of the reduction operations and not the corresponding intermediate values. This is a substantial asset when noticing the number of different reduction and division algorithms that can be found in the literature. Particularly because there is no obvious reason to favour one of these algorithms. As a second strength, the outcome of the reduction operation can be manipulated at different times during the full CRT-RSA execution. The area of leakage can be found from the reduction operation to the modular multiplication computed before and during the CRT exponentiation. Hence it offers potentially more leaking points that can be exploited and consequently it provides more chance to succeed in the attack.

One of the most significant differences between both attack resides in the number of traces necessary to run the attack. Whereas the side-channel treatment remains similar, the MRED technique requires the collection of a new set of traces for each part of the secret when the DSCA attack introduced in this paper only needs a unique set of traces. Cumulating the number of traces necessary to exhibit the full secret, the MRED attack would unlikely succeed if the total number of operations is limited on the device. In that case, the current attack would represent a clear advantage as the number of traces required to perform it is way lower and may not reach such a limit.

For the sake of comparison in the statistical processing, we implemented both techniques using simulated side-channel traces. As factor of comparison, the number of trace necessary to disclose the full secret is given for different key lengths and different levels of noise of variance σ. The results are shown on the table 2 and were obtained on simulated traces.

Table 2. Number of traces comparison between MRED and this paper Chosen CPA

Attack	σ	RSA bitlength					
		512	768	1024	1536	2048	4096
MRED	0.1	3840	5760	7680	11520	15360	30720
Alg.3.2	0.1	50	50	50	50	50	50
MRED	2	6400	9600	12800	19200	25600	51200
Alg.3.2	2	120	120	120	120	120	120
MRED	7	48000	72000	96000	144000	192000	384000
Alg.3.2	7	1500	1500	1500	1500	1500	1500

6   Countermeasures

First of all, as a chosen message attack, the threat does not concern the cryptographic protocols using message padding such as PKCS#1 [20]. Indeed this protocol removes control over the value of m from the attacker. Hence he will not be able to run a CRT-RSA processing specific messages. As both attacks presented require to send message of the form $n - u$ (with u of at most half of the bitlength) proper message padding is an already existing and efficient countermeasure.

However this attack suits well for protocols where any message can be processed. This is the case typically for a decryption using a CRT-RSA key. In that cases, the attack may represent a threat if the RSA implementation is not adequately protected. Different countermeasures have been published in the literature to strengthen the reduction operations.

In 2002 Joye and Villegas [13] introduced a protection for implementations of division algorithm. Whereas this idea represents a strong protection to tackle simple side-channel analyses within the reduction, it does not withstand the attacks estimating the outcome value of the reduction. As a result such implementation would remain susceptible to the simple attack and the statistical attack introduced in this paper.

Blinding methods like the additive and the multiplicative message randomisations are usually considered as the most effective countermeasures to defeat side-channel attacks. Indeed adding a random multiple $r \cdot n$ (or $r \cdot p$) of the modulus protects the reduction (or division) against side-channel techniques targeting the intermediate values. The operation $(m + r \cdot p) \bmod p$ cannot be threatened by classical attacks as all the intermediate values of the reduction are randomized differently at each execution with a new random r. However this countermeasure is not a good enough to prevent implementations from the attacks of this article as well as the MRED which aims at estimating the result of the modular reduction.

To be effective the additive randomization requires the reduced value to be combined with another technique. As an example a specific randomized reduction algorithms [10] can be chosen. In such a case the value m mod p would not be the exact reduced value and would render the attack not efficient. A second option would be to keep the secret modulus randomized during the whole CRT-RSA. Lastly multiplicative blinding techniques applied to the message would cover in the same way the outcome of the reduction operation. With one of this technique efficiently implemented both attack presented in this paper would be defeated.

7   Conclusion

This article introduced two new attacks targeting the reduction operations that are being performed before CRT-RSA exponentiations. As a result, the secret primes p or q may be exposed and consequently the whole secret key being disclosed. These attacks are applicable for cryptographic protocols when the message in input of the CRT-RSA can be chosen. Indeed both techniques exploit a specific format of messages with the aim to estimate intermediate values leading to the secret key. Their validity has been ascertained when successfully implemented on simulated and real hardware devices. The first attack makes use of a set of single traces to recover the secret prime with simple analysis. The second attack requires a statistical processing to expose the secret prime with a high number of traces. The latter can be compared to an existing attack using the same principle, named the MRED. Our study presents a comparison between both attacks and concluded that the technique represents a significant improvement compared to MRED. Indeed our techniques require much less traces to recover the whole secret which is of strong interest in practical attack considerations.

References

1. Amiel, F., Feix, B., Villegas, K.: Power analysis for secret recovering and reverse engineering of public Key algorithms. In: Adams, C., Miri, A., Wiener, M. (eds.) SAC 2007. LNCS, vol. 4876, pp. 110–125. Springer, Heidelberg (2007)
2. Avanzi, R.-M., Cohen, H., Doche, C., Frey, G., Lange, T., Nguyen, K., Vercauteren, F.: Handbook of Elliptic and Hyperelliptic Curve Cryptography. CRC Press (2006)
3. Barrett, P.: Implementing the Rivest Shamir and Adleman public key encryption algorithm on a standard digital signal processor. In: Odlyzko, A.M. (ed.) CRYPTO 1986. LNCS, vol. 263, pp. 311–323. Springer, Heidelberg (1987)
4. den Boer, B., Lemke, K., Wicke, G.: A DPA attack against the modular reduction within a CRT implementation of RSA. In: Kaliski Jr, B.S., Koç, Ç.K., Paar, C. (eds.) CHES 2002. LNCS, vol. 2523, pp. 228–243. Springer, Heidelberg (2003)
5. Brier, E., Clavier, C., Olivier, F.: Correlation Power Analysis with a Leakage Model. In: Joye, M., Quisquater, J.-J. (eds.) CHES 2004. LNCS, vol. 3156, pp. 16–29. Springer, Heidelberg (2004)
6. Clavier, C., Feix, B.: Updated recommendations for blinded exponentiation vs. single trace analysis. In: Prouff, E. (ed.) COSADE 2013. LNCS, vol. 7864, pp. 80–98. Springer, Heidelberg (2013)

7. Courrège, J.-C., Feix, B., Roussellet, M.: Simple Power Analysis on Exponentiation Revisited. In: Gollmann, D., Lanet, J.-L., Iguchi-Cartigny, J. (eds.) CARDIS 2010. LNCS, vol. 6035, pp. 65–79. Springer, Heidelberg (2010)
8. Diffie, W., Hellman, M.E.: New Directions in Cryptography. IEEE Transactions on Information Theory **22**(6), 644–654 (1976)
9. Doget, J., Prouff, E., Rivain, M., Standaert, F.-X.: Univariate side channel attacks and leakage modeling. IACR Cryptology ePrint Archive 2011:302 (2011)
10. Dupaquis, V., Venelli, A.: Redundant modular reduction algorithms. In: Prouff, E. (ed.) CARDIS 2011. LNCS, vol. 7079, pp. 102–114. Springer, Heidelberg (2011)
11. Garner, H.L.: The residue number system. In: Proceedings of the Western Joint Computer Conference, pp. 146–153 (1959)
12. Gierlichs, B., Batina, L., Tuyls, P., Preneel, B.: Mutual Information Analysis. In: Oswald, E., Rohatgi, P. (eds.) CHES 2008. LNCS, vol. 5154, pp. 426–442. Springer, Heidelberg (2008)
13. Joye, M., Villegas, K.: A protected division algorithm. In: Proceedings of the Fifth Smart Card Research and Advanced Application Conference, CARDIS 2002 (2002)
14. Kocher, P.C.: Timing Attacks on Implementations of Diffie-Hellman, RSA, DSS, and Other Systems. In: Koblitz, N. (ed.) CRYPTO 1996. LNCS, vol. 1109, pp. 104–113. Springer, Heidelberg (1996)
15. Kocher, P.C., Jaffe, J.M., June, B.C.: DES and Other Cryptographic Processes with Leak Minimization for Smartcards and other CryptoSystems. US Patent 6,278,783 (1998)
16. Mayer-Sommer, R.: Smartly analyzing the simplicity and the power of simple power analysis on smartcards. In: Koç, Ç.K., Paar, C. (eds.) CHES 2000. LNCS, vol. 1965, pp. 78–92. Springer, Heidelberg (2000)
17. Menezes, A., van Oorschot, P.C., Vanstone, S.A.: Handbook of Applied Cryptography. CRC Press (1996)
18. Messerges, T.S., Dabbish, E.A., Sloan, R.H.: Power analysis attacks of modular exponentiation in smartcards. In: Koç, Ç.K., Paar, C. (eds.) CHES 1999. LNCS, vol. 1717, pp. 144–157. Springer, Heidelberg (1999)
19. Montgomery, P.L.: Modular multiplication without trial division. Mathematics of Computation **44**(170), 519–521 (1985)
20. PKCS #1. RSA Cryptography Specifications Version 2.1. RSA Laboratories (2003)
21. Quisquater, J.-J., Couvreur, C.: Fast decipherment algorithm for RSA public-key cryptosystem. Electronic Letters **18**(21), 905–907 (1982)
22. Rivest, R.L., Shamir, A., Adleman, L.: A Method for Obtaining Digital Signatures and Public-Key Cryptosystems. Communications of the ACM **21**, 120–126 (1978)
23. Schindler, W., Lemke, K., Paar, C.: A stochastic model for differential side channel cryptanalysis. In: Rao, J.R., Sunar, B. (eds.) CHES 2005. LNCS, vol. 3659, pp. 30–46. Springer, Heidelberg (2005)
24. Walter, C.D.: Sliding Windows Succumbs to Big Mac Attack. In: Koç, Ç.K., Naccache, D., Paar, C. (eds.) CHES 2001. LNCS, vol. 2162, pp. 286–299. Springer, Heidelberg (2001)
25. Yen, S.-M., Lien, W.-C., Moon, S.-J., Ha, J.C.: Power Analysis by Exploiting Chosen Message and Internal Collisions – Vulnerability of Checking Mechanism for RSA-decryption. In: Dawson, E., Vaudenay, S. (eds.) Mycrypt 2005. LNCS, vol. 3715, pp. 183–195. Springer, Heidelberg (2005)

Theory

On Constant-Round Concurrent Zero-Knowledge from a Knowledge Assumption

Divya Gupta[✉] and Amit Sahai

Department of Computer Science, and Center for Encrypted Functionalities,
University of California, Los Angeles, USA
{divyag,sahai}@cs.ucla.edu

Abstract. In this work, we consider the long-standing open question of constructing constant-round concurrent zero-knowledge protocols in the plain model. Resolving this question is known to require non-black-box techniques. We consider non-black-box techniques for zero-knowledge based on knowledge assumptions, a line of thinking initiated by the work of Hada and Tanaka (CRYPTO 1998). Prior to our work, it was not known whether knowledge assumptions could be used for achieving security in the concurrent setting, due to a number of significant limitations. Nevertheless, we obtain the first constant round concurrent zero-knowledge argument for **NP** in the plain model based on a new variant of knowledge of exponent assumption. We give evidence that our new variant of knowledge of exponent assumption is in fact plausible. In particular, we show that our assumption holds in the generic group model.

1 Introduction

Zero-knowledge proofs [20] are fundamental and important tools in the design of cryptographic protocols. The original setting of zero-knowledge proofs contemplated a single prover and a single verifier executing a single protocol session in isolation. Concurrent zero-knowledge [16] ($c\mathcal{ZK}$) extends zero-knowledge to concurrent settings, where several protocol sessions are executed at the same time involving multiple provers and verifiers. Resolving the round complexity of concurrent zero-knowledge protocols has been a long standing open problem. There have been several negative results which give lower bounds for round complexity of black box simulation of $c\mathcal{ZK}$ [11,25,33]. The best result, which uses

Research supported in part from a DARPA/ONR PROCEED award, NSF grants 1228984, 1136174, 1118096, and 1065276, a Xerox Faculty Research Award, a Google Faculty Research Award, an equipment grant from Intel, and an Okawa Foundation Research Grant. This material is based upon work supported by the Defense Advanced Research Projects Agency through the U.S. Office of Naval Research under Contract N00014-11- 1-0389. The views expressed are those of the author and do not reflect the official policy or position of the Department of Defense, the National Science Foundation, or the U.S. Government.

© Springer International Publishing Switzerland 2014
W. Meier and D. Mukhopadhyay (Eds.): INDOCRYPT 2014, LNCS 8885, pp. 71–88, 2014.
DOI: 10.1007/978-3-319-13039-2_5

black box simulation, has $\omega(\log n)$ round complexity [31], where n is the security parameter. This nearly matches the best known lower bound for black box simulation [11], which states that any black-box concurrent zero-knowledge protocol must require at least $\tilde{\Omega}(\log n)$ rounds. Hence, our only hope of achieving constant round $c\mathcal{ZK}$ lies in non-black-box simulation. In his seminal work, Barak [3] introduced the first non-black-box simulation technique, but this technique or its variants have not yet yielded a concurrent zero-knowledge protocol with lower round complexity than the work of [31]. Indeed, Barak explicitly posed the problem of constructing constant-round concurrent zero-knowledge arguments as "an important open question" [3]. Despite many attempts in this direction, this is still a long-standing open problem in cryptography.

We stress that constant-round concurrent zero knowledge, according to the standard definition of polynomial-time simulation and in the standard plain model, is completely open. Prior to our work, there were no known complexity assumptions under which such constant-round concurrent zero knowledge protocols were known to exist, even under exponential hardness assumptions. On the other hand, concurrent zero-knowledge trivializes in alternative models like the CRS model or the random oracle model, where completely non-interactive zero-knowledge is possible. The only other model where constant-round concurrent zero-knowledge was known was in the timing model [16], where an assumption is needed that the network cannot introduce unbounded delays for messages. Constant-round arguments are known that achieve super-polynomial simulation [30] under super-polynomial hardness assumptions in the plain model, but this is meaningful only in scenarios where we must change our notion of efficient (adversarial) computation to beyond polynomial-time. In sum, before our work, no concrete assumption was known to be sufficient to build constant-round concurrent zero-knowledge protocols in the plain model. Our work opens up a new direction of understanding what kinds of concrete assumptions are sufficient for achieving novel concurrent *polynomial-time* simulations.

In this work, we consider whether non-black-box techniques based on knowledge assumptions can be applied to achieve constant round concurrent zero-knowledge protocols. We answer this question affirmatively, by giving the first constant-round concurrent zero-knowledge protocol based on a knowledge assumption, which is a novel variant of the knowledge of exponent assumption first introduced by Damgard [14] and used by Hada and Tanaka [23] in the context of ordinary zero-knowledge protocols.

Furthermore, our techniques allow us to avoid the inefficiency inherent in previous non-black-box techniques, such as those of Barak [3]. Indeed, we obtain our result by providing an efficient transformation from constant round stand alone protocols to constant round concurrently secure zero-knowledge protocols.

Recently, there has been an explosion of interest in knowledge assumptions. Knowledge assumptions became popular when these were applied to the construction of three round zero-knowledge arguments by [23]. This has led to a number of interesting research papers applying knowledge assumptions to a variety of different contexts [1, 5, 6, 8, 9, 13, 15, 18, 19, 21, 24, 32]. Prior to our work, to the best of our knowledge, knowledge assumptions have not been applied successfully to achieve concurrent security. As we explore below, this is because of a number of complications which arise when one applies knowledge assumptions to concurrent settings.

Our Contributions: We show the following:

1. We obtain the first constant round concurrent zero-knowledge argument for **NP** in plain model based on a new variant of knowledge of exponent assumption. Our compiler to get concurrently secure protocol is efficient and avoids the inefficiency inherent in other non-black-box techniques.
2. Unlike Hada and Tanaka [23], we do not require a knowledge assumption to argue the soundness of our protocol. Instead, we use a discrete log like assumption, which we call *DHLA* (See Section 2), to prove the soundness of our protocol.
3. We give evidence that our new variant of knowledge of exponent assumption is in fact plausible. In particular, we show that our assumption holds in the generic group model. Note that we provide the generic group proof only to show the plausibility of our *assumption*. We do not claim or argue that constant-round concurrent zero knowledge is interesting in the generic group model. Our result, on the other hand, shows that constant-round concurrent zero knowledge is possible under a *concrete assumption* in the *standard model*.
4. As we discuss in greater detail below, and as has been discussed throughout the history of knowledge assumptions in cryptography, knowledge assumptions are especially delicate assumptions whose plausibility may be hard to gauge. We give a novel framework to express knowledge assumptions in a more flexible way which may allow for formulation of plausible assumptions and exploration of their impact and application in cryptography.
5. Until now, it was quite plausible that impossibility techniques [2, 4, 10, 17, 26–28] might extend to show that constant-round concurrent zero knowledge is impossible to achieve. We consider a major contribution of our work to be providing the first solid evidence that such an impossibility result is unlikely to materialize. We do this by showing that constant-round concurrent zero knowledge is possible under a plausible concrete assumption in the plain model.

On Knowledge Assumptions and their Applications in Cryptography. Knowledge assumptions are inherently non-black-box. Informally speaking[1], knowledge assumptions can be expressed by assuming that there is a specific "Knowledge Commitment Protocol" such that we can efficiently extract the value committed by the adversary if he completes the commitment protocol successfully — in other words, we assume that any adversary that successfully completes the Knowledge Commitment Protocol must have "knowledge" of the value that it committed to. For the purpose of this introduction, assume that the Knowledge

---

[1] Our assumption is concrete. See Section 3.

Commitment Protocol is just a two message protocol in which first the Receiver sends a random message to the Committer and then the Committer responds with a commitment to a value[2]. Knowledge assumptions present a number of challenges to the research community from the point of view of the falsifiability rubric of Naor [29]: they do not fall in the desirable category of falsifiable assumptions in [29].

Furthermore, knowledge assumptions present challenges with regard to auxiliary inputs as is also pointed out in the early works of Hada and Tanaka [23]. Intuitively the problem arises if we consider what happens if an adversary is given as auxiliary input an obfuscated program. The adversary simply compiles and executes the obfuscated program to obtain the commitment message. Then a knowledge assumption, which is expected to hold *for all* auxiliary inputs, would imply an efficient extraction of the committed value. This would imply an efficient deobfuscation, which is problematic [7]. It was recently suggested by Bitansky et al [6] that it is more reasonable to assume that knowledge assumptions only hold with respect to "benign" auxiliary inputs. One of our contributions is to put forward a framework for formulating knowledge assumptions with respect to *Admissible Adversaries*. This allows us to specify a set of auxiliary inputs with respect to which the knowledge of exponent assumption would hold. For applications in cryptography we want this class to be as large as possible. Despite these drawbacks, the study of knowledge assumptions in cryptography has been thriving recently. This is evident by the long list of interesting research papers cited above. (See full version [22] for more details on related work).

Limitations of Knowledge Assumptions in the Setting of Concurrency. Undoubtedly, the reason that knowledge assumptions have attracted attention is because they are very useful to achieve important goals in cryptography. Indeed often it may seem that knowledge assumptions are so powerful that they can be used to achieve any plausible result that we want to achieve in cryptography. For example, when it comes to the simulation of protocols, intuitively it seems that whenever the adversary commits to some value, the simulator can use the knowledge assumption to extract the hidden value committed to. Hence, it seems this can become a universal technique for straight line simulation[3]. This intuition is false, as we describe below.

One way to see that the above intuition is false is by observing a long list of unconditional impossibility results for concurrent simulations in plain model [2,4,10,17,26–28] and observing that the above intuition seems to give a simulation technique applicable to any concurrent setting. Even in restricted models of concurrency, there are many natural protocol tasks that are impossible

---

[2] The commitment here does not refer to a semantically secure commitment scheme.

[3] For example, consider the following coin flipping protocol. Adversary commits to R, honest party sends R', adversary opens R. The result of coin flipping protocol would be $R \oplus R'$. Intuitively, knowledge assumption would allow the simulator to force the outcome of coin flipping to any string he wants since it would know R immediately after the adversary's commitment through extraction. Thus, one might conclude that with knowledge assumptions we can achieve the CRS model. This intuition is false.

even with knowledge assumptions. One of the most relevant examples is concurrently secure oblivious transfer (OT), in the "fixed roles" setting and with fixed inputs for honest parties. This setting is almost identical to concurrent zero-knowledge, with the only difference being that there one is trying to achieve OT as opposed to zero-knowledge, but there are no issues of "man-in-the-middle attacks" or adaptive choice of inputs. Nevertheless, a concurrently secure OT protocol in fixed roles setting and with fixed inputs for honest parties is impossible even with knowledge assumptions [2,17], yet as we show, there is a plausible assumption under which we achieve constant round concurrent zero-knowledge. The negative results show that potential specific knowledge assumptions, which would be powerful enough to allow for concurrently secure OT, *must be false.* (We stress that the novel knowledge of exponent assumption we formulate here would *not* naturally provide a simulation of a concurrent OT protocol.)

As suggested above, one of the main non-trivialities of our work is to formulate a plausible knowledge assumption that would allow us to achieve constant round concurrent zero-knowledge, while remaining plausible. We begin our discussion here with natural attempts to apply knowledge assumptions to the concurrent setting, and their limitations. We believe that this discussion will be useful to other researchers who would like to apply knowledge assumptions to other interesting problems in cryptography, while also illustrating the non-triviality of achieving concurrent security using a plausible knowledge assumption.

Perhaps the most promising idea would be to formulate an "interactive" knowledge assumption. Informally speaking, such an assumption would say that extraction is possible after an arbitrary interaction which took place prior to the final message in the Knowledge Commitment Protocol. However, any natural formulation of such an interactive knowledge assumption would be powerful enough to achieve concurrent realization of functionalities such as OT. Hence, we know that such an assumption must be false. Indeed such an assumption would be falsified by considering a scenario in which the actions of the adversary in the Knowledge Commitment Protocol are fully specified by messages that the adversary received in the past, and not directly by the adversary itself. (For example, the functionality being computed could provide the messages of the Knowledge Commitment Protocol as outputs to the adversary [2,4].) Intuitively, in such a situation, the adversary doesn't have any knowledge of the value he committed to, and hence the goal of extraction is untenable. Essentially the problem is that some "external knowledge" may find its way to the adversary by means of previous interactions and get used by it to generate its messages in Knowledge Commitment Protocol. Similar problems arise when trying to use auxiliary inputs to the extractor promised by a knowledge assumption in order to facilitate extraction. (See full version [22] for a brief discussion.)

Recursive Applications of Knowledge Assumptions and their Limitations. Another approach would be to apply a knowledge assumption *recursively* for each session. What we mean by this is as follows: Essentially, a knowledge assumption transforms an adversary circuit A into another (potentially polynomially larger) circuit A' that behaves just like A but also outputs an extracted value. If we apply a knowledge assumption recursively, then we would transform the original adversary circuit A into A', but then apply the knowledge assumption again to transform A' into A''. However, clearly if such a recursion is applied a super-constant number of times, then the final circuit might be super-polynomial in size. This problem was encountered by Bitansky et al [6] in the construction of succinct non-interactive adaptive arguments of knowledge (SNARKs) using extractable collision resistant hash functions (ECRH). To prove the property of proof of knowledge, the extractor needs to extract the full Probabilistically Checkable Proof (PCP) given only the root of a Merkel tree. The natural solution is to apply the knowledge extraction recursively at each level of the tree. But since each level of extraction potentially incurs a polynomial blow up, one can apply extraction only a constant number of times. One of the major contributions of [6] was to circumvent this problem by using Merkel trees with polynomial fan-in and constant depth. Note that, however, we do not have any such option while constructing a constant round concurrent zero-knowledge protocol because the number of concurrent sessions can be any unbounded polynomial.

One natural approach to avoid this blow-up with each recursive extraction would be to assume a stronger property on the running time of the extractor. For example, one can assume the existence of an extractor which only takes an additive $poly(n)$ (where n is the security parameter) factor more than the running time of the adversary. Note that the factor of $poly(n)$ is independent of the running time of the adversary. We call this the $+poly(n)$ assumption. However, this assumption seems too strong and in fact potentially implausible[4]. On the other hand, if we do not make such a strong assumption, the essence of the problem is that if we want to apply the knowledge extractor recursively, we cannot afford it to take even m^ϵ longer than the adversary, where m is the running time of the adversary and ϵ is an arbitrary constant. Note that we do not make the $+poly(n)$ assumption.

Intuition behind our assumption. Our first idea is to separate the process of extraction from the behavior of the adversary. More precisely, we will think of the adversary as a circuit M. If M completes the Knowledge Commitment

---

[4] Consider the following scenario: Given a random group element g from a special group G, the adversary is expected to output g^b (a commitment to b) and extractor's task is to output b. However, the Adversary applies a hash function on its input and gets a pseudorandom string $s = s_1 \ldots s_m$ of length m, where m depends on the running time of the adversary and is not a fixed polynomial in length. Now, it traverses the string s and recursively applies a special function A, such that $A(d, g^x) = g^{f(d,x)}$. In other words, the adversary computes $A(s_1, A(s_2, \ldots, A(s_m, g) \ldots))$. Now suppose A and f satisfy the following conditions: (1) $\mathsf{Time}(A) < \mathsf{Time}(f)$ (2) $\mathsf{Time}(f(s_1, f(s_2, \ldots, f(s_m, 1) \ldots))) = m \cdot \mathsf{Time}(f)$. Then, by the latter condition, the extractor needs to compute f iteratively. Thus, the extractor will need at least $O(m)$ more operations than the adversary, where m is decided by the adversary. We do not know if such an A and f exist. However, if such an A and f did exist, it would refute the $+poly(n)$ assumption.

Protocol, an application of our assumption to M gives us a *separate* extractor circuit E. The assumption states that the input wires of E can be any wire inside the circuit M, including input, intermediate, or output wires. The output of $E(x)$ is *only* the value committed to in the output of $M(x)$. Now that we have separated the extractor from the adversary, we make the following observation: It is reasonable to assume that when the assumption is applied to create an extractor circuit E, the assumption does not attempt to place any "external knowledge" into E or attempt to hide any knowledge in E. In other words, the extractor created by the assumption is *not* maliciously created. Hence, let us call it *benign* and denote it by B. Note that we will *only consider benign circuits that are created by the assumption*. The benign circuits are not assumed to remain benign if they are modified. Now we can state our assumption:

Assumption 1 (Informal knowledge assumption) *Consider a pair of malicious and benign circuits (M, B) such that M completes a Knowledge Commitment Protocol and outputs a commitment to a value. Then there exists a polysize benign extractor circuit E which takes as input a subset of wires of M, and outputs the value committed to by M. Moreover, the size of the extractor E is bounded by a fixed polynomial in the size of M and the security parameter n.*

Now consider a recursive application of our assumption. Recall that the recursive application is required for the following: Suppose we have an adversary and we execute it to obtain some number of messages until it completes a Knowledge Commitment Protocol. Then we apply the knowledge assumption to obtain an extractor that allows us to obtain the committed value. We then use the extracted value in order to execute the adversary for some additional number of messages until it finishes another Knowledge Commitment Protocol (and so on). Let us denote by M the execution of the adversary so far. Note that the inputs to M are essentially the original inputs to the adversary together with the outputs of the extractors so far. Denote by B the collection of extractors so far.

Now let us consider what happens when we apply our assumption to (M, B). We obtain an extractor E that extracts a value committed in the output of M. We observe that while B was involved in the execution of the adversary, only the outputs of B were ever used by M to compute its output commitment message. Furthermore, as argued above, B was benignly created by the assumption and thus has no external or hidden knowledge inside it. Thus we argue, that it is reasonable to assume that the size of the extractor E created by the assumption is a fixed polynomial in the size of only the malicious circuit M. Recall that M contains all the malicious computations done by the adversary. We now make the following observations about our assumption.

– We observe that without loss of generality, we can assume that in a recursive application of our assumption, the extractor created by the assumption in fact contains all of the extractors created previously inside of it. Namely the benign circuit B is a part[5] of the newly created extractor E. Thus E can make use of all of the intermediate wires of previously created extractors, without loss of generality. These intermediate values may contain useful knowledge which may help the extraction of the value committed in the output message of M.

- We also observe that the counter-example we contemplated in Footnote 4 to the $+poly(n)$ assumption is compatible with our assumption[6]. That is, the existence of the functions A and f specified in the counter-example would not refute our assumption. Essentially this is because our extractor E is *allowed to be polynomially larger* than the malicious circuit M.
- We further validate the plausibility of specific knowledge assumption that we make (see Section. 3) by providing a proof that the assumption holds in the generic group model (Section. 7).
- To understand what computational complexity limitations our assumption is placing on the Knowledge Commitment Protocol, let us first examine an important complexity limitation that the knowledge of exponent assumption of Hada and Tanaka (HTKEA) [23] places on the Knowledge Commitment Protocol. For simplicity of notation here, let us assume that the Knowledge Commitment Protocol is a non-interactive commitment denoted by $\mathsf{Com}(x)$. Consider a circuit A such that:

$$A(x) = \overbrace{\mathsf{Com}(\mathsf{Com}\ldots(\mathsf{Com}(f(x)))\ldots)}^{\ell}$$

where f is *not* efficiently computable. Then the HTKEA implies that there cannot exist such a polysize circuit A for any constant ℓ. This is because by making constant recursive invocations of HTKEA we will be able to extract $f(x)$ and generate a polysize circuit that computes f. Because our assumption admits further recursive invocations with efficient extractions, it would imply that such a polysize circuit A should not exist for larger values of ℓ. However, we note that the commitment we use is size increasing, namely $|,(x)| \geq 2|x|$. Therefore our assumption would imply that such a circuit A cannot exist for any ℓ which is $O(\log(n))$. We believe that if such a complexity assumption holds for a constant ℓ, as the HTKEA implies, then it is quite plausible that it holds for $\ell = O(log(n))$.

We describe two variants of our protocol: First, we provide a simpler protocol transformation that uses bilinear groups. This protocol is quite efficient and requires only 5 rounds. Our second protocol works with a knowledge assumption in general groups (without the need of a bilinear map), at the cost of a constant number of additional rounds, and is slightly less efficient.

---

[5] We stress that if all recursively created extractors contain all the previously created extractors inside it, then the last invocation of the assumption only needs to embed the previous extractor (since it already contains all previous extractors). This would prevent an exponential blow-up in size that a reader might otherwise worry would occur.

[6] On the other hand if the reader believes that the counter-example from Footnote 4 is not plausible, then it is easy to see that $+poly(n)$ assumption implies our assumption.

Comparison with independent concurrent work. Independently and concurrently[7] with our work, Chung et al. [12] have also constructed constant-round concurrent zero knowledge under a new concrete assumption. However, unlike our results which hold against non-uniform adversaries, their protocol only achieves soundness against uniform adversaries. To achieve their result, Chung et al. make an assumption which deals with the existence of succinct P-certificates, which is quite different from our knowledge assumption and their work uses quite different methods from ours. Their assumption can potentially be falsified by producing a valid P-certificate for a false statement corresponding to a polynomial-time computation. However, known constructions of P-certificates often make use of knowledge assumptions similar to ours in order to prove the soundness (see e.g. [6]) that Chung et al. assume to be true.

Organization. The paper is organized as follows: We discuss the technical sections beginning with background on zero-knowledge, canonical arguments and commitment schemes in Section 2. We describe the *DHLA* assumption and our knowledge assumption for bilinear groups in Section 3. We describe our protocol (which uses bilinear groups) in Section 4 and prove its soundness in Section 5. Next, we show that our protocol is zero-knowledge in a concurrent setting in Section 6. Then we prove that our knowledge assumption holds in the generic group model in Section 7. For general groups, the knowledge assumption and the protocol for concurrent zero-knowledge are described in the full version [22].

2 Definitions and Preliminaries

In the following sections, we will denote the security parameter by n. We denote a NP-complete language by L and if $x \in L$ then $W_L(x)$ returns a witness w to that fact.

Definition 1 (Bilinear Groups). *A bilinear group is a tuple $\mathcal{BG} = (q, \mathbb{G}, \mathbb{G}_T, e, g)$, where $\mathbb{G}$ and $\mathbb{G}_T$ are cyclic groups of prime order q, g generates $\mathbb{G}$, and $e : \mathbb{G} \times \mathbb{G} \to \mathbb{G}_T$ is an efficient non-degenerate bilinear map, i.e. $\forall X, Y \in \mathbb{G} \; \forall a, b \in \mathbb{Z}_q : e(X^a, Y^b) = e(X, Y)^{ab}$, and $e(g, g)$ generates $\mathbb{G}_T$. Let L_{QG} denote the set of $\{(q, g, e)\}$, where g generates a bilinear group of prime order q, where q is an n-bit prime, and e is an efficient non-degenerate bilinear map. For brevity, we will suppress the bilinear map, when it is obvious from the context, and simply write $(q, g) \in L_{QG}$. Also, we will assume that if q is an $n-$bit prime then any $x \in \mathbb{Z}_q$ can be represented by a unique $n-$bit string. For ease of notation, we just use x to denote this unique string.*

Definition 2 (Interactive Arguments). *Let P, V be two PPT interactive machines. We denote the probability that V accepts $x \in L$ on interacting with P by $\mathsf{Acc}\langle P(x, w), V(x) \rangle$. We say that $\langle P, V \rangle$ is an interactive argument for an NP-complete language L if the following two conditions are satisfied:*

---

[7] Our paper was first made public concurrently/independently of [12], see our full version [22].

- *Efficient Completeness: For every $x \in L$, there exists a string w, such that*

$$\mathsf{Acc}\langle P(x,w), V(x) \rangle = 1.$$

- *Computational Soundness: For every PPT machine P^* (cheating prover), every polynomial $poly(\cdot)$, all sufficiently long $x \notin L$ and all strings w,*

$$\mathsf{Acc}\langle P^*(x,w), V(x) \rangle < \tfrac{1}{poly(|x|)}.$$

Definition 3 (Non-Black-Box Zero-Knowledge protocol w.r.t. auxiliary input of length m). *Let m be a polynomial in n. Let P, V be two PPT interactive machines. We say that $\langle P, V \rangle$ is a non-black-box zero-knowledge protocol for L w.r.t. auxiliary input of length m if for every PPT machine V^* there exists a PPT machine S_{V^*} such that the following two distribution ensembles are indistinguishable:*

$$\{S_{V^*}(x,y)\}_{x \in L, y \in \{0,1\}^m} \ and \ \{\langle P(x,w), V^*(x,y) \rangle\}_{x \in L, w \in W_L(x), y \in \{0,1\}^m},$$

where $\{\langle P(x,w), V^*(x,y) \rangle\}_{x \in L, w \in W_L(x), y \in \{0,1\}^m}$ is a random variable taking the value of V^*'s random coins and the sequence of messages in the interaction between P and V^*.

2.1 Concurrent Zero-Knowledge ($c\mathcal{ZK}$)

Let $\langle P, V \rangle$ be an interactive proof system for a language L, and consider a concurrent adversary V^* that given an input $x \in L$ interacts with an unbounded number of copies of the prover P concurrently. Moreover, there is no restriction on the scheduling of the messages between P and V^* (in particular, V^* controls the scheduling of these messages).

The transcript of the concurrent session consists of the common input x, followed by a sequence of messages exchanged between the prover and the verifier. The view of V^* when it interacts with P consists of the random tape of V^* together with the transcript of the protocol.

To prove that any protocol is zero-knowledge in the concurrent setting, we show the existence of a simulator for every concurrent verifier V^* that interacts with m copies of P, where m is bounded by a polynomial in n.

Definition 4 (Non-Black-Box $c\mathcal{ZK}$ with auxiliary input of length m). *Let $\langle P, V \rangle$ be an interactive argument system for a language L. We say that $\langle P, V \rangle$ is non-black-box concurrent zero-knowledge if for every concurrent adversary V^* (with auxiliary input y of length m) that runs at most m concurrent sessions with P, where m is n^c for any constant c, then there exists a probabilistic polynomial time algorithm S_{V^*} that runs in time polynomial in the running time of V^* and n and satisfies that the following ensembles are computationally indistinguishable:*

$$\{S_{m,V^*}(x,y)\}_{x \in L, y \in \{0,1\}^m, m \leq n^c} \ and$$
$$\{\langle P(x,w), V^*(x,y) \rangle\}_{x \in L, w \in W_L(x), y \in \{0,1\}^m, m \leq n^c}$$

In the final constant round protocol for concurrent zero knowledge (Π) (see Section 4) using knowledge assumption in bilinear groups, we will use a discrete log based equivocal commitment scheme and three round canonical arguments as subroutines. Hence, we define and describe these next. Then we will describe the assumptions used to prove the soundness and the zero-knowledge properties of our protocol in bilinear groups. In the subsequent section, we will describe our protocol for concurrent zero-knowledge (Π) in detail. In this protocol, we also use a constant round statistically sound zero-knowledge protocol in stand alone setting.

2.2 Canonical Arguments

A three round canonical argument $\langle \overline{P}, \overline{V} \rangle$ for an NP-complete language L, proposed by [23], is described in Figure 1. CMT and RSP are the first and second messages of the prover and CH is the message sent by the verifier.

Definition 5. *An argument system $\langle \overline{P}, \overline{V} \rangle$ for an NP-complete language L is called a canonical argument system if it satisfies the following properties:*

B0 *The prover is a probabilistic polynomial time function which is given the NP-witness w. When this function is invoked with an incoming message M_{in} and its state, it outputs M_{out} and its updated state. The initial state of the prover is set to (x, w, R), where x is the common input, w is its auxiliary input and R is the random tape. When it is invoked with $(\epsilon, (x, w, R))$ it outputs the prover's first message which is a commitment CMT.*

B1 *The verifier selects the challenge CH uniformly at random from $\{0,1\}^n$.*

B2 *Strong-Soundness: For any $x \notin L$ and CMT, there exists at most one challenge CH $\in \{0,1\}^n$ for which there exists a RSP $\in \{0,1\}^*$ such that VER_x (CMT, CH, RSP) = 1.*

B3 *Honest Verifier Zero Knowledge (HVZK): There exists a probabilistic polynomial time Simulator S_{HV} such that following two ensembles are computationally indistinguishable:*

$$\{S_{HV}(x)\}_{x \in L} \text{ and } \{\langle \overline{P}(x, w), \overline{V}(x) \rangle\}_{x \in L, w \in W_L(x)},$$

where $\{\langle \overline{P}(x, w), \overline{V}(x) \rangle\}_{x \in L, w \in W_L(x)}$ is a random variable taking the value of $\overline{V}$'s internal coin tosses and the sequence of messages it receives in interaction between $\overline{P}$ (with auxiliary input w) and $\overline{V}$.

One of the ways to construct such a protocol, as described by Hada and Tanaka [23], is parallel composition of Blum's ZK protocol for Hamiltonicity.

2.3 Discrete Log based Equivocal Commitment Scheme Com$_{DL}$

The committer and the receiver are given a group $\mathbb{G}$ of prime order q, its generator g and an element $B \in \mathbb{G}$ such that q is an n-bit prime. To commit to $x \in \mathbb{Z}_q$, choose $r \xleftarrow{\$} \mathbb{Z}_q$ and send $Z = g^x \cdot B^r$. To open, the sender sends (x, r).

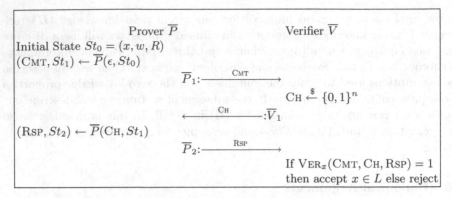

Fig. 1. Three Round Canonical Argument System $\langle \overline{P}, \overline{V} \rangle$

This commitment scheme is perfectly hiding i.e. $\mathsf{Com}_{DL}(x)$ and $\mathsf{Com}_{DL}(x')$ are identically distributed. If the committer does not know the discrete log of B, then Com_{DL} is computationally binding under discrete log assumption. We assume that discrete log assumption holds in all the groups we consider. Also, if Z is a commitment under Com_{DL}, then given two distinct openings of Z to (x, r) and (x', r') such that $x \neq x'$, one can easily solve for the discrete log of B, say b, as follows: $b = (x - x') \cdot (r' - r)^{-1}$. Also, if the simulator knows the discrete log of B, say b, it can open $Z = \mathsf{Com}_{DL}(x; r)$ as being a commitment to any $x' \in \mathbb{Z}_q$ by sending $r' = \mathsf{Open}_{DL}(x, x', r, b) = (x + r \cdot b - x') \cdot b^{-1}$.

3   Assumptions

We begin this section by describing an assumption which is very similar to the discrete logarithm assumption (DLA). Given a $(q, g) \in L_{QG}$, DLA says that given a random group element $A = g^a$, for any polysize circuit, it is hard to compute a with non negligible probability. *Diffie-Hellman Log Assumption* says that given a Diffie-Hellman tuple (g^a, g^b, g^{ab}), it is difficult to compute b even when a is chosen maliciously by the adversary. Let us denote Diffie-Hellman tuples by $\mathcal{DH}$.

Assumption 2 (Diffie-Hellman Log Assumption (*DHLA*)). *For every family of probabilistic polynomial size circuits* $I = \{I_n\}_{n \geq 1}$, *every* $poly(\cdot)$, *all sufficiently large* n's *and all* $(q, g) \in L_{QG}$ *such that* q *is of length* n, *consider the following probabilistic experiment:*

- I_n *on input ("Step 1", 1^n) outputs* (g, A), *where* $A \in \mathbb{G}$.
- *Given* (g, A) *as input, experiment chooses* $b \in \mathbb{Z}_q^*$ *and computes* $(B = g^b, X = A^b)$,

then DHLA says that if (g, A, B, X) *is a Diffie-Hellman tuple then the probability that* I_n, *given this tuple, outputs discrete log of* B *is negligible even when* A *is chosen maliciously by* I_n. *More formally,*

$$Pr[I_n(\text{ ``Step 2''}, g, A, B, X | (A, B, X) \in \mathcal{DH}) = b : B = g^b] < \tfrac{1}{poly(n)},$$

for any choice of A by I_n.

Knowledge Assumption: Below, by a circuit C we mean a collection of Boolean gates and wires. We use the non-standard convention that certain gates are specially marked as output gates.

Definition 6 (Admissible family of Adversaries). *An admissible family of adversaries $\mathcal{A}$ is a family of sets such that the following properties hold: Each set $S \in \mathcal{A}$ is such that $S = \{C_n, M_n, B_n, \text{aux}_n\}_{n \in \mathbb{N}}$. For each such set S, there exist constants $c, c' > 0$, such that C_n is a circuit with $|C_n| \leq n^c$, and $\text{aux} \subseteq \{0, 1\}^{n^{c'}}$. Furthermore, $\{M_n, B_n\}$ is a partition of the gates and the wires of the circuit C_n. If x is the input to C_n then by $M_n(x)$ we refer to the result of the computation $C_n(x)$ restricted to the output wires in M_n; we define $B_n(x)$ similarly.*

We will refer to M_n and B_n as the malicious and the benign parts respectively of the adversary circuit C_n.

Definition 7 ($\mathcal{A}$ admits polysize malicious extensions). *An admissible family of adversaries $\mathcal{A}$ admits polysize malicious extensions if the following holds: For any set of circuits $S \in \mathcal{A}$ where $S = \{C_n, M_n, B_n, \text{aux}_n\}_{n \in \mathbb{N}}$, and any polysize circuit family $\{F_n\}_{n \in \mathbb{N}}$ such that $\exists d > 0$, $|F_n| < n^d$ and the input wires to F_n are a subset of the wires in M_n (including both internal and output wires) and the output wires of B_n, we have that $S' = \{C_n \cup F_n, M_n \cup F_n, B_n, \text{aux}_n\} \in \mathcal{A}$.*

Next, based on the definition above, we define a variant of knowledge of exponent assumption based on the one described by Hada and Tanaka [23].

Assumption 3 *[m-Knowledge of Exponent Assumption (m-KEA) w.r.t. admissible adversaries]. We say that the m-Knowledge of Exponent Assumption holds with respect to a family of admissible adversaries $\mathcal{A}$, if for every $c > 0$, there exists a constant $c' > 0$ such that the following holds: For $m = n^c$, fix any $S = \{C_n, M_n, B_n, \text{aux}_n\}_{n \in \mathbb{N}} \in \mathcal{A}$. Then there exists a family of extraction circuits $\{E_n\}_{n \in \mathbb{N}}$ whose inputs are a subset of any wires in M_n, such that $|E_n| \leq (n \cdot |M_n|)^{c'}$. (Informally, this condition requires that the extraction only uses the internal wires of the malicious part of the adversary.) Furthermore, we require that the following conditions hold:*

1. *For all sufficiently large n, every polynomial $poly(\cdot)$, the following is true for all $aux \in \text{aux}_n$: Consider the following probabilistic experiment: For $i \in [1, m]$, primes q_i and generators g_i are chosen randomly such that $(q_i, g_i) \in L_{QG}$, where q_i is chosen to be of length n. Values $a_1, \ldots, a_m$ are chosen at random such that $a_i \in \mathbb{Z}_{q_i}^*$. Finally, R is chosen uniformly at random from sufficiently long strings so that the length of the tuple $x = ((q_1, g_1, g_1^{a_1}), \ldots, (q_m, g_m, g_m^{a_m}), aux, R)$ is exactly the length of the input to circuit C_n. If the input to C_n is not long enough to allow such an input then the assumption is vacuously true for this S. Now, we consider the output of $M_n(x)$, which*

we interpret as a tuple (j, B, X), where $j \in [m]$, and both B and X are in the group generated by g_j. Then, we interpret the output of $E_n(x)$ as the value $b_j \in \mathbb{Z}_{q_j}$, and require the following to be true:

$$Pr\left[X = B^{a_j} \wedge B \neq g_j^{b_j}\right] < \frac{1}{poly(n)}.$$

(Informally, this condition states that if the malicious part of the adversary outputs a tuple so that $(g_j, g_j^{a_j}, B, X)$ form a Diffie-Hellman tuple, then the extractor E_n successfully outputs the discrete log of B with respect to g_j.)

2. We have that $(C_n \cup E_n, M_n, B_n \cup E_n, \mathsf{aux}_n) \in \mathcal{A}$. (Informally, this means that the extraction circuit created by this assumption is benign.)

Definition 8. *An admissible set of adversaries $\mathcal{A}$ contains all polysize malicious adversaries if for all $c, c' > 0$, and for all circuit families $\{C_n\}_{n \in \mathbb{N}}$ such that $|C_n| \leq n^c$, for each n there exists some subset $\mathsf{aux}_n \subseteq \{0,1\}^{n^{c'}}$, such that $(C_n, C_n, \epsilon, \mathsf{aux}_n) \in \mathcal{A}$. We say that $\mathcal{A}$ contains all polysize malicious adversaries with all polysize auxiliary inputs if $\mathsf{aux}_n = \{0,1\}^{n^{c'}}$ for each circuit family above.*

Theorem 4 (Informal). *If the m-Knowledge of Exponent assumption holds with respect to an admissible adversary family $\mathcal{A}$ such that $\mathcal{A}$ contains all polysize malicious circuits and allow polysize malicious extension, and DHLA holds, then there exist constant-round concurrent zero-knowledge arguments for* **NP** *in the plain model.*

Furthermore, if $\mathcal{A}$ contains all polysize malicious adversaries with all polysize auxiliary inputs, then there exist constant-round concurrent zero-knowledge arguments for **NP** *in the plain model with respect to arbitrary auxiliary inputs.*

4   Constant Round Protocol for Concurrent Zero-Knowledge

The protocol starts by asking the verifier to use Knowledge Commitment Protocol to commit to a value b in $B = g^b$. We use equivocal commitments whose trapdoor is b to run a coin flipping protocol between the prover and the verifier. In parallel with the coin flipping protocol, we run a parallel repetition of Blum's Hamiltonicity protocol, where the result of coin flipping protocol determines the challenge message. We describe the 5-Round protocol for concurrent zero-knowledge argument in Figure 2. Note that the protocol execution does not make use of the bilinear map. It is only used by our zero-knowledge simulator to check that (A, B, X) forms a Diffie-Hellman tuple since it does not have access to the discrete log of A. We stress that this use of a bilinear map is not crucial, and that we eliminate the need for a bilinear map in our second protocol (See full version [22]).

This protocol uses the discrete log based commitment scheme Com_{DL} which is binding under the hardness of *DHLA*. The secret value b committed to by the verifier satisfies the following properties.

Here, P_i and V_i denote the ith prover and the ith verifier message respectively. $\langle \overline{P}, \overline{V} \rangle$ represents the three round canonical argument.

Prover P		Verifier V

Initial State $St_0 = (x, w, R)$

$(q, g) \xleftarrow{\$} L_{QG}$; $a \xleftarrow{\$} \mathbb{Z}_q^*$; $A \leftarrow g^a$

$$P_1:\xrightarrow{\quad (q,\ g,\ A) \quad}$$

If $(q, g) \notin L_{QG}$ then abort

else $b \xleftarrow{\$} \mathbb{Z}_q^*$; $B \leftarrow g^b$; $X \leftarrow A^b$

$$\xleftarrow{\quad (B,\ X) \quad}:V_1$$

Let Com_{DL} (defined in Section 2.3) be the commitment scheme using (g, B).

If $X \neq B^a$ then abort

else $(\mathrm{CMT}, St_1) \leftarrow \overline{P}(\epsilon, St_0)$

$\alpha \xleftarrow{\$} \mathbb{Z}_q$; $Z = \mathsf{Com}_{DL}(\alpha; \tilde{r})$

$$P_2:\xrightarrow{\quad \mathrm{CMT},\ Z \quad}$$

$$\beta \xleftarrow{\$} \{0,1\}^n$$

$$\xleftarrow{\quad \beta \quad}:V_2$$

$\mathrm{CH} \leftarrow \alpha \oplus \beta$

$(\mathrm{RSP}, St_2) \leftarrow \overline{P}(\mathrm{CH}, St_1)$

$$P_3:\xrightarrow{\quad (\alpha,\ \tilde{r},\ \mathrm{RSP}) \quad}$$

If $Z \neq \mathsf{Com}_{DL}(\alpha; \tilde{r})$ then abort

$\mathrm{CH} \leftarrow \alpha \oplus \beta$

If $\mathrm{VER}_x(\mathrm{CMT}, \mathrm{CH}, \mathrm{RSP}) = 1$

then accept else reject.

Fig. 2. Π: 5-Round Protocol for $c\mathcal{ZK}$ (P, V)

R1: For Soundness: Under $DHLA$ (Assumption 2), any cheating prover while interacting with the honest verifier cannot get the secret coins of the verifier. Hence, any cheating prover cannot output the discrete log of B sent by the verifier in Figure 2.

R2: For Zero-knowledge: Under $m\text{-}KEA$ (Assumption 3), our simulator will be able to output the discrete log of B no matter how the verifier behaves. Once the simulator gets the secret coins of V^*, which is the trapdoor to equivocal commitment scheme, the simulation is easy.

For **R2**, informally, it seems that even the cheating verifier must start by simply choosing b and computing (g^b, A^b) in order to pass the check $X = B^a$. That is, we assume that the verifier knows the secret coins b whenever it passes the check. $m\text{-}KEA$ defined in Section 3 captures this idea of knowledge and knowledge extraction formally. Under this variant of knowledge of exponent assumption, we will design a simulator which will extract the secret coins of the cheating verifier. Since, the simulator will have the trapdoor to Com_{DL}, it will be able to equivocate on its commitment to α and force the outcome of the coin flipping protocol to the challenge string output by the honest verifier simulator S_{HV}.

5 Π is Computationally Sound

Recall that *DHLA* says that given a Diffie-Hellman tuple (g, g^a, g^b, g^{ab}), even if a is chosen by the adversary, it is hard for it to guess b with non-negligible probability. We prove soundness of Π by the following two steps: Let P^* denote the cheating prover.

- If P^* succeeds in equivocating its commitment in coin flipping protocol then we can extract the trapdoor value b of Knowledge Commitment Protocol from P^*. This shows that P^* can be used to efficiently compute b and thereby break *DHLA*.
- We show that if P^* does not equivocate on its commitment in coin flipping protocol and convinces the verifier of a false statement, then such a P^* can be used to violate the underlying strong soundness of canonical arguments. In other words, it would violate the underlying soundness of Blum's Hamiltonicity protocol.

To prove the soundness of Π in the concurrent setting, it is sufficient to prove soundness of a single stand alone session. We prove the following theorem in the full version [22].

Theorem 5. *Under DHLA and strong soundness property (**B2**) of $\langle \overline{P}, \overline{V} \rangle$, protocol Π is computationally sound.*

6 Π is Concurrently Zero-Knowledge

To establish the zero-knowledge property, we build a sequence of extractors through recursive applications of *m-KEA*. Informally, each circuit uses the extractor provided by *m-KEA* to obtain the value b committed by V^* and then use this trapdoor value to equivocate in the coin flipping protocol. Through such an equivocation, the simulator can force the challenge message in Blum's Hamiltonicity protocol to be equal to the challenge the simulator received by calling the honest verifier simulator S_{HV} for Blum's Hamiltonicity. We prove that the simulation is computationally indistinguishable from the real execution through a sequence of hybrids.

Let $\mathcal{S}_{V^*}$ be the zero-knowledge simulator for Π which is described formally in the full version [22]. We prove the following theorem in [22].

Theorem 6. *If there are m concurrent sessions of Π and if our family of admissible adversaries $\mathcal{A}$ contains all polynomial size adversaries and allows poly-size malicious extensions, then under m-KEA and honest verifier zero-knowledge property of $\langle \overline{P}, \overline{V} \rangle$, the following distribution ensembles are computationally indistinguishable:*

$$\{\mathcal{S}_{V^*}(x, y)\}_{m, x \in L, y \in \{0,1\}^m} \text{ and } \{\langle P(x, w), V^*(x, y) \rangle\}_{m, x \in L, w \in W_L(x), y \in \{0,1\}^m},$$

In the full version [22] we also show that the circuit of our simulator $\mathcal{S}_{V^*}$ is a polynomial size circuit. In particular,

Theorem 7. *The size of the circuit of the simulator $\mathcal{S}_{V^*}$ is a fixed polynomial in the size of the circuit of V^* and the security parameter.*

7 m-KEA Holds in Bilinear Generic Group Model

In the full version [22] we formally show that m-KEA (Assumption 3) holds for any family of admissible adversaries (described below) that acts generically to the groups used in our protocol. In particular, we prove the following therem.

Theorem 8. m-KEA holds in the bilinear generic group model w.r.t. a family of admissible adversaries $\mathcal{A}$ which contains all polysize malicious circuits with all polysize auxiliary inputs.

References

1. Abe, M., Fehr, S.: Perfect NIZK with adaptive soundness. In: Vadhan, S.P. (ed.) TCC 2007. LNCS, vol. 4392, pp. 118–136. Springer, Heidelberg (2007)
2. Agrawal, S., Goyal, V., Jain, A., Prabhakaran, M., Sahai, A.: New impossibility results for concurrent composition and a non-interactive completeness theorem for secure computation. In: Safavi-Naini, R., Canetti, R. (eds.) CRYPTO 2012. LNCS, vol. 7417, pp. 443–460. Springer, Heidelberg (2012)
3. Barak, B.: How to go beyond the black-box simulation barrier. In: FOCS (2001)
4. Barak, B., Prabhakaran, M., Sahai, A.: Concurrent non-malleable zero knowledge. In: FOCS (2006)
5. Bellare, M., Palacio, A.: The knowledge-of-exponent assumptions and 3-round zero-knowledge protocols. In: Franklin, M. (ed.) CRYPTO 2004. LNCS, vol. 3152, pp. 273–289. Springer, Heidelberg (2004)
6. Bitansky, N., Canetti, R., Chiesa, A., Tromer, E.: From extractable collision resistance to succinct non-interactive arguments of knowledge, and back again. In: ITCS (2012)
7. Bitansky, N., Canetti, R., Paneth, O., Rosen, A.: On the existence of extractable one-way functions. Cryptology ePrint Archive, Report 2014/402 (2014). http://eprint.iacr.org/
8. Canetti, R., Dakdouk, R.R.: Extractable perfectly one-way functions. In: Aceto, L., Damgård, I., Goldberg, L.A., Halldórsson, M.M., Ingólfsdóttir, A., Walukiewicz, I. (eds.) ICALP 2008, Part II. LNCS, vol. 5126, pp. 449–460. Springer, Heidelberg (2008)
9. Canetti, R., Dakdouk, R.R.: Towards a theory of extractable functions. In: Reingold, O. (ed.) TCC 2009. LNCS, vol. 5444, pp. 595–613. Springer, Heidelberg (2009)
10. Canetti, R., Fischlin, M.: Universally composable commitments. In: Kilian, J. (ed.) CRYPTO 2001. LNCS, vol. 2139, pp. 19–40. Springer, Heidelberg (2001)
11. Canetti, R., Kilian, J., Petrank, E., Rosen, A.: Black-box concurrent zero-knowledge requires Omega~(log n) rounds. In: STOC (2001)
12. Chung, K.M., Lin, H., Pass, R.: Constant-round concurrent zero knowledge from falsifiable assumptions. Cryptology ePrint Archive, Report 2012/563 (2012). http://eprint.iacr.org/
13. Di Crescenzo, G., Lipmaa, H.: Succinct NP proofs from an extractability assumption. In: Beckmann, A., Dimitracopoulos, C., Löwe, B. (eds.) CiE 2008. LNCS, vol. 5028, pp. 175–185. Springer, Heidelberg (2008)

14. Damgård, I.: Towards practical public key systems secure against chosen ciphertext attacks. In: Feigenbaum, J. (ed.) CRYPTO 1991. LNCS, vol. 576, pp. 445–456. Springer, Heidelberg (1992)
15. Damgård, I., Faust, S., Hazay, C.: Secure two-party computation with low communication. In: Cramer, R. (ed.) TCC 2012. LNCS, vol. 7194, pp. 54–74. Springer, Heidelberg (2012)
16. Dwork, C., Naor, M., Sahai, A.: Concurrent zero-knowledge. In: STOC (1998)
17. Garg, S., Kumarasubramanian, A., Ostrovsky, R., Visconti, I.: Impossibility results for static input secure computation. In: Safavi-Naini, R., Canetti, R. (eds.) CRYPTO 2012. LNCS, vol. 7417, pp. 424–442. Springer, Heidelberg (2012)
18. Gennaro, R., Krawczyk, H., Rabin, T.: Okamoto-Tanaka revisited: Fully authenticated Diffie-Hellman with minimal overhead. In: Zhou, J., Yung, M. (eds.) ACNS 2010. LNCS, vol. 6123, pp. 309–328. Springer, Heidelberg (2010)
19. Goldwasser, S., Lin, H., Rubinstein, A.: Delegation of computation without rejection problem from designated verifier CS-proofs. IACR Cryptology ePrint Archive 2011 (2011)
20. Goldwasser, S., Micali, S., Rackoff, C.: The knowledge complexity of interactive proof systems. SIAM J. Comput. (1989)
21. Groth, J.: Short pairing-based non-interactive zero-knowledge arguments. In: Abe, M. (ed.) ASIACRYPT 2010. LNCS, vol. 6477, pp. 321–340. Springer, Heidelberg (2010)
22. Gupta, D., Sahai, A.: On constant-round concurrent zero-knowledge from a knowledge assumption. IACR Cryptology ePrint Archive (2012)
23. Hada, S., Tanaka, T.: On the existence of 3-round zero-knowledge protocols. In: Krawczyk, H. (ed.) CRYPTO 1998. LNCS, vol. 1462, pp. 408–423. Springer, Heidelberg (1998)
24. Ishai, Y., Kumarasubramanian, A., Orlandi, C., Sahai, A.: On invertible sampling and adaptive security. In: Abe, M. (ed.) ASIACRYPT 2010. LNCS, vol. 6477, pp. 466–482. Springer, Heidelberg (2010)
25. Kilian, J., Petrank, E., Rackoff, C.: Lower bounds for zero knowledge on the internet. In: FOCS (1998)
26. Kushilevitz, E., Lindell, Y., Rabin, T.: Information-theoretically secure protocols and security under composition. SIAM J. Comput. 39(5) (2010)
27. Lindell, Y.: General composition and universal composability in secure multi-party computation. In: FOCS (2003)
28. Lindell, Y.: Lower bounds for concurrent self composition. In: Naor, M. (ed.) TCC 2004. LNCS, vol. 2951, pp. 203–222. Springer, Heidelberg (2004)
29. Naor, M.: On cryptographic assumptions and challenges. In: Boneh, D. (ed.) CRYPTO 2003. LNCS, vol. 2729, pp. 96–109. Springer, Heidelberg (2003)
30. Pass, R.: Simulation in quasi-polynomial time, and its application to protocol composition. In: Biham, E. (ed.) EUROCRYPT 2003. LNCS, vol. 2656, pp. 160–176. Springer, Heidelberg (2003)
31. Prabhakaran, M., Rosen, A., Sahai, A.: Concurrent zero knowledge with logarithmic round-complexity. In: FOCS (2002)
32. Prabhakaran, M., Xue, R.: Statistically hiding sets. In: Fischlin, M. (ed.) CT-RSA 2009. LNCS, vol. 5473, pp. 100–116. Springer, Heidelberg (2009)
33. Rosen, A.: A note on the round-complexity of concurrent zero-knowledge. In: Bellare, M. (ed.) CRYPTO 2000. LNCS, vol. 1880, pp. 451–468. Springer, Heidelberg (2000)

Balancing Output Length and Query Bound in Hardness Preserving Constructions of Pseudorandom Functions

Nishanth Chandran[1](✉) and Sanjam Garg[2]

[1] Microsoft Research, Bangalore, India
nichandr@microsoft.com
[2] University of California, California, Berkeley
sanjamg@berkeley.edu

Abstract. We revisit hardness-preserving constructions of a pseudo-random function (PRF) from any length doubling pseudo-random generator (PRG) when there is a non-trivial upper bound q on the number of queries that the adversary can make to the PRF. Very recently, Jain, Pietrzak, and Tentes (TCC 2012) gave a hardness-preserving construction of a PRF that makes only $O(\log q)$ calls to the underlying PRG when $q = 2^{n^\epsilon}$ and $\epsilon \geq \frac{1}{2}$. This dramatically improves upon the efficiency of the construction of Goldreich, Goldwasser, and Micali (FOCS 1984). However, they explicitly left open the question of whether such constructions exist when $\epsilon < \frac{1}{2}$. In this work, we give constructions of PRFs that make only $O(\log q)$ calls to the underlying PRG when $q = 2^{n^\epsilon}$, for $0 < \epsilon < 1$; our PRF outputs $O(n^{2\epsilon})$ bits (on every input), as opposed to the construction of Jain et al. that outputs n bits. That is, our PRF is not length preserving; however it outputs more bits than the PRF of Jain et al. when $\epsilon > \frac{1}{2}$. We obtain our construction through the use of information-theoretic tools such as almost α-wise independent hash functions coupled with a novel proof strategy.

1 Introduction

Pseudo-random functions. Goldreich, Goldwasser, and Micali introduced the fundamental notion of pseudo-random functions in their seminal paper [12]. A pseudo-random function (PRF) is a keyed function $\mathsf{F} : \{0,1\}^\ell \times \{0,1\}^n \to \{0,1\}^m$ (with key length ℓ and input length u) such that no efficient adversary, that can make adaptive oracle queries to F, can distinguish between the outputs of F and a uniformly random function (from u bits to n bits). Pseudo-random functions have found extensive applications in cryptography - from symmetric-key cryptography to software protection to zero-knowledge proofs [9–11,16,17].

Work done while Nishanth Chandran was at Microsoft Research, Redmond.

Work done while Sanjam Garg was visiting Microsoft Research, Redmond. Part of this research conducted while at the IBM Research, T.J. Watson funded by NSF Grant No.1017660.

© Springer International Publishing Switzerland 2014
W. Meier and D. Mukhopadhyay (Eds.): INDOCRYPT 2014, LNCS 8885, pp. 89–103, 2014.
DOI: 10.1007/978-3-319-13039-2_6

Goldreich *et al.* showed how to construct a PRF from any length doubling *pseudo-random generator*, that is a function $G : \{0,1\}^n \to \{0,1\}^{2n}$ that takes as input a seed of n bits and outputs $2n$ bits that are computationally indistinguishable from the uniform distribution on $2n$ bits. We will refer to their construction as the GGM construction. The GGM construction gives rise to a PRF $GGM_G : \{0,1\}^\ell \times \{0,1\}^n \to \{0,1\}^m$ for $\ell = n$ and $m = n$ (in fact this construction works for any input length $u \in \mathbb{N}$ with other parameters remaining n).

Hardness preservation. Now, assume that the underlying PRG G has "σ bits of security", meaning that no adversary of size 2^σ can distinguish $G(U_n)$ from U_{2n} (where U_t denotes the uniform distribution on t bits) with advantage more than $2^{-\sigma}$. The GGM construction of a PRF is *hardness preserving*, i.e., if G has cn bits of security (for some constant $c > 0$ and all sufficiently large n), then GGM_G has $c'n$ bits of security for some $0 < c' \le c$. The domain size of the PRF can be arbitrary ($\{0,1\}^u$), but the GGM construction makes u calls to the underlying PRG, and hence the efficiency of GGM_G depends critically on u. Levin [15] suggested a trick that improves the efficiency when u is large; first hash the u bits down to v bits, using a universal hash function, and now apply the GGM construction to the v bits obtained. Now, the efficiency of the GGM construction can be reduced to v calls to G, and v can be set to $\omega(\log n)$, if we want security only against polynomial-size adversaries. However, if we care about preserving hardness (if G has cn bits of security), then we are forced to set $v = \Omega(n)$ and hence, the best hardness-preserving construction of a PRF, F, from a length-doubling PRG, G, requires $\Theta(n)$ calls to G (except in the case when $u = o(n)$ is sublinear, in which case one can use the GGM construction directly).

Constructing PRFs with bound on queries of adversary. Jain, Pietrzak, and Tentes [13] considered a setting, in which the size of the adversary is still 2^σ (or 2^{cn} when G is exponentially hard), but there is a bound q on the number of queries that the adversary can make to the PRF F. Surprisingly, by making such a restriction on the adversary, they can dramatically improve the efficiency of PRF constructions. In more detail, they consider an adversary of size 2^{cn} who can make only $q = 2^{n^\epsilon}$ (for constant $\frac{1}{2} \le \epsilon < 1$) queries to F; against such an adversary, they give a construction of a PRF F that only makes $\Theta(n^\epsilon)$ calls to the underlying PRG, but is still hardness preserving.

Hardness preservation and bounding queries in practice. We stress here that the notion of hardness preserving constructions of PRFs is important both in theory as well as in practice. For example, if we have a hardness preserving construction of a PRF (and the underlying PRG has exponential security) we can scale down the security parameter by almost a logarithmic factor in practical scenarios where security against only polynomial-size adversaries is required. This implies an almost exponential improvement in the efficiency. Furthermore, in most practical situations, we will be able to bound the number of queries that the adversary can make to the PRF; e.g., if we are using the PRF for constructing a symmetric encryption scheme, it is conceivable that we can bound the number of ciphertexts that the adversary can get to see (which inherently bounds the number of queries to the PRF that the adversary can get). In such situations, we can obtain much more efficient hardness preserving constructions of PRFs.

1.1   Our Results

Unfortunately, for the case when $\epsilon < \frac{1}{2}$, that is when $q = 2^{o(\sqrt{n})}$, the construction of Jain *et al.* [13] does not offer any further improvement in efficiency, and the best hardness preserving construction of F from G makes $\Theta(\sqrt{n})$ calls to G.

In this work, we are precisely interested in answering the following question: can we obtain a hardness preserving construction of a PRF from a PRG that makes $o(\sqrt{n})$ calls to G, when q, the bound on the number of queries that the adversary can make to the PRF, is $2^{o(\sqrt{n})}$?

Roughly speaking it seems that, as any hardness preserving construction of a PRF tries to make fewer calls to the underlying PRG, its ability to output pseudo-random bits reduces.[1] In the setting where a PRF from a PRG makes $o(\sqrt{n})$ calls to G and when q is bounded by $2^{o(\sqrt{n})}$ we present a tradeoff between the output length of the PRF, the bound q, and the level of hardness preserved. In particular:

- We provide a hardness preserving construction of a PRF F : $\{0,1\}^\ell \times \{0,1\}^n \to \{0,1\}^{n^{2\epsilon}}$ (where $\ell = \Theta(n)$ denotes the seed length, n denotes the input length, and $n^{2\epsilon}$ denotes the output length), that makes $\Theta(n^\epsilon)$ calls to G : $\{0,1\}^n \to \{0,1\}^{2n}$ for any $u \in \mathbb{N}$, where $q = 2^{n^\epsilon}$ and $0 < \epsilon < 1$. Note that the PRF in this construction outputs $n^{2\epsilon}$ bits as opposed to n bits as in the construction of Jain *et al.* [13]. For the case when $\epsilon > \frac{1}{2}$, our construction outputs more number of random bits than the construction of Jain *et al.* [13], while making the same number of calls to G.
- Next, we note that if we wished to output more number of bits (say n, as opposed to $n^{2\epsilon}$, when $\epsilon < \frac{1}{2}$), then we can obtain a construction that makes the same $\Theta(n^\epsilon)$ calls to G, but the resulting PRF F : $\{0,1\}^\ell \times \{0,1\}^n \to \{0,1\}^n$ has only $n^{2\epsilon}$ bits of security.
  Alternatively, we can obtain a construction of a PRF with cn bits of hardness and n bits of output by repeating the construction of F : $\{0,1\}^\ell \times \{0,1\}^n \to \{0,1\}^{n^{2\epsilon}}$, $n^{1-2\epsilon}$ times in parallel. While the total number of calls to the PRG made by this construction will be $\Theta(n^{1-\epsilon})$ (which is larger than $\Theta(n^\epsilon)$ for $0 < \epsilon < \frac{1}{2}$), the advantage of this construction is that the depth of the circuit evaluating our PRF will still be $\Theta(n^\epsilon)$.
- We also note that, just as in [13], our results extend to the case when we start with a PRG G that has only $\sigma = o(n)$ bits of hardness, and we have a bound $q = 2^{\sigma^\epsilon}$ on the number of queries that the adversary can make to the PRF, when $0 < \epsilon < 1$.[2]

---

[1] In fact with appropriate formalization Jain *et al.* conjecture this.
[2] For clarity of exposition, we present our results only in the case when G has exponential hardness.

- Finally, we mention that, similar to Jain *et al.* [13], our techniques can be used to give more efficient constructions in other settings; for example by applying it to the work of Naor and Reingold [21] who construct PRFs computable in low depth from pseudorandom synthesizers (objects stronger than PRGs but weaker than PRFs). The construction of [21] gives a hardness-preserving construction of a PRF from a pseudorandom synthesizer (PRS) with $\Theta(n)$ calls to the PRS in depth $\Theta(\log n)$. Our techniques can be used to improve the efficiency of their construction to $\Theta(\log q)$ whenever one can put an upper bound $q = 2^{n^\epsilon}, 0 < \epsilon < 1$ on the number of adversarial queries.

Concurrent and independent work. Concurrently and independently of our work, Berman *et al.* [4], using very different techniques based on cuckoo hashing, show how to construct a hardness preserving PRF from n bits to n bits that makes $\mathcal{O}(n^\epsilon)$ calls to the underlying PRG when the bound on the number of queries made by the adversary is 2^{n^ϵ}, for any $0 < \epsilon < 1$. We remark here, that our work first appeared online on the IACR Eprint Archive under a different title [7] in October 2012, while the work of Berman *et al.* [4] appeared on the IACR Eprint Archive in December 2012 [3] and was then published at TCC 2013. A technical comparison between our work and the work of Berman *et al.* [4] follows:

- In our construction the seed length of the PRF contructed, ℓ, is $\Theta(n)$, while the PRF construction of Berman *et al.* [4] requires a seed of length $\Theta(n^2)$.
- Secondly, in the case when $\epsilon < \frac{1}{2}$, their PRF outputs more random bits (n) than our construction $(n^{2\epsilon})$; however, when $\epsilon > \frac{1}{2}$, the situation is reversed, and our PRF outputs more random bits $(n^{2\epsilon})$ as opposed to their construction (as well as [13], since both constructions output n bits) while making the same number of GGM calls.

1.2   Technical Difficulties and New Ideas

In order to present the high level ideas behind our construction, we shall begin by describing Levin's trick and how Jain *et al.* build upon it to construct a PRF that makes fewer calls to the PRG.

Background. Recall that the key idea behind Levin's trick [15] is to first hash the input bits down using a (information-theoretically secure) universal hash function $h : \{0,1\}^u \to \{0,1\}^v$, and then applying the GGM construction to the obtained hashed bits. The efficiency of the GGM construction depends on the output length of this hash function. Now observe that that in order to use Levin's trick, to obtain a hardness preserving construction, one must set $v = \Theta(n)$. If v were to be $o(n)$, then the probability that two inputs x_i and x_j collide on the hash function's output would be $2^{-o(n)}$. In other words $\mathsf{F}(x_i) = \mathsf{F}(x_j)$ (since $\mathsf{F}(x) = \mathsf{GGM}_G(k, h(x))$) and this would prevent us from achieving exponential security. Hence we need that a hardness preserving construction of a PRF makes $\Theta(n)$ calls to the underlying PRG.

The key idea of Jain *et al.* is to exploit tools from information-theoretic t-wise independent hash functions and hash the bits using a hash function h_1 to $2t$ bits anyway (t is chosen to be $\log q$, where $q = 2^{\sqrt{n}}$ is the bound on the number of queries that the adversary can make to F) and then evaluate the GGM construction on it.[3] Of course there will be collisions and so they do not output the generated value directly. Instead they use the output in order to derive a key for another hash function h_2 which is then applied on the original un-hashed input. The resulting value is the output. The derivation of the key from the output of the GGM function involves stretching the output of the GGM function using the PRG. In particular, the output of GGM is stretched by a factor of t so that it is large enough to serve as the seed for h_2.

Roughly speaking if h_1 and h_2 are both t-wise independent hash functions, then Jain *et al.* can argue the security of their scheme as long as the adversary makes less than q queries to the PRF. The crucial idea is that since h_1 is t-wise independent not too many inputs (specifically no more than t) can collide on the first hashing. Furthermore the few (up to t) that do happen to collide on a specific value will ultimately lead to random outputs as h_2 is also t-wise independent. More formally, h_1 is a t-wise independent hash functions, and therefore the probability that one gets a $t+1$-wise collision after the first hashing (i.e., some $t+1$ of the q queries hash down to the same value) is upper bounded roughly by $\frac{1}{2^{t^2}}$, which is exponentially small, when $q \geq 2^{\sqrt{n}}$. This implies that with very high probability, at most t inputs will use the same seed for the t-wise independent hash function h_2 and hence the outputs will be pseudorandom.

Now, observe that the total number of calls that the above PRF construction makes to the underlying PRG is $\Theta(t)$. This is because the GGM construction is executed on a t-bit input (generated as the output of the hash function h_1) which makes t calls to the underlying PRG. The construction additionally makes t calls in deriving the key for h_2 from the output of GGM.

Overview of our construction: new ideas. In order to reduce the number of calls in the above construction, we need to reduce the number of calls that F makes in two places: the GGM part, as well as the PRG stretch (to sample the t-wise independent hash function h_2) part. We are interested in the case when $q = 2^{n^{\epsilon}}$, for $0 < \epsilon < \frac{1}{2}$ and would like to obtain a construction that makes $\Theta(n^{\epsilon})$ calls to the PRG. However in order to make just $\Theta(n^{\epsilon})$ calls to the PRG we need to reduce the output length of the first hash function h_1 to $\Theta(n^{\epsilon})$. Recall that the probability of a $t+1$-wise collision is bounded by roughly $\frac{1}{2^{t^2}}$, which is sub-exponential (when $\epsilon < \frac{1}{2}$) for our setting. Another problem is that far too many inputs to the F will collide in the key of h_2 than what h_2 is equipped to handle. The only option seems to be to make the hash functions much more resilient to collisions, or in other words increase the parameter t for both h_1 and h_2. That is, we use α-wise independent hash functions h_1 and h_2 for some parameter $\alpha = \omega(t)$. However this is fundamentally problematic since

---

[3] We consider only the case where $q = 2^{\sqrt{n}}$ in the discussion below, but the argument holds for $q = 2^{n^{\epsilon}}$, for $\frac{1}{2} \leq \epsilon < 1$.

the GGM part of the construction outputs n bits that need to be "stretched" to get a seed of length $\Theta(\alpha n)$ bits. This, unfortunately, would end up requiring $\omega(t)$ calls to G for stretching the output of GGM from n bits to $\Theta(\alpha n)$.

Our key idea here is to use approximate constructions of α-wise independent hash functions [1,14,20] for h_1 and h_2. The key advantage of these hash functions is that they can be constructed using roughly $\Theta(m\alpha)$ bits of randomness (where m is the *output* length), instead of the $\Theta(n\alpha)$ bits needed for perfect constructions (where n is the *input* length). Hence by decreasing the output length we can obtain the desired level of resilience to collisions. This allows us to obtain a tradeoff between the efficiency of the PRF and the length of the output it generates.

Even though our construction makes a seemingly simple tweak to the construction of Jain *et al.* [13], unfortunately their proof strategy does not work for us. The fundamental reason behind this is that having a perfect t-wise independent hash function allows them to reduce an adaptive distinguisher directly to a non-adaptive distinguisher. Intuitively speaking, this follows from the fact that the outputs of the t-wise independent hash function are uniformly random strings and hence when these values are used as the outputs of the PRF, they prove useless for the adaptive distinguisher. Formally, this follows from a claim of Maurer [18]. However, in our setting, the responses are not uniform and the slight bias could help the adversary choose its queries intelligently, triggering the events that ultimately allow it to distinguish the function from random. This prevents us from using the results of Maurer [18]. It is worth noting here, that the problem of constructing adaptively secure pseudorandom functions from non-adaptive pseudorandom functions [2,4,8,19,22,23] is a very important problem that has received plenty of attention.

For our construction, we prove security against adaptive distinguishers using a *step-ladder* approach. More specifically, consider an adaptive distinguisher that makes i queries to the PRF. Its distinguishing advantage can be used to upper bound the statistical distance between the distribution of the responses to i adaptive queries and the distribution of i uniform strings. This statistical difference gives us an upper bound on the advantage an adaptive distinguisher has in the choice it makes for the $i+1^{th}$ query over its non-adaptive counterpart. Given this we can evaluate the distinguishing advantage of an adaptive distinguisher that makes $i+1$ adaptive queries. Carefully applying this process repeatedly allows us to obtain a bound on the distinguishing advantage of an adaptive distinguisher making q queries.

Finally, we remark that our various results are obtained by setting α and m (output length of F) appropriately, according to the hardness of G and q.

1.3   Roadmap

We start by recalling the preliminary notions and definitions needed in Section 2. Then we provide our construction in Section 3 and the proof of our main theorem in Section 4. Finally we conclude in Section 5.

2   Preliminaries

In this section we recall and define some basic notions and setup notation. Let λ denote the security parameter. We say that a function is *negligible* in the security parameter if it is asymptotically smaller than the inverse of any fixed polynomial. Otherwise, the function is said to be *non-negligible* in λ. We say that an event happens with *overwhelming* probability if it happens with a probability $p(\lambda) = 1 - \nu(\lambda)$ where $\nu(\lambda)$ is a negligible function in λ.

Notation. We denote values and bit strings by lower case letters and random variables by uppercase letters. Sets are denoted by uppercase calligraphic letters. We use U_n to denote the random variable which takes values uniformly at random from the set of n bit long strings and $\mathcal{R}_{n,m}$ to denote the set of all functions $F : \{0,1\}^n \mapsto \{0,1\}^m$. For a set $\mathcal{X}$, $\mathcal{X}^t$ denotes the t'th direct product of $\mathcal{X}$, i.e., $(\mathcal{X}_1, \ldots, \mathcal{X}_t)$ of t identical copies of $\mathcal{X}$ and for a random variable X, $X^{(t)}$ denotes the random variable which consists of t independent copies of X. Let $x \leftarrow X$ denote the fact that x was chosen according to the random variable X and analogously let $x \leftarrow \mathcal{X}$ denote that x can chosen uniformly at random from set $\mathcal{X}$. For random variables X_0, X_1 distributed over a set $\mathcal{X}$, we use $X_0 \sim X_1$ to denote that the random variables are identically distributed, we use $X_0 \sim_\delta X_1$ to denote that they have a statistical distance δ, i.e. $\frac{1}{2} \sum_{x \in \mathcal{X}} |\Pr_{X_0}[x] - \Pr_{X_1}[x]| \leq \delta$, and finally we use $X_0 \sim_{(\delta,s)} X_1$ to denote that they are (δ, s) indistinguishable, i.e. for all distinguishers D of size at most $|D| \leq s$ we have $|\Pr_{X_0}[D(x) \to 1] - \Pr_{X_1}[D(x) \to 1]| \leq \delta$.

2.1   Pseudorandom Functions

We recall the definitions of pseudorandom generators (PRG) and pseudorandom functions (PRF). Subsequently we will describe the GGM construction of a pseudorandom function from a pseudorandom generator.

Definition 1. *[PRG [5, 25]] A length-increasing function* $G : \{0,1\}^n \mapsto \{0,1\}^m$ *where $m > n$ is a (δ, s)-hard pseudorandom generator if*

$$G(U_n) \sim_{(\delta,s)} U_m$$

*We say that G has σ bits of security if G is $(2^{-\sigma}, 2^\sigma)$-hard. G is **exponentially hard** if it has cn bits of security for some $c > 0$, and G is **sub-exponentially hard** if it has cn^ϵ bits of security for some $c > 0, 0 < \epsilon < 1$.*

Stretching a PRG. Let $G : \{0,1\}^n \to \{0,1\}^{2n}$ be a length doubling function. For $e \in \mathbb{N}$, let $G^e : \{0,1\}^n \to \{0,1\}^{en}$ be the function that takes an n bit string as input and expands it to an en bit string using $e - 1$ invocations of G. This can be done sequentially or via a more efficient parallel computation of depth $\lceil \log e \rceil$. We now have the following lemma.

Lemma 1. *As stated in* [13]. *Let* G *be a* (δ, s)*-hard PRG, then* G^e *is a* $(e \cdot \delta, s - e \cdot |\mathsf{G}|)$*-hard PRG.*

Definition 2 (PRF [12]**).**[4] *A function* $\mathsf{F} : \{0,1\}^\ell \times \{0,1\}^n \to \{0,1\}^m$ *is a* (q, δ, s)*-hard pseudorandom function (PRF) if for every oracle aided distinguisher* D^* *of size* $|D^*| \le s$ *making at most* q *oracle queries*

$$| \Pr_{k \leftarrow \{0,1\}^\ell}[D^{\mathsf{F}(k,\cdot)} \to 1] - \Pr_{f \leftarrow \mathcal{R}_{n,m}}[D^{f(\cdot)} \to 1] | \le \delta$$

F *has* σ *bits of security against* q *queries if* F *is* $(q, 2^{-\sigma}, 2^\sigma)$ *secure. If* q *is unspecified then it is assumed to be unbounded (the size* 2^σ *of the distinguisher is a trivial upper bound on* q*).*

The GGM Construction. Goldreich, Goldwasser and Micali [12] gave the first construction of a pseudorandom function from any length doubling PRG. Their construction is described below. For any length doubling PRG $\mathsf{G} : \{0,1\}^n \to \{0,1\}^{2n}$, $\mathsf{GGM}_\mathsf{G} : \{0,1\}^n \times \{0,1\}^n \to \{0,1\}^n$ is defined as a function that takes as input x along with a seed k. The output of the function $\mathsf{GGM}_\mathsf{G}(k, x)$ is k_x that can be obtained by recursive evaluation using $k_\epsilon = k$ and $k_{a||0}||k_{a||1} := \mathsf{G}(k_a)$ (where ϵ denotes the empty string).

Proposition 1 (PRF [12]**).** *If* G *is a* $(\delta_\mathsf{G}, s_\mathsf{G})$*-hard PRG, then for any* $n, q \in \mathbb{N}$, $\mathsf{GGM}_\mathsf{G} : \{0,1\}^n \times \{0,1\}^n \to \{0,1\}^n$ *is a* (q, δ, s)*-hard PRF where*

$$\delta = n \cdot q \cdot \delta_\mathsf{G} \qquad s = s_\mathsf{G} - q \cdot n \cdot |\mathsf{G}|$$

We remark, that in general, using the above same transformation, one can also obtain a PRF $\mathsf{GGM}_\mathsf{G} : \{0,1\}^n \times \{0,1\}^u \to \{0,1\}^n$ for any $u \in \mathbb{N}$. We will use GGM_G to refer to the PRF so obtained (with input length u) when the value of u is clear from context.

2.2 Information Theoretic Tools

The construction of pseudorandom functions presented in this paper relies on some well studied information theoretic tools. Next we recall these notions.

Definition 3 (α-wise independence [6,24]**).** *For* $\ell, m, n, \alpha \in \mathbb{Z}$, *a function* $h : \{0,1\}^\ell \times \{0,1\}^n \to \{0,1\}^m$ *is* α*-wise independent, if for every* α *distinct inputs* $x_1, \ldots, x_\alpha \in \{0,1\}^n$ *and a random key* $k \leftarrow \{0,1\}^\ell$ *the outputs are uniform, i.e.*

$$h_k(x_1)|| \ldots ||h_k(x_\alpha) \sim U_m^{(\alpha)}$$

Proposition 2. *For any* $\alpha, n, m \le n$ *there exists an* α*-wise independent hash function with key length* $\ell = n \cdot \alpha$.

---

[4] We use the specific definition of [13].

Definition 4 (Almost α-wise independence [20]). *For $\ell, m, n, \alpha \in \mathbb{Z}$, a function $h : \{0,1\}^\ell \times \{0,1\}^n \rightarrow \{0,1\}^m$ is (δ, α)-wise independent, if for every α distinct inputs $x_1, \ldots, x_\alpha \in \{0,1\}^n$ and a random key $k \leftarrow \{0,1\}^\ell$ the outputs are statistically close to uniform, i.e.*

$$h_k(x_1)\| \ldots \|h_k(x_\alpha) \sim_\delta U_m^{(\alpha)}$$

Proposition 3 ([1,14]). *For any α, n, m there exists a (δ, α)-wise independent hash function with key length $\ell = O(m\alpha + \log \frac{n}{\delta})$.[5]*

3 Our Construction

Our construction will use two parameters q, α. Recall that q represents the bound on the number of queries to the PRF that the adversary is allowed to make. We will use t as a shorthand for the value $\log q$. On the other hand α is a parameter that will depend on the other parameters in the system. Very roughly looking ahead α will need to increase as the desired level of security increases.

We use a $(\delta_1, \alpha + 1)$-wise independent hash function $h_1 : \{0,1\}^{\ell_1} \times \{0,1\}^n \rightarrow \{0,1\}^{2t}$ with appropriate seed length $\ell_1 = O(t\alpha + \log \frac{n}{\delta_1})$. (cf. Proposition 3) We will also need a $(\delta_2, \alpha + 1)$-wise independent hash function $h_2 : \{0,1\}^{\ell_2} \times \{0,1\}^n \rightarrow \{0,1\}^m$ with appropriate seed length $\ell_2 = O(m\alpha + \log \frac{n}{\delta_2})$.

Let $\mathsf{C}^\mathsf{G} : \{0,1\}^{\ell_1+n} \times \{0,1\}^n \rightarrow \{0,1\}^m$ be our PRF that on input a key $k = k_0\|k_1$(where $k_0 \in \{0,1\}^{\ell_1}$ and $k_1 \in \{0,1\}^n$) and $x \in \{0,1\}^n$, computes the output as:

$$\mathsf{C}(k, x) = h_2(\mathsf{G}^{2t}(\mathsf{GGM}_\mathsf{G}(k_1, h_1(k_0, x))), x).$$

Theorem 1 (Main Theorem). *If G is a $(\delta_\mathsf{G}, s_\mathsf{G})$-hard PRG, then C^G is a (q, δ, s)-secure PRF where*

$$\delta \le 4 \cdot q \cdot t \cdot \delta_\mathsf{G} + \frac{q^2}{2^n} + \frac{q^2}{2^{t\alpha}} + q^2 \cdot 2^{t\alpha}(\delta_1 + q \cdot \delta_2) + q \cdot \delta_2$$

$$m\alpha + \log \frac{n}{\delta_2} \le ctn$$

where $c > 0$ is an appropriately chosen constant and $t = \log q$. Finally note that the seed length, $\ell = O(t\alpha + \log \frac{n}{\delta_1} + n)$ and a total of $\Theta(t)$ calls are made to G.

$$s = s_\mathsf{G} - q \cdot |\mathsf{C}^\mathsf{G}| - 2q \cdot t \cdot |\mathsf{G}|$$

---

[5] To see that this is true, use Theorem 3 from [1] and consider the construction that outputs $2^n \cdot m$ bits which are δ-away (in L_1 norm) from $m\alpha$-wise independence using roughly $m\alpha + 2\log(\frac{\alpha \log(2^n \cdot m)}{2\delta})$ bits. This gives us the desired hash function.

Implications of Theorem 1. Now, suppose we want to obtain a hardness-preserving construction to obtain a PRF with $c'n$ bits of hardness, then in order to obtain $\delta = 2^{-c'n}$, we must set $t\alpha = n$. This means that $m \approx t^2$ (from the constraint $m\alpha + \log \frac{n}{\delta_2} \leq ctn$); in other words we obtain a hardness-preserving construction of a PRF that outputs $n^{2\epsilon}$ bits. At the other end of the spectrum, suppose we want the PRF to output n bits, then we must set $\alpha \approx t$, which gives us a PRF with $n^{2\epsilon}$ bits of hardness. To give a better perspective of the various results obtained by setting the values of α and m appropriately, we give some examples for the choices of these parameters in Figure 1.

Input	Query bound	Output	Hardness δ
n	q	$(\log q)^2$	2^{-cn}
n	q	$\sqrt{n}\log q$	$2^{-\sqrt{n}\log q}$
n	$q < 2^{\sqrt{n}}$	n	$2^{-(\log q)^2}$
n	$q \geq 2^{\sqrt{n}}$	$(\log q)^2$	2^{-cn}

Fig. 1. Security obtained for different settings of parameters

4   Proof of Theorem 1

We prove Theorem 1 by considering the following sequence of hybrids $\mathcal{H}_0, \mathcal{H}_1 \ldots, \mathcal{H}_4$. In hybrid $\mathcal{H}_i$ (for $i \in \{0, 1, \ldots 4\}$) samples are generated according to the circuit C_i (described in the sequel). In the first hybrid, C_0 corresponds to the execution of C itself and in the last hybrid $\mathcal{H}_4$, C_4 corresponds to the random function $\mathcal{R}_{n,m}$.

[-] **Hybrid $\mathcal{H}_0$:** This hybrid corresponds to actual evaluation of the function C. In other words $\mathsf{C}_0(k, x) = \mathsf{C}(k, x) = h_2(\mathsf{G}^{2t}(\mathsf{GGM}_\mathsf{G}(k_1, h_1(k_0, x))), x)$. For any machine D^{C_0} of size s and that makes q queries to C_0 let $p_0^D = \Pr[D^{\mathsf{C}_0(\cdot)} = 1]$.

[-] **Hybrid $\mathcal{H}_1$:** This hybrid is the same as the previous hybrid except that we use a random function $\mathcal{R}_{2t,n}$ instead of the GGM_G execution. More specifically, $\mathsf{C}_1(k, x) = h_2(\mathsf{G}^{2t}(\mathcal{R}_{2t,n}(h_1(k_0, x))), x)$. For any machine D^{C_1} of size s and that makes q queries to C_1 let $p_1^D = \Pr[D^{\mathsf{C}_1(\cdot)} = 1]$. We now show:

Lemma 2. *For every adversarial q-query distinguisher D of size $s \leq s_\mathsf{G} - 2q \cdot t \cdot |\mathsf{G}| - q \cdot |C^\mathsf{G}|$ we have that $|p_1^D - p_0^D| \leq 2q \cdot t \cdot \delta_\mathsf{G}$.*

Proof. Assume $|p_0^D - p_1^D| > 2q \cdot t \cdot \delta_\mathsf{G}$. Then we construct a distinguisher D' of size at most $s + q \cdot |C^\mathsf{G}|$ (which is equal to $s_\mathsf{G} - 2q \cdot t \cdot |\mathsf{G}|$), that distinguishes GGM_G from $\mathcal{R}_{2t,n}$ with a distinguishing advantage $> 2q \cdot t \cdot \delta_\mathsf{G}$, leading to a contradiction to Proposition 1. D' has access to an oracle $\mathcal{O}(\cdot)$ that generates a sample according to $\mathsf{GGM}_\mathsf{G}(k_1, \cdot)$ (for a random k_1) or according to $\mathcal{R}_{2t,n}$ and it needs to distinguish among the two.

Now we describe our distinguisher D'. D' samples a random seed k_0 of appropriate length. It executes D internally that makes queries for C. Consider a query x. Let y be the response of $\mathcal{O}$ on the query $h_1(k_0, x)$. D' responds with the value $h_2(G^{2t}(y, x))$ to D. Observe that the responses of D' to D correspond to evaluations of the circuit C_0 if the oracle $\mathcal{O}(\cdot)$ samples according to the distribution $GGM_G(k_1, \cdot)$ (for a random k_1) and to evaluations of the circuit C_1 if the oracle samples according to $\mathcal{R}_{2t,n}$. Hence the success of D is distinguishing between the two cases directly translates to the success of D' in distinguishing GGM_G from $\mathcal{R}_{2t,n}$. Note that D' makes q queries to $\mathcal{O}$ which is same as the number of queries D makes. Note that the size of our distinguisher D' is larger than the size of D by at most $q \cdot |C^G|$.

[-] **Hybrid $\mathcal{H}_2$:** This hybrid is the same as the previous hybrid except that we use a random function $\mathcal{R}_{n,2tn}$ instead of executing G^{2t}. More specifically, $C_2(k, x) = h_2(\mathcal{R}_{n,2tn}(\mathcal{R}_{2t,n}(h_1(k_0, x))), x)$. For any machine D^{C_2} of size s and that makes q queries to C_2, let $p_2^D = \Pr[D^{C_2(\cdot)} = 1]$. Now, we show:

Lemma 3. *For any adversarial q-query distinguisher D of size $s = s_G - 2q \cdot t \cdot |G| - q \cdot |C^G|$ we have that $|p_2^D - p_1^D| \leq 2q \cdot t \cdot \delta_G$.*

Proof. Assume $|p_0^D - p_1^D| > 2q \cdot t \cdot \delta_G$ for a distinguisher D of size s. Then we can construct a distinguisher D' of size at most $s + q \cdot |C^G|$ (which is equal to $s_G - 2q \cdot t \cdot |G|$), that distinguishes a q-tuple of samples of $G^{2t}(U_n)$ from a q-tuple of samples of U_{2tn} with a distinguishing advantage $> 2q \cdot t \cdot \delta_G$. Using a standard hybrid argument this distinguisher yields another distinguisher D'' that distinguishes between a single sample of $G^{2t}(U_n)$ from a single sample of U_{2tn} with a distinguishing advantage $> 2 \cdot t \cdot \delta_G$. This contradicts Lemma 1.

Now we describe our distinguisher D'. D' gets as input a q-tuple $(a_1, a_2 \ldots a_q)$ which has samples either from $G^{2t}(U_n)$ or from U_{2tn}. D' internally executes D and answers the oracle queries of D by executing C_1. However it uses a_i instead of generating values using G^{2t}. More specifically, it uses a fresh value of a_i for every query, except for repeat queries. In case of a repeat query it responds with the value that was returned previously (for the query being repeated). If the input tuple consists of samples from $G^{2t}(U_n)$, then the distribution corresponds to the circuit C_1. On the other hand if the samples are from U_{2tn}, then the distribution corresponds to the circuit C_2. Hence the success of D in distinguishing between the two cases directly translates to the success of D' in distinguishing q-tuple of $G^{2t}(U_n)$ from q-tuple of U_{2tn}. Note that the size of our distinguisher D' is larger than the size of D by at most $q \cdot |C^G|$.

[-] **Hybrid $\mathcal{H}_3$:** This hybrid is the same as the previous hybrid except that we use one random function $\mathcal{R}_{t,2nt}$ instead of two functions $\mathcal{R}_{n,2nt}$ and $\mathcal{R}_{2t,n}$. More specifically, $C_3(k, x) = h_2(\mathcal{R}_{n,2nt}(h_1(k_0, x)), x)$. For any machine D^{C_3} of size s and that makes q queries to C_3 let $p_3^D = \Pr[D^{C_3(\cdot)} = 1]$. We now show the following lemma:

Lemma 4. *For every adversarial q-query distinguisher D we have (unconditionally) that $|p_3^D - p_2^D| \leq \frac{q^2}{2^n}$.*

Proof. Observe that C_2 consists of two nested random functions $f(\cdot) = \mathcal{R}_{n,2nt}$ $(\mathcal{R}_{2t,n}(\cdot))$ and on the other hand C_3 consists of one random function $g(\cdot) = \mathcal{R}_{2t,2nt}(\cdot)$. Further, note that every time C_2 (resp., C_3) is called $f(\cdot)$ (resp., $g(\cdot)$) is executed exactly once. Hence, the distinguishing advantage of an unbounded q-query distinguisher D can be bounded by the distinguishing advantage of an unbounded distinguisher (that makes q queries) in distinguishing between $f(\cdot)$ and $g(\cdot)$.

Without loss of generality we assume that all the queries of the distinguisher are distinct. Let E be the event that the q queries (all distinct among themselves) of the distinguisher are such that all of the queries to the internal random function $\mathcal{R}_{2t,n}(\cdot)$ in $f(\cdot)$ are distinct. Observe that, conditioned on the event E, the distributions generated by $f(\cdot)$ and $g(\cdot)$ are the same. Hence the distinguishing advantage of an unbounded distinguisher (that makes q queries), in distinguishing between $f(\cdot)$ and $g(\cdot)$, can be upper bounded by the probability of event E failing to happen. This corresponds to the probability that q uniformly random values are such that there is a collision among two values. This value is $\frac{\binom{q}{2}}{2^n}$ which is upper bounded by $\frac{q^2}{2^n}$.

[-] **Hybrid $\mathcal{H}_4$:** This hybrid corresponds to a random function. More specifically, $C_4(k,x) = \mathcal{R}_{n,m}(x)$. For any machine D^{C_4} of size s that makes q queries to C_4, let $p_4^D = \Pr[D^{C_4(\cdot)} = 1]$.

Lemma 5. *For every adversarial q-query distinguisher D we (unconditionally) have that* $|p_4^D - p_3^D| \le \frac{q^2}{2^{t\alpha}} + q^2 \cdot 2^{t\alpha}(\delta_1 + q \cdot \delta_2) + q \cdot \delta_2.$

Proof. We prove this lemma using a step ladder approach. For an adaptive distinguisher D let E_i^D be the event such that D succeeds in making adaptive queries $x_1, x_2 \ldots x_i$ such that there exists a subset $\mathcal{I} \subseteq [i]$ of size $|\mathcal{I}| = \alpha + 1$ such that $h_1(k_0, x_j) = h_1(k_0, x_k)$ for all $j, k \in \mathcal{I}$. Intuitively speaking E_i^D is the event that D is able to force an $\alpha + 1$-wise collision on the output of the inner hash function in the i queries that it makes. At this point we claim the following lemma and prove it separately. This lemma will be used crucially in the rest of the analysis.

Lemma 6. *For every adversarial i-query distinguisher D we (unconditionally) have that* $|\Pr[D^{C_3(\cdot)} = 1|\neg E_i^D] - \Pr[D^{C_4(\cdot)} = 1|\neg E_i^D]| \le i \cdot \delta_2.$

Proof. We start by stressing that the distinguisher here is only allowed to make i queries and all the probabilities considered in this proof are for the setting where we condition on $\neg E_i^D$.

Consider a sequence of $i + 1$ hybrids – $Y_0, Y_1 \ldots Y_i$. In the hybrid Y_j (for $0 \le l \le i$) the adaptive query x_j for $j \in \{0, \ldots, i\}$ is answered as follows:

- If $j < l$ then return $\mathcal{R}_{n,m}(x_j)$.
- Else return $h_2(\mathcal{R}_{n,2tn}(h_1(k_0, x_j)), x_j)$.

We will next argue that for every $l \in \{0, 1, \ldots i - 1\}$ the statistical difference between the hybrids Y_l and Y_{l+1}, when restricted to the case $\neg E_i^D$, is bounded by δ_2. This directly implies the claimed lemma.

Now we will argue that any adaptive distinguisher D distinguishing between Y_l and Y_{l+1} with a probability greater that δ_2 can be used to construct a distinguisher D' that distinguishes the $(\alpha+1)$-wise independent hash function $h_2(k_1, \cdot)$ (for a random seed k_1) from $\mathcal{S}_{n,m}(\cdot)$ with probability at least δ_2 when making at most α adaptive queries to $\mathcal{O}(\cdot)$, where $\mathcal{O}(\cdot)$ is either $h_2(k_1, \cdot)$ (for a random seed k_1) or $\mathcal{S}_{n,m}(\cdot)$; here $\mathcal{S}_{n,m}(\cdot)$ is just a random oracle. D' internally executes D which makes i adaptive queries on inputs $x_1, \ldots x_i$. D' provides answers to the query x_j for $j \in \{0, \ldots, i\}$ as follows:

- If $j < l$ then return $\mathcal{R}_{n,m}(x_j)$.
- If $j = l$ then return $\mathcal{S}_{n,m}(x_j)$.
- Else return $h_2(\mathcal{R}_{n,2tn}(h_1(k_0, x_j)), x_j)$.

Observe that the view of D when $\mathcal{O}$ is $h_2(k_1, \cdot)$ corresponds to the hybrid Y_l. On the other hand the view of D when $\mathcal{O}$ is $\mathcal{S}_{n,m}(\cdot)$ corresponds to the hybrid Y_{l+1}. Note that both $\mathcal{R}$ and $\mathcal{S}$ are random oracles with identical output distributions Finally, note that, since we are conditioning on the event that $\neg E_i^D$, it follows that D' makes at most α adaptive queries to $\mathcal{O}(\cdot)$. Hence our claim follows.

We now complete the proof of Lemma 5. We start by evaluating the probability $\Pr[E_{i+1}^D | \neg E_i^D]$. First, consider the case where an adaptive distinguisher D is not given any of the responses. Now, note that the probability, for the $i+1^{th}$ query x_{i+1} made by adaptive D, to be such that for a particular subset $\mathcal{I} \subseteq [i]$ of size $|\mathcal{I}| = \alpha+1$, $h_1(k_0, x_j) = h_1(k_0, x_k)$ for all $j, k \in \mathcal{I}$, is $\frac{1}{2^{2 \cdot t \cdot \alpha}} + \delta_1$. By Lemma 6 given $\neg E_i^D$ we have that the statistical difference between the responses actually provided and uniformly random values is $i \cdot \delta_2$. Therefore, we can claim that given the responses, the success probability of D can increase by at most $i \cdot \delta_2$. Hence we have that the probability that the $i+1^{th}$ query made by adaptive D (when it is actually provided with the responses) such that for a particular subset $\mathcal{I} \subseteq [i]$ of size $|\mathcal{I}| = \alpha + 1$, $h_1(k_0, x_j) = h_1(k_0, x_k)$ for all $j, k \in \mathcal{I}$, is at most $\frac{1}{2^{2 t \cdot \alpha}} + \delta_1 + i\delta_2$. Taking union bound over all $\frac{(i+1)^{\alpha+1}}{(\alpha+1)!}$ possible $\alpha+1$-element subsets of the $i+1$ element set we get that:

$$\Pr[E_{i+1}^D | \neg E_i^D] \leq \frac{(i+1)^{\alpha+1}}{(\alpha+1)!} \cdot \left(\frac{1}{2^{2t\alpha}} + \delta_1 + i\delta_2 \right)$$

$$\leq 2^{(\alpha+1) \cdot \log{(i+1)}} \cdot \left(\frac{1}{2^{2t\alpha}} + \delta_1 + i\delta_2 \right) \qquad (1)$$

$$\leq 2^{t\alpha+t} \cdot \left(\frac{1}{2^{2t\alpha}} + \delta_1 + q\delta_2 \right) = \left(\frac{q}{2^{t\alpha}} + 2^{t\alpha} \cdot q \cdot (\delta_1 + q\delta_2) \right)$$

Next,

$$\Pr[\neg E_q^D] = \Pr[\neg E_q^D | \neg E_{q-1}^D] \Pr[\neg E_{q-1}^D] = (1 - \Pr[E_q^D | \neg E_{q-1}^D]) \Pr[\neg E_{q-1}^D]$$

$$= \prod_{i=0}^{q-1} (1 - \Pr[E_{i+1}^D | \neg E_i^D]) \geq \prod_{i=0}^{q-1} (1 - \frac{q}{2^{t\alpha}} - 2^{t\alpha} q (\delta_1 + q \cdot \delta_2)) \qquad (2)$$

$$\geq (1 - \frac{q}{2^{t\alpha}} - 2^{t\alpha} q (\delta_1 + q \cdot \delta_2))^q \geq 1 - \frac{q^2}{2^{t\alpha}} - q^2 \cdot 2^{t\alpha}(\delta_1 + q \cdot \delta_2)$$

Finally,

$$|\Pr[D^{C_3(\cdot)} = 1] - \Pr[D^{C_4(\cdot)} = 1]|$$
$$\leq |\Pr[D^{C_3(\cdot)} = 1|\neg E_q^D]\Pr[\neg E_q^D] + \Pr[D^{C_3(\cdot)} = 1|E_q^D]\Pr[E_q^D]$$
$$-\Pr[D^{C_4(\cdot)} = 1|\neg E_q^D]\Pr[\neg E_q^D] - \Pr[D^{C_4(\cdot)} = 1|E_q^D]\Pr[E_q^D]|$$
$$\leq (|\Pr[D^{C_3(\cdot)} = 1|\neg E_q^D] - \Pr[D^{C_4(\cdot)} = 1|\neg E_q^D]|)\Pr[\neg E_q^D] \qquad (3)$$
$$+(|\Pr[D^{C_3(\cdot)} = 1|E_q^D] - \Pr[D^{C_4(\cdot)} = 1|E_q^D]|)\Pr[E_q^D]$$
$$\leq |\Pr[D^{C_3(\cdot)} = 1|\neg E_q^D] - \Pr[D^{C_4(\cdot)} = 1|\neg E_q^D]| + \Pr[E_q^D]$$
$$\leq q \cdot \delta_2 + \frac{q^2}{2^{t\alpha}} + q^2 \cdot 2^{t\alpha}(\delta_1 + q \cdot \delta_2)$$

This completes the proof of the claimed lemma.

The proof of Theorem 1 follows from Lemmas 2, 3, 4 and 5.

5   Conclusion

Pseudorandom functions (PRF) are one of the most fundamental primitives in cryptography, both from a theoretical and a practical standpoint. However unfortunately, many of the known black-box constructions from PRGs are inefficient. Recently, Jain, Pietrzak, and Tentes [13] gave a hardness-preserving construction of a PRF that makes only $O(\log q)$ calls to the underlying PRG when $q = 2^{n^\epsilon}$ and $\epsilon \geq \frac{1}{2}$. This dramatically improves upon the efficiency of the GGM construction. However, they explicitly left open the question of whether such constructions exist when $\epsilon < \frac{1}{2}$. In this work, we give constructions of PRFs that make only $O(\log q)$ calls to the underlying PRG even when $q = 2^{n^\epsilon}$ for $0 < \epsilon < \frac{1}{2}$.

References

1. Alon, N., Goldreich, O., Håstad, J., Peralta, R.: Simple construction of almost k-wise independent random variables. Random Struct. Algorithms **3**(3), 289–304 (1992)
2. Berman, I., Haitner, I.: From non-adaptive to adaptive pseudorandom functions. In: Cramer, R. (ed.) TCC 2012. LNCS, vol. 7194, pp. 357–368. Springer, Heidelberg (2012)
3. Berman, I., Haitner, I., Komargodski, I., Naor, M.: Hardness preserving reductions via cuckoo hashing. IACR Cryptology ePrint Archive 2012: 722 (2012)
4. Berman, I., Haitner, I., Komargodski, I., Naor, M.: Hardness preserving reductions via cuckoo hashing. In: Sahai, A. (ed.) TCC 2013. LNCS, vol. 7785, pp. 40–59. Springer, Heidelberg (2013)
5. Blum, M., Micali, S.: How to generate cryptographically strong sequences of pseudo random bits. In: 23rd Annual Symposium on Foundations of Computer Science, Chicago, Illinois, USA, November 3-5, pp. 112–117 (1982)

6. Carter, L., Wegman, M.N.: Universal classes of hash functions (extended abstract). In: Proceedings of the 9th Annual ACM Symposium on Theory of Computing, Boulder, Colorado, USA, May 4-6, pp. 106–112 (1977)
7. Chandran, N., Garg, S.: Hardness preserving constructions of pseudorandom functions, revisited. IACR Cryptology ePrint Archive 2012: 616 (2012)
8. Cho, C., Lee, C.-K., Ostrovsky, R.: Equivalence of uniform key agreement and composition insecurity. In: Rabin, T. (ed.) CRYPTO 2010. LNCS, vol. 6223, pp. 447–464. Springer, Heidelberg (2010)
9. Goldreich, O.: Towards a theory of software protection. In: Odlyzko, A.M. (ed.) CRYPTO 1986. LNCS, vol. 263, pp. 426–439. Springer, Heidelberg (1987)
10. Goldreich, O.: Two remarks concerning the goldwasser-micali-rivest signature scheme. In: Odlyzko, A.M. (ed.) CRYPTO 1986. LNCS, vol. 263, pp. 104–110. Springer, Heidelberg (1987)
11. Goldreich, O., Goldwasser, S., Micali, S.: On the cryptographic applications of random functions. In: Blakely, G.R., Chaum, D. (eds.) CRYPTO 1984. LNCS, vol. 196, pp. 276–288. Springer, Heidelberg (1985)
12. Goldreich, O., Goldwasser, S., Micali, S.: How to construct random functions. J. ACM **33**(4), 792–807 (1986)
13. Jain, A., Pietrzak, K., Tentes, A.: Hardness preserving constructions of pseudorandom functions. In: Cramer, R. (ed.) TCC 2012. LNCS, vol. 7194, pp. 369–382. Springer, Heidelberg (2012)
14. Kurosawa, K., Johansson, T., Stinson, D.R.: Almost k-wise independent sample spaces and their cryptologic applications. In: Fumy, W. (ed.) EUROCRYPT 1997. LNCS, vol. 1233, pp. 409–421. Springer, Heidelberg (1997)
15. Levin, L.A.: One-way functions and pseudorandom generators. Combinatorica **7**(4), 357–363 (1987)
16. Luby, M.: Pseudorandomness and cryptographic applications. Princeton computer science notes. Princeton University Press (1996)
17. Luby, M., Rackoff, C.: A study of password security. In: Pomerance, C. (ed.) CRYPTO 1987. LNCS, vol. 293, pp. 392–397. Springer, Heidelberg (1988)
18. Maurer, U.M.: Indistinguishability of random systems. In: Knudsen, L.R. (ed.) EUROCRYPT 2002. LNCS, vol. 2332, pp. 110–133. Springer, Heidelberg (2002)
19. Myers, S.: Black-box composition does not imply adaptive security. In: Cachin, C., Camenisch, J.L. (eds.) EUROCRYPT 2004. LNCS, vol. 3027, pp. 189–206. Springer, Heidelberg (2004)
20. Naor, J., Naor, M.: Small-bias probability spaces: Efficient constructions and applications. SIAM J. Comput. **22**(4), 838–856 (1993)
21. Naor, M., Reingold, O.: Synthesizers and their application to the parallel construction of psuedo-random functions. In: 36th Annual Symposium on Foundations of Computer Science, Milwaukee, Wisconsin, October 23-25, pp. 170–181 (1995)
22. Pietrzak, K.: Composition does not imply adaptive security. In: Shoup, V. (ed.) CRYPTO 2005. LNCS, vol. 3621, pp. 55–65. Springer, Heidelberg (2005)
23. Pietrzak, K.: Composition implies adaptive security in minicrypt. In: Vaudenay, S. (ed.) EUROCRYPT 2006. LNCS, vol. 4004, pp. 328–338. Springer, Heidelberg (2006)
24. Wegman, M.N., Carter, L.: New classes and applications of hash functions. In: 20th Annual Symposium on Foundations of Computer Science, San Juan, Puerto Rico, October 29-31, pp. 175–182 (1979)
25. Yao, A.C.-C.: Theory and applications of trapdoor functions (extended abstract). In: 23rd Annual Symposium on Foundations of Computer Science, Chicago, Illinois, USA, November 3-5, pp. 80–91 (1982)

Block Ciphers

Linear Cryptanalysis
of the PP-1 and PP-2 Block Ciphers

Michael Colburn and Liam Keliher[✉]

AceCrypt Research Group, Mount Allison University,
Sackville, NB, Canada
{mgcolburn,lkeliher}@mta.ca

Abstract. PP-1 and PP-2 are scalable SPN-based block ciphers introduced in 2008 and 2013, respectively. PP-2 was intended as an improvement to PP-1, which was broken in 2011 using differential cryptanalysis. The designers of PP-2 claim that it is comparable to the Advanced Encryption Standard (AES) in its resistance to linear and differential cryptanalysis. However, we demonstrate that both PP-1 and PP-2 with 64-bit and 128-bit block sizes are vulnerable to linear cryptanalysis. Specifically, we find high probability linear hulls that allow us to break each cipher faster than exhaustive search of the keyspace. This is the first use of linear cryptanalysis against PP-1, and the first successful attack of any kind against PP-2. We confirm our theoretical results by experimentally breaking a reduced-round version of PP-2.

Keywords: PP-1 · PP-2 · Block cipher · SPN · Linear cryptanalysis

1 Introduction

PP-1 and PP-2 are block ciphers with flexible block and key sizes introduced in 2008 and 2013, respectively [3,4]. Both are based on the substitution-permutation network (SPN) structure [15]. PP-1, which is involutional (self-inverting) and targeted to limited resource environments, has been broken by multiple flavours of differential cryptanalysis [9,12,13]. In contrast, the designers of PP-2 claim that it is comparable to the Advanced Encryption Standard (AES) [5] in its resistance to differential and linear cryptanalysis. However, we demonstrate that both PP-1 and PP-2 with 64-bit and 128-bit block sizes are vulnerable to linear cryptanalysis. Specifically, we find high probability linear hulls that allow us to break each cipher faster than exhaustive search of the keyspace. This is the first use of linear cryptanalysis against PP-1, and the first successful attack of any kind against PP-2. We confirm our theoretical results by experimentally breaking a reduced-round version of PP-2.

We focus initially on PP-2, since it is a new cipher, and then we adapt our analysis to PP-1. Overall, we find that PP-2 exhibits higher resistance to linear cryptanalysis than PP-1, especially when the block size is 128 bits, but neither cipher has a security level close to that of the AES.

© Springer International Publishing Switzerland 2014
W. Meier and D. Mukhopadhyay (Eds.): INDOCRYPT 2014, LNCS 8885, pp. 107–123, 2014.
DOI: 10.1007/978-3-319-13039-2_7

The rest of the paper is organized as follows. In Section 2 we give the structure of PP-2. In Section 3 we outline the relevant aspects of linear cryptanalysis, and in Section 4 we describe the application of linear cryptanalysis to PP-2. We present the results of our experimental attacks on reduced-round PP-2 in Section 5. In Section 6 we briefly describe the structure of PP-1, and then explain how it can also be attacked using linear cryptanalysis. We conclude in Section 7.

2   Structure of PP-2

PP-2, designed by Bucholc et al., is based on a standard SPN structure in which each round consists of three layers: *key mixing, substitution,* and *permutation.*[1] Let n denote the block size (i.e., the size in bits of each plaintext, ciphertext, or intermediate data block), let r denote the number of rounds, and let $|\boldsymbol{k}|$ denote the key size (where $\boldsymbol{k}$ is the key). The specified values of n, r, and $|\boldsymbol{k}|$ are given in Table 1. We will focus on the two most standard block sizes, namely $n = 64$ and $n = 128$.

Table 1. Number of PP-2 rounds (r) for various values of n and $|\boldsymbol{k}|$

Key length |k| \ Block size n	64	128	192	256	...
64	11				
128	13	22			...
192	15	24	32		...
256	17	26	34	43	...

The PP-2 *key-scheduling algorithm* uses the original key $\boldsymbol{k}$ to generate r subkeys $\boldsymbol{k}_1, \boldsymbol{k}_2, \ldots, \boldsymbol{k}_r$, each of which is n bits long. Since the structure of the key-scheduling algorithm does not affect our attack, we do not consider it further. For more details, see [3].

For each round i ($1 \leq i \leq r$), the n-bit input block $\boldsymbol{x}_i$ is partitioned into J 64-bit subblocks $\boldsymbol{x}_{i,1}, \boldsymbol{x}_{i,2}, \ldots, \boldsymbol{x}_{i,J}$ (where $J = n/64$). Similarly, the n-bit subkey $\boldsymbol{k}_i$ is partitioned into $\boldsymbol{k}_{i,1}, \boldsymbol{k}_{i,2}, \ldots, \boldsymbol{k}_{i,J}$.

Key-Mixing Layer

In the key-mixing layer, each byte of $\boldsymbol{x}_{i,j}$ is combined with the corresponding byte of $\boldsymbol{k}_{i,j}$ (for $1 \leq j \leq J$) using the operations specified in Fig. 1 (taken from [3]), where $\oplus$, $\boxplus$, and $\boxminus$ denote bitwise XOR, addition mod 256, and subtraction mod 256, respectively.

---

[1] The designers of PP-2 refer to the key-mixing and substitution layers collectively as the *nonlinear layer (NL).*

Substitution Layer

In the substitution layer, each byte produced by the key-mixing layer is processed through the same 8×8 s-box. The PP-2 s-box is given in Appendix A.

Permutation Layer

In the permutation layer, the 64-bit outputs from the previous two layers are rejoined, and a bitwise permutation is applied to the resulting n-bit block. The bits of the block are indexed $0, \ldots, (n-1)$ from left to right (so index 0 corresponds to the most-significant bit of the first byte), and bits are grouped into index classes based on their position mod 4. The bits in each index class are circularly right rotated (mod n) by a fixed amount: class 0 rotates 12 bits, class 1 rotates 28 bits, class 2 rotates 44 bits, and class 3 rotates 60 bits.

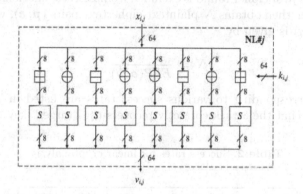

Fig. 1. Key-mixing and substitution layers of PP-2

3 Linear Cryptanalysis

Linear cryptanalysis, introduced by Matsui in 1993 [10], is a known-plaintext attack that exploits relatively large *expected linear probability* (ELP) values over consecutive cipher rounds.

Definition 1. *Let* $F : \{0,1\}^m \rightarrow \{0,1\}^m$ *be bijective, and let* $a, b \in \{0,1\}^m$ *be fixed. If* $\mathbf{X} \in \{0,1\}^m$ *is a uniformly distributed random variable, then the* linear probability $LP(a, b)$ *is defined as*

$$LP(a, b) = (2 \cdot \text{Prob}_{\mathbf{X}} \{a \bullet \mathbf{X} = b \bullet F(\mathbf{X})\} - 1)^2 \qquad (1)$$

where $\bullet$ *denotes the inner (dot) product over* $\{0,1\}$. *If* F *is parameterized by a key* k, *we write* $LP(a, b; k)$, *and the* expected linear probability $ELP(a, b)$ *is defined as*

$$ELP(a, b) = E_{\mathbf{K}} [LP(a, b; \mathbf{K})] \qquad (2)$$

where **K** *is a random variable uniformly distributed over the space of keys, and* $E[\,]$ *denotes expectation.*

The values $a\ /\ b$ in Definition 1 are called input/output *masks*. A nonzero LP value quantifies a correlation between the inputs and outputs of a mapping. Note that $LP(\mathbf{0},\mathbf{0}) = 1$, and $LP(\mathbf{z},\mathbf{0}) = LP(\mathbf{0},\mathbf{z}) = 0$ for $\mathbf{z} \neq \mathbf{0}$.

LP values can be computed for unkeyed mappings such as individual s-boxes, and ELP values can be computed for keyed mappings such as single cipher rounds or multiple consecutive cipher rounds, although exact computation is typically infeasible in the multiple-round case.

An attacker uses linear cryptanalysis (specifically, Matsui's Algorithm 2 [10]) by pre-computing masks $a \neq \mathbf{0}$, $b \neq \mathbf{0}$ such that $ELP(a, b)$ is relatively large over $t < r$ consecutive "core" cipher rounds. For much of this paper we will assume that the *first* cipher round is excluded from ELP calculations, so $t = r-1$.

The attacker then obtains N plaintext-ciphertext pairs (p, c), where N (the *data complexity*) is given by

$$N = \frac{d}{ELP(a, b)} \tag{3}$$

Values of d corresponding to various success rates are listed in Table 2 [10]. Success means that the attack correctly recovers the "target" key bits.

Table 2. Success rates for linear cryptanalysis

d	8	16	32	64
Success rate	48.6%	78.5%	96.7%	99.9%

We omit most of the details of linear cryptanalysis here. In brief, the attacker guesses sufficiently many target bits from the first subkey to be able to encrypt each plaintext p through the first round, producing an intermediate block p', and uses this to evaluate an *experimental* LP value

$$LP_{\text{exp}}(a, b) = (2 \cdot \text{Prob}\{a \bullet p' = b \bullet c\} - 1)^2 \tag{4}$$

where the probability is computed over all N (p, c) pairs. A correct guess of the relevant subkey bits should yield an LP_{exp} value that is higher than the corresponding values for incorrect guesses. Ideally this attack can be repeated with other mask pairs (using the same (p, c) data) to recover the entire first subkey, and ultimately all r subkeys.

For the core cipher rounds under consideration, analysts often consider the *maximum expected linear probability* (MELP):

$$MELP = \max_{a \neq \mathbf{0}, b \neq \mathbf{0}} ELP(a, b) \tag{5}$$

A block cipher is said to have *provable security*[2] against linear cryptanalysis [6] if the MELP is sufficiently small that the resulting data complexity is prohibitively large, especially if it is greater than the total number of plaintexts available (2^n).

3.1   Characteristics and Hulls

A *linear characteristic* (or *linear trail*) for rounds $2 \ldots r$ of a block cipher is a vector of masks $\Omega = \langle a_2, a_3, \ldots, a_r, a_{r+1} \rangle$. The associated *linear characteristic probability* (LCP) is defined to be the product of the corresponding one-round ELP values:

$$LCP(\Omega) = \prod_{i=2}^{r} ELP_i(a_i, a_{i+1}) \tag{6}$$

where ELP_i denotes an ELP value for round i. For an SPN, the one-round value $ELP_i(a_i, a_{i+1})$ is computed as follows:

- process a_{i+1} through the *inverse* permutation[3] to obtain b_i
- partition a_i / b_i into input/output masks for the s-boxes
- set $ELP_i(a_i, a_{i+1})$ equal to the product of the s-box LP values

For a given characteristic, any s-box with nonzero input/output masks is called *active*. Clearly a characteristic with an active s-box whose LP value is zero will have an LCP of zero. In general, characteristics with small numbers of active s-boxes have larger LCP values, and therefore are of interest to attackers.

Nyberg [14] defined the *linear hull* $H(a, b)$ for the core rounds under consideration as the set of all characteristics with first mask a and final mask b, i.e, all characteristics of the form $\Omega = \langle a, -, -, \ldots, -, -, b \rangle$. The main result in [14] is the following equation linking ELP values and LCP values:

$$ELP(a, b) = \sum_{\Omega \in H(a,b)} LCP(\Omega) = \sum_{\Omega \in H(a,b)} \prod_{i=2}^{r} ELP_i(a_i, a_{i+1}) \tag{7}$$

If a characteristic has a large LCP value, it is often used as an approximation to the ELP of the linear hull to which it belongs, but this may significantly underestimate the true ELP value (a situation called the *linear hull effect*). This is the approach taken by Bucholc et al. in their preliminary analysis of PP-2 [3]. However, since we are interested in finding large (accurate) ELP values, we work with linear hulls.

4   Applying Linear Cryptanalysis to PP-2

4.1   Relevant PP-2 Design Features

There are several PP-2 design features that affect our attack. Items 1–3 below facilitate linear cryptanalysis, while Item 4 is a complicating factor.

---

[2] The term *provable security* has other well established meanings in cryptography, typically involving computationally hard problems [15].

[3] For a linear transformation that is not a permutation, this step may need to be modified; see [5].

1. The permutation layer is in fact a bitwise permutation. This was traditionally the case for SPNs, but in many newer ciphers it has been replaced by a more general linear transformation layer. A bitwise permutation exhibits very poor diffusion: a mask with low Hamming weight prior to the permutation will yield a mask with the same Hamming weight after the permutation. This potentially allows the construction of characteristics with small numbers of active s-boxes, in particular characteristics with one active s-box per round.[4] In contrast, characteristics over 4 or more AES rounds always involve at least 25 active s-boxes [5].

2. PP-2 has no final *whitening key* [15], i.e., no subkey following the substitution and permutation layers of the last round. Since these layers are not key-dependent, for the purpose of cryptanalysis they can be stripped off by partially decrypting each ciphertext, effectively leaving an $(r-1)$-round cipher with a final whitening key (the subkey from the original last round). This means that the numbers of rounds given in Table 1 can all be reduced by 1. (Interestingly, PP-1 does not have this flaw.)

3. The largest nontrivial LP value for the PP-2 s-box is $\left(\frac{9}{64}\right)^2$. A more important feature, however, in light of the low-diffusion permutation layer, is the fact that almost all choices of weight-1 input/output masks for the s-box yield nonzero LP values. These values are given in Table 3, where rows (resp., columns) are indexed by input (resp., output) masks.

4. Unlike the AES, which uses XOR for all key mixing, PP-2 employs the three byte-level operations in Fig. 1. The presence of modular addition and subtraction potentially complicates the calculation of one-round ELP values, since these operations essentially behave like additional (key-dependent) nonlinear functions (see [16]). We bypass this complicating factor by considering only characteristics for which the indices of all active s-boxes (for each 64-bit subblock of the n-bit block) belong to the set $\{1, 3, 4, 6\}$, with the convention that s-boxes are numbered $0 \ldots 7$, left to right — i.e., every active s-box is associated with XOR key mixing.

4.2   High Probability Linear Hulls for PP-2

To find high probability linear hulls for PP-2, we adapt the algorithm used by Keliher to find high probability hulls for the Q cipher [7]. This straightforward algorithm computes values $ELP(a, b)$ for $wt(a) = wt(b) = 1$ (where $wt(\ )$ denotes Hamming weight). It does so by summing the LCP values of all characteristics in the linear hull $H(a, b)$, limiting consideration to characteristics in which every intermediate mask also has Hamming weight 1. (Technically, it deals with a subset of the linear hull, and therefore computes a lower bound on

---

[4] Interestingly, a number of recently introduced lightweight ciphers such as PRESENT [1] incorporate bitwise permutations. However, some (including PRESENT) have other design features that appear to give them a high level of resistance to cryptanalysis.

Table 3. PP-2 s-box LP values for weight-1 input/output masks (in hex)

	0x01	0x02	0x04	0x08	0x10	0x20	0x40	0x80
0x01	$(\frac{6}{64})^2$	$(\frac{5}{64})^2$	$(\frac{6}{64})^2$	$(\frac{6}{64})^2$	$(\frac{1}{64})^2$	$(\frac{2}{64})^2$	$(\frac{6}{64})^2$	$(\frac{2}{64})^2$
0x02	$(\frac{4}{64})^2$	$(\frac{4}{64})^2$	$(\frac{5}{64})^2$	$(\frac{4}{64})^2$	$(\frac{4}{64})^2$	$(\frac{9}{64})^2$	$(\frac{1}{64})^2$	0
0x04	$(\frac{2}{64})^2$	$(\frac{3}{64})^2$	$(\frac{2}{64})^2$	$(\frac{4}{64})^2$	$(\frac{4}{64})^2$	$(\frac{8}{64})^2$	$(\frac{1}{64})^2$	$(\frac{7}{64})^2$
0x08	$(\frac{1}{64})^2$	$(\frac{2}{64})^2$	$(\frac{1}{64})^2$	$(\frac{4}{64})^2$	$(\frac{6}{64})^2$	$(\frac{6}{64})^2$	$(\frac{6}{64})^2$	$(\frac{6}{64})^2$
0x10	$(\frac{1}{64})^2$	$(\frac{2}{64})^2$	$(\frac{9}{64})^2$	$(\frac{8}{64})^2$	$(\frac{7}{64})^2$	$(\frac{1}{64})^2$	$(\frac{6}{64})^2$	$(\frac{3}{64})^2$
0x20	$(\frac{1}{64})^2$	0	$(\frac{5}{64})^2$	$(\frac{7}{64})^2$	$(\frac{2}{64})^2$	$(\frac{2}{64})^2$	$(\frac{2}{64})^2$	$(\frac{3}{64})^2$
0x40	$(\frac{1}{64})^2$	$(\frac{6}{64})^2$	$(\frac{5}{64})^2$	$(\frac{2}{64})^2$	$(\frac{3}{64})^2$	$(\frac{3}{64})^2$	$(\frac{1}{64})^2$	$(\frac{1}{64})^2$
0x80	$(\frac{3}{64})^2$	$(\frac{4}{64})^2$	$(\frac{4}{64})^2$	$(\frac{4}{64})^2$	$(\frac{7}{64})^2$	$(\frac{1}{64})^2$	$(\frac{1}{64})^2$	$(\frac{6}{64})^2$

the relevant ELP value, but in general this is a much better (i.e., higher) lower bound than would be obtained by using the LCP of a single characteristic.) The algorithm also counts the number of characteristics included in each (partial) linear hull. The restriction to weight-1 masks is practical for a couple of reasons:

- it reduces computational complexity, making the calculation of ELP values feasible
- the characteristics with the highest LCP values generally have one active s-box per round, and these are guaranteed to be included

We make the following two modifications to the algorithm in [7]:

1. We omit characteristics that activate any s-box *not* associated with XOR key mixing, consistent with Item 4 in Section 4.1.
2. We loosen the weight-1 restriction on masks other than the first in a characteristic so that one *or two* s-boxes can be active in any intermediate round. We also allow the final mask to have weight greater than 1 as long as it contains only one nonzero byte. The goal of loosening the weight-1 restriction (which does not increase the number of subkey bits we have to guess) is to test whether or not it results in larger ELP values.

For any mask pair (a, b) produced by our algorithm, we process a through the inverse permutation and note the byte in which the single 1 now lies. The associated byte of subkey k_1 is the "target," i.e., the subkey byte that we need to guess to process each p through Round 1 to obtain p', as described in Section 3. For each target byte, we use the (a, b) pair with the highest ELP. In theory the resulting set of mask pairs will enable a successful linear cryptanalytic attack to reconstruct all of k_1.

Note that the hull-finding algorithm is very fast. For any parameters used in this paper, the execution time was under 2 seconds on a standard PC.

4.3   Linear Hulls for PP-2 with $n = 64$, $|k| = 64$

PP-2 with $n = 64$, $|k| = 64$ has 11 rounds (Table 1), corresponding to 10 "real" rounds (see Item 2 in Section 4.1), so we form linear hulls over $r - 1 = 9$ core rounds. The results are given in Table 4. The number of characteristics in each (partial) linear hull was found to be at least 2.0×10^{10}. When we re-introduced the restriction of weight-1 masks, the number of characteristics per hull dropped to approximately 1.5×10^4, but this had almost no effect on the ELP values for the masks in Table 4 (an increase of at most 0.2 in the absolute value of each exponent), which indicates that characteristics with weight-1 masks dominate each hull.

Table 4. Linear hulls over 9 rounds for PP-2 with $n = 64$, $|k| = 64$ (masks in hex)

Target byte	Mask a	Mask b	ELP
0	00 08 00 00 00 00 00 00	00 00 00 02 00 00 00 00	$2^{-55.0}$
1	00 00 00 00 04 00 00 00	00 00 00 02 00 00 00 00	$2^{-54.0}$
2	00 01 00 00 00 00 00 00	00 00 00 02 00 00 00 00	$2^{-54.1}$
3	00 00 00 00 08 00 00 00	00 00 00 02 00 00 00 00	$2^{-54.4}$
4	00 02 00 00 00 00 00 00	00 00 00 02 00 00 00 00	$2^{-55.0}$
5	00 00 00 00 00 00 08 00	00 00 00 02 00 00 00 00	$2^{-55.3}$
6	00 00 00 00 00 00 10 00	00 00 00 02 00 00 00 00	$2^{-55.5}$
7	00 00 00 00 02 00 00 00	00 00 00 02 00 00 00 00	$2^{-53.6}$

All ELP values in Table 4 are larger than 2^{-56}, so if we choose $d = 16$ from Table 2 (corresponding to a 78.5% success rate), the resulting data complexity is $N = 2^{60}$. The time complexity, typically expressed as the number of full encryptions, is less straightforward to calculate. If we focus on a single mask pair, the work to process each (p, c) pair primarily involves a single s-box lookup. (The step in which the guess of the target subkey byte is mixed with the corresponding plaintext byte can be incorporated into a modified s-box.) If we assume that the substitution layer is $\frac{1}{2}$ the work of a single round, and that a single s-box lookup is $\frac{1}{8}$ of the substitution layer for $n = 64$ (without the use of parallelism), the work per (p, c) pair is $\frac{1}{16}$ of a one-round encryption, and therefore $\frac{1}{16 \cdot 11} = \frac{1}{176}$ of a full 11-round encryption. There are 2^8 subkey byte guesses and $N = 2^{60}$, so we estimate the time complexity to be

$$\frac{2^8 \cdot 2^{60}}{176} \approx 2^{60.5} \text{ encryptions}$$

Table 5. Linear hulls over 20 rounds for PP-2 with $n = 128$, $|k| = 128$ (masks in hex)

Target byte	Mask a	Mask b	ELP
0	00 08 00 00 00 00 00 00 ...	... 00 00 00 02 00 00 00 00	$2^{-120.6}$
1	... 00 10 00 00 00 00 00 00	00 00 00 02 00 00 00 00 ...	$2^{-120.4}$
2	... 00 01 00 00 00 00 00 00	00 00 00 02 00 00 00 00 ...	$2^{-119.9}$
3	00 00 00 00 08 00 00 00 ...	00 00 00 02 00 00 00 00 ...	$2^{-120.9}$
4	... 00 02 00 00 00 00 00 00	00 00 00 02 00 00 00 00 ...	$2^{-120.7}$
5	00 00 00 00 00 00 08 00 ...	... 00 00 00 02 00 00 00 00	$2^{-121.8}$
6	... 00 04 00 00 00 00 00 00	00 00 00 02 00 00 00 00 ...	$2^{-121.3}$
7	... 00 00 00 00 02 00 00 00	... 00 00 00 02 00 00 00 00	$2^{-120.2}$
8	... 00 08 00 00 00 00 00 00	00 00 00 02 00 00 00 00 ...	$2^{-120.6}$
9	00 10 00 00 00 00 00 00 ...	... 00 00 00 02 00 00 00 00	$2^{-120.4}$
10	00 01 00 00 00 00 00 00 ...	... 00 00 00 02 00 00 00 00	$2^{-119.9}$
11	... 00 00 00 00 08 00 00 00	... 00 00 00 02 00 00 00 00	$2^{-120.9}$
12	00 02 00 00 00 00 00 00 ...	... 00 00 00 02 00 00 00 00	$2^{-120.7}$
13	... 00 00 00 00 00 00 08 00	00 00 00 02 00 00 00 00 ...	$2^{-121.8}$
14	00 04 00 00 00 00 00 00 ...	... 00 00 00 02 00 00 00 00	$2^{-121.3}$
15	00 00 00 00 02 00 00 00 ...	00 00 00 02 00 00 00 00 ...	$2^{-120.2}$

This is better than exhaustive search of the keyspace, which requires 2^{64} encryptions in the worst case, and 2^{63} on average.[5]

4.4   Linear Hulls for PP-2 with $n = 128$, $|k| = 128$

PP-2 with $n = 128$, $|k| = 128$ has 22 rounds (Table 1), corresponding to 21 "real" rounds, and therefore we form linear hulls over 20 core rounds. The results are given in Table 5. To save space, we only include 64 bits of each 128-bit mask, using an ellipsis (...) to indicate the remaining zero bits as appropriate. When we re-introduced the restriction of weight-1 masks, there was little effect on the ELP values for the masks in Table 5 (an increase of at most 0.4 in the absolute value of each exponent).[6]

---

[5] Admittedly, a successful exhaustive search recovers the entire original key.

[6] We do not report the number of characteristics in each (partial) linear hull for PP-2 (or for PP-1) when $n = 128$, since the values overflow a 64-bit unsigned integer data type. (We did not implement a higher precision counter.)

All the ELP values in Table 5 are lower bounded by 2^{-122}, so again choosing $d = 16$ from Table 2 gives a data complexity of $N = 2^{126}$ (which is close to the entire codebook). However, some of the mask pairs allow us to reduce this to 2^{124}. Following an argument analogous to that in Section 4.3, we estimate the data complexity of an attack using a mask pair with ELP $\approx 2^{-120}$ to be

$$\frac{1}{2} \cdot \frac{1}{16} \cdot \frac{1}{22} \cdot 2^8 \cdot 2^{124} \approx 2^{122.5} \text{ encryptions}$$

This is faster than exhaustive search, which requires 2^{128} encryptions in the worst case, and 2^{127} on average.

Remark 1. It is significant that PP-2 with $n = 128$ needs 20 core rounds to achieve this level of resistance to linear cryptanalysis. In contrast, the AES requires only 4 core rounds in order for all ELP values to be upper bounded by 1.8×2^{-110} [8].

5   Experimental Results for Reduced-Round PP-2

For our experimental linear cryptanalysis of PP-2, we focused on $n = 64$ with 6 rounds. Since this corresponds to 5 "real" rounds, we require linear hulls over rounds $2 \ldots 5$, i.e., over 4 core rounds. We used our search algorithm to find the linear hull with the highest ELP value for each potential target byte in the first subkey. The resulting masks and ELP values are given in Table 6.

Table 6. Linear hulls over 4 rounds for experimental attack on PP-2 (masks in hex)

Target byte	Mask a	Mask b	ELP
0	00 08 00 00 00 00 00 00	00 00 00 02 00 00 00 00	$2^{-23.9}$
1	00 10 00 00 00 00 00 00	00 00 00 02 00 00 00 00	$2^{-23.9}$
2	00 01 00 00 00 00 00 00	00 00 00 02 00 00 00 00	$2^{-23.9}$
3	00 00 00 00 08 00 00 00	00 20 00 00 00 00 00 00	$2^{-25.3}$
4	00 02 00 00 00 00 00 00	00 00 00 00 00 00 02 00	$2^{-25.0}$
5	00 00 00 00 01 00 00 00	00 00 00 02 00 00 00 00	$2^{-25.8}$
6	00 04 00 00 00 00 00 00	00 00 00 80 00 00 00 00	$2^{-25.6}$
7	00 00 00 00 02 00 00 00	00 20 00 00 00 00 00 00	$2^{-24.6}$

Each ELP value in Table 6 is lower bounded by 2^{-26}. We chose $d = 32$ from Table 2 for a data complexity of 2^{31} and a theoretical success rate of 96.7%. We then ran multiple trials of linear cryptanalysis. For convenience, we grouped trials into clusters of 8 such that each trial in a cluster attacked a distinct target byte. For each cluster, we randomly generated a new key for PP-2, and also randomly generated and encrypted 2^{31} plaintexts.

The results for 10 clusters are given in Table 7, where Ti denotes trial i, and a checkmark ($\checkmark$) denotes a successful trial, i.e., one that recovered its target byte. The experimental success rate is $\frac{54}{80} = 67.5\%$. For our purposes, this confirms that linear cryptanalysis with appropriately chosen masks and data complexity succeeds against (reduced-round) PP-2. The discrepancy with Matsui's predicted success rate is interesting, and may be explained in part by recent results re-examining Matsui's underlying theory (for example, see [2]).

Table 7. Results of linear cryptanalysis trials for reduced-round PP-2

Target byte	T1	T2	T3	T4	T5	T6	T7	T8	T9	T10
0		$\checkmark$	$\checkmark$	$\checkmark$		$\checkmark$	$\checkmark$	$\checkmark$	$\checkmark$	
1		$\checkmark$	$\checkmark$			$\checkmark$			$\checkmark$	
2	$\checkmark$	$\checkmark$	$\checkmark$		$\checkmark$	$\checkmark$	$\checkmark$	$\checkmark$	$\checkmark$	$\checkmark$
3		$\checkmark$		$\checkmark$	$\checkmark$	$\checkmark$	$\checkmark$	$\checkmark$	$\checkmark$	$\checkmark$
4	$\checkmark$	$\checkmark$	$\checkmark$		$\checkmark$		$\checkmark$	$\checkmark$	$\checkmark$	$\checkmark$
5	$\checkmark$		$\checkmark$		$\checkmark$	$\checkmark$	$\checkmark$		$\checkmark$	
6	$\checkmark$	$\checkmark$	$\checkmark$	$\checkmark$						
7	$\checkmark$	$\checkmark$	$\checkmark$			$\checkmark$	$\checkmark$	$\checkmark$	$\checkmark$	$\checkmark$

6 Linear Cryptanalysis of PP-1

PP-1, introduced in 2008 by Chmiel et al. [4], was designed to be involutional (self-inverting) in order to reduce resource requirements, especially in hardware. It is the predecessor of PP-2, and therefore the two ciphers have many elements in common:

- SPN-based structure
- flexible block size that is a multiple of 64 (for PP-1, the legal sizes are $n = 64, 128, 192, 256$)
- key mixing via byte-level XOR, addition mod 256, or subtraction mod 256
- single 8×8 s-box used throughout every substitution layer
- permutation layer that is in fact a bitwise permutation, not a more general linear transformation

6.1 Structural Differences

For the purposes of our analysis, there are several significant differences between PP-1 and PP-2:

1. Each PP-1 round mixes *two* n-bit subkeys, one immediately before the sub-sitution layer, and one immediately after. This is depicted for one 64-bit subblock unit in Fig. 2 (taken from [4]).
2. PP-1 and PP-2 use different 8×8 s-boxes. The (involutional) PP-1 s-box is given in Appendix A.
3. PP-1 and PP-2 have different permutation layers. The (involutional) PP-1 permutation is given algorithmically in [4]; we do not reproduce it here.
4. The number of PP-1 rounds (r) depends on the block size, but not on the key size. These values are listed in Table 8. Chmiel et al. specify that for a given block size n, the key size $|k|$ can equal n or $2n$.
5. PP-1 and PP-2 have different key-scheduling algorithms, but apart from the fact that PP-1 requires two subkeys per round, we ignore key-scheduling details.

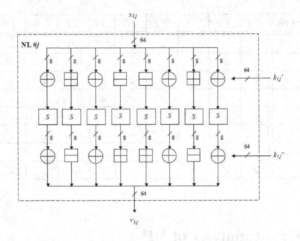

Fig. 2. Key-mixing and substitution layers of PP-1

Table 8. Number of PP-1 rounds (r) for various values of n

Block size (n)	64	128	192	256
Number of rounds (r)	11	22	32	43

6.2 High Probability Linear Hulls for PP-1

To apply linear cryptanalysis to PP-1, we take the following into account:

1. The second subkey in the last round of PP-1 acts as a final whitening key, so we cannot simply strip off the last round as we did when attacking PP-2 (we can only strip off the final permutation). However, we will guess subkey bytes from the first *and* last rounds simultaneously, so we form linear hulls over $r - 2$ core rounds.

2. The s-box LP table entries corresponding to weight-1 input/output masks are mostly nonzero. These are given in Table 9. (As is the case for PP-2, the maximum s-box LP value is $\left(\frac{9}{64}\right)^2$, and this appears in the table.)

3. We limit consideration to characteristics in which every active s-box is associated with XOR key mixing, i.e., the index of every active s-box (within its 64-bit subblock) belongs to the set $\{0, 2, 5, 7\}$.

Table 9. PP-1 s-box LP values for weight-1 input/output masks (in hex)

	0x01	0x02	0x04	0x08	0x10	0x20	0x40	0x80
0x01	$\left(\frac{4}{64}\right)^2$	$\left(\frac{7}{64}\right)^2$	$\left(\frac{3}{64}\right)^2$	$\left(\frac{1}{64}\right)^2$	$\left(\frac{5}{64}\right)^2$	$\left(\frac{7}{64}\right)^2$	$\left(\frac{6}{64}\right)^2$	$\left(\frac{5}{64}\right)^2$
0x02	$\left(\frac{7}{64}\right)^2$	0	$\left(\frac{5}{64}\right)^2$	$\left(\frac{3}{64}\right)^2$	$\left(\frac{7}{64}\right)^2$	$\left(\frac{1}{64}\right)^2$	$\left(\frac{3}{64}\right)^2$	$\left(\frac{6}{64}\right)^2$
0x04	$\left(\frac{3}{64}\right)^2$	$\left(\frac{5}{64}\right)^2$	0	$\left(\frac{2}{64}\right)^2$	$\left(\frac{4}{64}\right)^2$	$\left(\frac{8}{64}\right)^2$	$\left(\frac{4}{64}\right)^2$	$\left(\frac{1}{64}\right)^2$
0x08	$\left(\frac{1}{64}\right)^2$	$\left(\frac{3}{64}\right)^2$	$\left(\frac{2}{64}\right)^2$	$\left(\frac{4}{64}\right)^2$	$\left(\frac{9}{64}\right)^2$	$\left(\frac{3}{64}\right)^2$	$\left(\frac{7}{64}\right)^2$	0
0x10	$\left(\frac{5}{64}\right)^2$	$\left(\frac{7}{64}\right)^2$	$\left(\frac{4}{64}\right)^2$	$\left(\frac{9}{64}\right)^2$	$\left(\frac{2}{64}\right)^2$	$\left(\frac{8}{64}\right)^2$	$\left(\frac{4}{64}\right)^2$	$\left(\frac{2}{64}\right)^2$
0x20	$\left(\frac{7}{64}\right)^2$	$\left(\frac{1}{64}\right)^2$	$\left(\frac{8}{64}\right)^2$	$\left(\frac{3}{64}\right)^2$	$\left(\frac{8}{64}\right)^2$	$\left(\frac{4}{64}\right)^2$	$\left(\frac{2}{64}\right)^2$	$\left(\frac{3}{64}\right)^2$
0x40	$\left(\frac{6}{64}\right)^2$	$\left(\frac{3}{64}\right)^2$	$\left(\frac{4}{64}\right)^2$	$\left(\frac{7}{64}\right)^2$	$\left(\frac{4}{64}\right)^2$	$\left(\frac{2}{64}\right)^2$	$\left(\frac{6}{64}\right)^2$	$\left(\frac{6}{64}\right)^2$
0x80	$\left(\frac{5}{64}\right)^2$	$\left(\frac{6}{64}\right)^2$	$\left(\frac{1}{64}\right)^2$	0	$\left(\frac{2}{64}\right)^2$	$\left(\frac{3}{64}\right)^2$	$\left(\frac{6}{64}\right)^2$	$\left(\frac{6}{64}\right)^2$

We again adapt the algorithm in [7] to find high probability linear hulls over rounds $2 \ldots (r-1)$, also loosening the weight-1 restriction on masks as described in Section 4.2. For any resulting mask pair $(\boldsymbol{a}, \boldsymbol{b})$:

- We process $\boldsymbol{a}$ through the inverse permutation (which is identical to the forward permutation) to determine the "target" byte in the first round. If this byte is associated with XOR keying, we only need to guess the *first* (outermost) associated subkey byte in the first round. If the target byte is *not* associated with XOR keying, we need to guess *both* associated subkey bytes in the first round (i.e., 16 bits).
- The mask $\boldsymbol{b}$ guarantees that the target byte in the last round is associated with XOR keying, so we only need to guess the second (outermost) associated subkey byte in the last round.

6.3   Linear Hulls for PP-1 with $n = 64$, $|k| = 64, 128$

PP-1 with $n = 64$ has 11 rounds, so we form linear hulls over $r - 2 = 9$ core rounds. The results are given in Table 10. Note that the ELP values are significantly higher than those in Table 4. The number of characteristics in each (partial) linear hull was found to be at least 3.1×10^{14}.

Table 10. Linear hulls over 9 rounds for PP-1 with $n = 64$, $|k| = 64, 128$ (masks in hex)

Target byte	Mask a	Mask b	ELP
0	00 00 00 00 00 00 00 02	00 00 00 00 00 00 00 09	$2^{-48.2}$
1	00 00 00 00 00 00 00 40	00 00 00 00 00 00 00 09	$2^{-47.7}$
2	20 00 00 00 00 00 00 00	00 00 00 00 00 00 00 09	$2^{-48.6}$
3	00 00 00 00 00 00 00 10	00 00 00 00 00 00 00 09	$2^{-47.7}$
4	02 00 00 00 00 00 00 00	00 00 00 00 00 00 00 09	$2^{-49.1}$
5	00 00 00 00 00 00 00 80	00 00 00 00 00 00 00 09	$2^{-48.8}$
6	00 00 00 00 00 00 00 20	00 00 00 00 00 00 00 09	$2^{-48.2}$
7	00 00 00 00 00 00 00 01	00 00 00 00 00 00 00 09	$2^{-48.3}$

Using an ELP value of 2^{-49} with $d = 16$ from Table 2, we get a data complexity of $N = 2^{53}$. Following the reasoning in Section 4.3, if we need to guess a total of 16 subkey bits (8 in the first round and 8 in the last round), then the estimated time complexity is

$$\frac{2 \cdot 2^{16} \cdot 2^{53}}{176} \approx 2^{62.5} \text{ encryptions}$$

where the extra factor of 2 results from performing a partial encryption through the first round *and* a partial decryption through the last round. This time complexity is better than exhaustive search when $|k| = 64$, and significantly better when $|k| = 128$. If we need to guess a total of 24 subkey bits (16 in the first round and 8 in the last round), then the time complexity jumps to $2^{70.5}$, which is only better than exhaustive search when $|k| = 128$.

6.4   Linear Hulls for PP-1 with $n = 128$, $|k| = 128, 256$

PP-1 with $n = 128$ has 22 rounds, so we form linear hulls over $r - 2 = 20$ core rounds. The results are given in Table 11.

Using a conservative ELP value of 2^{-106} with $d = 16$ from Table 2, we get a data complexity of $N = 2^{110}$. Following the reasoning in Section 4.4 and Section 6.3, if we need to guess 16 subkey bits, the estimated time complexity is

$$\frac{1}{2} \cdot \frac{1}{16} \cdot \frac{1}{22} \cdot 2 \cdot 2^{16} \cdot 2^{110} \approx 2^{117.5} \text{ encryptions}$$

which is better than exhaustive search when $|k| = 128$, and significantly better when $|k| = 256$. If we need to guess 24 subkey bits, then the time complexity

Table 11. Linear hulls over 20 rounds for PP-1 with $n = 128$, $|k| = 128, 256$ (masks in hex)

Target byte	Mask a	Mask b	ELP
0	00 00 00 00 00 00 00 02 ...	00 00 00 00 00 00 00 08 ...	$2^{-104.0}$
1	00 00 00 00 00 00 00 40 ...	00 00 00 00 00 00 00 08 ...	$2^{-103.9}$
2	20 00 00 00 00 00 00 00 ...	00 00 00 00 00 00 00 08 ...	$2^{-104.2}$
3	... 00 00 00 00 00 00 00 40	00 00 00 00 00 00 00 08 ...	$2^{-103.7}$
4	02 00 00 00 00 00 00 00 ...	00 00 00 00 00 00 00 08 ...	$2^{-104.5}$
5	00 00 00 00 00 00 00 10 ...	00 00 00 00 00 00 00 08 ...	$2^{-103.8}$
6	... 00 00 00 00 00 20 00 00	00 00 00 00 00 00 00 08 ...	$2^{-104.6}$
7	... 00 00 00 00 00 00 00 10	00 00 00 00 00 00 00 08 ...	$2^{-103.7}$
8	00 00 00 00 00 02 00 00 ...	00 00 00 00 00 00 00 08 ...	$2^{-104.2}$
9	00 00 00 00 00 00 00 04 ...	00 00 00 00 00 00 00 08 ...	$2^{-104.9}$
10	00 00 20 00 00 00 00 00 ...	00 00 00 00 00 00 00 08 ...	$2^{-105.3}$
11	... 00 00 00 00 00 00 00 04	00 00 00 00 00 00 00 08 ...	$2^{-105.1}$
12	00 00 01 00 00 00 00 00 ...	00 00 00 00 00 00 00 08 ...	$2^{-105.5}$
13	00 00 00 00 00 00 00 01 ...	00 00 00 00 00 00 00 08 ...	$2^{-104.6}$
14	00 00 00 00 00 00 00 20 ...	00 00 00 00 00 00 00 08 ...	$2^{-104.2}$
15	... 00 00 00 00 00 00 00 01	00 00 00 00 00 00 00 08 ...	$2^{-104.1}$

jumps to $2^{125.5}$, which is slightly better than exhaustive search when $|k| = 128$, and significantly better when $|k| = 256$.

7   Conclusion

In this paper we have examined the scalable block ciphers PP-1 and PP-2, and have shown how both ciphers (with 64-bit or 128-bit block sizes) can be attacked using linear cryptanalysis. Specifically, we give high probability linear hulls that allow us to break each cipher faster than exhaustive search, and we confirm our analysis by experimentally breaking a reduced-round version of PP-2. This contradicts the security claims made by the designers, in particular the claim that PP-2 has resistance to linear cryptanalysis comparable to that of the AES.

Appendix A: PP-1 and PP-2 S-boxes

The PP-1 s-box is given in Fig. 3, and the PP-2 s-box is given in Fig. 4. For each s-box, the high-end nibble of the input is used as the row index, and the low-end nibble is used as the column index.

```
     0   1   2   3   4   5   6   7   8   9   A   B   C   D   E   F
   ----------------------------------------------------------------
0 | 9E  BC  C3  82  A2  7E  41  5A  51  36  3F  AC  E3  68  2D  2A
1 | EB  9B  1B  35  DC  1E  56  A5  B2  74  34  12  D5  64  15  DD
2 | B6  4B  8E  FB  CE  E9  D9  A1  6E  DB  0F  2C  2B  0E  91  F1
3 | 59  D7  3A  F4  1A  13  09  50  A9  63  32  F5  C9  CC  AD  0A
4 | 5B  06  E6  F7  47  BF  BE  44  67  7B  B7  21  AF  53  93  FF
5 | 37  08  AE  4D  C4  D1  16  A4  D6  30  07  40  8B  9D  BB  8C
6 | EF  81  A8  39  1D  D4  7A  48  0D  E2  CA  B0  C7  DE  28  DA
7 | 97  D2  F2  84  19  B3  B9  87  A7  E4  66  49  95  99  05  A3
8 | EE  61  03  C2  73  F3  B8  77  E0  F8  9C  5C  5F  BA  22  FA
9 | F0  2E  FE  4E  98  7C  D3  70  94  7D  EA  11  8A  5D  00  EC
A | D8  27  04  7F  57  17  E5  78  62  38  AB  AA  0B  3E  52  4C
B | 6B  CB  18  75  C0  FD  20  4A  86  76  8D  5E  01  ED  46  45
C | B4  FC  83  02  54  D0  DF  6C  CD  3C  6A  B1  3D  C8  24  E8
D | C5  55  71  96  65  1C  58  31  A0  26  6F  29  14  1F  6D  C6
E | 88  F9  69  0C  79  A6  42  F6  CF  25  9A  10  9F  BD  80  60
F | 90  2F  72  85  33  3B  E7  43  89  E1  8F  23  C1  B5  92  4F
```

Fig. 3. PP-1 s-box

```
     0   1   2   3   4   5   6   7   8   9   A   B   C   D   E   F
   ----------------------------------------------------------------
0 | 9E  81  F7  2F  CC  F1  46  6E  B7  CE  29  76  14  42  E6  BB
1 | EC  39  B6  EF  A3  D5  EA  91  3D  37  F0  51  44  C6  0C  BC
2 | 41  80  AB  1D  6C  D2  C0  ED  00  FE  3B  F9  A4  24  FF  DB
3 | 4F  7A  4A  B8  A9  4E  79  A8  15  9B  B2  A7  31  52  69  58
4 | 71  DC  77  99  04  A6  B9  25  E7  92  5E  62  57  89  C1  61
5 | D1  66  48  C2  AA  38  2D  8E  65  8C  C3  6A  C8  7B  DA  95
6 | 90  0E  0A  6B  F4  5B  8D  A1  05  0B  10  03  8B  9D  85  E1
7 | BD  19  FA  6D  88  22  E4  4C  AF  49  F8  BE  83  07  FD  D3
8 | 8F  A5  BF  E8  C4  E5  FB  16  35  3C  64  2C  0D  C5  43  02
9 | 59  C7  7E  E3  18  CF  06  4B  9C  D0  F3  70  D7  33  87  B1
A | DF  20  E2  EE  F5  32  56  B3  84  74  2B  34  47  36  96  DD
B | 63  0F  97  28  D6  5F  7D  9F  53  09  8A  12  5A  AE  1B  3A
C | 7F  40  30  A0  D4  27  82  3E  4D  08  7C  1C  17  2E  01  CD
D | B5  CB  54  A2  D9  EB  50  93  F2  3F  1F  9A  13  CA  21  94
E | E9  23  5D  1A  AC  B0  67  86  73  1E  26  6F  45  98  11  BA
F | E0  D8  75  72  AD  55  68  78  F6  2A  B4  5C  C9  60  DE  FC
```

Fig. 4. PP-2 s-box

Acknowledgments. Funding for this research was provided by the Natural Sciences and Engineering Research Council of Canada (NSERC). We are grateful to the anonymous reviewers for comments that improved the final version of this paper.

References

1. Bogdanov, A., Knudsen, L.R., Leander, G., Paar, C., Poschmann, A., Robshaw, M.J.B., Seurin, Y., Vikkelsoe, C.: PRESENT: An Ultra-Lightweight Block Cipher. In: Paillier, P., Verbauwhede, I. (eds.) CHES 2007. LNCS, vol. 4727, pp. 450–466. Springer, Heidelberg (2007)
2. Bogdanov, A., Tischhauser, E.: On the Wrong Key Randomisation and Key Equivalence Hypotheses in Matsui's Algorithm 2. In: Moriai, S. (ed.) FSE 2013. LNCS, vol. 8424, pp. 19–38. Springer, Heidelberg (2014)
3. Bucholc, K., Chmiel, K., Grocholewska-Czurylo, A., Stoklosa, J.: PP-2 Block Cipher. In: 7th International Conference on Emerging Security Information Systems and Technologies (SECURWARE 2013), pp. 162–168. XPS Press, Wilmington (2013)
4. Chmiel, K., Grocholewska-Czurylo, A., Stoklosa, J.: Involutional Block Cipher for Limited Resources. In: IEEE Global Telecommunications Conference (GLOBECOM 2008). IEEE Press (2008)
5. Daemen, J., Rijmen, V.: The Design of Rijndael: AES–The Advanced Encryption Standard. Springer, Heidelberg (2002)
6. Hong, S., Lee, S., Lim, J., Sung, J., Cheon, D., Cho, I.: Provable Security against Differential and Linear Cryptanalysis for the SPN Structure. In: Schneier, B. (ed.) FSE 2000. LNCS, vol. 1978, pp. 273–283. Springer, Heidelberg (2001)
7. Keliher, L., Meijer, H., Tavares, S.: High Probability Linear Hulls in Q. In: Preneel, B., et al. (eds.) Final Report of European Project IST-1999-12324: New European Schemes for Signatures, Integrity, and Encryption (NESSIE), pp. 749–761 (2004)
8. Keliher, L., Sui, J.: Exact Maximum Expected Differential and Linear Probability for Two-Round Advanced Encryption Standard. IET Information Security 1(2), 53–57 (2007)
9. Lee, Y., Jeong, K., Sung, J., Lee, C., Hong, S., Chang, K.-Y.: Security Analysis of Scalable Block Cipher PP-1 Applicable to Distributed Sensor Networks. Int. J. Distr. Sens. Net. 2013, 1–9 (2013)
10. Matsui, M.: Linear Cryptanalysis Method for DES Cipher. In: Helleseth, T. (ed.) EUROCRYPT 1993. LNCS, vol. 765, pp. 386–397. Springer, Heidelberg (1994)
11. Menezes, A.J., van Oorschot, P.C., Vanstone, S.A.: Handbook of Applied Cryptography. CRC Press, Boca Raton (1997)
12. Misztal, M.: Differential Cryptanalysis of PP-1 Cipher. Annales UMCS Informatica AI XI 2, 9–24 (2011)
13. Misztal, M., Courtois, N.T.: Aggregated Differentials and Cryptanalysis of PP-1 and GOST. Periodica Mathematica Hungarica 65(2), 177–192 (2012)
14. Nyberg, K.: Linear Approximation of Block Ciphers. In: De Santis, A. (ed.) EUROCRYPT 1994. LNCS, vol. 950, pp. 439–444. Springer, Heidelberg (1995)
15. Stinson, D.R.: Cryptography: Theory and Practice, 3rd edn. Chapman & Hall/CRC, Boca Raton (2006)
16. Wallén, J.: Linear Approximations of Addition Modulo 2^n. In: Johansson, T. (ed.) FSE 2003. LNCS, vol. 2887, pp. 261–273. Springer, Heidelberg (2003)

On the Key Schedule of Lightweight Block Ciphers

Jialin Huang[1,2]($\boxtimes$), Serge Vaudenay[1], and Xuejia Lai[2]

[1] EPFL, Lausanne, Switzerland
[2] Shanghai Jiao Tong University, Shanghai, China
serge.vaudenay@epfl.ch,
{xuejia.lai,jlhuang.cn}@gmail.com

Abstract. Key schedules in lightweight block ciphers are often highly simplified, which causes weakness that can be exploited in many attacks. Today it remains an open problem on how to use limited operations to guarantee enough diffusion of key bits in lightweight key schedules. Also, there are few tools special for detecting weakness in the key schedule.

In 2013 Huang et al. pointed out that insufficient actual key information (AKI) in computation chains is responsible for many attacks especially the meet-in-the-middle (MITM) attacks. Motivated by this fact, in this paper we develop an efficient (with polynomial time complexity) and effective tool to search the computation chains which involve insufficient AKI for iterated key schedules of lightweight ciphers. The effectiveness of this tool is shown by an application on TWINE-80.

Then, we formulate the cause of key bits leakage phenomenon, where the knowledge of subkey bits is leaked or overlapped by other subkey bits in the same computation chain. Based on the interaction of diffusion performed by the key schedule and by the round function, a necessary condition is thus given on how to avoid key bits leakage.

Therefore, our work sheds light on the design of lightweight key schedules by guiding how to quickly rule out unreasonable key schedules and maximize the security under limited diffusion.

Keywords: Automatic tool · Meet-In-The-Middle · PRESENT · TWINE · Actual key information

1 Introduction

Today the demands of secure communication on source constrained environments such as RFID tags and sensor networks motivate the development of many lightweight block ciphers. In these lightweight ciphers, the security margin that conventional block ciphers are equipped with is

This work was supported by the National Natural Science Foundation of China (61073149 and 61272440 and 61472251).

© Springer International Publishing Switzerland 2014
W. Meier and D. Mukhopadhyay (Eds.): INDOCRYPT 2014, LNCS 8885, pp. 124–142, 2014.
DOI: 10.1007/978-3-319-13039-2_8

reduced as much as possible in order to optimize the software and hardware efficiency. One obvious sacrifice is that the key schedules are highly simplified for saving memory. Some key schedules have round-by-round iteration with low diffusion [3,20,21]. Some key schedules do simple permutation or linear operations on master keys [10,17]. Some have no key schedule, and just use master keys directly in each round [9,14]. These key schedules are succinct but responsible for many attacks, especially related-key attacks [13,18], MITM attacks and their variants [5,12,19], and special attacks such as the invariant subspace attack on PRINTcipher [15].

Although causing so many risks, the key schedules in lightweight block ciphers are still designed in a heuristic and ad-hoc way, especially when a tradeoff between security and memory constraints should be made. Because of the lack of systematic guidelines, research on design principles of lightweight key schedules is pressing.

1.1 Related Work

Avoiding MITM attacks and key bits leakage are two of the goals in the design of key schedules [11,16]. Designers tend to exploit relatively fast diffusion or avalanche to achieve these goals, which is infeasible for lightweight key schedules. The resistance to MITM attacks is usually claimed by ensuring that all master key bits are used within several rounds. Huang et al. show that considering the diffusion of key bits in the "round" level is not enough by investigating the interaction of diffusion between the key schedule and the round function [11]. They also propose a measure called actual key information (AKI) to evaluate the effective speed of diffusing key bits and claim that a computation path should have as high AKI as possible.

From the other aspect, automatic tools have been developed to search cryptanalytic properties relevant to key schedules. Most of these tools focus on searching related-key differential characteristics [1,2,8], while the tools aiming at other weaknesses of key schedules are much fewer. Bouillaguet et al. propose automatic algorithms for searching guess-and-determine and MITM attacks on round-reduced AES, using the linear relations in the key schedule [4]. Derbez et al. give a way to automatically search the kind of Demirci-Selçuk MITM attacks against AES [7]. Huang et al. present a method to calculate the AKI for 2-round iterated key schedules whose key size is double of the block size [11].

1.2 Our Contribution

While the concept of AKI that considers the minimal information of involved key bits in a computation path is meaningful, the algorithm described in [11] targets a too particular type of key schedules. Firstly, the master key size is double of the block size. Secondly, the number of searched rounds is fixed to two. Additionally, the time complexity of this 2-round search is exponential to the key size. If this algorithm is extended to R rounds, the time complexity is exponential to not only the key size but also R, making practical applications infeasible.

Instead of computing AKI for conventional block ciphers, in this paper we consider computing AKI for lightweight block ciphers. We target iterated key schedules with low diffusion, as they exist in a certain amount of lightweight block ciphers. Computing AKI for the static or permutation-based key schedules is much easier, and thus not our concern here. We generalize the problem of computing AKI to any number of rounds in lightweight block ciphers. Then, we develop an efficient and effective algorithm to solve it (Sect. 3). Based on the observation of characters of lightweight iterated key schedules, we use a greedy strategy, resulting in an algorithm with polynomial time complexity. The tools in [7] focus on discovering a special kind of MITM attacks for AES (the Demirci-Selçuk attack that depends on algebraic relations in AES), while our tool aims to present a simple, efficient and generic approach to evaluate the design of iterated key schedules in lightweight ciphers. We show the feasibility of this algorithm by applying it to TWINE-80.

Although the key bits leakage has been mentioned in [11,16], it does not achieve deserving attention in cipher design as other properties. This phenomenon is revisited in Sect. 4, and found in almost all the computation paths that contain insufficient information of key bits. We analyze the major cause of key bits leakage by explicitly relating the incidence matrix of the diffusion in key schedules to that of round functions. Furthermore, a formulated necessary criterion for key schedule design is proposed to guide to avoid key bits leakage within a given number of rounds. This necessary criterion can be a guidance of quickly ruling out unreasonable key schedules at first step, and then our tool can be used to do further examination.

2   Preliminary

2.1   Notations

The lightweight block ciphers studied here have a block size of n bits and a master key size of k bits. Each round function includes a key extraction and incorporation layer (AK), and a cipher permutation layer F that updates the n-bit internal state with confusion and diffusion after adding key bits. The output internal state in round i is denoted as s_i. The iterated key schedule updates each k-bit key state round-by-round from the previous one. That is, $k_i = \mathcal{KS}(k_{i-1})$, where k_i is the output *subkey* bits in the ith round of the key schedule, and k_0 is the master key. Normally $\mathcal{KS}$ is invertible, thus the size of all k_i is k. An rk-bit *round key* rk_i is extracted from k_i in the key extraction phase and added to the internal state. The most common case is to apply parts of the subkey k_i directly to the round key rk_i. Other notations include:

- $k_i[j]$ (resp. $s_i[j]$) denotes the jth bit of bit-string k_i (resp. s_i).
- M_s is the incidence matrix representing the diffusion of F. In the encryption direction, $M_s[j', j] = 1$ means that the $s_i[j']$ depends on $s_{i-1}[j]$, where $M_s[j', j]$ denotes the (j', j) entry in matrix M_s. In the decryption direction, $M_s[j', j] = 1$ means that $s_{i-1}[j']$ depends on $s_i[j]$.
- M_k is the incidence matrix for the diffusion of $\mathcal{KS}$, defined in a similar way with M_s.
- e_i is a vector with the ith bit equal to 1 and other bits equal to 0.
- $|\cdot|$ denotes the size of a set, or the Hamming weight of a bit-string, depending on the context.

2.2 Definitions

Definition 1. *Given an input set $\Delta_{input} \subset \{x_1, x_2, ..., x_n\}$, and the set Δ_{sk} of all subkey bits that are generated from the master key, a computation chain[1] in a block cipher is a sequence $o^1, o^2, ..., o^h$, in which each o^j is either an element of Δ_{input}, or the output of $f_j(o^{i1}, o^{i2}, ..., o^{ip}, \Delta_{jk})$, where $\Delta_{jk} \subset \Delta_{sk}$, for $1 \le i1, i2, ..., ip < j$, and f_j depends on the cipher structure and round function. Consider $\Delta_{output} \subset \{o^j | j = 1, ..., h, o^j \notin \Delta_{input}\}$ as the output set of this chain.*

Definition 2. *([11, Definition 3]) According to a key schedule, the minimal amount of subkey bits from which one can derive all subkey bits in a computation chain is called actual key information (AKI).*

The n-bit $\{x_1, x_2, ..., x_n\}$ could be n-bit plaintexts ($\mathcal{P}$) or ciphertexts ($\mathcal{C}$).

A forward computation chain has a calculation direction consistent with the encryption process, while the backward chain is consistent with the decryption process. The length of a chain is denoted as r if it covers r rounds of the block cipher. AKI($\mathcal{K}$) is the AKI of a set $\mathcal{K}$ of subkey bits. Normally, AKI($\mathcal{K}$) is the key length, but it could be smaller. We give a simple example to explain the above definitions.

In Fig.1[2], $n = k = 6$, $\Delta_{input} = \{s_0[0], s_0[1]\}$, and $\Delta_{output} = \{o^h\}$, where $o^h = s_4[1] = f_h(s_0[0], s_0[1], s_1[4], s_1[5], s_2[0], s_2[1], s_3[4], \Delta_{hk})$. Obviously, f_h is specifically defined by M_s. Also, $\Delta_{hk} = \{k_0[0], k_0[1], k_1[4], k_1[5], k_2[0], k_2[1], k_3[4]\}$. According to the key schedule (defined by M_k), with the knowledge of $\{k_0[0], k_0[1], k_0[2], k_0[3], k_0[5]\}$, we can deduce all 7 subkey bits in Δ_{hk}.

---

[1] For convenient explanation in our context, we use this terminology similar but not the same as [11].

[2] In this example, we don't need to give the detailed description of F, and just need to concern its input, output and diffusion pattern M_s.

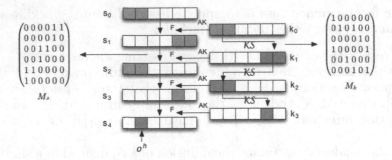

Fig. 1. The 4-round forward computation chain determined by M_s is marked by grey boxes, both in the round function and the key schedule. Here F represents a round transformation.

Obviously AKI is always not larger than k since the knowledge of all master key bits is enough to produce all subkey bits. For a static key schedule or a permutation-based key schedule, since each subkey is directly derived from some master key bits, it is relatively simple to find the AKI.

2.3   A Brief Description of Meet-In-The-Middle (MITM) Attack

Among the risks caused by the insufficient AKI in a computation chain, the most serious and relevant ones are MITM attacks and their variants. A basic MITM attack uses the fact that the cipher can be decomposed into two consecutive parts and the computation of each part only involves partial master key (or its mapping). Then, a check point is calculated by each part independently, and used to filter a number of wrong key pairs.

Thus, for mounting an $(r_f + r_b)$-round MITM attack, we need at least: an r_f-round forward computation chain with $o^h = E_f(\Delta_f, \mathcal{K}_f)$ where $\Delta_f \subset \mathcal{P}$, and $\mathcal{K}_f$ is all involved subkey bits in this forward chain; an r_b-round backward computation chain with $o^h = E_b(\Delta_b, \mathcal{K}_b)$ where $\Delta_b \subset \mathcal{C}$, and $\mathcal{K}_b$ is all subkey bits in this backward chain. The total time complexity C_{comp} using N plaintext/ciphertext pairs is:

$$C_{comp} = \max(2^{\mathrm{AKI}(\mathcal{K}_f)}, 2^{\mathrm{AKI}(\mathcal{K}_b)}) \cdot N + 2^{\mathrm{AKI}(\mathcal{K}_f)+\mathrm{AKI}(\mathcal{K}_b)-N\cdot|o^h|-|\mathcal{K}_{common}|}$$

where $\mathcal{K}_{common}$ is the common key bits shared by $\mathcal{K}_f$ and $\mathcal{K}_b$. The MITM attack works faster than the brute force attack when C_{comp} is less than 2^k, which is the required computations for the brute force attack in the worst case.

3   Automatic Search Tool for Detecting Weakness on Iterated Key Schedules

3.1   AKI Computation Problem

The core of our algorithm is to calculate the AKI of one r-round computation chain. With this algorithm, all r-round computation chains are examined (given a fixed r), so that the one with the least AKI can be found. And by repeating this algorithm with different r, the longest chain with AKI less than k can be determined.

Let V be the set of all subkey bits, and they are linked according to the diffusion of key schedules. Record all subkey bits involved in an r-round computation chain as a set $AV_r \subset V$. Our aim is **to find a set $BV_r \subset V$ such that the subkey bits in BV_r can derive all subkey bits in AV_r, and find the BV_r with the smallest size**. More precisely, BV_r should satisfy the following property: for each bit $b_i \in AV_r$, either $b_i \in BV_r$, or b_i can be reached by the bits in BV_r, where we define that a subkey bit can be reached iff all its parent subkey bits can be reached. $\text{AKI}(AV_r)$ is equal to the size of the smallest BV_r.

AV_r is determined by the cipher structure, the round function and the key extraction phase. BV_r is decided based on AV_r and the key schedule. Our tool contains two phases: the precomputation phase and the online phase, which are described in the next two sections. All explanation is for the forward direction, and for the backward direction we only need to exchange the terms of i and $i - 1$ in our analysis.

3.2   Precomputation

The precomputation phase determines AV_r of r-round computation chains. In a forward r-round chain, AV_r is the set of involved subkey bits (see $\mathcal{K}_f$ in Sect. 2.3), and decided by the diffusion of round functions (i.e., M_s) and the key extraction phase. Based on our representation of M_s, the direction of deducing AV_r is opposite to the direction of its corresponding computation chain. That is, a forward computation chain uses Δ_f (plaintext bits) and $\mathcal{K}_f$ as inputs to compute the output o^h, while in the process of computing AV_r, we start from the output o^h and date back round-by-round to find all related inputs of o^h so that $\mathcal{K}_f$ can be figured out. An r-round computation chain is obtained by multiplying M_s r times with an initial state determined by o^h. Note that in this way, what we obtain is the involved round key bits in a chain, and they should be traced back to the corresponding subkey bits in the key schedule according to the key extraction phase. For byte or nibble-oriented ciphers, one byte or one nibble can be represented with one bit.

For a fixed length r, there are 2^n possible initial states, hence corresponding to 2^n r-round computation chains. Normally, all these 2^n chains require to be checked to find the one with minimal AKI. However, in fact we only need to enumerate n computation chains based on the following observation: the minimum AKI will be found among the chains whose output are only one bit. Since the AV_r starting from one bit in round r are always subsets of the remaining, the initial states with only one active bit (i.e., o^h is one bit) always lead to computation chains with less AKI than the chains with more than one active bit in the initial states. In practical attacks, exploitable chains are often among these n ones. The pseudocode that computes AV_r starting from the ith bit of the rth round internal state is as Algorithm 1, which will be called n times for different i, $i = 1, \ldots, n$. Hereafter we use $AV_r[j]$ ($AV_r[j_1..j_2]$) to represent the set of related subkey bits in the jth (from j_1th to j_2th) round of the r-round chain.

Algorithm 1. Compute AV_r

1: **procedure** COMPUTEAV$_r(r, M_s, i)$
2:     BlockState$[r] = e_i$;   ▷ *Start from the initial state with the ith bit active.*
3:     **for** $j = r - 1 \to 1$ **do**   ▷ *Deduce all related input bits round-by-round.*
4:         BlockState$[j] = $ BlockState$[j + 1] \times M_s$; ▷ *Derive the state bits in round j.*
5:         $AV_r[j] = $ KeyExtraction(BlockState$[j]$);   ▷ *Relate round key to subkey.*
6:     return AV_r;

3.3 Online Phase

Given AV_r and M_k, the online phase looks for BV_r for AV_r. The simplest approach is to brute-force search all predecessors of all subkey bits in AV_r as well as themselves, and among these bits find the smallest subset that can derive AV_r. Using such method, the time complexity is exponential to the number of all bits, about $2^{r \cdot k}$. For a lightweight block cipher, if the diffusion in the round function is not very fast, the size of AV_r is limited by r to a great extent. Also, the low diffusion in the key schedule results in a sparse M_k. Thus, the exhaustive search can be pruned efficiently. However, the time complexity is still too high.

In practical situations, most of the time a discovery of weakness is already able to demonstrate insecurity of the design. Hence, instead of exhaustively searching to guarantee finding the optimal solution, we can relax our goal to find sub-optimal results which still can satisfy our requirements. This inspires us to exploit a greedy algorithm that makes the best choice at each step with the hope of approximating the optimum finally. It is easier to find a BV_r satisfying $|BV_r| \leq |AV_r|$ at each step. We expect that with our greedy strategy, the size of final BV_r is small enough, and always not larger than the size of AV_r and k. Thanks to the characters of iterated lightweight key schedules, our greedy method generates sub-optimal results, while consumes much less time compared with the brute-force searching. If the sub-optimal results we find are already enough to suggest weaknesses in the design, then the real situation of the design may be worse. Since the optimal AKI may indicate fewer key bits involved in a chain, which is even more undesirable for a secure design.

To obtain the final solution, we divide the problem into a series of subproblems, make a choice that is best at the moment, and then keep on applying the greedy strategy to the current result. The choice made each time depends on the choices made so far but not on the future choices. Our greedy choice is based on the following assumption and characteristics of lightweight key schedules. For a forward computation chain:

- **Property 1.** The orientation of key information is along the encryption direction, i.e., from k_{i-1} to k_i. Thus, we assume that the knowledge of a bits of subkey k_i can bring about the knowledge of b bits of k_j iff $i < j$, and $a \geq b^3$.
- **Property 2.** $\mathcal{KS}$ is a bijective and invertible mapping, and each round of subkey has k bits information. A subkey bit cannot be derived from other subkey bits in the same round.
- **Property 3.** A majority of rows in M_k have Hamming weight 1, and the number of these rows is denoted as N_{one}. In most lightweight key schedules, N_{one} is close to k. The remaining tiny part of the rows in M_k have weight larger than 1, while still very small. Denote the maximal Hamming weight among all rows as MAX_{weight}.

Note that Property 1 is made for conveniently explaining our greedy strategy, instead of placing restrictions on the key schedules. Without this assumption, the results may be better but more time complexity is required, i.e., a smaller BV_r for the same AV_r may be found, since more relations in the key schedule can be utilized among the subkey bits. What we obtain with Algorithm 2 is in fact an upper bound of real AKI. If the upper bound of real AKI is insufficient, then real AKI is more insufficient.

We explain how our approach works (see also Algorithm 2) according to the general components required in a greedy algorithm. We start from the given AV_r, and each time we try to reduce the size of current AV_r while making sure that we still can derive all subkey bits in the original AV_r from it according to M_k. The subkey bits in the top round[4] of AV_r require no process due to Property 1. Thus, we only need to deduce the subkey bits within 2 to r rounds, denoting the number of these bits as n_a. At each step we trace the currently remaining n_a subkey bits as back as possible. That is, for the bits in round i of AV_r (i.e., $AV_r[i]$), we try to deduce them with the least amount of bits in round j, where $j < i$ and j should be as small as possible. If any subset of AV_r can be represented with fewer (then the size of AV_r is reduced) or equal subkey bits in former rounds[5], then we remove the original subset from AV_r and add these new

---

[3] This is because of the diffusion direction we consider here.

[4] In an r-round forward chain, w.l.o.g., the top round is the round 1, and the bottom round is the round r.

[5] In a forward chain, round j is a former round of round i iff $j < i$.

bits of former rounds to AV_r. According to Property 1 and Property 2, deducing the same amount of subkey information to former rounds will increase the chance of leaking other subkey bits, while the total number of subkey bits in AV_r doesn't increase. The above process continues until no bits in the current state of AV_r can be deduced to former rounds without increasing $|AV_r|$. The final state of AV_r is the BV_r we want to find. In Algorithm 2, the function GenNextCombination(num) is to generate a combination of num subkey bits from AV_r, and each time a subset of num subkey bits selected from the current n_a bits is considered.

Algorithm 2. Compute AKI

1: **procedure** PROCESSAV$_r$(num)   ▷ *Deduce num key bits to former rounds.*
2:     S = GenNextCombination(num); ▷ *S is the set of currently chosen num bits.*
3:     **while** S != NULL **do**
4:         MIN_{round} = ChooseMinRound(S);
5:         **for** $i = 1 \rightarrow MIN_{round}$ **do**
6:             $S[i]$ = DeduceS(i);   ▷ *Deduce the bits in S to round i according to M_k.*
7:             $tmpAV_r[i] = AV_r[i] \cup S[i]$;
8:             **if** $(|tmpAV_r[i]| - |AV_r[i]|) \leq num$ **then**
9:                 $AV_r[i] = tmpAV_r[i]$;   ▷ *Include the new bits in round i to AV_r.*
10:                $AV_r = AV_r - S$;   ▷ *Remove the original num bits from AV_r.*
11:                break;
12:        S = GenNextCombination(num);
13:    $n_a = |AV_r[2..r]|$;   ▷ *Update the number of remaining bits to be processed.*
14:    **return** AV_r;

15: **procedure** COMPUTEAKI(MAX_{weight}, N_{one}, AV_r)
16:    $num = 1$;   ▷ *Consider all subsets with size one at first.*
17:    **while** (AV_r is reducible) **do**   ▷ *Process until AV_r is unchangable.*
18:        ProcessAV$_r$(num);   ▷ *Select num bits from current $AV_r[2..r]$ to process.*
19:        Choose num according to n_a, MAX_{weight}, N_{one};
20:    $BV_r = AV_r$;   ▷ *The final state of AV_r is what we want.*
21:    AKI = $|BV_r|$;
22:    **return** AKI;

Our method is suitable for both forward and backward directions by using corresponding matrices. We compute the AKI of all computation chains with the same length to find the minimal AKI.

3.4   The Complexity

The memory complexity of matrix M_k is k^2. The time complexity to build n AV_r is $O(nr)$. The memory cost to store AV_r corresponds to the diffusion of the round function M_s. The total time complexity is dominant by the online phase. The main trick in our algorithm is that we process the subkey bits which can be derived by former one bit (i.e., the in-degree is one) at first. Since most bits in AV_r have in-degree one (Property 3), these subkey bits can be deduced to the top rounds directly. Hence, the number of remaining subkey bits that require processing (i.e., n_a) reduces greatly. Because of this sharp decrease of n_a, as well as limited MAX_{weight} in lightweight key schedules, GenNextCombination(num) only enumerates the combinations with a time complexity polynomial to n_a. When deducing any combinations of current AV_r cannot help to reduce the size of AV_r, AV_r is non-reducible. The total time complexity to obtain the final AKI for one AV_r is polynomial to $n_a r^2$. Examples to explain the time complexity are given in Appendix B. Our results listed in the following sections are obtained within seconds on a PC.

3.5   Application to TWINE-80

To show the strength of our tool, we apply it to TWINE-80 [20] and mount a MITM attack based on the found results. The number of rounds we can reach is 14 rounds, while the designers confirmed that the subkeys for the first 3 rounds contain all 80 master key bits. This result is far from reaching the number of rounds of the best known attack in a single key model, which is a 23-round impossible differential attack. However, our attack is the first one with very low data complexity on TWINE-80. Indeed, as pointed out in [4,6], it is important to determine the number of rounds that can be attacked with only a few available data, for a better security understanding. Since the attack is not our main concern in this paper and just an example to show the application of the tool, we refer to more details about TWINE-80 and our attack in Appendix A.

4   Key Bits Leakage

4.1   The Least AKI in TWINE and PRESENT

As we mentioned before, low data complexity attack is an important research direction to make us understand the security of block ciphers in a "real-life" scenario. If a forward and a backward computation chain with insufficient AKI are discovered, then an adversary is likely to mount an MITM attack even the available data is restricted to only a few pairs of plaintext/ciphertext. Therefore, the least AKI among all forward (backward) chains provides a probable security bound for the low data complexity attack, particularly, the MITM attack. For example, if we know the smallest AKI is 58 bits in all 6-round forward chains, and the smallest AKI is 60 bits in all 6-round backward chains, then basically, we cannot mount any 12-round MITM attacks with time complexity less than about 2^{60}. Also, if the least AKI has reached to k bits after r rounds for both

Table 1. The least AKI of the forward computation chains in TWINE and PRESENT, and corresponding theoretical AKI

	TWINE-128(nibbles)		PRESENT-128(bits)		TWINE-80(nibbles)		PRESENT-80(bits)	
round	theoretical	AKI	theoretical	AKI	theoretical	AKI	theoretical	AKI
1	0	0	4	4	0	0	4	4
2	1	1	20	20	1	1	20	20
3	2	2	84	77	2	2	80	64
4	4	4	128	125	4	4	80	80
5	7	7	128	128	7	6	80	80
6	12	11	128	128	12	10	-	-
7	19	17	-	-	19	15	-	-
8	27	23	-	-	20	18	-	-
9	32	27	-	-	20	19	-	-
10	32	30	-	-	20	20	-	-
11	32	31	-	-	-	-	-	-
12	32	32	-	-	-	-	-	-

forward and backward directions, then the cipher is resistant against $2r$-round basic MITM attacks.

In this section, we concern the approximately smallest AKI given the number of rounds, for TWINE and PRESENT. Both of these two ciphers have 64 bits of block size, and 80/128-bit master key. For PRESENT, a 64-bit round key is extracted from the 80/128-bit subkey in each round, while for TWINE a 32-bit (8 nibbles) round key is extracted from the 80/128-bit subkey in each round. Again, we use one bit to represent one nibble for TWINE.

Our search covers different rounds until the AKI is larger than k. The results are in Table 1. Details of some chains with the least AKI in Table 1 are given in Appendix B. The theoretical value of AKI can be computed by the diffusion of round function. Indeed, it is the total number of round key bits we expect to have in a computation chain, i.e., $|AV_r|$, and is not larger than k. Take Fig.1 as an example again, the number of involved round key bits (the grey boxes) in the chain is 7. However, all 6 bits of the master key should be enough to derive these 7 bits, which means that the theoretical AKI of this chain is 6 bits.

Remark. As we mentioned before, the real AKI of a chain could be less than what we find, thus our results in fact indicate a necessary but not sufficient security bound.

The authors in [11] mention three causes responsible for poor key schedules: the size of round keys is too small, no avalanche effect, and key bits leakage. However, no details about these causes are discussed carefully. In the next, we will focus on the key bits leakage.

4.2   Key Bits Leakage and Three Examples

We find that, almost all chains whose practical AKI are less than the theoretical one have a particular phenomenon, where the knowledge of subkey bits are leaked by some other subkey bits in the same chain. This phenomenon is denoted as *key bits leakage*.

Besides the structure of the cipher, for iterated key schedules the interplay of diffusion patterns between key schedules and round functions plays a crucial role on this phenomenon. In this section, we use three examples to demonstrate different levels of key bits leakage caused by different interaction of M_s and M_k. Recall that M_s and M_k are the incidence matrices of diffusion in round functions and key schedules, respectively.

$$
\begin{pmatrix}
0&0&0&0&1&1\\
0&0&0&0&1&0\\
0&0&1&1&0&0\\
0&0&1&0&0&0\\
1&1&0&0&0&0\\
1&0&0&0&0&0
\end{pmatrix}
\quad
\begin{pmatrix}
0&0&0&0&1&1\\
0&0&0&0&1&0\\
0&0&1&1&0&0\\
0&0&1&0&0&0\\
1&1&0&0&0&0\\
1&0&0&0&0&0
\end{pmatrix}
\quad
\begin{pmatrix}
1&0&0&0&0&0\\
0&1&0&1&0&0\\
0&0&0&0&1&0\\
1&0&0&0&0&1\\
0&0&1&0&0&0\\
0&0&0&1&0&1
\end{pmatrix}
\quad
\begin{pmatrix}
1&0&1&0&0&0\\
0&0&0&1&0&0\\
0&1&0&0&1&0\\
1&0&0&0&0&0\\
0&0&1&0&0&1\\
0&0&0&0&1&0
\end{pmatrix}
$$

(a) M_s (b) Case 1: M_k (c) Case 2: M_k (d) Case 3: M_k

Fig. 2. The M_s and different M_k of these three cases

Assume that we already have the round transformation for a block cipher, and now we need to compare the security of three candidate key schedules that have the same level of diffusion (i.e., to some extent they have the same efficiency). Here, we use the same cipher structure as Fig.1. The block size is equal to the key size and every internal state bit is XORed by one bit of the round key. All these three cases share the same M_s, as in (a) of Fig.2, and the three key schedules correspond to different M_k as in (b), (c) and (d) of Fig.2. They have the same level of diffusion: three subkey bits are respectively derived from two subkey bits of the previous round, and the remaining three subkey bits only depend on one bit in the previous round. Therefore, a common property in three M_k is that they all have three rows with weight 2 and three rows with weight 1.

The first case (replace the M_k in Fig.1 with (b) of Fig.2) is a typical and extreme example for key bits leakage, where M_s and M_k are identical. Recall that according to the diffusion of round function F (decided by M_s), a 4-round computation chain for calculating $s_4[1]$ (i.e., we can use e_1 to represent the involved and non-involved bits of s_4 in this chain) is marked by the grey boxes. That is, $k_0 = e_1{\cdot}M_s^4 = (1, 1, 0, 0, 0, 0)$, $k_1 = e_1{\cdot}M_s^3 = (0, 0, 0, 0, 1, 1)$, $k_2 = e_1{\cdot}M_s^2 = (1, 1, 0, 0, 0, 0)$, $k_3 = e_1{\cdot}M_s = (0, 0, 0, 0, 1, 0)$. The vector k_i represents the involved and non-involved subkey bits with 1 and 0, respectively. M_s^4 is M_s to the power of 4. The involved subkey bits are: $k_0[0]$, $k_0[1]$, $k_1[4]$, $k_1[5]$, $k_2[0]$, $k_2[1]$ and $k_3[4]$, 7 bits in total. Thus, what we expect is that 6 bits of the master key are all exploited in this 4-round chain. However, $k_3 \cdot M_k^3 = k_0$, $k_2 \cdot M_k^2 = k_0$, $k_1 \cdot M_k = k_0$. This reveals the situation far from our anticipation: once $k_0[0]$ and $k_0[1]$ are known, the remaining five key bits are not required to guess. Therefore, this computation chain actually contains only two key bits, and the AKI is 2. The ratio between AKI and the theoretical value (6 bits) now is $\frac{1}{3}$, meaning that only 33.3% of the key bits in this chain are effective.

Above is the worst case of key bits leakage in a block cipher. Apparently, this is a bad example for cipher design when M_s and M_k have incomplete diffusion. More generally, we should avoid constructing homologous M_s and M_k in lightweight block ciphers, since the low diffusion will make the consequence of key bits leakage worse.

For the second case, similarly we compute AKI according to M_k in (c) of Fig.2 and find that all 4-round computation chains have 5 bits of AKI. Again for the third case, the AKI of all 4-round chains are 6 bits, meaning that the whole key space is covered now. The security of these two cases is much better than the first case. Thus, we can conclude that not only the amount of diffused bits but also the positions of diffused bits can affect the number of actual key information in a computation chain. With the same amount of diffusion operations, some diffusing patterns will cause severe key bits leakage, while others will not.

5   How to Avoid Key Bits Leakage

In practical design, we hope that key bits leakage does not occur in any round and the AKI can be guaranteed as large as possible in every chain. Hereafter, in our multiplication between matrices and vectors, bit-addition is OR and bit-multiplication is AND. The sum of binary matrices or vectors, denoted as $\sum$, is a bitwise OR of these vectors or matrices.

5.1   The Cause of Key Bits Leakage

After examining the computation chains with least AKI in Table 1, as well as other chains having much lower AKI compared to the theoretical value, we find that almost all of them have the following property: for the computation chain starting from e_j, we can always derive subkeys in this chain from fewer key bits in the top round[6]. That is,

$$|\sum_{i=1}^{r} e_j \cdot M_s^i \cdot M_k^{r-i}| < \sum_{i=1}^{r} |e_j \cdot M_s^i|. \tag{1}$$

---

[6] In a forward computation chain, the top round is the round with the smallest index. In a backward computation chain, the top round is the round with the biggest index.

Recall that $|\cdot|$ is the Hamming weight of the binary vector, M_s^i is the matrix of M_s to the power of i, the sum of weight ($\sum$ in the right side) is a sum over integers. The same notations are used hereafter.

Also, the number of finally involved key bits in the top round that can derive all subkey bits in the chain is less than the master key size. i.e.,

$$|\sum_{i=1}^{r} e_j \cdot M_s^i \cdot M_k^{r-i}| < k. \tag{2}$$

For example, in Table 1, Eq.(1) and Eq.(2) hold for the 6 to 11 rounds of TWINE-128. The theoretical value is $\min(\sum_{i=1}^{r} |e_j \cdot M_s^i|, k)$. Thus, in these rounds,

$$\text{AKI} \leq |\sum_{i=1}^{r} e_j \cdot M_s^i \cdot M_k^{r-i}| < \min(\sum_{i=1}^{r} |e_j \cdot M_s^i|, k), \tag{3}$$

where the ratio of AKI and the theoretical value is smaller than 1.

Note that $\sum_{i=1}^{r} e_j \cdot M_s^i \cdot M_k^{r-i}$ is just the upper bound of AKI, since we may be able to find a set of fewer subkey bits from different rounds (instead of only from the top round) that also can be used to compute all key bits in the chain. However, the first equality sign in Eq.(3) can be obtained for most of the cases we checked for PRESENT and TWINE. This means that the smallest set of subkey bits that can derive all key bits in the chain is selected from the top round. There are two probable reasons responsible for this. Firstly, a large number of rows have weight 1 in M_k so that most subkey bits can directly date back to previous rounds without diffusing. Also, even for the remaining small quantity of rows having weight larger than 1, these weights are low and sometimes neighboring (e.g. the key schedule of PRESENT).

5.2   Necessary Condition for Aoiding Key Bits Leakage

Based on the above cause, we explicitly formulate a necessary condition for avoiding key bits leakage, i.e., gaining the maximal AKI in each computation chain. The tool in Sect. 3 can compute the AKI for each chain and check the ratio. However, this only can be done after the key schedule has been designed. Here we give a design principle only in terms of M_s and M_k, so that we can use it as a guideline when designing the key schedule for a known M_s. For convenience and w.l.o.g., we take the case that M_s and M_k have a equal size k again. In the case where they have different sizes, M_s just needs to be merged or converted according to the key extraction phase.

Corollary 1. *For a block cipher, if there is no key bits leakage within R rounds, then, $\forall r \in [2 \ldots R]$ and $\forall j \in [0 \ldots k-1]$*

$$\begin{cases} |\sum_{i=1}^{r} M_s^i \cdot M_k^{r-i}|_j \geq \sum_{i=1}^{r} |M_s^i|_j, & if \ \sum_{i=1}^{r} |M_s^i|_j < k \\ |\sum_{i=1}^{r} M_s^i \cdot M_k^{r-i}|_j = k, & if \ \sum_{i=1}^{r} |M_s^i|_j \geq k \end{cases} \tag{4}$$

where $| \cdot |_j$ represents the Hamming weight of the jth row of the matrix.

Proof. In order to prove that Eq.(4) is a necessary condition for avoiding key bits leakage, we need to prove that once Eq.(4) doesn't hold, there is key bits leakage within R rounds.

We discuss the case that $\sum_{i=1}^{r} |M_s^i|_j < k$ firstly. When Eq.(4) doesn't hold, then $\exists r$, $\exists j$ so that $|\sum_{i=1}^{r} M_s^i \cdot M_k^{r-i}|_j < \sum_{i=1}^{r} |M_s^i|_j$. Since $e_j \cdot M_s^i \cdot M_k^{r-i}$ is equal to the jth row of $M_s^i \cdot M_k^{r-i}$, and $e_j \cdot M_s^i$ is equal to the jth row of M_s^i, hence this r and j satisfies that $|\sum_{i=1}^{r} e_j \cdot M_s^i \cdot M_k^{r-i}| < \sum_{i=1}^{r} |e_j \cdot M_s^i|$, which is the same as Eq.(1). This means that, considering the rounds from $R-r+1$ (resp. $R+r-1$) to R in a forward (resp. backward) chain which computes the jth bit of round R, the involved subkey bits between these rounds can leak each other through some subkey bits in round $R-r+1$ (resp. $R+r-1$).

The second case is that $\sum_{i=1}^{r} |M_s^i|_j \geq k$. Note that since M_k and M_s are $k \times k$ matrices, $|\sum_{i=1}^{r} M_s^i \cdot M_k^{r-i}|_j$ is always not larger than k, for all j. Hence, when Eq.(4) doesn't hold, then $\exists r$, $\exists j$ so that $|\sum_{i=1}^{r} M_s^i \cdot M_k^{r-i}|_j < k$. Similarly, this r and j satisfies that $|\sum_{i=1}^{r} e_j \cdot M_s^i \cdot M_k^{r-i}| < k$. In this case, we can find that the r-round chain, which computes the jth bit in the beginning round, contains fewer than k bits of AKI. However, this chain should have contained k bits of AKI.   □

Note that in a computation chain that calculates the jth bit of the internal state, indeed AKI can be smaller than $|\sum_{i=1}^{r} M_s^i \cdot M_k^{r-i}|_j$, i.e., $|\sum_{i=1}^{r} M_s^i \cdot M_k^{r-i}|_j$ only provides an upper bound for AKI. This is why Corollary 1 is not a sufficient condition for key bits leakage.

Based on Corollary 1, we can design diffusion of a lightweight key schedule in a more targeting way when the round function has been decided, so that a computation chain is able to diffuse as many key bits as possible even though the number of diffusion operations is limited.

6 Conclusion

In this paper, we develop a effective and efficient tool to automatically search AKI in iterated key schedules of lightweight block ciphers. Also, we formulate the cause of key bits leakage phenomenon from the point of the relation of incidence matrices that represent diffusion of round functions and key scheduling. Based on this cause, a necessary condition on how to avoid key bits leakage in design of lightweight key schedules is given, which can be a guidance of quickly ruling out unreasonable key schedules and maximizing security under low diffusion. In further research we will consider using our algorithm and the necessary condition to examine more designs of lightweight key schedules.

Appendix

A   A MITM Attack on 14 Rounds of TWINE-80

TWINE is a lightweight block cipher using a variant of Type-2 generalized Feistel scheme. TWINE-80 iterates 36 rounds to encrypt 64-bit block message with 80-bit master key. We refer to [20] for more details about this cipher.

Here we choose similar notations as in [20], $X^i = X_0^i \| X_1^i \| \ldots \| X_{14}^i \| X_{15}^i$ and $X^{i+1} = X_0^{i+1} \| X_1^{i+1} \| \ldots \| X_{14}^{i+1} \| X_{15}^{i+1}$ as the input and output of round i, and $P = X^1$. RK^i is the round key used in round i, extracted from WK^i, where WK^i is the key state after running i rounds of the key schedule. Note that $WK_0^0 \| \ldots \| WK_{19}^0$ is the master key, and $RK^i = WK_1^{i-1} \| WK_3^{i-1} \| WK_4^{i-1} \| WK_6^{i-1} \| WK_{13}^{i-1} \| WK_{14}^{i-1} \| WK_{15}^{i-1} \| WK_{16}^{i-1}$.

Here we use a forward computation chain which computes X_5^8 from the plaintext (X^1), and a backward chain computing X_5^8 from the ciphertext (X^{15}). The AKI of this forward chain is 17 nibbles, i.e., $\mathcal{K}_f$ can be decided by all nibbles in WK^0 except WK_{10}^0, WK_{11}^0 and WK_{12}^0. The AKI of the backward chain is 18 nibbles, that is, $\mathcal{K}_b$ can be known from WK_4^{12}, and all nibbles in WK^{13} except WK_0^{13}, WK_7^{13} and WK_{12}^{13}. Note that according to the key schedule, after guessing all nibbles in WK^{13} except WK_0^{13}, WK_7^{13} and WK_{12}^{13}, 10 nibbles in WK^0 can be known (WK_0^0, WK_1^0, WK_3^0, WK_6^0, WK_8^0, WK_{10}^0, WK_{11}^0, WK_{13}^0, WK_{15}^0, WK_{19}^0). Thus, these two computation chains share 8 nibbles of key information, which means that we can filter 32 bits of information from the key. The attack process is as Algorithm 3. We use 9 pairs of plaintext/ciphertext, providing a 36-bit filter from the internal state.

The time complexity of the meeting phase is $9 \cdot 2^{18 \cdot 4} \cdot 0.5 = 2^{74.17}$ encryptions. The number of remaining candidate keys is $2^{(17+18) \cdot 4 - 32 - 36} = 2^{72}$. Thus, 2^{72} trivial encryptions are needed in the search phase. The total time complexity is about $2^{74.46}$ encryptions of 14-round TWINE-80. For simplicity, we omit the cost of memory access for finding a match in T_1, assuming that the time complexity of one table look-up is negligible compared with that of one encryption. This assumption is quite natural and reasonable in most cases. However, strictly speaking, those costs should be considered. The memory complexity is $2^{17 \cdot 4} = 2^{68}$ blocks.

Algorithm 3. The 14-round MITM attack on TWINE-80

1: **for** each possible value of 17 nibbles for $\mathcal{K}_f$ **do**
2:     Encrypt $P_{(i)}$ to get $X^8_{5(i)}$, $i = 1 \ldots 9$.
3:     Store corresponding key values for $\mathcal{K}_f$ in T_1 indexed by $X^8_{5(1)} \| \ldots \| X^8_{5(9)}$.
4: **for** each possible values of 18 nibbles for $\mathcal{K}_b$ **do**
5:     Decrypt $C_{(i)}$ to get $X^8_{5(i)}$, $i = 1 \ldots 9$.
6:     **if** this $X^8_{5(1)} \| \ldots \| X^8_{5(9)}$ exists in T_1 **then**
7:         Take the corresponding key values for $\mathcal{K}_f$ as well as current key value for $\mathcal{K}_b$ as candidate keys.
8: Exhaustively search each remaining candidate key.

B   Details of Computation Chains in Table 1

For TWINE-128, o^h is the output nibble of the chain, $s_i[j]$ here is the jth nibble of the internal state in round i, and $k_i[j]$ ($k_i[j_1 - j_2]$) is the jth (j_1 to j_2) output subkey nibble(s) in the ith round of key schedule. For PRESENT-128, the results are bit-oriented. The subkey bits (nibbles) in BV_r are enough to compute the output o^h of the r-round chain.

Table 2. The BV_r corresponding to TWINE128 for $r = 7, 8, 9$ in Table 1

round	the output o^h	BV_r	AKI
7	$s_7[10]$	$k_5[23], k_6[2-3], k_6[8-15], k_6[17], k_6[22],$ $k_6[24-25], k_6[28], k_6[28], k_6[31]$	17
8	$s_8[14]$	$k_6[23], k_7[0], k_7[2-4], k_7[7-15], k_7[17-18],$ $k_7[22], k_7[24-25], k_7[27-28], k_7[30-31]$	23
9	$s_9[4]$	$k_8[0], k_8[2-3], k_8[5-15], k_8[17-18], k_8[20], k_8[22-31]$	27

Table 3. The BV_r corresponding to PRESENT128 for $r = 3, 4$ in Table 1

round	the output o^h	BV_r	AKI
3	$s_3[3]$	$k_2[51-127]$	77
4	$s_4[0]$	$k_2[3-127]$	125

For PRESENT-128, 120 rows of M_k have weight 1, and 8 rows have weight 4. We consider the 4-round chain with the least AKI, which is 125 bits. The size of original AV_4 is 148 bits, and n_a is 84. After we deduce all one in-degree cases, n_a decreases to 4. Then we can quickly deduce these 4 bits 2-by-2, 3-by-3, and 4-by-4. $s_i[j]$ is the jth bit of the internal state in round i.

References

1. Biryukov, A., Nikolić, I.: Automatic Search for Related-Key Differential Character-istics in Byte-Oriented Block Ciphers: Application to AES, Camellia, Khazad and Others. In: Gilbert, H. (ed.) EUROCRYPT 2010. LNCS, vol. 6110, pp. 322–344. Springer, Heidelberg (2010)
2. Biryukov, A., Nikolić, I.: Search for Related-Key Differential Characteristics in DES-Like Ciphers. In: Joux, A. (ed.) FSE 2011. LNCS, vol. 6733, pp. 18–34. Springer, Heidelberg (2011)
3. Bogdanov, A., Knudsen, L.R., Leander, G., Paar, C., Poschmann, A., Robshaw, M.J.B., Seurin, Y., Vikkelsoe, C.: PRESENT: An Ultra-Lightweight Block Cipher. In: Paillier, P., Verbauwhede, I. (eds.) CHES 2007. LNCS, vol. 4727, pp. 450–466. Springer, Heidelberg (2007)
4. Bouillaguet, C., Derbez, P., Fouque, P.-A.: Automatic Search of Attacks on Round-Reduced AES and Applications. In: Rogaway, P. (ed.) CRYPTO 2011. LNCS, vol. 6841, pp. 169–187. Springer, Heidelberg (2011)
5. Boztaş, Ö., Karakoç, F., Çoban, M.: Multidimensional Meet-in-the-Middle Attacks on Reduced-Round TWINE-128. In: Avoine, G., Kara, O. (eds.) LightSec 2013. LNCS, vol. 8162, pp. 55–67. Springer, Heidelberg (2013)
6. Canteaut, A., Naya-Plasencia, M., Vayssière, B.: Sieve-in-the-Middle: Improved MITM Attacks. In: Canetti, R., Garay, J.A. (eds.) CRYPTO 2013. LNCS, vol. 8042, pp. 222–240. Springer, Heidelberg (2013)
7. Derbez, P., Fouque, P.-A.: Exhausting Demirci-Selçuk Meet-in-the-Middle Attacks Against Reduced-Round AES. In: Moriai, S. (ed.) FSE 2013. LNCS, vol. 8424, pp. 541–560. Springer, Heidelberg (2014)
8. Fouque, P.-A., Jean, J., Peyrin, T.: Structural Evaluation of AES and Chosen-Key Distinguisher of 9-Round AES-128. In: Canetti, R., Garay, J.A. (eds.) CRYPTO 2013. LNCS, vol. 8042, pp. 183–203. Springer, Heidelberg (2013)
9. Guo, J., Peyrin, T., Poschmann, A., Robshaw, M.: The LED Block Cipher. In: Preneel, B., Takagi, T. (eds.) CHES 2011. LNCS, vol. 6917, pp. 326–341. Springer, Heidelberg (2011)
10. Hong, D., et al.: HIGHT: A New Block Cipher Suitable for Low-Resource Device. In: Goubin, L., Matsui, M. (eds.) CHES 2006. LNCS, vol. 4249, pp. 46–59. Springer, Heidelberg (2006)
11. Huang, J., Lai, X.: Revisiting Key Schedule's Diffusion in Relation with Round Function's Diffusion. In: Designs, Codes and Cryptography, pp. 1–19 (2013)
12. Isobe, T., Shibutani, K.: Security Analysis of the Lightweight Block Ciphers XTEA, LED and Piccolo. In: Susilo, W., Mu, Y., Seberry, J. (eds.) ACISP 2012. LNCS, vol. 7372, pp. 71–86. Springer, Heidelberg (2012)
13. Jean, J., Nikolić, I., Peyrin, T., Wang, L., Wu, S.: Security Analysis of PRINCE. In: Moriai, S. (ed.) FSE 2013. LNCS, vol. 8424, pp. 92–111. Springer, Heidelberg (2014)
14. Knudsen, L., Leander, G., Poschmann, A., Robshaw, M.J.B.: PRINTcipher: A Block Cipher for IC-Printing. In: Mangard, S., Standaert, F.-X. (eds.) CHES 2010. LNCS, vol. 6225, pp. 16–32. Springer, Heidelberg (2010)
15. Leander, G., Abdelraheem, M.A., AlKhzaimi, H., Zenner, E.: A Cryptanalysis of PRINTcipher: The Invariant Subspace Attack. In: Rogaway, P. (ed.) CRYPTO 2011. LNCS, vol. 6841, pp. 206–221. Springer, Heidelberg (2011)
16. May, L., Henricksen, M., Millan, W., Carter, G., Dawson, E.: Strengthening the Key Schedule of the AES. In: Batten, L.M., Seberry, J. (eds.) ACISP 2002. LNCS, vol. 2384, pp. 226–240. Springer, Heidelberg (2002)

17. Needham, R.M., Wheeler, D.J.: TEA Extensions. Report, Cambridge University, Cambridge, UK (October 1997)
18. Özen, O., Varıcı, K., Tezcan, C., Kocair, Ç.: Lightweight Block Ciphers Revisited: Cryptanalysis of Reduced Round PRESENT and HIGHT. In: Boyd, C., González Nieto, J. (eds.) ACISP 2009. LNCS, vol. 5594, pp. 90–107. Springer, Heidelberg (2009)
19. Sekar, G., Mouha, N., Velichkov, V., Preneel, B.: Meet-in-the-Middle Attacks on Reduced-Round XTEA. In: Kiayias, A. (ed.) CT-RSA 2011. LNCS, vol. 6558, pp. 250–267. Springer, Heidelberg (2011)
20. Suzaki, T., Minematsu, K., Morioka, S., Kobayashi, E.: TWINE: A Lightweight Block Cipher for Multiple Platforms. In: Knudsen, L.R., Wu, H. (eds.) SAC 2012. LNCS, vol. 7707, pp. 339–354. Springer, Heidelberg (2013)
21. Wu, W., Zhang, L.: LBlock: A Lightweight Block Cipher. In: Lopez, J., Tsudik, G. (eds.) ACNS 2011. LNCS, vol. 6715, pp. 327–344. Springer, Heidelberg (2011)

Cryptanalysis of Reduced-Round SIMON32 and SIMON48

Qingju Wang[1,2], Zhiqiang Liu[1,2]([✉]), Kerem Varıcı[2,3]([✉]), Yu Sasaki[4]([✉]),
Vincent Rijmen[2]([✉]), and Yosuke Todo[4]([✉])

[1] Department of Computer Science and Engineering,
Shanghai Jiao Tong University, Shanghai, China
zhiqiang.liu@esat.kuleuven.be
[2] ESAT/COSIC and iMinds, KU Leuven, Leuven, Belgium
kerem.varici@uclouvain.be
[3] ICTEAM-Crypto Group, Universite catholique de Louvain, Louvain, Belgium
{sasaki.yu,todo.yosuke}@lab.ntt.co.jp
[4] NTT Secure Platform Laboratories, Tokyo, Japan
vincent.rijmen@esat.kuleuven.be

Abstract. SIMON family is one of the recent lightweight block cipher designs introduced by NSA. So far there have been several cryptanalytic results on this cipher by means of differential, linear and impossible differential cryptanalysis. In this paper, we study the security of SIMON32, SIMON48/72 and SIMON48/96 by using integral, zero-correlation linear and impossible differential cryptanalysis. Firstly, we present a novel experimental approach to construct the best known integral distinguishers of SIMON32. The small block size, 32 bits, of SIMON32 enables us to experimentally find a 15-round integral distinguisher, based on which we present a key recovery attack on 21-round SIMON32, while previous best results only achieved 19 rounds. Moreover, we attack 20-round SIMON32, 20-round SIMON48/72 and 21-round SIMON48/96 based on 11 and 12-round zero-correlation linear hulls of SIMON32 and SIMON48 respectively. Finally, we propose new impossible differential attacks which improve the previous impossible differential attacks. Our analysis shows that SIMON maintains enough security margin.

Keywords: SIMON · Integral · Zero-correlation · Impossible differential

1 Introduction

Lightweight primitives are designed to be efficient for limited resource environments, but they should also ensure that the message is transmitted confidentially. Therefore, the vital design motivation is to maintain a reasonable trade-off between the security and performance. During recent years, many

---

Due to page limitations, several details are omitted in this proceedings version. In particular, impossible differential attacks are only described in the full version [1].

© Springer International Publishing Switzerland 2014
W. Meier and D. Mukhopadhyay (Eds.): INDOCRYPT 2014, LNCS 8885, pp. 143–160, 2014.
DOI: 10.1007/978-3-319-13039-2_9

lightweight ciphers have been designed. Prominent examples are included but not limited to these: ICEBERG [2], mCrypton [3], HIGHT [4], PRESENT [5], KATAN [6], LED [7], Piccolo [8], KLEIN [9], EPCBC [10], PRINCE [11] and TWINE [12].

In 2013, NSA also proposed two families of highly-optimized block ciphers, SIMON and SPECK [13], which are flexible to provide excellent performance in both hardware and software environments. Moreover both families offer large variety of block and key sizes such that the users can easily match the security requirements of their applications without sacrificing the performance. However, no cryptanalysis results are included in the specification of these algorithms.

Related Work and Our Contributions. On the one hand, several external cryptanalysis results on SIMON and SPECK were published. In [14,15], differential attacks are presented on various state sizes of SIMON and SPECK, while the best linear attacks on SIMON are given in [16]. In [17] Biryukov et al. exploit the threshold search technique [18], where they showed better differential characteristics and proposed attacks with better results on several versions of SIMON and SPECK. Very recently, there are some differential attack results about SIMON32 and SIMON48 in ePrint [19]. These results need to be further verified although they seem intriguing.

In this paper, we investigate the security of SIMON32, SIMON48/72 and SIMON48/96 by using integral, zero-correlation linear and impossible differential cryptanalysis. We firstly apply integral cryptanalysis. Regarding SIMON32, because the block size is only 32 bits, we can experimentally observe the behaviors of all the plaintexts under a fixed key. Our experiments show that the number of distinguished rounds rapidly increases when the number of active bits becomes close to the block size. On the contrary, exploiting integral distinguishers with a large number of active bits for recovering the key is hard in general. Indeed, our distinguisher needs 31 active bits. To make the data complexity smaller than the code book, we cannot iterate the analysis even for two sets of the distinguisher. We then exploit the fact that the key schedule consists of simple linear equations, and show that reducing any fraction of subkey space can immediately reduce the main key space by solving the linear equations with Gaussian elimination. By combining several known cryptanalytic techniques we present an attack on 21-round SIMON32/64. As for SIMON48, the approach cannot be applied due to the large search space. However, according to the experimental results for SIMON32, we may expect that there exist good integral distinguishers of SIMON48 when the number of active bits is near the block size.

Moreover, we construct 11 and 12-round zero-correlation linear hulls of SIMON32 and SIMON48 respectively by using miss-in-the-middle technique. Then based on these distinguishers, we mount attacks on 20-round SIMON32, 20-round SIMON48/72 and 21-round SIMON48/96 delicately with the help of divide-and-conquer technique. Finally, we demonstrate impossible differential attacks on 18-round SIMON32, 18-round SIMON48/72 and 19-round SIMON48/96. Although these results are not better than the ones achieved by

Table 1. Summary of Attack Results on SIMON

Cipher	Full Rounds	Attack	Attacked Rounds	Complexity Time(EN)	Data	Memory(Bytes)	Source
SIMON32/64	32	Imp. Diff.	13	$2^{50.1}$	$2^{30.0}$KP	$2^{20.0}$	[20]
		Imp.Diff.	**18**	$2^{61.14}$	2^{32}KP	$2^{47.67}$	[1]
		Diff.	16	$2^{26.481}$	$2^{29.481}$CP	2^{16}	[15]
		Diff.	18	$2^{46.0}$	$2^{31.2}$CP	$2^{15.0}$	[14]
		Diff.	19	2^{32}	2^{31}CP	-	[17]
		Zero-Corr.	**20**	$2^{56.96}$	2^{32}KP	$2^{41.42}$	**Subsec 4.2**
		Integral	**21**	$2^{63.00}$	2^{31}CP	2^{54}	**Subsec 3.2**
SIMON48/72	36	**Imp. Diff.**	**18**	$2^{61.87}$	2^{48}KP	$2^{42.12}$	[1]
		Diff.	18	$2^{43.253}$	$2^{46.426}$CP	2^{24}	[15]
		Diff.	19	$2^{52.0}$	$2^{46.0}$CC	$2^{20.0}$	[14]
		Diff.	20	2^{52}	2^{46}CP	-	[17]
		Zero-Corr.	**20**	$2^{59.7}$	2^{48}KP	2^{43}	**Subsec 4.3**
SIMON48/96	36	Imp. Diff.	15	$2^{53.0}$	$2^{38.0}$KP	$2^{20.6}$	[20]
		Imp.Diff.	**19**	$2^{85.82}$	2^{48}KP	$2^{66.68}$	[1]
		Diff.	18	$2^{69.079}$	$2^{50.262}$CP	$2^{45.618}$	[15]
		Diff.	19	$2^{76.0}$	$2^{46.0}$CC	$2^{20.0}$	[14]
		Diff.	20	2^{75}	2^{46}CP	-	[17]
		Zero-Corr.	**21**	$2^{72.63}$	2^{48}KP	$2^{46.73}$	**Subsec 4.3**

CP: Chosen Plaintext; KP: Known Plaintext; CC: Chosen Ciphertext; EN: Encryptions

using differential, integral and zero-correlation linear cryptanalysis, they are the currently best impossible differential attacks for SIMON32 and SIMON48. Our improvements upon the state-of-the-art cryptanalysis for SIMON are given in Table 1.

Organization. The remainder of this paper is organized as follows. In Section 2, we give a brief description of SIMON. Section 3 covers the integral attack. In Section 4, zero-correlation cryptanalysis is studied. Finally, we conclude the paper in Section 5. Impossible differential attacks are shown in [1]. Table 2 contains the notations that we use throughout this paper.

2 Brief Description of SIMON

We denote the SIMON block cipher using n-bit words by SIMON$2n$, with $n \in \{16, 24, 32, 48, 64\}$. SIMON$2n$ with an m-word key is referred to SIMON$2n/mn$.

SIMON is a two-branch balanced Feistel network with simple round functions consisting of three operations: AND (&), XOR ($\oplus$) and rotation ($\lll$). In round $i-1$, by using a function $F(x) = (x \lll 1)\&(x \lll 8)\oplus(x \lll 2)$, (L_{i-1}, R_{i-1}) are updated to (L_i, R_i) by $L_i = F(L_{i-1}) \oplus R_{i-1} \oplus k_{i-1}$ and $R_i = L_{i-1}$. The output of the last round (L_r, R_r) (r is the number of rounds) yields the ciphertext. The structure of the round function of SIMON is depicted in Figure 6 in Appendix A.

The key schedule of SIMON processes three different procedures depending on the key size. The first mn round keys are directly initialized with the main

Table 2. Notations: Top 8 are for general and bottom 4 are for integral attack.

L_r, R_r	left and right branches of the input state to the r-th round
$L_{r,\{i\sim j\}}, R_{r,\{i\sim j\}}$	the bits from bit i to bit j of L_r and R_r
$\Delta L_r, \Delta R_r$	left and right branches of the input difference of state to the r-th round
$\Gamma L_r, \Gamma R_r$	left and right branches of the input linear mask of state to the r-th round
$\Delta F(\cdot)$	the output difference after round function F
k_r	the subkey in the r-th round
$k_{r,\{i\sim j\}}$	the bits from bit i to bit j of k_r
?	an undetermined difference or linear mask

Let Λ be a collection of state vectors $X = (x_0, \ldots, x_{n-1})$ where $x_i \in \mathbb{F}_2$ is the the i-th word of X:	
A	if all i-th words x_i in Λ are distinct, x_i is called active
B	if the sum of all i-th words x_i in Λ can be predicted, x_i is called balanced
C	if the values of all i-th words x_i in Λ are equal, x_i is called passive/constant
*	if the sum of all i-th words x_i in Λ can not be predicted

key, while the remaining key words are generated by three slightly different procedures depending on the key words value m:

$$k_{i+m} = c \oplus (z_j)_i \oplus k_i \oplus Y_m \oplus (Y_m \lll 1), \quad Y_m = \begin{cases} k_{i+1} \lll 3, & \text{if } m = 2, \\ k_{i+2} \lll 3, & \text{if } m = 3, \\ k_{i+3} \lll 3 \oplus k_{i+1}, & \text{if } m = 4. \end{cases}$$

Here, the value c is constant `0xff...fc`, and $(z_j)_i$ denotes the i-th (least significant) bit from one of the five constant sequences z_j ($0 \le j \le 4$). The main key can be derived if any sequence of m consecutive subkeys are known.

3 Integral Cryptanalysis of SIMON

The integral attack [21,22] first constructs an integral distinguisher, which is a set of plaintexts such that the states after several rounds have a certain property, e.g. the XOR sum of all states in the set is 0. Then, several rounds are appended to the distinguisher for recovering subkeys. In this section, we investigate the integral properties and present integral attacks on 21-round SIMON32/64.

3.1 Integral Distinguishers of SIMON32

We experimentally find integrals of SIMON32. The results are shown in Table 3. Here the active bits are the ones in the input of round 1. An interesting observation is that the number of rounds increases rapidly when the number of active bits becomes close to the block size. Giving a theoretical reasoning for this observation seems hard. In other words, experimental approaches are useful for a small block size such that all plaintexts can be processed in a practical time. We explain the algorithm of our experiments as follows:

1. Firstly, we generate 2^t plaintexts ($t \ge 16$) by setting the right half (16 bits) and $(t - 16)$ bits of the left half of the input in round 1 to be active, while keeping the remaining bits as constant.

Table 3. The Number of Rounds of SIMON32 Integral Distinguishers

Num. of Active Bits	16	17	18	19	20	21	22	23	24	25	26	27	28	29	30	31
Num. of Rounds	9	9	9	9	10	10	10	10	11	11	11	12	13	13	14	15

2. (a) Choose the main key randomly. Encrypt 2^t plaintexts r rounds and check whether certain bits of the output are balanced (i.e., for each of these bits, the XOR sum of the bit over 2^t output states is 0). If yes, keep this as an integral candidate.
   (b) Repeat (a) 2^{13} times and verify if the integral candidate always holds. If not, discard it.
3. If there is an integral candidate for all the structures with the same pattern (i.e., with the same t active bits), we regard this as an r-round integral distinguisher of SIMON32.

As a result, we obtain a 15-round distinguisher (Figure 1) with 31 active bits:

$$(CAAA, AAAA, AAAA, AAAA,\quad AAAA, AAAA, AAAA, AAAA)$$
$$\rightarrow (****, ****, ****, ****,\quad *B**, ****, B**, ***B). \quad (1)$$

The distinguisher in (1) is not ensured for all of 2^{64} keys. Because our experiment did not return any failure, we expect that the success probability of this distinguisher is at least $1 - 2^{-13}$.

3.2 21-round Integral Attack of SIMON32/64

We use a 15-round integral distinguisher shown in Figure 1. We first prepare 2^{31} internal state values $(X_L\|X_R)$ in which 31 bits are active, then compute the corresponding plaintext $(L_0\|R_0)$ as $L_0 \leftarrow X_R$ and $R_0 \leftarrow F(X_R) \oplus X_L$. Those 2^{31} plaintexts yield balanced bits in 3 positions after 15 rounds, i.e. (L_{15}, R_{15}). Moreover, the subsequent subkey XOR to R_{15} in round 16 never breaks the balanced property as long as the number of plaintexts in a set is even. We then mount a key recovery attack on 21-round SIMON-32/64 by adding six rounds after the distinguisher, which is illustrated in Figure 2.

3.2.1 Overall Strategy

The attacker guesses a part of the last 5-round subkeys $k_{16}, k_{17}, \ldots, k_{20}$. Then he partially decrypts the 2^{31} ciphertexts up to the state $R_{15} \oplus k_{15}$, and computes their XOR sum at the balanced bits. The 15-round distinguisher in Figure 1 has 3 balanced bits. Because the partial decryption up to all of those 3 bits requires too much subkey guesses, we only use 1 balanced bit at position 0. Thus, the subkey space can be reduced by 1 bit per set of 2^{31} plaintexts. In Figure 2, bit-position 0 of $(R_{15} \oplus k_{15})$ is circled and the related bits to the partial decryption are shown. 3 bits of k_{16}, 6 bits of k_{17}, 10 bits of k_{18}, 14 bits of k_{19}, 16 bits of k_{20}, in total 49 subkey bits are related. Because the block size is 32 bits, the analysis

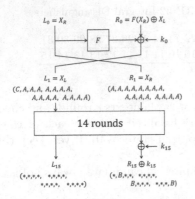

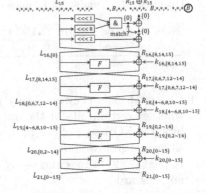

Fig. 1. 15-round Integral Distinguisher   **Fig. 2.** 6-round Key-Recovery

with 2^{31} plaintexts can be iterated at most twice, which implies that the 49-bit subkey space can be reduced at most 2 bits.

To detect the correct key, we further utilize the key schedule. 4 consecutive subkey values can reveal the main key value. We aim to recover 64 bits of $k_{17}, \ldots, k_{20}$. Among 64 bits, 46 bits are suggested from the 6-round partial decryption. Moreover, because 5 subkeys $k_{16}, \ldots, k_{20}$ are linked only with linear equations, 3 bits of $k_{16,\{8,14,15\}}$ can be converted to 3-bit information for the remaining 18 bits of $k_{17}, \ldots, k_{20}$ by solving linear equations with Gaussian elimination. Thus, for each of 49 subkey bits suggested by the 6-round partial decryption, the attacker can obtain 64 bits of $k_{17}, \ldots, k_{20}$ only by guessing 15-bit information of $k_{17}, \ldots, k_{20}$, which leads to a faster key recovery attack than the exhaustive search.

3.2.2 Efficient Subkey Recovery

To perform the 6-round partial decryption with 49-bit subkey guess with a straight-forward method, partial decryption for 2^{31} ciphertexts with 2^{49} guesses are performed, which requires 2^{80} computations i.e. more than the exhaustive search. Several methods are known to reduce the complexity. Here, we use partial-sum [23], meet-in-the-middle match [24], and exploiting linearity for meet-in-the-middle match [25].

The attack finds 49 subkey bits satisfying $\bigoplus(R_{15} \oplus k_{15})_{\{0\}} = 0$, which is $\bigoplus((L_{15,\{15\}} \& L_{15,\{8\}}) \oplus L_{15,\{14\}} \oplus L_{16,\{0\}}) = 0$. This is further converted to

$$\bigoplus(L_{15,\{15\}} \& L_{15,\{8\}}) = \bigoplus(L_{15,\{14\}} \oplus L_{16,\{0\}}). \tag{2}$$

Hence, we can compute the left-hand side and right-hand side of Equation (2) independently, and later find the match between two independent computations as the meet-in-the-middle attack. The computation of the left-hand and right-hand side of Equation (2) is shown in the left and right part of Figure 3, in

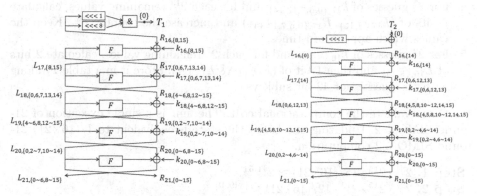

Fig. 3. Computations of $\bigoplus(L_{15,\{15\}}\&L_{15,\{8\}})$ and $\bigoplus(L_{15,\{14\}} \oplus L_{16,\{0\}})$

which 42 bits of subkeys are involved respectively. Compared to the original 6-round partial decryption in Figure 2, the number of related subkey bits are reduced from 49 to 42, which contributes to reduce the attack complexity. The complexity is further reduced by the partial-sum technique. Namely, every time subkey bits are guessed and state values are updated, we compress the amount of data only by keeping the state values appearing odd times.

3.2.2.1 Computation of $\bigoplus(L_{15,\{15\}}\&L_{15,\{8\}})$
Given a set including 2^{31} plaintexts, $\bigoplus(L_{15,\{15\}}\&L_{15,\{8\}})$ for 2^{42} distinct subkey values can be computed with $2^{50.55}$ 21-round SIMON32 computations. The computed results along with 42-bit guessed subkeys are stored in a table T_1. We first initialize the following counters which remembers the parity of internal state values.

- 2^{27} counters T_{20}^x, each corresponding to $x = (L_{20,\{0,2\sim7,10\sim14\}}, R_{20,\{0\sim6,8\sim15\}})$.
- 2^{20} counters T_{19}^x, each corresponding to $x = (L_{19,\{4\sim6,8,12\sim15\}}, R_{19,\{0,2\sim7,10\sim14\}})$.
- 2^{13} counters T_{18}^x, each corresponding to $x = (L_{18,\{0,6,7,13,14\}}, R_{18,\{4\sim6,8,12\sim15\}})$.
- 2^7 counters T_{17}^x, each corresponding to $x = (L_{17,\{8,15\}}, R_{17,\{0,6,7,13,14\}})$.

We then compute $\bigoplus(L_{15,\{15\}}\&L_{15,\{8\}})$ by the following procedure.

1. For 2^{15} guesses of $k_{20,\{0\sim6,8\sim15\}}$ and for each 2^{31} ciphertext values, calculate 27 bits of $(L_{20,\{0,2\sim7,10\sim14\}}, R_{20,\{0\sim6,8\sim15\}})$, and increase the relevant counter T_{20}^x by 1. Keep the values of $(L_{20,\{0,2\sim7,10\sim14\}}, R_{20,\{0\sim6,8\sim15\}})$ which appear odd times.
2. For 2^{12} guesses of $k_{19,\{0,2\sim7,10\sim14\}}$ and for each 2^{27} remaining values, calculate 20 bits of $(L_{19,\{4\sim6,8,12\sim15\}}, R_{19,\{0,2\sim7,10\sim14\}})$ and increase the counter T_{19}^x. Keep the values which appear odd times.
3. For 2^8 guesses of $k_{18,\{4\sim6,8,12\sim15\}}$ and for each 2^{20} remaining values, calculate 13 bits of $(L_{18,\{0,6,7,13,14\}}, R_{18,\{4\sim6,8,12\sim15\}})$ and increase the counter T_{18}^x. Keep the values which appear odd times.

4. For 2^5 guesses of $k_{17,\{0,6,7,13,14\}}$ and for each 2^{13} remaining values, calculate 7 bits of $(L_{17,\{8,15\}}, R_{17,\{0,6,7,13,14\}})$ and increase the counter T_{17}^x. Keep the values which appear odd times.

5. For 2^2 guesses of $k_{16,\{8,15\}}$ and for each 2^7 remaining values, calculate 2 bits of $L_{15,\{8,15\}}$ and then 1-bit of $(L_{15,\{15\}}\&L_{15,\{8\}})$. Store it in a table T_1 along with the guesses for 42-bit subkeys.

We then evaluate the computational cost. The unit is a single execution of 21-round SIMON32. Updating one bit of the state is equivalent to $1/(16 \cdot 21)$ 21-round SIMON32 computation.

Step 1. $2^{31} \cdot 2^{15} \cdot 15/(16 \cdot 21) \approx 2^{41.51}$.
Step 2. $2^{27} \cdot 2^{15} \cdot 2^{12} \cdot 12/(16 \cdot 21) \approx 2^{49.19}$.
Step 3. $2^{20} \cdot 2^{15} \cdot 2^{12} \cdot 2^8 \cdot 8/(16 \cdot 21) \approx 2^{49.61}$.
Step 4. $2^{13} \cdot 2^{15} \cdot 2^{12} \cdot 2^8 \cdot 2^5 \cdot 5/(16 \cdot 21) \approx 2^{46.93}$.
Step 5. $2^7 \cdot 2^{15} \cdot 2^{12} \cdot 2^8 \cdot 2^5 \cdot 2^2 \cdot 3/(16 \cdot 21) \approx 2^{42.19}$.

The sum of the above 5 steps is $2^{50.55}$ 21-round SIMON32 computations. The table T_1 contains 2^{42} elements of 43-bit information, which is less than 2^{45} bytes.

3.2.2.2   Computation of $\bigoplus(L_{15,\{14\}} \oplus L_{16,\{0\}})$
For each of 2^{31} plaintexts set, $\bigoplus(L_{15,\{14\}} \oplus L_{16,\{0\}})$ for distinct 2^{42} subkey values can be computed with $2^{54.01}$ 21-round SIMON32 computations. The computed results along with 42-bit guessed subkeys are stored in a table T_2. Because the procedure is similar to the computation of T_1, the attack is explained shortly.

1. For 2^{16} guesses of $k_{20,\{0\sim15\}}$ and 2^{31} ciphertext values, calculate 29 bits of $(L_{20,\{0,2\sim4,6\sim14\}}, R_{20,\{0\sim15\}})$. The complexity of this step is $2^{31} \cdot 2^{16} \cdot 16/(16 \cdot 21) \approx 2^{42.61}$.

2. For 2^{13} guesses of $k_{19,\{0,2\sim4,6\sim14\}}$ and 2^{29} remaining values, calculate 21 bits of $(L_{19,\{4,5,8,10\sim12,14,15\}}, R_{19,\{0,2\sim4,6\sim14\}})$. The complexity of this step is $2^{29} \cdot 2^{16} \cdot 2^{13} \cdot 13/(16 \cdot 21) \approx 2^{53.31}$.

3. For 2^8 guesses of $k_{18,\{4,5,8,10\sim12,14,15\}}$ and 2^{21} remaining values, calculate 12 bits of $(L_{18,\{0,6,12,13\}}, R_{18,\{4,5,8,10\sim12,14,15\}})$. The complexity of this step is $2^{21} \cdot 2^{16} \cdot 2^{13} \cdot 2^8 \cdot 8/(16 \cdot 21) \approx 2^{52.61}$.

4. For 2^4 guesses of $k_{17,\{0,6,12,13\}}$ and 2^{12} remaining values, calculate 5 bits of $(L_{17,\{14\}}, R_{17,\{0,6,12,13\}})$. The complexity of this step is $2^{12} \cdot 2^{16} \cdot 2^{13} \cdot 2^8 \cdot 2^4 \cdot 4/(16 \cdot 21) \approx 2^{46.61}$.

5. For 2 guesses of $k_{16,\{14\}}$ and 2^5 remaining values, calculate 2 bits of $L_{15,\{14\}}$ and then 1-bit of $(L_{15,\{14\}} \oplus L_{16,\{0\}})$. Store it in a table T_2 along with the guesses for 42-bit subkeys. The complexity of this step is $2^5 \cdot 2^{16} \cdot 2^{13} \cdot 2^8 \cdot 2^4 \cdot 2 \cdot 2/(16 \cdot 21) \approx 2^{39.61}$.

Table T_2 contains 2^{42} elements of 43-bit information, which is less than 2^{45} bytes.

3.2.2.3   Matching T_1 and T_2

After T_1 and T_2 are independently generated, we derive valid 49-bit subkey candidates. Because both of T_1 and T_2 contain 2^{42} elements, the number of pairs is 2^{84}. From Equation (2), the valid candidates will match the 1-bit result in T_1 and T_2. Moreover, 42-bit subkeys used in T_1 and 42-bit subkeys in T_2 overlap in 35 bits. Thus, $2^{84-1-35} = 2^{48}$ valid candidates are generated, which reduces the entire 49-bit space by one bit.

3.2.3   Entire Attack Procedure and Complexity Evaluation

1. Represent the three subkey bits $k_{15,\{8,14\sim15\}}$ by using $k_{16}\|k_{17}\|k_{18}\|k_{19}$ according to the key schedule of SIMON32 and keep the three linear equations.
2. Generate a set of 2^{31} plaintexts.
3. For each of 2^{31} plaintexts, compute T_1 and T_2 as explained before, and identify the correct key candidates to reduce the subkey space of 49 bits in the last 6 rounds.
4. For each of remaining subkey candidates, guess the 15 bits $k_{19,\{1,15\}}\|$ $k_{18,\{0\sim3,7,9\}}\|k_{17,\{1\sim5,11,15\}}$ and obtain three bits of $k_{17,\{8\sim10\}}$ by solving the linear equations with Gaussian elimination. Then compute all bits of the original key by inverting the key schedule, and check the correctness of the guess by using two plaintext-ciphertext pairs.

The data complexity of the attack is 2^{31} chosen-plaintexts. The time complexity for Step 3 is $2^{50.55} + 2^{54.01} \approx 2^{54.13}$ 21-round SIMON32 computations. After Step 3, 2^{48} subkey candidates remain. In Step 4, the cost of Gaussian elimination is much smaller than 21-round SIMON32, and thus is ignored. The check with two plaintext-ciphertext pairs can be done one by one, that is, the check for the second pair is performed only with the first check is passed with probability 2^{-32}. Hence, the time complexity is $2^{48} \cdot 2^{15}(1 + 2^{-32}) \approx 2^{63}$ 21-round SIMON32 computations. In total, the time complexity is $2^{54.13} + 2^{63} \approx 2^{63.00}$ 21-round SIMON32 computations. The memory complexity is $2 \cdot 2^{45}$ bytes for constructing T_1 and T_2 and 2^{48} 49-bit subkey candidates after analyzing a plaintext set, which is less than 2^{51} bytes. The success probability is $1 - 2^{-13}$ due to the probability of the 15-round distinguisher.

4   Zero-Correlation Linear Cryptanalysis of SIMON

The zero-correlation attack is one of the recent cryptanalytic method introduced by Bogdanov and Rijmen [26]. The attack is based on linear approximations with zero correlation (i.e. linear approximations with probability exactly 1/2). We introduce 11 and 12-round zero-correlation linear approximations of SIMON32 and SIMON48, based on which we present key recovery attacks on 20-round SIMON32, 20-round SIMON48/72 and 21-round SIMON48/96 respectively.

4.1   Zero-Correlation Linear Distinguishers of SIMON

By applying miss-in-the-middle technique, we construct 11-round zero-correlation linear hull for SIMON32 (see Figure 4). More specifically, this distinguisher consists

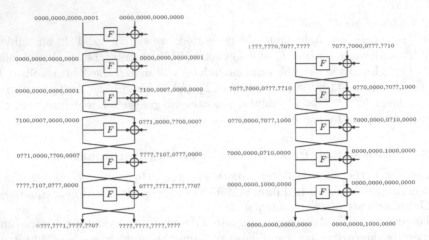

Fig. 4. Zero-Correlation Linear Approximations of 11-round SIMON32. The '0' at bottom left and the '1' at top right (in red) constitute the contradiction that ensures correlation zero.

of two parts: forward part (along the encryption direction) and backward part (along the decryption direction). For the forward part, we find that for any 6-round non-zero correlation linear hull with input mask being $(0x0001, 0x0000)$, the most significant bit of the left half of its output mask must be 0. As to the backward part, we observe that for any 5-round non-zero correlation linear hull with input mask being $(0x0000, 0x0080)$, the most significant bit of the left half of its output mask must be 1. Combining the above two parts, we can deduce that an 11-round linear hull with input and output masks being $(0x0001, 0x0000)$ and $(0x0000, 0x0080)$ must be a zero-correlation linear hull. Similarly, a 12-round zero-correlation linear hull for SIMON48 can be derived (see Table 4 in Appendix B).

4.2 Zero-Correlation Linear Attack on 20-round SIMON32

Let E denote the 20-round SIMON32 from round 0 to round 19. Suppose that the 11-round zero-correlation linear distinguisher given in Figure 4 covers from round 5 to round 15. We now present an attack on E based on this distinguisher by adding five rounds before the distinguisher and four rounds after the distinguisher, which is illustrated in Figure 5.

4.2.1 Overall Strategy

For each of the 2^{32} plaintext-ciphertext pairs, the attacker first guesses a part of the last 4-round subkeys $k_{16}, k_{17}, k_{18}, k_{19}$ and partially decrypts the ciphertext up to the state $R_{16,\{7\}}$. Then he guesses a part of the first 5-round subkeys $k_0, k_1, \ldots, k_4$ and partially encrypts the plaintext up to the state $L_{5,\{0\}}$. Finally, the attacker computes the value of $L_{5,\{0\}} \oplus R_{16,\{7\}}$. The subkey bits related to the above partial encryption and partial decryption are shown in Figure 5. We can see that 14 bits of k_0, 10 bits of k_1, 6 bits of k_2, 3 bits of k_3, one bit of k_4,

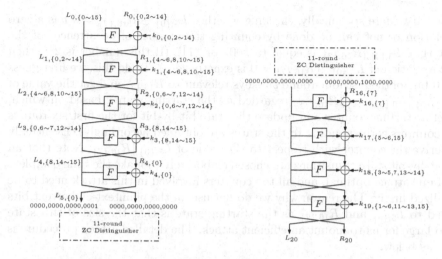

Fig. 5. Add 5 rounds before and 4 rounds after the Distinguisher

one bit of k_{16}, 3 bits of k_{17}, 6 bits of k_{18}, 10 bits of k_{19}, in total 54 subkey bits are related.

For a guessed value of the 54 subkey bits, if the event that $L_{5,\{0\}} \oplus R_{16,\{7\}}$ is equal to 0 happens 2^{31} times (i.e., the correlation of the linear equation $L_{5,\{0\}} \oplus R_{16,\{7\}} = 0$ is exactly 0), then we take this guessed subkey information as a correct subkey candidate. According to [26] and the Wrong-Key Randomization Hypothesis given in [27], for a wrong subkey candidate, the probability that the correlation of $L_{5,\{0\}} \oplus R_{16,\{7\}} = 0$ is 0 can be estimated as $\frac{1}{\sqrt{2\pi}} 2^{\frac{4-32}{2}} \approx 2^{-15.33}$. Thus the 54-bit subkey space can be reduced by a factor of $2^{15.33}$ approximately.

In order to recover the master key value (i.e., k_0, k_1, k_2, k_3), we further exploit the key schedule. Among 64 bits of the master key, 33 bits are suggested from the above procedure. Moreover, $k_4, k_{16}, k_{17}, k_{18}, k_{19}$ can be derived from the master key by using linear equations, therefore, one bit of k_4, one bit of k_{16}, 3 bits of k_{17}, 6 bits of k_{18} and 10 bits of k_{19} (totally 21 subkey bits) can be converted to 21-bit information for the remaining 31 bits of the master key. More specifically, for each of the 33 master key bits suggested above, the attacker can guess 10-bit information of the master key and then obtain 21 linear equations of 21 variables (i.e., the remaining 21 bits of the master key). By solving these linear equations with Gaussian elimination, the attacker can retrieve the master key value.

4.2.2 Efficient Subkey Recovery

We now explain the strategy for efficiently performing 4-round partial decryption and 5-round partial encryption with 54-bit subkey guess. By using a straightforward approach, we need to do the partial decryption and partial encryption for 2^{32} plaintext-ciphertext pairs with 2^{54} subkey guesses. This requires $2^{32+54} = 2^{86}$ computations, which is much more than the exhaustive key search. In our attack, we adopt the divide-and-conquer technique delicately to reduce the time complexity. More specifically, checking whether $L_{5,\{0\}} \oplus R_{16,\{7\}} = 0$ has a zero correlation or not can be done by counting the number of occurrences of the event that $L_{5,\{0\}} \| R_{16,\{7\}}$ is equal to "00" or "11" (If this number is 2^{31}, then the correlation of $L_{5,\{0\}} \oplus R_{16,\{7\}} = 0$ is exactly zero). To do this, we first guess the 20 bits of the last four-round subkeys relevant to $R_{16,\{7\}}$ and get the value of $L_{0,\{0\sim15\}} \| R_{0,\{0,2\sim14\}} \| R_{16,\{7\}}$ (regarded as the starting state), based on which, we set a starting counter and update the state bit-by-bit for the first six rounds (the counters corresponding to the states are obtained accordingly). Eventually we derive the counter with respect to the value of $L_{5,\{0\}} \| R_{16,\{7\}}$. Note that all the bit-by-bit state transitions are chosen elaborately to make the time complexity of our attack optimal, and all the counters involved in this attack need to be initialized firstly. The reason why we do not use all the plaintext-ciphertext bits related to $L_{5,\{0\}}$ and $R_{16,\{7\}}$ as the starting state is that the size of this state is too large for us to mount an efficient attack. The detailed attack procedure is given as below.

1. Collect all the 2^{32} plaintext-ciphertext pairs of E. Let T_1 be a vector of 2^{31} counters correspond to all possible values of $L_{0,\{0\sim15\}} \| R_{0,\{0,2\sim14\}} \| R_{16,\{7\}}$ (denoted as S_1^1). Guess the 20 subkey bits $k_{16,\{7\}} \| k_{17,\{5\sim6,15\}} \| k_{18,\{3\sim5,7,13\sim14\}}$ $\| k_{19,\{1\sim6,11\sim13,15\}}$. Then for each plaintext-ciphertext pair:
   (a) Do partial decryption to get the value of $R_{16,\{7\}}$ and increase the corresponding counter T_{1,S_1^1} by one according to the value of S_1^1. After that, we will do bit-by-bit state transitions based on S_1^1 and update the counters corresponding to the intermediate states.
   (b) Let T_2 be a vector of 2^{30} counters which correspond to all possible values of $L_{0,\{1\sim15\}} \| R_{0,\{0,3\sim7,9\sim14\}} \| L_{1,\{2,8\}} \| R_{16,\{7\}}$ (denoted as S_2^1). Guess the subkey bits $k_{0,\{2,8\}}$. Encrypt partially for each possible value of S_1^1 to obtain the value of $L_{1,\{2,8\}}$, then add T_{1,S_1^1} to the relevant counter T_{2,S_2^1} according to the value of S_2^1.
   (c) Guess subkey bits $k_{0,\{9\}}$, $k_{0,\{3\}}$, $k_{0,\{4,10\}}$, $k_{0,\{11\}}$, $k_{0,\{5\}}$ and $k_{0,\{0,6\sim7,12\sim14\}}$ step by step (see Table 5 in Appendix B).[1] Do similarly to the above and finally get the values of the counters corresponding to the state $L_{1,\{0,2\sim14\}} \| R_{1,\{4\sim6,8,10\sim15\}} \| R_{16,\{7\}}$ (denoted as S_0^2).
2. Let X_1 be a vector of 2^{24} counters which correspond to all possible values of $L_{1,\{0,2\sim7,9\sim14\}} \| R_{1,\{4\sim6,8,11\sim15\}} \| L_{2,\{10\}} \| R_{16,\{7\}}$ (denoted as S_1^2). Guess the subkey bit $k_{1,\{10\}}$. For each possible value of S_0^2, do partial encryption to derive the value of $L_{2,\{10\}}$ and add T_{8,S_0^2} to the corresponding counter X_{1,S_1^2} according to the value of S_1^2. After that, guess the subkey bits $k_{1,\{4\}}$, $k_{1,\{11\}}$, $k_{1,\{12\}}$, $k_{1,\{13\}}$, $k_{1,\{5\}}$, $k_{1,\{6\}}$ and $k_{1,\{8,14\sim15\}}$ sequentially. Do similarly to the above and eventually obtain the values of the counters corresponding to the state $L_{2,\{4\sim6,8,10\sim15\}} \| R_{2,\{0,6\sim7,12\sim14\}} \| R_{16,\{7\}}$ (denoted as S_0^3).
3. Let Y_1 be a vector of 2^{16} counters which correspond to all possible values of $L_{2,\{4\sim6,8,11\sim15\}} \| R_{2,\{0,6\sim7,13\sim14\}} \| L_{3,\{12\}} \| R_{16,\{7\}}$ (denoted as S_1^3). Guess the subkey bit $k_{2,\{12\}}$. For each possible value of S_0^3, do partial encryption to gain

---

[1] Please refer to the full version for more details of the subsequential attack procedures.

the value of $L_{3,\{12\}}$ and add X_{8,S_0^3} to the relevant counter Y_{1,S_1^3} according to the value of S_1^3. Then guess the subkey bits $k_{2,\{13\}}$, $k_{2,\{14\}}$, $k_{2,\{6\}}$, $k_{2,\{7\}}$ and $k_{2,\{0\}}$ step by step. Do similarly to the above and finally derive the values of the counters corresponding to the state $L_{3,\{0,6\sim7,12\sim14\}}\|R_{3,\{8,14\sim15\}}\|R_{16,\{7\}}$ (denoted as S_0^4).

4. Let Z_1 be a vector of 2^9 counters which correspond to all possible values of $L_{3,\{0,6\sim7,12\sim13\}}\| R_{3,\{8,15\}}\|L_{4,\{14\}}\|R_{16,\{7\}}$ (denoted as S_1^4). Guess the subkey bit $k_{3,\{14\}}$. For each possible value of S_0^4, do partial encryption to get the value of $L_{4,\{14\}}$ and add Y_{6,S_0^4} to the corresponding counter Z_{1,S_1^4} according to the value of S_1^4. After that, guess the subkey bits $k_{3,\{15\}}$ and $k_{3,\{8\}}$ step by step. Do similarly to the above and eventually get the values of the counters corresponding to the state $L_{4,\{8,14\sim15\}}\|R_{4,\{0\}}\|R_{16,\{7\}}$ (denoted as S_0^5).

5. Let W be a vector of 2^2 counters which correspond to all possible values of $L_{5,\{0\}}\|R_{16,\{7\}}$. Guess the subkey bit $k_{4,\{0\}}$. For each possible value of S_0^5, do partial encryption to obtain the value of $L_{5,\{0\}}$ and add Z_{3,S_0^5} to the relevant counter in W according to the value of $L_{5,\{0\}}\|R_{16,\{7\}}$. If $W_0 + W_3 = 2^{31}$ (Note that W_0, W_3 are the counters corresponding to the cases that $L_{5,\{0\}}\|R_{16,\{7\}} = $ "00" and $L_{5,\{0\}}\|R_{16,\{7\}} = $ "11", respectively), keep the guessed 54-bit subkey information (i.e., $k_{0,\{0,2\sim14\}}\|k_{1,\{4\sim6,8,10\sim15\}}$ $\|k_{2,\{0\sim7,12\sim14\}}\|k_{3,\{8,14\sim15\}}\|k_{4,\{0\}}\|k_{16,\{7\}}\|k_{17,\{5\sim6,15\}}\|k_{18,\{3\sim5,7,13\sim14\}}\|$ $k_{19,\{1\sim6,11\sim13,15\}}$, denoted as η) as a possible subkey candidate, and discard it otherwise.

According to [26] and the Wrong-Key Randomization Hypothesis given in [27], the probability that a wrong subkey candidate for η is kept after Step 5 can be approximated as $\frac{1}{\sqrt{2\pi}}2^{-14} \approx 2^{-15.33}$, thus about $2^{54} \times 2^{-15.33} = 2^{38.67}$ subkey candidates for η will be left after the above procedure.

4.2.3   Master Key Recovery

1. Represent the subkey bits $k_{4,\{0\}}$, $k_{16,\{7\}}$, $k_{17,\{5\sim6,15\}}$, $k_{18,\{3\sim5,7,13\sim14\}}$ and $k_{19,\{1\sim6,11\sim13,15\}}$ by using $k_{0,\{0\sim15\}}$, $k_{1,\{0\sim15\}}$, $k_{2,\{0\sim15\}}$ and $k_{3,\{0\sim15\}}$ according to the key schedule of SIMON32 and keep these 21 linear equations.

2. For each of the remaining $2^{38.67}$ values of η, do the following to recover the 64-bit master key:
   (a) Guess the 10 subkey bits $k_{0,\{1,15\}}$, $k_{1,\{0\sim3,7,9\}}$ and $k_{2,\{1\sim2\}}$ and obtain 21 linear equations with respect to $k_{2,\{3\sim5,8\sim11,15\}}$ and $k_{3,\{0\sim7,9\sim13\}}$.
   (b) Solve the linear equations by means of Gaussian elimination so as to get the value of $k_{2,\{3\sim5,8\sim11,15\}} \|k_{3,\{0\sim7,9\sim13\}}$, thus all bits of master key can be gained. Verify whether the master key is correct or not by using two plaintext-ciphertext pairs (do the verification for one pair firstly, if the master key can pass the test, do the verification for the other pair).

4.2.4   Complexity of the Attack

The data complexity of this attack is 2^{32} known plaintexts. The memory complexity is primarily owing to keeping the remaining subkey candidates for η

in Step 5 of the *Efficient subkey recovery* phase, thus it can be estimated as $2^{38.67} \cdot 54/8 \approx 2^{41.42}$ bytes.

Regarding the time complexity of this attack, it is mainly dominated by Steps 1–4 of the *Efficient subkey recovery* phase and Step 2(b) of the *Master key recovery* phase, which can be derived as follows.

1. In Step 1 of the *Efficient subkey recovery* phase, the time complexity can be estimated as $2^{52}/5 + 3 \cdot 2^{48}/5 + 2 \cdot 2^{47}/5 + 2^{49}/5 + 2^{54} \cdot 3/5 \approx 2^{53.42}$ 20-round SIMON32 encryptions (See Table 5 in Appendix B).

2. In Step 2 of the *Efficient subkey recovery* phase, the time complexity can be estimated as $7 \cdot 2^{54}/5 + 2^{55} \cdot 3/5 \approx 2^{55.38}$ 20-round SIMON32 encryptions.

3. In Step 3 of the *Efficient subkey recovery* phase, the time complexity can be measured as $3 \cdot 2^{56}/5 + 2 \cdot 2^{55}/5 + 2^{54}/5 \approx 2^{55.77}$ 20-round SIMON32 encryptions.

4. In Step 4 of the *Efficient subkey recovery* phase, the time complexity can be measured as $2 \cdot 2^{55}/5 + 2^{54}/5 = 2^{54}$ 20-round SIMON32 encryptions.

5. In Step 2(b) of the *Master key recovery* phase, solving 21 linear equations with 21 variables by using Gaussian elimination needs about $\frac{1}{3} \cdot 21^3 \approx 3087$ bit-XOR operations, which can be measured by $\frac{3087}{16 \cdot 4 \cdot 20} \approx 2^{1.27}$ 20-round SIMON32 encryptions (Note that there are three XOR operations and one AND operation in the round function of SIMON. For simplicity, we approximate them as four XOR operations in our analysis), thus the time complexity of this step can be approximated as $2^{38.67} \cdot 2^{10} \cdot 2^{1.27} + 2^{38.67} \cdot 2^{10} \approx 2^{50.44}$ 20-round SIMON32 encryptions.

Therefore, the total time complexity of this attack is about $2^{53.42} + 2^{55.38} + 2^{55.77} + 2^{54} + 2^{50.44} \approx 2^{56.96}$ 20-round SIMON32 encryptions.

4.3   Zero-Correlation Linear Attacks on SIMON48

Similarly, by using the 12-round zero-correlation linear distinguisher in Table 4 in Appendix, we can mount key recovery attacks on 20-round SIMON48/72 and 21-round SIMON48/96. For the former, the data, memory and time complexities are about 2^{48} known plaintexts, 2^{43} bytes and $2^{59.7}$ 20-round SIMON48/72 encryptions, respectively. As to the latter, the data, memory and time complexities are about 2^{48} known plaintexts, $2^{46.73}$ bytes and $2^{72.63}$ 21-round SIMON48/96 encryptions, respectively.

5   Discussion and Conclusion

Discussion. As mentioned before, applying our experiments to SIMON48 is hard due to the large block size especially when the number of active bits is close to the block size. We then did experiments in which the number of active bits is

24 (i.e., half of the state) and 30 (i.e., 5/8 of the state), and found 9 and 10-round distinguishers, respectively. Interestingly, according to the experimental results for SIMON32 in Table 3, we observed that if half of the state (16 bits) are active, 9-round distinguishers can be found, and if 5/8 of the state (20 bits) are active, 10-round distinguishers can be derived. It seems that the ratio between the number of active bits and the block size for SIMON48 matches with SIMON32 well, thus we may find 13-round distinguisher with 7/8 of the state (42 bits) being active and 15-round distinguisher with 47 active bits for SIMON48. It remains an open problem to apply this experimental approach efficiently to block ciphers with larger block size.

Conclusion. In this paper, we investigated the security of SIMON32 and SIMON48 by using integral, zero-correlation linear and impossible differential cryptanalysis, and obtained some new results on these ciphers. Firstly, we introduced a novel approach to find a 15-round integral distinguisher of SIMON32, with which an efficient attack was mounted on 21-round SIMON32. This approach gives a new way of constructing integral distinguishers for block ciphers with small block size. Secondly, we presented attacks on 20-round SIMON32, 20-round SIMON48/72 and 21-round SIMON48/96 delicately based on 11 and 12-round zero-correlation linear hulls of SIMON32 and SIMON48 respectively. Our attacks improved the previous best results (appeared in FSE 2014) in terms of the number of attacked rounds. Moreover, we proposed improved impossible differential attacks on SIMON32 and SIMON48. It is expected that our results could be beneficial to the security evaluation of SIMON.

A   Round Function of SIMON

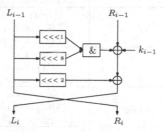

Fig. 6. The Round Function of SIMON

B   Details of Zero-Correlation Linear Cryptanalysis

Table 4. Zero-Correlation Linear Approximations of 12-round SIMON48

	Round	Left	Right
Forward	0	0000,0000,0000,0000,0000,0001	0000,0000,0000,0000,0000,0000
	1	0000,0000,0000,0000,0000,0000	0000,0000,0000,0000,0000,0001
	2	0000,0000,0000,0000,0000,0001	?100,000?,0000,0000,0000,0000
	3	?100,000?,0000,0000,0000,0000	0??1,0000,??00,000?,0000,0001
	4	0??1,0000,??00,000?,0000,0001	?0??,?10?,0???,0000,??00,000?
	5	?0??,?10?,0???,0000,??00,000?	????,????,????,??0?,0???,0001
	6	????,????,????,??0?,0???,0001	????,????,????,????,????,??0?
	7	????,????,????,????,????,??0?	????,????,????,????,????,????
Backward	5	????,????,????,?0?0,???0,001?	0???,10?0,???0,000?,?000,00??
	4	0???,10?0,???0,000?,?000,00??	??10,000?,?000,00?0,0000,0010
	3	??10,000?,?000,00?0,0000,0010	1000,00?0,0000,0000,0000,000?
	2	1000,00?0,0000,0000,0000,000?	0000,0000,0000,0000,0000,0010
	1	0000,0000,0000,0000,0000,0010	0000,0000,0000,0000,0000,0000
	0	0000,0000,0000,0000,0000,0000	0000,0000,0000,0000,0000,0010

Table 5. Attack Procedure in Step 1

i	Input state (S_i^1)	Guessed subkey bit	Output state S_{i+1}^1	Counters related to S_{i+1}^1
0	$L_{0,\{0\sim15\}}\|R_{0,\{0\sim15\}}$	$k_{16,\{7\}}\|k_{17,\{5\sim6,15\}}$ $\|k_{18,\{3\sim5,7,13\sim14\}}$ $\|k_{19,\{1\sim6,11\sim13,15\}}$	$L_{0,\{0\sim15\}}\|R_{0,\{0,2\sim14\}}$ $\|R_{16,\{7\}}$	T_{1,S_1^1}
1	$L_{0,\{0\sim15\}}\|R_{0,\{0,2\sim14\}}$ $\|R_{16,\{7\}}$	$k_{0,\{2,8\}}$	$L_{0,\{1\sim15\}}\|R_{0,\{0,3\sim7,9\sim14\}}$ $\|L_{1,\{2,8\}}\|R_{16,\{7\}}$	T_{2,S_2^1}
2	$L_{0,\{1\sim15\}}\|R_{0,\{0,3\sim7,9\sim14\}}$ $\|L_{1,\{2,8\}}\|R_{16,\{7\}}$	$k_{0,\{9\}}$	$L_{0,\{1\sim6,8\sim15\}}$ $\|R_{0,\{0,3\sim7,10\sim14\}}$ $\|L_{1,\{2,8\sim9\}}\|R_{16,\{7\}}$	T_{3,S_3^1}
3	$L_{0,\{1\sim6,8\sim15\}}$ $\|R_{0,\{0,3\sim7,10\sim14\}}$ $\|L_{1,\{2,8\sim9\}}\|R_{16,\{7\}}$	$k_{0,\{3\}}$	$L_{0,\{2\sim6,8\sim15\}}$ $\|R_{0,\{0,4\sim7,10\sim14\}}$ $\|L_{1,\{2\sim3,8\sim9\}}\|R_{16,\{7\}}$	T_{4,S_4^1}
4	$L_{0,\{2\sim6,8\sim15\}}$ $\|R_{0,\{0,4\sim7,10\sim14\}}$ $\|L_{1,\{2\sim3,8\sim9\}}\|R_{16,\{7\}}$	$k_{0,\{4,10\}}$	$L_{0,\{3\sim6,8\sim15\}}$ $\|R_{0,\{0,5\sim7,11\sim14\}}$ $\|L_{1,\{2\sim4,8\sim10\}}\|R_{16,\{7\}}$	T_{5,S_5^1}
5	$L_{0,\{3\sim6,8\sim15\}}$ $\|R_{0,\{0,5\sim7,11\sim14\}}$ $\|L_{1,\{2\sim4,8\sim10\}}\|R_{16,\{7\}}$	$k_{0,\{11\}}$	$L_{0,\{3\sim6,8,10\sim15\}}$ $\|R_{0,\{0,5\sim7,12\sim14\}}$ $\|L_{1,\{2\sim4,8\sim11\}}\|R_{16,\{7\}}$	T_{6,S_6^1}
6	$L_{0,\{3\sim6,8,10\sim15\}}$ $\|R_{0,\{0,5\sim7,12\sim14\}}$ $\|L_{1,\{2\sim4,8\sim11\}}\|R_{16,\{7\}}$	$k_{0,\{5\}}$	$L_{0,\{4\sim6,8,10\sim15\}}$ $\|R_{0,\{0,6\sim7,12\sim14\}}$ $\|L_{1,\{2\sim5,8\sim11\}}\|R_{16,\{7\}}$	T_{7,S_7^1}
7	$L_{0,\{4\sim6,8,10\sim15\}}$ $\|R_{0,\{0,6\sim7,12\sim14\}}$ $\|L_{1,\{2\sim5,8\sim11\}}\|R_{16,\{7\}}$	$k_{0,\{0,6\sim7,12\sim14\}}$	$L_{1,\{0,2\sim14\}}\|R_{1,\{4\sim6,8,10\sim15\}}$ $\|R_{16,\{7\}}$ (also denoted as S_0^2)	T_{8,S_8^1} (i.e., T_{8,S_0^2})

The time complexities of substeps $0-7$ are estimated as follows:

substep 0: $2^{20} \cdot 2^{32} \cdot 4/20 = 2^{52}/5$;   substep 1: $2^{20} \cdot 2^{31} \cdot 2^2 \cdot 2/(16 \cdot 20) = 2^{48}/5$;

substep 2: $2^{20} \cdot 2^{30} \cdot 2^3/(16 \cdot 20) = 2^{47}/5$;   substep 3: $2^{20} \cdot 2^{29} \cdot 2^4/(16 \cdot 20) = 2^{47}/5$;

substep 4: $2^{20} \cdot 2^{28} \cdot 2^6 \cdot 2/(16 \cdot 20) = 2^{49}/5$;   substep 5: $2^{20} \cdot 2^{27} \cdot 2^7/(16 \cdot 20) = 2^{48}/5$;

substep 6: $2^{20} \cdot 2^{26} \cdot 2^8/(16 \cdot 20) = 2^{48}/5$;   substep 7: $2^{20} \cdot 2^{25} \cdot 2^{14} \cdot 6/(16 \cdot 20) = 2^{54} \cdot 3/5$.

Acknowledgments. The authors are grateful to all anonymous reviewers for their valuable comments. We also thank Lauren De Meyer, Tomer Ashur and Andras Boho for helping with the integral distinguishers. Moreover, the authors are supported by the National Natural Science Foundation of China (no. 61202371), Major State Basic Research Development Program (973 Plan, no. 2013CB338004), China Postdoctoral Science Foundation (no. 2012M521829) and Shanghai Postdoctoral Research Funding Program (no. 12R21414500).

References

1. Wang, Q., Liu, Z., Varıcı, K., Sasaki, Y., Rijmen, V., Todo, Y.: Cryptanalysis of Reduced-round SIMON32 and SIMON48. Cryptology ePrint Archive, Report 2014/761 (2014). http://eprint.iacr.org/
2. Standaert, F.-X., Piret, G., Rouvroy, G., Quisquater, J.-J., Legat, J.-D.: ICEBERG: An Involutional Cipher Efficient for Block Encryption in Reconfigurable Hardware. In: Roy, B., Meier, W. (eds.) FSE 2004. LNCS, vol. 3017, pp. 279–299. Springer, Heidelberg (2004)
3. Lim, C.H., Korkishko, T.: mCrypton – A Lightweight Block Cipher for Security of Low-Cost RFID Tags and Sensors. In: Song, J., Kwon, T., Yung, M. (eds.) WISA 2005. LNCS, vol. 3786, pp. 243–258. Springer, Heidelberg (2006)
4. Hong, D., et al.: HIGHT: A New Block Cipher Suitable for Low-Resource Device. In: Goubin, L., Matsui, M. (eds.) CHES 2006. LNCS, vol. 4249, pp. 46–59. Springer, Heidelberg (2006)
5. Bogdanov, A., Knudsen, L.R., Leander, G., Paar, C., Poschmann, A., Robshaw, M.J.B., Seurin, Y., Vikkelsoe, C.: PRESENT: An Ultra-Lightweight Block Cipher. In: Paillier, P., Verbauwhede, I. (eds.) CHES 2007. LNCS, vol. 4727, pp. 450–466. Springer, Heidelberg (2007)
6. De Cannière, C., Dunkelman, O., Knežević, M.: KATAN and KTANTAN — A Family of Small and Efficient Hardware-Oriented Block Ciphers. In: Clavier, C., Gaj, K. (eds.) CHES 2009. LNCS, vol. 5747, pp. 272–288. Springer, Heidelberg (2009)
7. Guo, J., Peyrin, T., Poschmann, A., Robshaw, M.J.B.: The LED Block Cipher. In: Preneel, Takagi [28], pp. 326–341
8. Shibutani, K., Isobe, T., Hiwatari, H., Mitsuda, A., Akishita, T., Shirai, T.: Piccolo: An Ultra-Lightweight Blockcipher. In: Preneel, Takagi [28], pp. 342–357
9. Gong, Z., Nikova, S., Law, Y.W.: KLEIN: A New Family of Lightweight Block Ciphers. In: Juels, A., Paar, C. (eds.) RFIDSec 2011. LNCS, vol. 7055, pp. 1–18. Springer, Heidelberg (2012)
10. Yap, H., Khoo, K., Poschmann, A., Henricksen, M.: EPCBC - A Block Cipher Suitable for Electronic Product Code Encryption. In: Lin, D., Tsudik, G., Wang, X. (eds.) CANS 2011. LNCS, vol. 7092, pp. 76–97. Springer, Heidelberg (2011)
11. Borghoff, J., Canteaut, A., Güneysu, T., Kavun, E.B., Knezevic, M., Knudsen, L.R., Leander, G., Nikov, V., Paar, C., Rechberger, C., Rombouts, P., Thomsen, S.S., Yalçın, T.: PRINCE – A Low-Latency Block Cipher for Pervasive Computing Applications - Extended Abstract. In: Wang, X., Sako, K. (eds.) ASIACRYPT 2012. LNCS, vol. 7658, pp. 208–225. Springer, Heidelberg (2012)
12. Suzaki, T., Minematsu, K., Morioka, S., Kobayashi, E.: TWINE: A Lightweight Block Cipher for Multiple Platforms. In: Knudsen, Wu [29], pp. 339–354

13. Beaulieu, R., Shors, D., Smith, J., Treatman-Clark, S., Weeks, B., Wingers, L.: The SIMON and SPECK Families of Lightweight Block Ciphers. Cryptology ePrint Archive, Report 2013/404 (2013)
14. Abed, F., List, E., Wenzel, J., Lucks, S.: Differential Cryptanalysis of round-reduced Simon and Speck. In: Cid, C., Rechberger, C. (eds.) International Workshop on Fast Software Encryption - FSE 2014. LNCS. Springer (2104)
15. Alkhzaimi, H.A., Lauridsen, M.M.: Cryptanalysis of the SIMON Family of Block Ciphers. Cryptology ePrint Archive, Report 2013/543 (2013). http://eprint.iacr.org/
16. Alizadeh, J., Bagheri, N., Gauravaram, P., Kumar, A., Kumar Sanadhya, S.: Linear Cryptanalysis of Round Reduced SIMON. Cryptology ePrint Archive, Report 2013/663 (2013). http://eprint.iacr.org/
17. Biryukov, A., Roy, A., Velichkov, V.: Differential Analysis of Block Ciphers SIMON and SPECK. In: Cid, C., Rechberger, C. (eds.) International Workshop on Fast Software Encryption - FSE 2014. LNCS. Springer (2104)
18. Biryukov, A., Velichkov, V.: Automatic Search for Differential Trails in ARX Ciphers. In: Benaloh, J. (ed.) CT-RSA 2014. LNCS, vol. 8366, pp. 227–250. Springer, Heidelberg (2014)
19. Wang, N., Wang, X., Jia, K., Zhao, J.: Improved Differential Attacks on Reduced SIMON Versions. Cryptology ePrint Archive, Report 2014/448 (2014). http://eprint.iacr.org/
20. Abed, F., List, E., Lucks, S., Wenzel, J.: Differential and Linear Cryptanalysis of Reduced-Round Simon. Cryptology ePrint Archive, Report 2013/526 (2013). http://eprint.iacr.org/
21. Daemen, J., Knudsen, L.R., Rijmen, V.: The Block Cipher SQUARE. In: Biham, E. (ed.) FSE 1997. LNCS, vol. 1267, pp. 149–165. Springer, Heidelberg (1997)
22. Knudsen, L.R., Wagner, D.: Integral Cryptanalysis. In: Daemen, J., Rijmen, V. (eds.) FSE 2002. LNCS, vol. 2365, pp. 112–127. Springer, Heidelberg (2002)
23. Ferguson, N., Kelsey, J., Lucks, S., Schneier, B., Stay, M., Wagner, D., Whiting, D.: Improved Cryptanalysis of Rijndael. In: Schneier, B. (ed.) FSE 2000. LNCS, vol. 1978, pp. 213–230. Springer, Heidelberg (2001)
24. Sasaki, Y., Wang, L.: Meet-in-the-Middle Technique for Integral Attacks against Feistel Ciphers. In: Knudsen, Wu [29], pp. 234–251
25. Sasaki, Y., Wang, L.: Bitwise Partial-sum on HIGHT: A New Tool for Integral Analysis against ARX Designs. In: ICISC 2013, vol. 8565. LNCS, pp. 1–14. Springer, Heidelberg (2013)
26. Bogdanov, A., Rijmen, V.: Linear hulls with correlation zero and linear cryptanalysis of block ciphers. Des. Codes Cryptography 70(3), 369–383 (2014)
27. Harpes, C., Kramer, G.G., Massey, J.L.: A Generalization of Linear Cryptanalysis and the Applicability of Matsui's Piling-Up Lemma. In: Guillou, L.C., Quisquater, J.-J. (eds.) EUROCRYPT 1995. LNCS, vol. 921, pp. 24–38. Springer, Heidelberg (1995)
28. Preneel, B., Takagi, T. (eds.): CHES 2011. LNCS, vol. 6917. Springer, Heidelberg (2011)
29. Knudsen, L.R., Wu, H. (eds.): SAC 2012. LNCS, vol. 7707. Springer, Heidelberg (2013)

General Application of FFT in Cryptanalysis and Improved Attack on CAST-256

Long Wen[1], Meiqin Wang[1](✉), Andrey Bogdanov[2](✉), and Huaifeng Chen[1]

[1] Key Laboratory of Cryptologic Technology and Information Security,
Ministry of Education, Shandong University, Jinan 250100, China
longwen@mail.sdu.edu.cn, mqwang@sdu.edu.cn
[2] Technical University of Denmark, Kongens Lyngby, Denmark
anbog@dtu.dk

Abstract. Fast Fourier Transformation (FFT) technique was used to reduce the time complexity of linear cryptanalysis by Collard *et al.* at ICISC 2007. This powerful technique has been used to improve the time complexity of zero-correlation linear cryptanalysis as well as integral attack by Bogdanov *et al.* and Todo respectively. Yet whether FFT is applicable when multiple modular additions with subkeys are involved during the partial encryption and decryption phase remains unknown, which has limited its application to some degree. In this paper, we give a general scheme to use FFT technique in linear cryptanalysis, zero-correlation or integral attack where multiple modular additions (together with multiple XORs) with subkeys are involved in the key recovery process. Based on this scheme, we can attack one more round of CAST-256 than the zero-correlation attack on 28-round CAST-256 at ASIACRYPT 2012 by Bogdanov *et al.*, which also becomes the best attack against CAST-256 without any weak key assumption.

Keywords: FFT · Modular addition · CAST-256 · Zero-correlation

1 Introduction

Fast Fourier Transformation (FFT) technique was introduced to the cryptanalysis domain of block cipher by Collard *et al.* at ICISC 2007 [11], where they pointed out that under certain circumstances the time complexity of linear cryptanalysis could be reduced with FFT technique. Moreover, the FFT technique has been applied in correlation attacks against stream ciphers [10,13,15]. As multiple zero-correlation linear cryptanalysis proposed in [9] shares a similar key recovery phase with classic linear cryptanalysis, Bogdanov *et al.* showed that FFT technique is applicable to multiple zero-correlation linear cryptanalysis as well, and with which they achieved better cryptanalysis results against Camellia [6]. Later, Todo showed that the FFT technique can also be integrated into the integral attack [18]. Yet, only the case where XOR subkey or one modular addition with subkey is included in the key recovery phase has been demonstrated

© Springer International Publishing Switzerland 2014
W. Meier and D. Mukhopadhyay (Eds.): INDOCRYPT 2014, LNCS 8885, pp. 161–176, 2014.
DOI: 10.1007/978-3-319-13039-2_10

as FFT compatible in [11] while the common scenario in some ciphers such as CAST-256, Twofish and Safer family ciphers, *etc.*, where the operation between the subkeys and intermediate states is modular addition and multiple modular additions with subkeys are involved in the key recovery phase, has not been investigated. In this paper, by analyzing the properties of circulant matrices we give a general scheme to use FFT in the key recovery phase where multiple modular additions (together with XOR) with subkeys are involved. Basing on the proposed scheme, we improve the zero-correlation attack on CAST-256 from 28-round in [7] to 29-round.

CAST-256 is designed by Adams [1] and was one of the fifteen candidates in the first round of AES project. The block size of CAST-256 is 128 bits, the key size could be 128, 160, 192, 224 or 256 bits and the number of rounds is 48. There have been quite a few results on the security of CAST-256. For CAST-256 with 128-bit key, in 2007 Nakahara *et al.* identified 12-round linear approximations and gave the distinguishing attack on 12-round CAST-256 [14]. At FSE 1999, Wagner proposed the boomerang attack against 16-round CAST-256 [19]. As to attacks on CAST-256 with 256-bit key, Seki *et al.* gave the differential attack on 36-round CAST-256 under 2^{-35} weak key assumption [16] in 2001. Then using linear cryptanalysis, Wang *et al.* [20] presented a key recovery attack on 24-round CAST-256 with 256-bit key at SAC 2008. And at ASIACRYPT 2012, Bogdanov *et al.* [7] constructed zero-correlation linear approximations over 24-round CAST-256 and recovered the key for 28-round CAST-256 by adding four rounds before the zero-correlation linear approximations. We here revisited the attack proposed by Bogdanov *et al.* Using the same zero-correlation linear distinguisher, we manage to attack one more round (29-round) of CAST-256 due to the application of FFT technique.

Zero-correlation (ZC) linear cryptanalysis is introduced by Bogdanov and Rijmen [8]. The new technique is based on the availability of numerous key-independent unbiased (i.e., of correlation zero) linear approximations for many ciphers. (If p is the probability for a linear approximation to hold, its correlation is defined as $c = 2p - 1$). A more data-efficient distinguisher than the original one was proposed in [9] utilizing the existence of multiple linear approximations with correlation zero in target ciphers, referred to as *multiple ZC linear cryptanalysis*. At ASIACRYPT'12 [7], fundamental links of integral cryptanalysis to ZC cryptanalysis were revealed, and a multidimensional distinguisher was constructed for the ZC property, referred to as *multidimensional ZC linear cryptanalysis*. Later, the multidimensional ZC linear cryptanalysis model was verified with the experiments on small variant of LBlock with block length 32-bit [17]. ZC linear cryptanalysis has been demonstrated to improve the state-of-the-art attacks on TEA, XTEA, CAST-256, Camellia, CLEFIA, LBlock, HIGHT, and E2 [6,7,9,17,22,23]. Other improvements in the domain of linear cryptanalysis include the proposition of a new linear cryptanalysis in the related-key setting [5], the extension of integral ZC to the ARX block ciphers [21], as well as the links between ZC and impossible differential (ID) distinguisher [3,4].

1.1   Our Contributions

General scheme of FFT-based technique in cryptanalysis. Although FFT has been introduced to the domain of cryptanalysis for block cipher since 2007, its application has been limited to the cases where only XOR subkey or one modular addition with subkey is involved in the key recovery phase. In this paper, by observing the property of circulant matrices, we present a general scheme to use FFT technique to reduce the time complexity when multiple modular additions with subkeys are involved in the partial encryption/decryption phase. This scheme is useful to apply FFT technique to the cryptanalysis of block ciphers with modular addition with subkeys. The cases with only XOR subkey or one modular addition considered in [11] belong to the general scheme.

Multiple ZC attack with FFT on 29-round CAST-256. Bogdanov *et al.* presented a multidimensional ZC attack on 28-round CAST-256 with 24-round ZC linear distinguisher at ASIACRYPT 2012 [7]. Basing on the same ZC linear distinguisher and the general scheme of the FFT technique, we can attack 29-round CAST-256. This is the best attack against CAST-256 without weak key assumption. The attacks on CAST-256 are summarized in Table 1.

Table 1: Summary of Attacks on CAST-256

Attack	#Rounds	Data	Time	Memory	Weak key rate	Ref.
boomerang	16	$2^{49.3}$CP	-	-	1	[19]
linear	24	$2^{124.1}$KP	$2^{156.52}$	-	1	[14]
differential	36	2^{123}CP	2^{182}	-	2^{-35}	[16]
multidimensional ZC	28	$2^{98.8}$KP	$2^{246.9}$	2^{68}bytes	1	[7]
multiple ZC	**29**	$\mathbf{2^{123.2}}$**KP**	$\mathbf{2^{218.1}}$	$\mathbf{2^{113}}$**bytes**	**1**	**Sec.4**

[1]CP: Chosen Plaintext; KP: Known Plaintext.

Organization of the paper. The remainder of this paper is organized as follows. Section 2 recalls the multiple ZC linear cryptanalysis. In Section 3 we present a general scheme to use FFT technique when multiple modular additions with subkeys are involved in the partial encryption/decryption phase. Based on the proposed scheme from Section 3, improved attack against CAST-256 is illustrated in Section 4. We conclude the paper in Section 5.

2   Preliminaries

2.1   Basis of Zero-Correlation Cryptanalysis

A linear approximation $\alpha \to \beta$ of a n-bit vectorial function $f : \ _2^n \to \ _2^n$ has a *correlation* defined as $C_{\alpha,\beta}^f = 2\varepsilon_{\alpha,\beta}^f - 1$, where $\varepsilon_{\alpha,\beta}^f = \Pr\{\beta \cdot f(x) \oplus \alpha \cdot x\} - 1/2$

is the *bias* of $\alpha \to \beta$ and '·' means the scalar product of two n-bit binary vectors, $\alpha \cdot x = \bigoplus_{i=0}^{n} \alpha_i x_i$. In ZC linear cryptanalysis, the distinguisher uses linear approximations with correlation zero for all keys while the classical linear cryptanalysis utilizes linear approximations whose correlation is far from zero. Given a distinguisher of ZC linear approximations over part of a cipher, the key recovery can be done with a procedure similar to that of Matsui's Algorithm 2.

To be convenient, we describe a n-bit block cipher E as a cascade $E = E_f \circ E_d \circ E_b$, where E_d is covered by distinguisher. Suppose that we can obtain ℓ ZC linear approximations over E_d and the number of the known plaintext-ciphertext (P, C) pairs is N.

2.2 Multiple ZC Linear Cryptanalysis

We briefly recall the multiple ZC linear cryptanalysis proposed at FSE 2012 [9] in this section. For each of the ℓ given linear approximations, by partially encrypting and decrypting the N (P, C) pairs over E_f and E_b under guessed key values, the adversary computes $T_i, 1 \leq i \leq \ell$, which is the number of (P, C) pairs that fulfill the i-th linear approximation among the N (P, C) pairs. Each T_i suggests an empirical correlation value $\hat{c}_i = 2\frac{T_i}{N} - 1$. Then, the adversary evaluates the statistic:

$$\sum_{i=1}^{\ell} \hat{c}_i^2 = \sum_{i=1}^{\ell} \left(2\frac{T_i}{N} - 1 \right)^2.$$

Under a statistical independency assumption, the value $\sum_{i=1}^{\ell} \hat{c}_i^2$ for the right key approximately follows a normal distribution with mean $\mu_0 = \frac{\ell}{N}$ and standard deviation $\sigma_0 = \frac{\sqrt{2\ell}}{N}$ while for the wrong key the distribution is approximately a normal distribution with mean $\mu_1 = \frac{\ell}{N} + \frac{\ell}{2^n}$ and standard deviation $\sigma_1 = \frac{\sqrt{2\ell}}{N} + \frac{\sqrt{2\ell}}{2^n}$.

We denote the type-I error probability (the probability to wrongfully discard the right key guess) as β_0 and the type-II error probability (the probability to accept a wrong key guess as the right key) as β_1. Consider the decision threshold $\tau = \mu_0 + \sigma_0 z_{1-\beta_0} = \mu_1 - \sigma_1 z_{1-\beta_1}$ ($z_{1-\beta_0}$ and $z_{1-\beta_1}$ are the quantiles of the standard normal distribution), then the number of known plaintexts N should be approximately:

$$N \approx \frac{2^n (z_{1-\beta_0} + z_{1-\beta_1})}{\sqrt{\ell/2} - z_{1-\beta_1}}. \tag{1}$$

3 General Scheme of FFT Technique in Cryptanalysis

Collard *et al.* has applied FFT technique to improve the time complexity of key recovery of linear cryptanalysis [11]. Later Bogdanov *et al.* took advantage of FFT technique in zero-correlation cryptanalysis [6] and Todo showed that the FFT technique can also be integrated into the integral attack [18]. However, only the case where XOR subkey or one modular addition with subkey is included in the partial decryption or encryption phase has been demonstrated as FFT compatible. For some ciphers with modular additions with subkeys such as CAST-256, Twofish and SAFER family ciphers, the key recovery process involves multiple modular additions with subkeys (together with XOR with subkeys), but the application of FFT for them has not been investigated. In this section, we will focus to give a more generalized application of FFT-based technique in linear cryptanalysis, zero-correlation cryptanalysis or integral attack.

Firstly, we give some notations used in this section as follows.

- $a|b$ means the concatenation of a and b,
- $a \boxplus b = a + b \mod 2^k$, for $a, b \in \mathbb{F}_2^k$,
- $a \boxminus b = a - b \mod 2^k$, for $a, b \in \mathbb{F}_2^k$.

3.1 Some Definitions and Results from [12]

In this subsection, we will introduce some definitions and results related to circulant matrix from[12].

Definition 1. *[12, Subsection 3.1] A circulant matrix of order n, or circulant for short, is meant a square matrix of the form*

$$C = \begin{pmatrix} c_1 & c_2 & \cdots & c_n \\ c_n & c_1 & \cdots & c_{n-1} \\ \vdots & \vdots & \ddots & \vdots \\ c_2 & c_3 & \cdots & c_1 \end{pmatrix},$$

where the elements of each row are identical to those of the previous row, but are moved one position to the right and wrapped around.

Definition 2. *[12, Subsection 5.1] A left circulant matrix of order n has each successive row moved one place to the left with the form*

$$C = \text{left-circ}(c_1, c_2, \cdots, c_n) = \begin{pmatrix} c_1 & c_2 & \cdots & c_n \\ c_2 & c_3 & \cdots & c_1 \\ \vdots & \vdots & \ddots & \vdots \\ c_n & c_1 & \cdots & c_{n-1} \end{pmatrix}.$$

Definition 3. *[12, Subsection 5.6] Let $A_1, A_2, \ldots, A_m$ be square matrices each of order n, a block circulant matrix of type (m, n) is meant an $mn \times mn$ matrix of the form*

$$\begin{pmatrix} A_1 & A_2 & \cdots & A_m \\ A_m & A_1 & \cdots & A_{m-1} \\ \vdots & \vdots & \ddots & \vdots \\ A_2 & A_3 & \cdots & A_1 \end{pmatrix},$$

and a left block circulant matrix of type (m, n) has the following form

$$\begin{pmatrix} A_1 & A_2 & \cdots & A_m \\ A_2 & A_3 & \cdots & A_1 \\ \vdots & \vdots & \ddots & \vdots \\ A_m & A_1 & \cdots & A_{m-1} \end{pmatrix}.$$

Definition 4. *[12, Subsection 5.8] A (left) circulant of level-1 is an ordinary (left) circulant. Let A be a (left) block circulant of type (m, n) and if each block of A is a (left) circulant, then it is a level-2 (left) circulant of type (m, n). In general, a level-q (left) circulant of type $(n_q, n_{q-1}, \ldots, n_1), q \geq 2$, is a block (left) circulant of order n_q whose blocks are level-$(q-1)$ (left) circulant of type $(n_{q-1}, \ldots, n_1)$.*

Corollary 1. *[12, Subsection 5.1] A is a left circulant matrix if and only if it has the form $A = \Gamma C$ where C is a circulant matrix and the first rows of A and C are identical, where*

$$\Gamma = \begin{pmatrix} 1 & 0 & 0 & \cdots & 0 & 0 \\ 0 & 0 & 0 & \cdots & 0 & 1 \\ 0 & 0 & 0 & \cdots & 1 & 0 \\ \vdots & \vdots & \vdots & \ddots & \vdots & \vdots \\ 0 & 1 & 0 & \cdots & 0 & 0 \end{pmatrix}.$$

Theorem 1. *[12, Subsection 5.8] A level-q circulant of type $(n_q, n_{q-1}, \ldots, n_1)$ C is diagonalizable by the unitary matrix $F = F_{n_q} \otimes F_{n_{q-1}} \otimes F_{n_{q-2}} \otimes \ldots \otimes F_1$:*

$$C = F^* diag(\lambda)F,$$

where λ is the vector of eigenvalues of C, the symbol $\otimes$ is the Kronecker product and $F_{n_l}, 1 \leq l \leq q$ is the Fourier matrix of size $n_l * n_l$ defined by:

$$F_{n_l}(i, j) = \frac{1}{\sqrt{n_l}} \omega^{i \cdot j} \quad (0 \leq i, j \leq n_l - 1),$$

with $\omega = e^{\frac{2\pi\sqrt{-1}}{n_l}}$ and F^* is the transpose of F.

As is shown in [11], the eigenvalues of a circulant matrix can be obtained by a Fourier transform. From $C = F^* diag(\lambda)F$ and $FF^* = I$, we can get $FC = diag(\lambda)F$. The first column of FC is

$$(FC)(: 1) = (diag(\lambda)F)(: 1) = diag(\lambda)F(: 1).$$

From the definition of F, $F(: 1) = \frac{1}{\sqrt{n_1 n_2 \cdots n_q}}(1, 1, 1, \ldots, 1)^T$, so

$$diag(\lambda) = (FC)(: 1)\sqrt{n_1 n_2 \cdots n_q} = FC(: 1)\sqrt{n_1 n_2 \cdots n_q}.$$

Therefore, the eigenvalues vector λ of a level-1 circulant matrix C of and type $(n_1, n_2, \ldots, n_q)$ can be computed by performing one FFT on the first column of C. Multiplying C by a vector x can be denoted as

$$Cx = F^* diag(\lambda) Fx = F^* (diag(\lambda)(Fx)),$$

which means that Cx can be obtained by first computing an FFT on x, then multiplying the resulting vector by the eigenvalues of C element wise, and performing an FFT on the resulting vector. Hence, the time cost of computing the matrix-vector product is $3n_1 n_2 \cdots n_q \log(n_1 n_2 \cdots n_q)$.

3.2 General Scheme of FFT-Based Technique in Cryptanalysis

If the key recovery process of linear cryptanalysis, zero-correlation cryptanalysis and integral attack involves several modular additions (together with XOR) with subkeys, the evaluation of linear approximation can be implemented with the matrix-vector product $X \cdot V$ where V is counter vector computed from plaintext-ciphertext pairs and the element of row $(K_{i1}|K_{i2}|\ldots|K_{j1}|K_{j2}|\ldots)$ and column $(X_{i1}|X_{i2}|\ldots|X_{j1}|X_{j2}|\ldots)$ for the matrix X is computed as

$$g((X_{i1} \boxplus K_{i1}), (X_{i2} \boxplus K_{i2}), \ldots, (X_{j1} \oplus K_{j1}), (X_{j2} \oplus K_{j2}), \ldots), \qquad (2)$$

where g is the decryption function and X_t corresponds to some value derived from the plaintext, ciphertext or intermediate state, and K_t is the subkey ($t = i1, i2, \ldots, j1, j2, \ldots$). In order to show that the matrix-vector product $X \cdot V$ can be implemented with FFT, we will first introduce how to convert a level-q left circulant matrix to a level-q circulant matrix.

Let Γ_1 be the left circulant matrix of size n_1 $\Gamma_1 = left\text{-}circ(1, 0, \cdots, 0)$, Γ_2 be the level-2 left circulant of type (n_2, n_1) which is shown as follows,

$$\Gamma_2 = \begin{pmatrix} \Gamma_1 & 0 & 0 & \cdots & 0 & 0 \\ 0 & 0 & 0 & \cdots & 0 & \Gamma_1 \\ 0 & 0 & 0 & \cdots & \Gamma_1 & 0 \\ \vdots & \vdots & \vdots & \ddots & \vdots & \vdots \\ 0 & \Gamma_1 & 0 & \cdots & 0 & 0 \end{pmatrix},$$

and Γ_q be the level-q left circulant of type $(n_q, n_{q-1}, \cdots, n_1)$ $q > 1$ which has the following form,

$$\Gamma_q = \begin{pmatrix} \Gamma_{q-1} & 0 & \cdots & 0 & 0 \\ 0 & 0 & \cdots & 0 & \Gamma_{q-1} \\ 0 & 0 & \cdots & \Gamma_{q-1} & 0 \\ \vdots & \vdots & \ddots & \vdots & \vdots \\ 0 & \Gamma_{q-1} & \cdots & 0 & 0 \end{pmatrix},$$

where Γ_{q-1} is the level-$(q-1)$ left circulant of type $(n_{q-1}, n_{q-2}, \cdots, n_1)$.

Proposition 1. *A level-q left circulant of type $(n_q, n_{q-1}, \cdots, n_1)$ A_q can be converted to a level-q circulant of type $(n_q, n_{q-1}, \cdots, n_1)$ A'_q by $A'_q = \Gamma_q A_q$. Vice verse, $A_q = \Gamma_q A'_q$.*

Proof. Let A_q be

$$
A_q = \begin{pmatrix}
A_{q-1,1} & A_{q-1,2} & A_{q-1,3} & \cdots & A_{q-1,n_q} \\
A_{q-1,2} & A_{q-1,3} & A_{q-1,4} & \cdots & A_{q-1,1} \\
A_{q-1,3} & A_{q-1,4} & A_{q-1,5} & \cdots & A_{q-1,2} \\
\vdots & \vdots & \vdots & \ddots & \vdots \\
A_{q-1,n_q} & A_{q-1,1} & A_{q-1,2} & \cdots & A_{q-1,n_q-1}
\end{pmatrix},
$$

where $A_{q-1,i}$ is a level-$(q-1)$ left circulant of type $(n_{q-1}, n_{q-2}, \cdots, n_1)$ for each $i \in \{1, 2, \cdots, n_q\}$, then we have

$$
\begin{aligned}
A'_q &= \begin{pmatrix}
\Gamma_{q-1} & 0 & 0 & \cdots & 0 & 0 \\
0 & 0 & 0 & \cdots & 0 & \Gamma_{q-1} \\
0 & 0 & 0 & \cdots & \Gamma_{q-1} & 0 \\
\vdots & \vdots & \vdots & \ddots & \vdots & \vdots \\
0 & \Gamma_{q-1} & 0 & \cdots & 0 & 0
\end{pmatrix}
\times
\begin{pmatrix}
A_{q-1,1} & A_{q-1,2} & A_{q-1,3} & \cdots & A_{q-1,n_q} \\
A_{q-1,2} & A_{q-1,3} & A_{q-1,4} & \cdots & A_{q-1,1} \\
A_{q-1,3} & A_{q-1,4} & A_{q-1,5} & \cdots & A_{q-1,2} \\
\vdots & \vdots & \vdots & \ddots & \vdots \\
A_{q-1,n_q} & A_{q-1,1} & A_{q-1,2} & \cdots & A_{q-1,n_q-1}
\end{pmatrix} \\[2mm]
&= \begin{pmatrix}
\Gamma_{q-1}A_{q-1,1} & \Gamma_{q-1}A_{q-1,2} & \Gamma_{q-1}A_{q-1,3} & \cdots & \Gamma_{q-1}A_{q-1,n_q} \\
\Gamma_{q-1}A_{q-1,n_q} & \Gamma_{q-1}A_{q-1,1} & \Gamma_{q-1}A_{q-1,2} & \cdots & \Gamma_{q-1}A_{q-1,n_q-1} \\
\Gamma_{q-1}A_{q-1,n_q-1} & \Gamma_{q-1}A_{q-1,n_q} & \Gamma_{q-1}A_{q-1,1} & \cdots & \Gamma_{q-1}A_{q-1,n_q-2} \\
\vdots & \vdots & \vdots & \ddots & \vdots \\
\Gamma_{q-1}A_{q-1,2} & \Gamma_{q-1,3}A_{q-1,3} & \Gamma_{q-1}A_{q-1,4} & \cdots & \Gamma_{q-1}A_{q-1,1}
\end{pmatrix} \\[2mm]
&= \begin{pmatrix}
A'_{q-1,1} & A'_{q-1,2} & A'_{q-1,3} & \cdots & A'_{q-1,n_q} \\
A'_{q-1,n_q} & A'_{q-1,1} & A'_{q-1,2} & \cdots & A'_{q-1,n_q-1} \\
A'_{q-1,n_q-1} & A_{q-1,n_q} & A'_{q-1,1} & \cdots & A'_{q-1,n_q-2} \\
\vdots & \vdots & \vdots & \ddots & \vdots \\
A'_{q-1,2} & A'_{q-1,3} & A'_{q-1,4} & \cdots & A'_{q-1,1}
\end{pmatrix},
\end{aligned}
$$

where $A'_{q-1,i} = \Gamma_{q-1}A_{q-1,i}$ for $i \in \{1, 2, \cdots, n_q\}$. So we can get that A'_q is a block circulant of type $(n_q, \prod_{j=1}^{j=q-1} n_j)$. In the similar way, we can derive that $A'_{q-1,i}$ is a block circulant of type $(n_{q-1}, \prod_{j=1}^{j=q-2} n_j)$ for $i \in \{1, 2, \cdots, n_q\}$. Repeating the deriving process, we can obtain that A'_q is a level-q circulant of type $(n_q, n_{q-1}, \cdots, n_1)$. Using the similar proof method, the second part of the proposition can be proved. $\square$

From Proposition 1, the matrix-vector product $C_{left-q}x$ can be written as $\Gamma_q C_q x$ where C_{left-q} is a level-q left circulant and C_q is a level-q circulant. As the matrix-vector product $C_q x$ can be performed using multidimensional FFT computation and the result of $C_{left-q}x$ can be obtained by $\Gamma_q(C_q x)$. Note that left-multiplying a matrix by Γ_q can be implemented with some simple row permutations.

Proposition 2. *Let the element of row $(i_1|i_2)$ and column $(j_1|j_2)$ for the matrix M be $M_{i_1|i_2,j_1|j_2} = f(i_1 \boxplus j_1, i_2 \boxplus j_2), (i_1, j_1 \in \mathbb{F}_2^{k_1}, i_2, j_2 \in \mathbb{F}_2^{k_2})$, then the matrix M is a level-2 left circulant of type $(2^{k_1}, 2^{k_2})$. The time complexity to multiply M with a vector is $3 \cdot (k_1 + k_2) \cdot 2^{k_1+k_2}$.*

Proof. The matrix M is as follows

$$
M = \begin{pmatrix} M_{0|0,0|0} & \cdots & M_{0|0,2^{k_1}-1|2^{k_2}-1} \\ \vdots & \ddots & \vdots \\ M_{2^{k_1}-1|2^{k_2}-1,0|0} & \cdots & M_{2^{k_1}-1|2^{k_2}-1,2^{k_1}-1|2^{k_2}-1} \end{pmatrix}.
$$

Let $M^{(i_1,j_1)}$ be the following square matrix,

$$
M^{(i_1,j_1)} = \begin{pmatrix} M_{i_1|0,j_1|0} & M_{i_1|0,j_1|1} & \cdots & M_{i_1|0,j_1|2^{k_2}-1} \\ M_{i_1|1,j_1|0} & M_{i_1|1,j_1|1} & \cdots & M_{i_1|1,j_1|2^{k_2}-1} \\ \vdots & \vdots & \ddots & \vdots \\ M_{i_1|2^{k_2}-1,j_1|0} & M_{i_1|2^{k_2}-1,j_1|1} & \cdots & M_{i_1|2^{k_2}-1,j_1|2^{k_2}-1} \end{pmatrix}.
$$

From $f(i_1 \boxplus j_1, i_2 \boxplus j_2) = f(i_1 \boxplus j_1, (i_2 \boxplus 1) \boxplus (j_2 \boxminus 1))$, we know that $M_{i_2,j_2}^{(i_1,j_1)} = M_{i_2 \boxplus 1,j_2 \boxminus 1}^{(i_1,j_1)}$ for any i_2, j_2. It means that the successive row moved one place to the left, so $M^{(i_1,j_1)}$ is a left circulant for each i_1, j_1.

Besides, we have

$$
M = \begin{pmatrix} M^{(0,0)} & M^{(0,1)} & \cdots & M^{(0,2^{k_1}-1)} \\ M^{(1,0)} & M^{(1,1)} & \cdots & M^{(1,2^{k_1}-1)} \\ \vdots & \vdots & \ddots & \vdots \\ M^{(2^{k_1}-1,0)} & M^{(2^{k_1}-1,1)} & \cdots & M^{(2^{k_1}-1,2^{k_1}-1)} \end{pmatrix}.
$$

For any i_2, j_2, the elements of row i_2 and column j_2 for $M^{(i_1,j_1)}$ and $M^{(i_1 \boxplus 1,j_1 \boxminus 1)}$ are $f(i_1 \boxplus j_1, i_2 \boxplus j_2)$ and $f((i_1 \boxplus 1) \boxplus (j_1 \boxminus 1), i_2 \boxplus j_2)$, respectively. As

$$
f((i_1 \boxplus 1) \boxplus (j_1 \boxminus 1), i_2 \boxplus j_2) = f(i_1 \boxplus j_1, i_2 \boxplus j_2),
$$

we know that $M^{(i_1,j_1)} = M^{(i_1 \boxplus 1,j_1 \boxminus 1)}$. Therefore M is a level-2 left circulant of type $(2^{k_1}, 2^{k_2})$.

From the result of Subsection 3.1, the time complexity to multiply M with a vector is $3 \cdot (k_1 + k_2) \cdot 2^{k_1+k_2}$. $\square$

Corollary 2. *Let the element of row $(i_1|\cdots|i_m)$ and column $(j_1|\cdots|j_m)$ of the matrix M be*

$$
M_{i_1|\cdots|i_m,j_1|\cdots|j_m} = f(i_1 \boxplus j_1, \cdots, i_m \boxplus j_m),
$$

where $i_t, j_t \in \mathbb{F}_2^{k_t}, 1 \le t \le m$. Then M is a level-m left circulant of type $(2^{k_1}, 2^{k_2}, \cdots, 2^{k_m})$. The time complexity to multiply M with a vector is $3(\sum_{t=1}^{t=m} k_t) \cdot 2^{\sum_{t=1}^{t=m} k_t}$.

Proof. We prove it inductively.

As $m = 2$, M is a level-2 left circulant of type $(2^{k_1}, 2^{k_2})$ according to Proposition 2.

For $m = s > 2$, consider the block matrix $M^{(i_1, j_1)}$,

$$M^{(i_1,j_1)}_{i_2|\cdots|i_s,j_2|\cdots|j_s} = f(i_1 \boxplus j_1, \cdots, i_s \boxplus j_s) = f_{i_1 \boxplus j_1}(i_2 \boxplus j_2, \cdots, i_s \boxplus j_s).$$

We have

$$M = \begin{pmatrix} M^{(0,0)} & M^{(0,1)} & \cdots & M^{(0,2^{k_1}-1)} \\ M^{(1,0)} & M^{(1,1)} & \cdots & M^{(1,2^{k_1}-1)} \\ \vdots & \vdots & \ddots & \vdots \\ M^{(2^{k_1}-1,0)} & M^{(2^{k_1}-1,1)} & \cdots & M^{(2^{k_1}-1,2^{k_1}-1)} \end{pmatrix}.$$

With the proof technique for Proposition 2, we can derive that M is a left block circulant of type $(2^{k_1}, 2^{\sum_{t=2}^{t=s} 2^{k_t}})$. Suppose that the corollary is right when $m = s - 1$, it means that $M^{(i_1,j_1)}$ is a level-$(s-1)$ left circulant of type $(2^{k_2}, \cdots, 2^{k_s})$. So M is a level-s left circulant of type $(2^{k_1}, 2^{k_2}, \cdots, 2^{k_s})$. Then from the conclusion of Subsection 3.1, we get the time to multiply M with a vector is about $3(\sum_{t=1}^{t=m} k_t) \cdot 2^{\sum_{t=1}^{t=m} k_t}$ using multidimensional FFT technique. $\square$

If there are XOR operations with subkey in the evaluation of linear approximation, the XOR operation for k bits can be denoted as k modular additions in $\mathbb{F}_2$, then it can be also transformed to evaluate a linear approximation where only modular additions are involved.

In [11], Collard *et al.* consider that only XOR operation or one modular addition with subkey is included in the partial decryption of linear cryptanalysis. If only one modular addition with subkey is involved, the matrix is level-1 left circulant matrix of type (2^{k_1}). For XOR operation with subkey $f(X \oplus K), X, K \in \mathbb{F}_2^k$, the evaluation of linear approximation can be implemented with matrix-vector product where the matrix is level-k circulant of type $\underbrace{(2, 2, \ldots, 2)}_{k \quad times}$. The two cases belong to the above general scheme in Corollary 2.

4   Multiple ZC Attack on 29-round CAST-256 with FFT

4.1   Description of CAST-256

CAST-256 is designed by Adams [1] and is one of the fifteen candidates in the first round of AES project. It belongs to the CAST family symmetric ciphers which are constructed using the CAST design procedure by Adams [2]. CAST-256 employees the generalized Feistel structure with four 32-bit branches and uses three different round functions. The block size of CAST-256 is 128 bits and the key size could be 128, 160, 192, 224 or 256 bits. The number of rounds is 48 consisting of six forward quad-rounds followed with six reverse quad-rounds. Let's denote the three different round functions as F_1, F_2 and F_3. Exclusive-or, modulo-2^{32} addition, modulo-2^{32} subtraction and left rotation operations are used in all round functions, and they are denoted as $\oplus$, $\boxplus$, $\boxminus$ and $\lll$ respectively. Four 8×32 S-boxes are used in CAST-256's round functions, denoted as S_1, S_2, S_3 and S_4. F_1, F_2 and F_3 are shown in Figure 1 and can be presented as follows:

$$F_1 : I = ((k_m \boxplus I) \lll k_r), \ O = ((S_1[I_1] \oplus S_2[I_2] \boxminus S_3[I_3]) \boxplus S_4[I_4]),$$

$$F_2 : I = ((k_m \oplus I) \lll k_r), \ O = ((S_1[I_1] \boxminus S_2[I_2] \boxplus S_3[I_3]) \oplus S_4[I_4]),$$

$$F_3 : I = ((k_m \boxminus I) \lll k_r), \ O = ((S_1[I_1] \boxplus S_2[I_2] \oplus S_3[I_3]) \boxminus S_4[I_4]),$$

where I is a 32-bit (4-byte) intermediate variable and O is the 32-bit return value of functions F_1, F_2 and F_3. The four bytes of I is denoted as I_1, I_2, I_3, I_4. Details of round functions, F_1, F_2 and F_3, are shown in Figure 1, where k_r and k_m are the 5-bit rotation subkey and the 32-bit masking subkey for each round, respectively. Forward-quad and reverse-quad are shown in Figure 2.

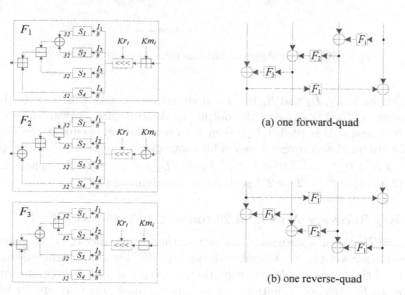

Fig. 1: Round Functions   Fig. 2: Forward-Quad and Reverse-Quad

4.2   ZC Linear Approximations of 24-round CAST-256

In this paper, we utilize the same $24r$ ZC linear approximations given by [7] and also add four rounds before the linear approximations. The details of the ZC linear approximations is illustrated in Figure 3. However, besides the four rounds added before the linear approximations, we can still append one round after the ZC linear approximations due to the power of the FFT technique described in Section 3. Note that, the FFT technique [6] is not suitable to cooperate with multidimensional ZC linear cryptanalysis because normally the key recovery process in such cryptanalysis cannot be denoted as a matrix-vector product. Yet, the model proposed by Bogdanov *et al.* at FSE 2012 [9] works well with the FFT technique. Anyway, the FFT technique enables us to attack 29-round CAST-256, round 9 to round 37.

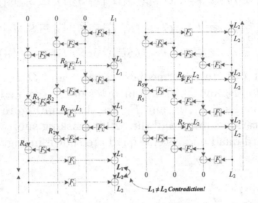

Fig. 3: ZC Linear Approximations over 24-round CAST-256

In Figure 3, L_1, L_2 and $R_i, 0 \le i \le 6$ are nonzero masks and $L_1 \ne L_2$. If the input mask is $(0,0,0,L_1)$ and the output mask after three forward-quads and three reverse-quads is $(0,0,0,L_2)$, then the contradiction occurs at the joint of the forward-quad and reverse-quad. This contradiction makes the linear approximations $(0,0,0,L_1) \xrightarrow{24r} (0,0,0,L_2)$, $L_1 \ne 0, L_2 \ne 0, L_1 \ne L_2$ ZC. There are at most $(2^{32} - 1) \cdot (2^{32} - 2) \approx 2^{64}$ such linear approximations.

4.3   Key Recovery Attack on 29-round CAST-256 with FFT

The more ZC linear approximations we use, the less number of (P, C) pairs are required to recover the key following the multiple ZC linear cryptanalysis model. To control the time complexity of partially encryption and decryption, we will use the ZC linear approximations whose nonzero mask bits only appear at the least significant byte of L_1 and L_2. This gives us $(2^8 - 1) \cdot (2^8 - 2) = 64770$ ZC linear approximations.

As we wrote before, by adding four rounds before and one round after the 24-round ZC linear approximations, we could attack 29-round CAST-256, see Figure 4. Suppose that we have N (P, C) pairs, then the detailed key recovery process is proceeded as Algorithm 1, where we use $P_4[0]$ and $C_4[0]$ to denote the least significant byte of P_4 and C_4. Note that the matrix M after Step 23 in Algorithm 1 is defined as:

$$M(km_{12}|km_{37}, X_2|C_1) = (-1)^{(F_1(kr_{12}|km_{12}, X_2) \diamond L_1) \oplus (F_1(kr_{37}|km_{37}, C_1) \diamond L_2)}.$$

In function F_1, subkeys km_{12} and km_{37} are involved with states X_2 and C_1 through modular addition, then according to Proposition 2, this matrix is a level-2 left circulant of type $(2^{32}, 2^{32})$.

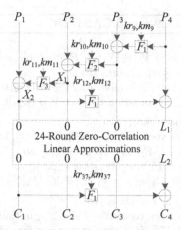

Fig. 4: Attack on 29-round CAST-256

Complexity estimation. The time complexity for Step 4 is about $2^{74} \cdot N \cdot 2/29 \approx N \cdot 2^{70.1}$ encryptions. If we set $\alpha_0 = 2^{-2.7}$ and $\alpha_1 = 2^{-50}$, then $N \approx 2^{123.2}$ and $\tau \approx 2^{15.9}$. So the time complexity for Step 4 is about $2^{193.3}$ encryptions. The time complexity for Step 9 is about $2^{74} \cdot 2^{37} \cdot 2^{112} \cdot 1/29 \approx 2^{218.1}$ encryptions. The time complexity for Step 16 is about $2^{74} \cdot 2^{37} \cdot 2^{10} \cdot 64770 \cdot 2^{80} \cdot 2 \approx 2^{218.0}$ scalar products of two 8-bit binary vectors. The time complexity for Step 21 and 23 is about $2^{74} \cdot 2^{37} \cdot 2^{10} \cdot 64770 \cdot 2^{66} \cdot 2 \approx 2^{204.0}$ scalar products of two 8-bit binary vectors. The time complexity for Step 25 is about $2^{74} \cdot 2^{37} \cdot 2^{10} \cdot 64770 \cdot 2^{64} \cdot 2/29 \approx 2^{197.1}$ encryptions. The time complexity for Step 26 is about $2^{74} \cdot 2^{37} \cdot 2^{10} \cdot 64770 \approx 2^{137.0}$ FFT operations because the matrix M is level-2 left block circulant with left circulant block matrix with type $(2^{32}, 2^{32})$. According to Proposition 3, the time complexity for the matrix-vector product is about $3 \cdot 64 \cdot 2^{64} \approx 2^{71.6}$. If we assume that such an operation equals to one 29-round CAST-256 encryption, then the time complexity for Step 26 is about $2^{208.6}$ encryptions. The time complexity for Step 31, which is the exhaustive search of the right key, is about 2^{206} encryptions. Thus the dominant time complexities for our attack are $2^{218.1}$ encryptions in Step 9 and $2^{218.0}$ scalar products of two 8-bit binary vectors in Step 16. However, compared with one 29-round CAST-256 encryptions, the scalar products of two 8-bit binary vectors can be ignored. Thus, the time complexity of our attack on 29-round CAST-256 is about $2^{218.1}$ encryptions.

The memory requirements of our attack are to store the counter vectors in Algorithm 1. The dominant part is to store V_1 in Step 2. As we have $2^{123.2}$ (P, C) pairs, each element in V_1 should be 16-bit length. So the memory requirements are about 2^{113} bytes.

In summary, 29-round CAST-256 with 256-bit key size could be attacked using multiple ZC linear cryptanalysis with data complexity $2^{123.2}$ known plaintexts, time complexity $2^{218.1}$ encryptions and about 2^{113} bytes.

Algorithm 1. Attack on 29-round CAST-256

---

1   **for** *all 2^{74} values of $kr_9, km_9, kr_{10}, km_{10}$* **do**

2   Allocate a counter vector $V_1[P_1|X_1|C_1|P_4[0]|C_4[0]]$, initialize to 0.

3   **for** *all N plaintext-ciphertext pairs* **do**

4   Compute $X_1 = P_2 \oplus F_2(kr_{10}|km_{10}, P_3 \oplus F_1(kr_9|km_9, P_4))$.

5   Add one to the corresponding counter $V_1[P_1|X_1|C_1|P_4[0]|C_4[0]]$.

6   **for** *all 2^{37} values of kr_{11}, km_{11}* **do**

7   Allocate a counter vector $V_2[X_2|C_1|P_4[0]|C_4[0]]$, initialize to 0.

8   **for** *all 2^{112} values of $P_1|X_1|C_1|P_4[0]|C_4[0]$* **do**

9   Compute $X_2 = P_1 \oplus F_3(kr_{11}|km_{11}, X_1)$.

10          $V_2[X_2|C_1|P_4[0]|C_4[0]] += V_1[P_1|X_1|C_1|P_4[0]|C_4[0]]$.

11   **for** *all 2^{10} values of kr_{12}, kr_{37}* **do**

12   Allocate a counter vector C of size 2^{64}, initialize to 0.

13   **for** *each of 64770 ZC linear approximations* **do**

14   Allocate a counter vector $V_3[X_2|C_1|P_4[0] \diamond L_1|C_4[0] \diamond L_2]$, initialize to 0.

15   **for** *all 2^{80} values of $X_2|C_1|P_4[0]|C_4[0]$* **do**

16   Compute $X_2|C_1|P_4[0] \diamond L_1|C_4[0] \diamond L_2$.

17              $V_3[X_2|C_1|P_4[0] \diamond L_1|C_4[0] \diamond L_2] += V_2[X_2|C_1|P_4[0]|C_4[0]]$.

18   Allocate a counter vector $V_4[X_2|C_1]$, initialize to 0.

19   **for** *all 2^{66} values of $X_2|C_1|P_4 \diamond L_1|C_4 \diamond L_2$* **do**

20   **if** $(P_4[0] \diamond L_1) \oplus (C_4[0] \diamond L_2) == 0$ **then**

21                $V_4[X_2|C_1] += V_2[X_2|C_1|P_4[0] \diamond L_1|C_4[0] \diamond L_2]$.

22   **if** $(P_4[0] \diamond L_1) \oplus (C_4[0] \diamond L_2) == 1$ **then**

23                $V_4[X_2|C_1] -= V_2[X_2|C_1|P_4[0] \diamond L_1|C_4[0] \diamond L_2]$.

             // $M(km_{12}|km_{37}, X_2|C_1) = (-1)^{(F_1(kr_{12}|km_{12}, X_2) \diamond L_1) \oplus (F_1(kr_{37}|km_{37}, C_1) \diamond L_2)}$

             // *M is a level-2 left circulant of type $(2^{32}, 2^{32})$.*

24   **for** *all 2^{64} values of $X_2|C_1$* **do**

               // *Compute M's the first column , then M is known.*

25                $M(0|0, X_2|C_1) = (-1)^{(F_1(kr_{12}|0, X_2) \diamond L_1) \oplus (F_1(kr_{37}|0, C_1) \diamond L_2)}$

26   Evaluate the vector $\epsilon = M \cdot V_3[X_2|C_1]$ using FFT technique.

27              $C += (\epsilon/N)^2$

28   **for** *all 2^{64} values of $km_{12}|km_{37}$* **do**

29   **if** $C_{km_{12}|km_{37}} < \tau$ **then**

30   Current value of $kr_{9,10,11,12,37}$ and $km_{9,10,11,12,37}$ is a key candidate.

31   Exhaustive search the right key.

---

5   Conclusion

In this paper, a more general scheme to use FFT technique in the key recovery phase is put forward, with which we can reduce the time complexity of key recovery phase when multiple modular additions with subkeys are involved in the partial encryption/decryption. We hope that this scheme would be useful for future cryptanalysis against block ciphers with modular addition with subkeys. As an application, we use the 24-round ZC linear distinguisher proposed at ASIACRYPT 2012 and improve the zero-correlation attack on CAST-256 by one round. Our attack is the best key recovery attack (without weak key assumption) on CAST-256.

Acknowledgments. This work has been supported by the National Basic Research 973 Program of China under Grant No. 2013CB834205, the National Natural Science Foundation of China under Grant Nos. 61133013, 61103237, the Program for New Century Excellent Talents in University of China under Grant No. NCET-13-0350, as well as the Interdisciplinary Research Foundation of Shandong University of China under Grant No. 2012JC018.

References

1. Adams, C.M.: The CAST-256 Encryption Algorithm. In: AES Proposal (1998)
2. Adams, C.M.: Constructing Symmetric Ciphers Using the CAST Design Proce-dure. Designs, Codes and Cryptography **12**(3), 283–316 (1997)
3. Blondeau, C., Bogdanov, A., Wang, M.: On the (In)Equivalence of Impossible Differential and Zero-Correlation Distinguishers for Feistel- and Skipjack-Type Ciphers. In: Boureanu, I., Owesarski, P., Vaudenay, S. (eds.) ACNS 2014. LNCS, vol. 8479, pp. 271–288. Springer, Heidelberg (2014)
4. Blondeau, C., Nyberg, K.: New Links Between Differential and Linear Cryptanal-ysis. In: Johansson, T., Nguyen, P.Q. (eds.) EUROCRYPT 2013. LNCS, vol. 7881, pp. 388–404. Springer, Heidelberg (2013)
5. Bogdanov, A., Boura, C., Rijmen, V., Wang, M., Wen, L., Zhao, J.: Key Difference Invariant Bias in Block Ciphers. In: Sako, K., Sarkar, P. (eds.) ASIACRYPT 2013, Part I. LNCS, vol. 8269, pp. 357–376. Springer, Heidelberg (2013)
6. Bogdanov, A., Geng, H., Wang, M., Wen, L., Collard, B.: Zero-Correlation Linear Cryptanalysis with FFT and Improved Attacks on ISO Standards Camellia and CLEFIA. In: Lange, T., Lauter, K., Lisoněk, P. (eds.) SAC 2013. LNCS, vol. 8282, pp. 306–323. Springer, Heidelberg (2014)
7. Bogdanov, A., Leander, G., Nyberg, K., Wang, M.: Integral and Multidimensional Linear Distinguishers with Correlation Zero. In: Wang, X., Sako, K. (eds.) ASI-ACRYPT 2012. LNCS, vol. 7658, pp. 244–261. Springer, Heidelberg (2012)
8. Bogdanov, A., Rijmen, V.: Linear Hulls with Correlation Zero and Linear Crypt-analysis of Block Ciphers. Designs, Codes and Cryptography **70**(3), 369–383 (2014)
9. Bogdanov, A., Wang, M.: Zero Correlation Linear Cryptanalysis with Reduced Data Complexity. In: Canteaut, A. (ed.) FSE 2012. LNCS, vol. 7549, pp. 29–48. Springer, Heidelberg (2012)
10. Chose, P., Joux, A., Mitton, M.: Fast Correlation Attacks: An Algorithmic Point of View. In: Knudsen, L.R. (ed.) EUROCRYPT 2002. LNCS, vol. 2332, pp. 209–221. Springer, Heidelberg (2002)

176   L. Wen et al.

11. Collard, B., Standaert, F.-X., Quisquater, J.-J.: Improving the Time Complexity of Matsui's Linear Cryptanalysis. In: Nam, K.-H., Rhee, G. (eds.) ICISC 2007. LNCS, vol. 4817, pp. 77–88. Springer, Heidelberg (2007)
12. Davis, P.J.: Circulant Matrices, pp. 176–191. Wiley-Interscience, Chichester (1979)
13. Lu, Y., Meier, W., Vaudenay, S.: The Conditional Correlation Attack: A Practical Attack on Bluetooth Encryption. In: Shoup, V. (ed.) CRYPTO 2005. LNCS, vol. 3621, pp. 97–117. Springer, Heidelberg (2005)
14. Nakahara Jr., J., Rasmussen, M.: Linear Analysis of Reduced-round CAST-128 and CAST-256. In: SBSEG 2007, pp. 45–55 (2007)
15. Naya-Plasencia, M.: Cryptanalysis of Achterbahn-128/80. In: Biryukov, A. (ed.) FSE 2007. LNCS, vol. 4593, pp. 73–86. Springer, Heidelberg (2007)
16. Seki, H., Kaneko, T.: Differential Cryptanalysis of CAST-256 Reduced to Nine Quad-rounds. IEICE Transactions on Fundamentals of Electronics Communications and Computer Sciences $E84-A(4)$, 913–918 (2001)
17. Soleimany, H., Nyberg, K.: Zero-Correlation Linear Cryptanalysis of Reduced-Round LBlock. IACR Cryptology ePrint Archive, 2012:570 (2012). http://eprint.iacr.org/2012/570
18. Todo, Y.: FFT-Based Key Recovery for the Integral Attack. IACR Cryptology ePrint Archive, 2014:187 (2014). http://eprint.iacr.org/2014/187
19. Wagner, D.: The Boomerang Attack. In: Knudsen, L. (ed.) FSE 1999. LNCS, vol. 1636, pp. 156–170. Springer, Heidelberg (1999)
20. Wang, M., Wang, X., Hu, C.: New Linear Cryptanalytic Results of Reduced-Round of CAST-128 and CAST-256. In: Avanzi, R., Keliher, L., Sica, F. (eds.) SAC 2008. LNCS, vol. 5381, pp. 429–441. Springer, Heidelberg (2009)
21. Wen, L., Wang, M.: Integral Zero-Correlation Distinguisher for ARX Block Cipher, with Application to SHACAL-2. In: Susilo, W., Mu, Y. (eds.) ACISP 2014. LNCS, vol. 8544, pp. 454–461. Springer, Heidelberg (2014)
22. Wen, L., Wang, M., Bogdanov, A.: Multidimensional Zero-Correlation Attacks on Lightweight Block Cipher HIGHT: Improved Cryptanalysis of an ISO Standard. Information Processing Letters $114(6)$, 322–330 (2014)
23. Wen, L., Wang, M., Bogdanov, A.: Multidimensional Zero-Correlation Linear Cryptanalysis of E2. In: Pointcheval, D., Vergnaud, D. (eds.) AFRICACRYPT 2004. LNCS, vol. 8469, pp. 147–164. Springer, Heidelberg (2014)

Side Channel Analysis - II

Cryptanalysis of the Double-Feedback XOR-Chain Scheme Proposed in Indocrypt 2013

Subhadeep Banik[1]($\boxtimes$), Anupam Chattopadhyay[2], and Anusha Chowdhury[3]

[1] DTU Compute, Technical University of Denmark, 2800 Kongens Lyngby, Denmark
subb@dtu.dk
[2] School of Computer Engineering, Nanyang Technological University,
Singapore, Singapore
[3] Department of Computer Science and Engineering, Indian Institute of Technology
Kanpur, Kanpur, India
anushac@iitk.ac.in

Abstract. For any modern chip design with a considerably large portion of logic, design for test (DFT) is a mandatory part of the design process which helps to reduce the complexity of testing sequential circuits. Scan-chains are one of the most commonly-used DFT techniques. However, the presence of scan-chains makes the device vulnerable to scan-based attacks from a cryptographic point of view. Techniques to cryptanalyze stream ciphers like Trivium, with additional hardware for scan-chains, are already available in literature (Agrawal et al. Indocrypt 2008). Such ideas were extended to more complicated stream ciphers like MICKEY 2.0 in the paper by Banik et al. at Indocrypt 2013. In this paper, we will look at the Double-Feedback XOR-Chain based countermeasure that was proposed by Banik et al. in Indocrypt 2013, to protect scan-chains from such scan-based attacks. We will show that such an XOR-Chain based countermeasure is vulnerable to attack. As an alternative, we propose a novel countermeasure based on randomization of XOR gates, that can protect scan-chains against such attacks.

Keywords: Scan-based attack · MICKEY 2.0 · Double-Feedback XOR-Chain scheme

1 Introduction

In today's world, cryptographic algorithms are being implemented in hardware in order to meet high throughput requirements. These cryptographic chips and associated systems need to be tested at fabrication and in-field. DFT(Design for Testability) techniques have been used since the early days of electronic data processing equipment. Scan-based DFT is the most widely used DFT scheme for integrated circuit testing as it is simple and yields high fault coverage. The advantage of scan-based testing is that it provides full observability and controllability of the internal nodes of the IC. Moreover, the scan-chains can be connected to an external five-pin JTAG interface during chip packaging to provide

© Springer International Publishing Switzerland 2014
W. Meier and D. Mukhopadhyay (Eds.): INDOCRYPT 2014, LNCS 8885, pp. 179–196, 2014.
DOI: 10.1007/978-3-319-13039-2_11

on-chip debug capability in-field. On-chip debug capability eases the development and maintenance of software running on the chips and is very important for microprocessors. Scan tests cover stuck-at-faults, caused by manufacturing problems. From a security point of view, the flip-side of this design paradigm is that using scan in secure chip, for instance in a cryptographic one, can reduce the security level, which makes the technique unacceptable. Scan-based attacks have already been reported against block ciphers like AES [25] and DES [26] and stream-ciphers like RC4 [22]. The eStream cipher Trivium [11] was successfully cryptanalyzed in [2], using scan-based attack. Thereafter, in [4] a scan-based side channel attack on the stream cipher MICKEY 2.0 [5] was presented.

Some countermeasures have been proposed in literature to secure the scan technique. As the traditional Scan DFT scheme can compromise the secret key, the idea of a secure scan DFT was proposed in [25] where neither the secret key nor the testability is compromised. A single feedback XOR-Chain scheme was proposed in [2] to protect the scan-chain from attack. However, in [4] it was shown that by carefully observing the scanned out vector, the attacker can find out the positions where the XOR gates have been inserted. As a countermeasure, a Double-Feedback XOR-Chain scheme was proposed. So, testability vs security is indeed non-trivial. In this paper we will show that the Double-Feedback XOR-Chain scheme that was proposed in [4] is also vulnerable to attack.

1.1   Scan-Based Attack

On one hand, scan-based test is a powerful test technique. On the other hand, it is an equally powerful attack tool because intermediate results can be scanned out and analyzed. The scan-chain can be used as a back door for accessing secret information, thereby jeopardizing the overall security.

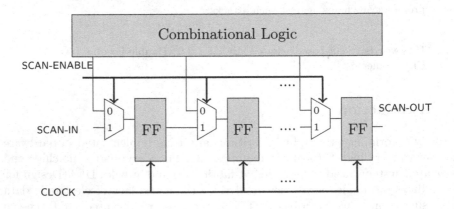

Fig. 1. Example of a Scan-chain using Multiplexers

As can be seen in Figure 1, a scan-based test involves construction of one or more scan-chains in a chip by connecting the internal registers and flip-flops of a device and by making either ends of the chain available to the boundary scan interface, via the SCAN-IN and SCAN-OUT ports. During the testing/SCAN phase, all flip-flops are disconnected from the combinatorial digital logic of the device and connected in single or multiple connected chains. This is done by placing a multiplexer in front of the D input of each flip-flop controlled by the SCAN-ENABLE signal. In normal mode of operation, the SCAN-ENABLE signal is set to 0, so the flip-flop accepts the D-input and the device behaves normally. In test mode, the SCAN-ENABLE signal is set to 1 and in this event the flip-flop accepts the SCAN-IN input. However, it is assumed that an outsider does not have any knowledge of the structure of the scan-chain that ties the flip-flops of the device together. The flip-flops in a scan-chain are generally not connected according to their positions in their respective registers. Rather, a Computer-aided tool optimizes the scan-chain according to the physical locations of the individual flip-flops.

The most commonly employed attack model used in literature is one in which the attacker knows the algorithmic design details of the cryptosystem being implemented in the device and in which the SCAN-ENABLE, SCAN-IN, SCAN-OUT pins can be controlled by him. That is to say, he has the ability to run the device under normal mode or test mode interchangeably. He can alter the public variables, scan-in values, stop the normal mode of operation of the device after any given number of clock rounds, drive the device into test mode and then scan-out the contents of the flip-flops in the scan-chain. The only unknown entity for the attacker, is the arrangement of the state bits of the cryptosystem inside the scan-chain, and the task of cryptanalysis is to determine this function. Any scan-based attack usually proceeds in two phases:

(a) Pre-processing stage → In this stage, the attacker performs various tests on the device to gain information about the structure of the scan-chain and deduce the internal correspondence of each flip-flop in the scan-chain with each flip-flop in the design of the cryptosystem implemented in the device. In other words, the attacker attempts to ascertain the position of each bit in the internal state of the cryptosystem in the scan-chain. Once the correspondence has been found out, the adversary proceeds to the online stage.

(b) Online stage → The adversary lets the device get initialized with some unknown Key and IV, and halts the device at some suitable clock round. He then scans out the content of the flip-flops in the scan-chain through the SCAN-OUT port and thereby gets a vector (call it $\mathbf{V}$). Since the position of each bit of the internal state in the scan-chain is now known to him, he is able to reconstruct the internal state of the cipher from $\mathbf{V}$ and this completes the attack. For further details, refer to [4].

1.2   Protecting Scan-Chains

The Flipped-Scan technique to protect scan-chains was proposed in [22]. This involved placing inverters at random points in the scan-chain. Security stemmed from the fact that an adversary could not guess the number and positions of the inverters. Hence an attacker would not be able to reconstruct the internal state of any cryptosystem from the scanned out vector.

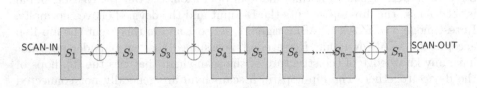

Fig. 2. The XOR-Chain scheme proposed in [2]

This technique was cryptanalyzed in [2] using a technique called RESET attack. It was shown that if all flip-flops in the scan-chain are initially RESET, then the positions of the inverters can be completely determined by the $0 \to 1$ and $1 \to 0$ transitions in the scanned-out vector. As an alternative, the XOR-Chain structure was proposed in [2] (See Figure 2). The technique involves placing XOR gates at random points of the chain. Security again stems from the fact that an adversary is unable to guess the number and positions of the XOR gates.

It was shown in [4] that the XOR-Chain structure would be vulnerable to attack if primitives like FDP, FDPE, FDS (of the Xilinx Virtex 6 library) were used in the scan-chain. The INIT value of these primitives is 1 by default, i.e., a GLOBAL SET/RESET (GSR) signal would set their states to 1. In [4], a detailed procedure is outlined as to how the positions of the XOR gates can be determined if the flip-flops of the scan-chain are initially SET. In the same paper, the authors proposed the Double-Feedback XOR-Chain structure that would resist both the aforementioned attacks (See Figure 3). Its structure is similar to the XOR-Chain scheme, except for the fact that the outputs of two successive flip-flops are fed back to each XOR gate. Security again comes from the fact that an adversary can not guess the number and positions of the XOR gates. In this paper we propose a generic attack strategy that can be used to cryptanalyze both the XOR-Chain and Double-Feedback XOR-Chain structures. As a countermeasure we propose the rXOR-Chain countermeasure scheme.

1.3   Contribution and Organization of the Paper

We will propose a generic strategy to attack the both the XOR-Chain scheme [2] and the Double-Feedback XOR-Chain scheme that was proposed in [4]. In this paper, to explain the details of the attack algorithm, we will concentrate on the Double-Feedback XOR-Chain structure. Our contribution is twofold:

- We will show that the attacker can find out the positions of the XOR gates and also the position of each internal state bit in the scan-chain. The attacker can use this information to reconstruct the internal state of any cryptosystem from the scanned out vector (see Section 3.1) and thus this is as good as breaking the cryptosystem.
- As an alternative, we propose a novel countermeasure based on randomization of the XOR gates, that can protect scan-chains against such attacks.

The organization of the paper is as follows. In Section 2 we will give some details about the Double-Feedback XOR-Chain scheme that was proposed in [4] to prevent scan-based side channel attacks. In Section 3, we will outline the details of the attack against this structure. We will also give a runtime analysis for the algorithm which is used to attack the previous countermeasure. In Section 4, we propose randomization of the XOR gates in order to protect scan-chains. Furthermore, we give robustness analysis and overhead analysis for this new architecture. Section 5 concludes the paper.

2   The Double-Feedback XOR-Chain Scheme

We assume that there are n flip-flops in the scan-chain. We also define the sequence a_i, $(1 \leq i \leq n)$ over $GF(2)$ as follows. If there is an XOR gate between the $(i-1)^{th}$ and the ith flip-flop then we assign the variable $a_i = 1$ else we assign $a_i = 0$. Therefore, if a_i is 1 $(\forall \, i \in [1, n-1])$, then the output of the ith and the $(i+1)^{th}$ flip-flop would be fed back to the XOR gate placed in front of the ith flip-flop (see Figure 3). For $i = n$, (i.e. for the last flip-flop in the chain) $a_n = 0$. This is the "Double-Feedback XOR-Chain" structure proposed in [4].

The state of the ith flip-flop $(1 \leq i \leq n)$ at clock round t $(t \geq 0)$ after the SCAN-ENABLE signal is asserted, is given by the symbol S_i^t. The tth $(t \geq 1)$ round input to the scan-chain is given as x_t. Due to the structure of the Double-Feedback XOR-Chain, the flip-flops get updated by the following recursive equation (See Figure 3).

$$S_i^t = \begin{cases} S_{i-1}^{t-1} \oplus a_i \cdot (S_i^{t-1} \oplus S_{i+1}^{t-1}), & \text{if} \quad 1 < i < n, \\ S_{i-1}^{t-1}, & \text{if} \quad i = n, \\ x_t \oplus a_i \cdot (S_i^{t-1} \oplus S_{i+1}^{t-1}), & \text{if} \quad i = 1. \end{cases} \quad (1)$$

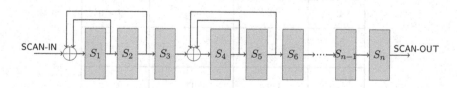

Fig. 3. Diagram of the Double Feedback XOR-Chain scheme proposed in [4]

3   Cryptanalysis

The goal of the attacker is to determine the value of the vector $[a_1, a_2, \ldots, a_n]$ and also deduce the correspondence between each flip-flop in the scan-chain with the individual state bits of the cryptosystem. We will show in Section 3.1, this is sufficient for the attacker to deduce the internal state of any cryptosystem. We will begin with the following Lemma.

Lemma 1. *Let S_i denote the ith flip-flop in the scan-chain. Consider the first two flip-flops. It can be shown that the attacker can easily find out a_1 and the correspondence of S_1 and S_2 with the internal state bit by appropriate assertion of the* SCAN-ENABLE *signals and scanned in vector.*

Proof. n order to attack the Double-Feedback XOR-Chain scheme, the attacker should follow the steps given below.

1. RESET all the flip-flops in the chain.
2. Assert the SCAN-ENABLE signal and scan-in a test vector **X** of appropriate length.
3. After the entire vector **X** has been scanned-in the attacker de-asserts the SCAN-ENABLE signal and allows the device to run in normal mode and observes the output that the device produces.
4. The attacker then makes deductions on the basis of the observed output.

After resetting the chain, the attacker enters the unit length vector 1 via SCAN-IN port. As can be seen in Figure 4, at time $t = 1$, the flip-flop S_1 will get the value 1 irrespective of the value of a_1 (Note that although we have shown XOR gates between all flip-flops in the diagram, $a_i = 0$ implies that the corresponding gate is absent).

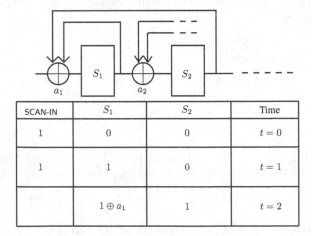

SCAN-IN	S_1	S_2	Time
1	0	0	$t = 0$
1	1	0	$t = 1$
	$1 \oplus a_1$	1	$t = 2$

Fig. 4. First two flip-flops in the scan-chain

This will give rise to an internal state of a cryptosystem where only one internal state bit is 1 and all the other bits are 0. As stated earlier, the attacker stops the SCAN mode and invokes the normal mode at this point of time. Now, the cryptosystem is executed in the normal operational mode on this internal state to generate an output vector $\mathbf{Y}_1$ of appropriate length.

We used the stream cipher MICKEY 2.0 to carry out experiments and check whether the internal state has a one-to-one correspondence with the generated keystream. Every time a different internal state bit was made 1 and all the other internal state bits were kept 0. Since the keystream generating algorithm of MICKEY 2.0 is public (see Appendix A), we were able to generate keystream vectors for each of these special internal states and tabulate them. Now the attacker would simply compare the vector $\mathbf{Y}_1$ with the entries of the table. The entry where a match is obtained reveals the state bit which is at the head of the scan-chain.

In the next round of experiments, the attacker again resets the chain and inputs the all 1 vector of length 2 via the SCAN-IN port. Since the chain was initially reset, at time $t = 2$, the flip-flop S_2 gets value 1, irrespective of the value of a_2 (see Figure 4). At this point of time, the logic in flip-flop S_1 is algebraically equal to $1 \oplus a_1$. This gives rise to two situations, the value of S_1 at $t = 2$ is therefore equal to:

$$\begin{cases} 0, & \text{if} \quad a_1 = 1, \\ 1, & \text{if} \quad a_1 = 0, \end{cases}$$

This will give rise to an internal state of a cryptosystem where only either one or two internal state bits are 1 and all the rest 0. As before, the attacker stops the SCAN mode of operation at this point, invokes the normal operational mode and generates the output vector $\mathbf{Y}_2$.

At this point the attacker knows that the initial state at the beginning of the normal mode contains at most two ones. The state bit corresponding to S_1 may be either zero or one and one other bit, that corresponding to S_2 is definitely 1. Again taking MICKEY 2.0, as an example, the attacker prepares two sets of internal states:

1. The first in which the state bit corresponding to S_1 is 0, all other state bits are zero except one state bit which is one. If the internal state consists of n bits, this will give rise to $n - 1$ internal states.
2. The second set is the same as the first except that the state bit corresponding to S_1 is set to 1. This gives rise to another $n - 1$ internal states.

As before the attacker generates $2n - 2$ keystream vectors corresponding to each of the internal states and tabulates them. Thereafter the attacker compares the vector $\mathbf{Y}_2$ with the entries of this table. A match will therefore reveal both the values of a_1 and the state bit corresponding to S_2. The process has been explained pictorially in Figure 5.   □

We will now prove that the attacker can, in a manner similar to the procedure outlined above, proceed to deduce the values of the remaining $a_i's$ and the correspondence of the remaining flip-flops $S_3, S_4, \ldots, S_n$ with the internal state bits

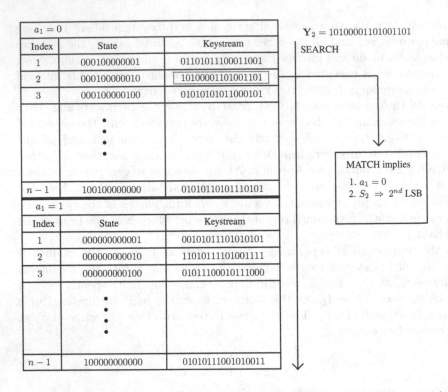

Fig. 5. Searching Table to find a_1 and correspondence of S_2 (An example)

of the cryptosystem. Central to the proof, is an induction step that computes the value of a_i and the correspondence of the flip-flop S_{i+1} with the prior knowledge of the values of $a_1, a_2, \ldots, a_{i-1}$ and the correspondence to the internal state bits of the flip-flops $S_1, S_2, \ldots, S_i$. We formalize the proof in Theorem 1.

Theorem 1. *At any point of time, if the values of $a_1, a_2, \cdots a_{i-1}$ and the internal state bits corresponding to $S_1, S_2, \ldots, S_i$ are already known, then the value of a_i and the state bit corresponding to S_{i+1} can be found out efficiently. Hence by Mathematical Induction on i, the attacker can deduce the values of all the $a_i's$ and the correspondence to the internal state bits of all the flip-flops in the scan-chain.*

Proof. or $i = 1$, the above statement is obviously true and the proof is given in Lemma 1. As before the attacker employs a strategy of resetting the scan-chain, enabling SCAN mode, scanning in an input vector **X** of his choice, and finally disabling SCAN mode and obtaining the output produced by the device. The input vector chosen by the attacker in this case is the all 1 vector of length $i + 1$.

Now as per the induction step, let us assume that the attacker knows the values of $a_1, a_2, \ldots, a_{i-1}$ and the internal correspondence of $S_1, S_2, S_3, \ldots, S_i$. If the all one vector of length of $i + 1$ is scanned in, then the knowledge of

$a_1, a_2, \ldots, a_{i-1}$ allows the attacker to calculate the state of scan-chain at time $t = i$. At $t = i$, S_i holds the logic 1 as the chain was initially RESET. The logic values at flip-flops S_j ($\forall j < i$) can be computed easily via Equation (1). For conciseness, let us denote by the symbols s_j the logic states of the flip-flops S_j at time $t = i$ (please refer to Figure 6).

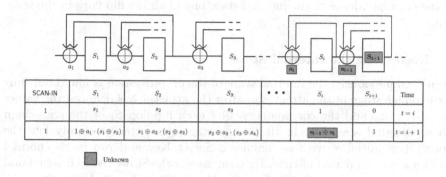

SCAN-IN	S_1	S_2	S_3	$\cdots$	S_i	S_{i+1}	Time
1	s_1	s_2	s_3		1	0	$t = i$
1	$1 \oplus a_1 \cdot (s_1 \oplus s_2)$	$s_1 \oplus a_2 \cdot (s_2 \oplus s_3)$	$s_2 \oplus a_3 \cdot (s_3 \oplus s_4)$		$s_{i-1} \oplus a_i$	1	$t = i+1$

Unknown

Fig. 6. First $i + 1$ flip-flops in the scan-chain

At $t = i + 1$, by applying Equation (1), the attacker can compute the logic value at all the flip-flops $S_1, S_2, \ldots, S_{i-1}, S_{i+1}$. S_{i+1} holds the logic 1. The algebraic expression for the logic at S_i is however equal to $s_{i-1} \oplus a_i$, and since a_i is unknown to the attacker at this point the attacker can not conclusively identify the logic state at this flip-flop.

Now at this point the attacker de-asserts the SCAN-ENABLE signal and drives the device into normal operational mode and records the output vector $\mathbf{Y}_i$. At the point of time that the device is driven into operational mode, the internal state of the cryptosystem would look as follows

1. The internal state bit corresponding to the flip-flop S_{i+1} would have the logic value 1.
2. The internal state bits corresponding to the flip-flop S_j ($\forall j < i$) would equal certain algebraic expressions in $s_1, s_2, \ldots, s_{i-1}$ and $a_1, a_2, \ldots, a_{i-1}$. Since the attacker knows the value of all these variables, he can compute their values.
3. The internal state bit corresponding to the flip-flop S_i would have the logic value $s_{i-1} \oplus a_i$. This would be equal to s_{i-1} if $a_i = 0$ and $1 \oplus s_{i-1}$ otherwise.
4. The internal state bits corresponding to the rest of the flip-flops S_j ($\forall j > i+1$, whose correspondences to internal bits are undetermined at this point) would all be equal to zero.

As before the attacker prepares two sets of internal states. In the first set the state bits corresponding to S_j ($\forall j < i$) are set to the values the attacker computes as described above. The state bit corresponding to S_i is set to s_{i-1}, and the remaining bits are set to zero except one bit which is set to 1, This gives rise to $n - i$ internal states. The second set is the same as the first except for the fact that the bit corresponding to S_i is set to $1 \oplus s_{i-1}$. The attacker computes the output generated by each of these $2n - 2i$ states and tabulates them. Thereafter the attacker compares the vector $\mathbf{Y}_i$ with each of these entries. A match will reveal the internal correspondence of S_{i+1} as well as the the value of a_i.

Proceeding via the principle of Mathematical Induction on the variable i, we can see that the attacker would manage to deduce the values of all the $a_i's$ and the correspondence to the internal state bits of all the flip-flops in the scan-chain. □

3.1   Recovering the Internal State

As outlined previously, the internal state of the cryptosystem is found out during the online stage of the attack, i.e. after the attacker has deduced the values of $a_1, a_2, \ldots, a_n$ and the correspondence of each flip-flop S_i of the scan-chain with an internal state bit. In the online stage, the attacker simply lets the cryptosystem initialize with an unknown Secret Key and run in the normal mode for a certain period of time. He then asserts the SCAN-ENABLE signal and scans out the contents of the flip-flops via the SCAN-OUT port. Using the notation of [4], let $Y = [y_1, y_2, \ldots, y_n]^T$ be the first n scanned out bits. Also let $S_t = [S_1^t, S_2^t, \ldots, S_n^t]^T$, denote the state of the scan-chain t clock rounds after the assertion of the SCAN-ENABLE signal. The task at hand is to find out S_0 (the state of the scan-chain at the time of assertion of the SCAN-ENABLE signal) from the knowledge of Y. Since the attacker already knows the position of each internal state bit in the scan-chain, once he knows the value of S_0 he can easily reconstruct the internal state. Now as per Equation (1), if all the inputs x_i are equal to 0, then S_t is related to S_{t-1}:

$$S_t = A \cdot S_{t-1}$$

where A is the tridiagonal matrix given by

$$A = \begin{pmatrix} a_1 & a_1 & 0 & 0 & 0 & \cdots & 0 & 0 \\ 1 & a_2 & a_2 & 0 & 0 & \cdots & 0 & 0 \\ 0 & 1 & a_3 & a_3 & 0 & \cdots & 0 & 0 \\ 0 & 0 & 1 & a_4 & a_4 & \cdots & 0 & 0 \\ \vdots & \vdots & \vdots & \vdots & \vdots & \ddots & \vdots & \vdots \\ 0 & 0 & 0 & 0 & 0 & \cdots & 1 & 0 \end{pmatrix}$$

This gives us the identity $S_t = A^t \cdot S_0$. Note that once the attacker computes the values of $a_1, a_2, \ldots, a_n$ he can construct the matrix A. Let $\mathbf{e}_n = [0, 0, \ldots, 0, 1]$. Now since $y_i = S_n^{i-1}$ and since $S_n^{i-1} = \mathbf{e}_n \cdot S_{i-1} = \mathbf{e}_n \cdot A^{i-1} \cdot S_0$ for all $1 \leq i \leq n$, we can write the following identity:

$$Y = [y_1, y_2, \ldots, y_n]^T = \begin{pmatrix} \mathbf{e}_n \cdot I_n \\ \mathbf{e}_n \cdot A \\ \vdots \\ \mathbf{e}_n \cdot A^{n-1} \end{pmatrix} \cdot S_0 = \mathcal{C} \cdot S_0.$$

Here $\mathcal{C}$ is the matrix given by $[\mathbf{e}_n \cdot I_n, \ \mathbf{e}_n \cdot \mathcal{A}, \ \ldots \ , \ \mathbf{e}_n \cdot \mathcal{A}^{n-1}]^T$. It was proven in [4, Theorem 3], that the matrix $\mathcal{C}$ is invertible and so we can easily calculate $\mathcal{S}_0 = \mathcal{C}^{-1} \cdot Y$. This concludes the attack.

3.2 Runtime Analysis for the Attack

A strategy to attack the Double-Feedback XOR-Chain scheme has been outlined above. To compute the complexity of the algorithm, note that every time the attacker compares the scanned out vector $\mathbf{Y_i}$ with a certain number of entries and a match is found, he comes to know the corresponding a_i and S_{i+1}. Now the number of entries for each case is just a polynomial in n (for the ith case, $2n - 2i$ comparisons are done), where n is the number of flip-flops in the scan-chain. So, to find $(a_1, a_2, \ldots, a_{n-1})$ and the correspondence for $(S_1, S_2, \ldots S_n)$, the total time required is a polynomial in n and is roughly of the order of n^2. Therefore, the runtime complexity of the algorithm given in the first part of Section 3 is $O(n^2)$.

Thereafter in order to find the internal state of the cryptosystem the attacker first needs to compute the matrix $\mathcal{C}$ and then its inverse. Computation of the matrix $\mathcal{C}$ requires the computation of the matrices $A, A^2, A^3, \ldots, A^{n-1}$. Even if the attacker employs a naive $O(n^3)$ approach to multiply 2 matrices, computing $\mathcal{C}$ should not take more than $O(n^4)$ time. Inverting $\mathcal{C}$ using Gaussian elimination also takes $O(n^3)$ time. Thus the complexity of the algorithm is $O(n^4)$.

Note that the advantage of this polynomial time algorithmic strategy is that it can be used to cryptanalyze any cryptosystem in general. For instance, it can be applied to stream ciphers like Grain, Trivium etc. In this paper we have taken MICKEY 2.0 as an example to illustrate the procedure of the attack. Also it has been assumed in the paper that the scan-chain is initially reset. However if the scan-chain is implemented with primitives like FDP, FDPE, FDS which are set to 1 on application of a GSR signal, one can design a similar attack strategy to cryptanalyze such systems in which the attacker would scan-in all 0 vectors in place of all 1's.

4   Countermeasure : Randomization of XOR Gates

A technique of preventing scan-based attack is presented in [25] via introducing a so-called *secure* mode of operation. During secure mode, user is not permitted to scan in or scan out. In the same paper, it was also noted that the security of the scheme completely relies on the preserving the integrity of keeping the enable_scan_in/out signals, which can be hijacked by fault attacks. To protect it against such possibility, multi-stage buffer between the test inputs and the crypto core is proposed. Despite these protections, it is recently reported [3] that, an AES implementation is vulnerable in presence of such mode-reset countermeasure. In [3], the mapping between the scan cells and the AES input words are identified establishing an attack. No countermeasure is provided.

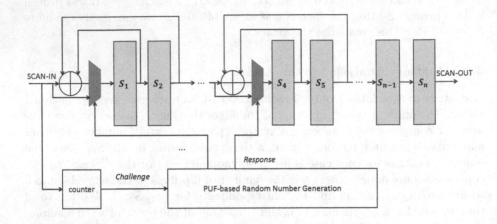

Fig. 7. rXOR-Chain Schematic

To hide the mapping between the scan cells and the actual variables of a cipher is what drove the previous single-feedback and Double-Feedback XOR-Chain schemes. As this is also falling prey to cryptanalysis, as shown in the previous section, we move towards a further secure architecture, named as random XOR-Chain (**rXOR-Chain**) structure.

The basic idea of this architecture is shown in the Figure 7. The Double-Feedback XOR-Chain as proposed in the previous work is retained. On top of that, we introduce a multiplexer at the end of every Double-Feedback XOR gate. The multiplexer receives either the previous input or the input from the XOR gate depending on a control bit. The control bits for each of the multiplexer comes from an internal key, which is fixed at the design time. For randomization of the XOR-Chain structure, a random internal key is needed. This can be achieved by two different techniques.

- Utilizing the intrinsic randomness of the final hardware implementation to establish a Physically Unclonable Function (PUF). PUF is a physical entity that is embodied in a physical structure and is easy to evaluate but hard to predict. When a physical stimulus is applied to the structure, it reacts in an unpredictable (but repeatable) way due to the complex interaction of the stimulus with the physical microstructure of the device. Here, the PUF response for a given challenge will act as the internal random key.
- Designing a smaller stream cipher with internal keystream feeding to the multiplexers' control inputs. This gives rise to a *nested cipher* scheme though, at the cost of increased design overhead. On the other hand, it is expected that due to authentication purposes, most modern ICs will have some kind of PUF implementation.

The schematic of PUF-based rXOR-Chain is also captured in Figure 7. From the bits provided in the scan input, a counter is driven. The counter, in turn, acts as the challenge of the PUF. The PUF response is used to control the multiplexers. The proposed PUF-based rXOR-Chain arrangement is only one of many possible variations. The key idea is that the actual locations of the XOR gates are controlled by PUF. Since the PUF-generated bits are known to a valid user as exepected responses to their challenges, he/she can always determine the position of the XORs and therefore, test the chip.

4.1   Robustness Analysis

In [4], it was claimed that in order to find the positions of the XOR gates in a Double-Feedback XOR-Chain structure, the attacker had to solve a system of n equations of algebraic degree close to n over the polynomial ring $GF(2)[a_1, a_2, \ldots, a_n]$. Solving such a system of equations does not appear to be easier than simply guessing the values $[a_1, a_2, \ldots, a_{n-1}]$ or breaking the cipher itself. The above fact is indeed true if the attacker has the ability to operate the scan-chain only in SCAN mode. However, if he can make use of the NORMAL mode of operation as well, then the Double-Feedback XOR-Chain scheme fails to protect the scan-chain as shown in section 3.

The rXOR-Chain is however not vulnerable to such an attack because of the randomness it introduces into the scan-chain structure each time the device is driven into SCAN mode. The rXOR-Chain structure completely prevents the attack we have outlined in Section 3, as randomization implies that every time the attacker drives the device into SCAN mode, he encounters a different Double-Feedback structure. So any algorithm that the attacker devises to cryptanalyze the structure must work with a single instance of the device being driven into SCAN mode. In such a situation the robustness analysis of [4] holds, and the only way the attacker can break the system is to solve a system of multivariate equations over $GF(2)$ of high algebraic degree.

4.2   Overhead Analysis

The overhead of the proposed rXOR-Chain structure is introduced due to the following components, for which analytical overhead estimation is provided.

- Double-Feedback XOR-Chain : This overhead is presented to prevent straight-forward attacks on the basic feedback-free scan-chain structure. From a basic logic-wise estimation [21], 2.5 gates per 2-input XOR logic is accounted. Assuming that each 3-input XOR gate requires two 2-input XOR gates, the Double-Feedback XOR-Chain requires $5n$ gates, where n is the number of feedbacks introduced.
- MUX : Given the multiplexers are 2×1 1-bit structure, 2 AND and 1 OR gate is needed. Accounting for 1.5 gates per each of these gates, $4.5n$ gates are needed.

- Counter : Assuming the counter range of $0 \cdots 128$, 7 Flip-Flops are needed. This is roughly equivalent to 84 gates, considering 12 gates per Flip-Flop [21]. For a 7-bit incrementer logic, 7 half-adder circuits are needed. Each half-adder circuit consumes 1 XOR gate and 1 AND gate, totalling 4 gates. Hence, altogether $7 \times 4 + 84 = 112$ gates are needed.
- PUF : Clearly, the largest area contribution comes from the PUF structure. To make the PUF responses reliable, significant area overhead is introduced by employing error correcting codes (ECC). However, in a recent sense-amplifier-based PUF construction [8], Built-In-Self-Test (BIST) is used to select reliable PUF bits, which is used for the key generation. The generated key is fed to an AES implementation for constructing a strong-PUF, with large number of challenge-response pairs. The overall PUF area, considering the AES logic, SRAM and sense-amplifier array is reported to be $118.8k\mu m^2$. The PUF provides 128-bit key. In our case, it is possible to have a smaller PUF construction generating less number of keys or with smaller number of challenge-response pairs.

5   Conclusion

Scan-chains are a powerful tool to test any electrical device for faults, but in the absence of proper countermeasures, it provides a potent side channel to cryptanalyze the underlying cryptosystem. In this paper we showed that the XOR-Chain mechanism which was proposed in [4], is vulnerable to attack. We also showed that the attack has only polynomial runtime complexity and it is applicable to any stream cipher in general. As a countermeasure, we proposed randomization of the XOR gates in the scan-chain structure so that it can resist such scan-based side channel attacks. We also gave robustness analysis and overhead analysis of the proposed architecture. Note that determining the number of feedbacks for robust cryptosystem is an open problem and so, is the exploration of nested ciphering schemes. In particular, it is known that several PUF structures are vulnerable to attacks, as shown in [16] and this introduces new security backdoors (for further details see [19]). Applying the countermeasure will ensure that at an acceptable cost, testability and security can live together.

Appendix A: A Description of MICKEY 2.0

MICKEY 2.0 uses an 80-bit key and a variable length IV, the length of which may be between 0 and 80 bits. The physical structure of the cipher consists of two 100 bit registers R and S that exercise mutual control over each other's evolution. Let $r_0, r_1, r_2, \ldots, r_{99}$ denote the contents of the register R and $s_0, s_1, s_2, \ldots, s_{99}$ denote the contents of the register S. In order to describe the structure of the cipher and its working let us first define the following routines:- (Note that the description given here is based on [5].)

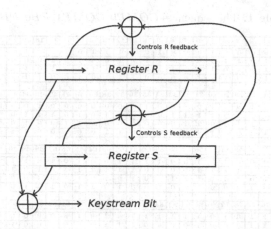

Fig. 8. The variable clocking architecture of MICKEY

Clocking Register R. Let $r_0, r_1, \ldots, r_{99}$ be the state of the register R before clocking, and let $r'_0, r'_1, \ldots, r'_{99}$ be the state of the register R after clocking. Define the integer array $RTAPS$ as follows

$$RTAPS = \{\ 0, 1, 3, 4, 5, 6, 9, 12, 13, 16, 19, 20, 21, 22, 25, 28, 37, 38, 41, 42,$$
$$45, 46, 50, 52, 54, 56, 58, 60, 61, 63, 64, 65, 66, 67, 71, 72, 79, 80,$$
$$81, 82, 87, 88, 89, 90, 91, 92, 94, 95, 96, 97\}$$

Now define an operation

$$CLOCK\_R(R, INPUT\_BIT\_R, CONTROL\_BIT\_R)$$

1. Define $FEEDBACK\_BIT = r_{99} + INPUT\_BIT\_R$
2. For $1 \leq i \leq 99 : r'_i = r_{i-1}.\ r'_0 = 0.$
3. For $0 \leq i \leq 99 :$ if $i \in RTAPS$, $r'_i = r'_i + FEEDBACK\_BIT$.
4. If $CONTROL\_BIT\_R = 1$:
     For $0 \leq i \leq 99 : r'_i = r'_i + r_i$

Clocking Register S. Let $s_0, s_1, \ldots, s_{99}$ be the state of the register S before clocking, and let $s'_0, s'_1, \ldots, s'_{99}$ be the state of the register S after clocking. Let $\hat{s}_0, \hat{s}_1, \ldots, \hat{s}_{99}$ be intermediate variables. Define the four sequences $COMP0_i$, $1 \leq i \leq 98$; $COMP1_i$, $1 \leq i \leq 98$; $FB0_i$, $0 \leq i \leq 99$ and $FB1_i$, $0 \leq i \leq 99$ over $GF(2)$ as in Table 1: Now define an operation

$$CLOCK\_S(S, INPUT\_BIT\_S, CONTROL\_BIT\_S)$$

1. Define $FEEDBACK\_BIT = s_{99} + INPUT\_BIT\_S$
2. For $1 \leq i \leq 98 : \hat{s}_i = s_{i-1} + \big((s_i + COMP0_i) \cdot (s_{i+1} + COMP1_i)\big).\ \hat{s}_0 = 0,\ \hat{s}_{99} = s_{98}.$
3. If $CONTROL\_BIT\_S = 0$:
     For $0 \leq i \leq 99 : s'_i = \hat{s}_i + (FB0_i \cdot FEEDBACK\_BIT)$
   Else If $CONTROL\_BIT\_S = 1$:
     For $0 \leq i \leq 99 : s'_i = \hat{s}_i + (FB1_i \cdot FEEDBACK\_BIT)$

Table 1. The sequences $COMP0, COMP1, FB0, FB1$

i	0	1	2	3	4	5	6	7	8	9	10	11	12	13	14	15	16	17	18	19	20	21	22	23	24
$COMP0_i$	0	0	0	1	1	0	0	0	1	0	1	1	1	1	1	0	1	0	0	1	0	1	0	1	0
$COMP1_i$	1	0	1	1	0	0	1	0	1	1	1	1	1	0	0	1	0	1	0	0	0	1	1	0	1
$FB0_i$	1	1	1	1	0	1	0	1	1	1	1	1	1	1	0	0	1	0	1	1	1	1	1	1	1
$FB1_i$	1	1	1	0	1	1	1	1	0	0	0	0	1	1	1	0	1	0	0	1	1	0	0	0	1

i	25	26	27	28	29	30	31	32	33	34	35	36	37	38	39	40	41	42	43	44	45	46	47	48	49
$COMP0_i$	1	0	1	0	1	1	0	1	0	0	1	0	0	0	0	0	0	0	1	0	1	0	1	0	1
$COMP1_i$	0	1	1	1	0	1	1	1	1	0	0	0	1	1	0	1	0	1	1	1	0	0	0	0	1
$FB0_i$	1	1	1	1	0	0	1	1	0	0	0	0	0	1	1	1	0	0	1	0	0	1	0	0	1
$FB1_i$	0	1	1	0	0	1	0	1	1	0	0	0	0	1	1	0	0	0	0	1	1	0	1	1	0

i	50	51	52	53	54	55	56	57	58	59	60	61	62	63	64	65	66	67	68	69	70	71	72	73	74
$COMP0_i$	0	0	0	0	1	0	1	0	0	1	1	1	1	1	0	0	1	0	1	0	1	1	1	1	1
$COMP1_i$	0	0	0	1	0	1	1	1	0	0	0	1	1	1	1	1	1	0	1	0	1	1	1	0	1
$FB0_i$	0	1	0	0	1	1	1	1	0	1	0	0	1	1	1	1	0	1	0	1	0	0	0	0	0
$FB1_i$	0	0	1	0	0	0	1	0	0	1	0	0	1	0	1	1	0	1	0	1	0	0	1	0	1

i	75	76	77	78	79	80	81	82	83	84	85	86	87	88	89	90	91	92	93	94	95	96	97	98	99
$COMP0_i$	1	1	1	0	1	0	1	1	1	1	1	1	0	1	0	1	0	0	0	0	0	0	1	1	
$COMP1_i$	1	1	1	0	0	0	1	0	0	0	0	1	1	1	0	0	0	0	1	1	1	0	0		
$FB0_i$	1	1	0	1	0	0	0	1	1	0	1	1	1	0	0	1	1	0	0	1	1	0	0	0	
$FB1_i$	0	0	0	1	1	1	1	0	1	1	1	1	1	0	0	0	0	0	0	1	0	0	0	1	

The $CLOCK\_KG$ Routine. We define another operation

$$CLOCK\_KG(R, S, MIXING, INPUT\_BIT)$$

1. $CONTROL\_BIT\_R = s_{34} + r_{67}$, $CONTROL\_BIT\_S = s_{67} + r_{33}$
2. If $MIXING = 1$:
   $INPUT\_BIT\_R = INPUT\_BIT + s_{50}$
   Else If $MIXING = 0$:
   $INPUT\_BIT\_R = INPUT\_BIT$
3. $INPUT\_BIT\_S = INPUT\_BIT$
4. $CLOCK\_R(R, INPUT\_BIT\_R, CONTROL\_BIT\_R)$
5. $CLOCK\_S(S, INPUT\_BIT\_S, CONTROL\_BIT\_S)$

Working of the Cipher. We will now describe the algorithm governing the functioning of the cipher. Let $K = k_0, k_1, \ldots, k_{79}$ be the 80 bit key used by the cipher. Let $IV = iv_0, iv_1, \ldots, iv_{v-1}$ be the v-bit IV ($0 \leq v \leq 80$). Then the cipher operates in the 4 stages as described below.

STAGE 1. IV loading
   Initialize both R and S to the all-zero state.
   For $0 \leq i \leq v - 1 : CLOCK\_KG(R, S, 1, iv_i)$

STAGE 2. Key loading
   For $0 \leq i \leq 79 : CLOCK\_KG(R, S, 1, k_i)$

STAGE 3. Preclock Stage
   For $0 \leq i \leq 99 : CLOCK\_KG(R, S, 1, 0)$

STAGE 4. PRGA(Pseudo-Random stream generation algorithm)

$i \leftarrow 0$

While key-stream is required

$\quad z_i = r_0 + s_0$

$\quad CLOCK\_KG(R, S, 0, 0)$

$\quad i \leftarrow i + 1$

References

1. The ECRYPT Stream Cipher Project. eSTREAM Portfolio of Stream Ciphers (revised on September 8, 2008)
2. Agrawal, M., Karmakar, S., Saha, D., Mukhopadhyay, D.: Scan Based Side Channel Attacks on Stream Ciphers and Their Counter-Measures. In: Chowdhury, D.R., Rijmen, V., Das, A. (eds.) INDOCRYPT 2008. LNCS, vol. 5365, pp. 226–238. Springer, Heidelberg (2008)
3. Ali, S.S., Saeed, S.M., Sinanoglu, O., Karri, R.: Scan attack in presence of mode-reset countermeasure. In: 2013 IEEE 19th International On-Line Testing Symposium (IOLTS), pp. 230–231 (2013)
4. Banik, S., Chowdhury, A.: Improved Scan-Chain Based Attacks and Related Countermeasures. In: Paul, G., Vaudenay, S. (eds.) INDOCRYPT 2013. LNCS, vol. 8250, pp. 78–97. Springer, Heidelberg (2013)
5. Babbage, S., Dodd, M.: The stream cipher MICKEY 2.0. ECRYPT Stream Cipher Project Report. http://www.ecrypt.eu.org/stream/p3ciphers/mickey/mickey_p3.pdf
6. Babbage, S., Dodd, M.: The stream cipher MICKEY-128 2.0. ECRYPT Stream Cipher Project Report. http://www.ecrypt.eu.org/stream/p3ciphers/mickey/mickey128_p3.pdf
7. Banik, S., Maitra, S.: A Differential Fault Attack on MICKEY 2.0. IACR Cryptology ePrint Archive 2013: 29
8. Bhargava, M., Mai, K.: An efficient reliable PUF-based cryptographic key generator in 65nm CMOS. In Design, Automation and Test in Europe Conference and Exhibition (DATE), pp. 1–6 (2014)
9. Bulens, P., Kalach, K., Standaert, F.X., Quisquater, J.J.: FPGA Implementations of eSTREAM Phase-2 Focus Candidates with Hardware Profile. http://www.ecrypt.eu.org/stream/papersdir/2007/024.pdf
10. Cid, C., Robshaw, M., (eds.), Babbage, S., Borghoff, J., Velichkov, V. (contributors): The eSTREAM Portfolio in 2012. Version 1.0 (16 January 2012). http://www.ecrypt.eu.org/documents/D.SYM.10-v1.pdf
11. De Cannière, C., Preneel, B.: Trivium Specifications. http://www.ecrypt.eu.org/stream/p3ciphers/trivium/trivium_p3.pdf
12. Floyd, T.L.: Digital Fundamentals, 10th edn. Prentice Hall (2009)
13. Gaj, K., Southern, G., Bachimanchi, R.: Comparison of hardware performance of selected Phase II eSTREAM candidates. http://www.ecrypt.eu.org/stream/papersdir/2007/026.pdf
14. Gierlichs, B., Batina, L., Clavier, C., Eisenbarth, T., Gouget, A., Handschuh, H., Kasper, T., Lemke-Rust, K., Mangard, S., Moradi, A., Oswald, E.: Susceptibility of eSTREAM Candidates towards Side Channel Analysis. In: Proceedings of SASC 2008 (2008). http://www.ecrypt.eu.org/stvl/sasc2008/
15. Good, T., Benaissa, M.: Hardware performance of eStream phase-III stream cipher candidates. http://www.ecrypt.eu.org/stream/docs/hardware.pdf

16. Helfmeier, C., Boit, C., Nedospasov, D., Seifert, J.-P.: Cloning Physically Unclonable Functions. In: IEEE International Symposium on Hardware-Oriented Security and Trust (HOST) (2013)
17. Helleseth, T., Jansen, C.J.A., Kazymyrov, O., Kholosha, A.: State space cryptanalysis of the MICKEY cipher. In: 2013 Information Theory and Applications Workshop, February 10-15. Catamaran Resort, San Diego (2013)
18. Hong, J., Kim, W.-H.: TMD-Tradeoff and State Entropy Loss Considerations of Streamcipher MICKEY. In: Maitra, S., Veni Madhavan, C.E., Venkatesan, R. (eds.) INDOCRYPT 2005. LNCS, vol. 3797, pp. 169–182. Springer, Heidelberg (2005)
19. Katzenbeisser, S., Kocabaş, U., Rožić, V., Sadeghi, A.-R., Verbauwhede, I., Wachsmann, C.: PUFs: myth, fact or busted? a security evaluation of physically unclonable functions (PUFs) cast in silicon. In: Prouff, E., Schaumont, P. (eds.) CHES 2012. LNCS, vol. 7428, pp. 283–301. Springer, Heidelberg (2012)
20. Kitsos, P.: On the Hardware Implementation of the MICKEY-128 Stream Cipher. http://www.ecrypt.eu.org/stream/papersdir/2006/059.pdf
21. Lano, J., Mentens, N., Preneel, B., Verbauwhede, I.: Power Analysis of Synchronous Stream Ciphers with Resynchronization Mechanism. In: ECRYPT Workshop, SASC - The State of the Art of Stream Ciphers, pp. 327–333 (2004)
22. Sengar, G., Mukhopadhyay, D., Chowdhury, D.R.: Secured flipped scan-chain model for crypto-architecture. IEEE Transactions on CAD of Integrated Circuits and Systems 26(11), 2080–2084 (2007)
23. Rogawski, M.: Hardware evaluation of eSTREAM Candidates: Grain, Lex, Mickey128, Salsa20 and Trivium. http://www.ecrypt.eu.org/stream/papersdir/2006/015.pdf
24. Tischhauser, E.: Nonsmooth cryptanalysis, with an application to the stream cipher MICKEY. Journal of Mathematical Cryptology 4(4), 317–348 (2011)
25. Yang, B., Wu, K., Karri, R.: Secure scan-chain: a design-for-test architecture for crypto-chips. IEEE Transactions on CAD of Integrated Circuits and Systems 25(10), 2287–2293 (2006)
26. Yang, B., Wu, K., Karri, R.: Scan based side channel attack on dedicated hardware implementations of data encryption standard. In: ITC 2004, pp. 339–344 (2004)

EscApe: Diagonal Fault Analysis of APE

Dhiman Saha[1]([✉]), Sukhendu Kuila[2], and Dipanwita Roy Chowdhury[1]

[1] Department of Computer Science and Engineering, IIT Kharagpur,
Kharagpur, India
{dhimans,drc}@cse.iitkgp.ernet.in
[2] Department of Mathematics, Vidyasagar University, Medinipur, India
babu.sukhendu@gmail.com

Abstract. This work presents an adaptation of the classical diagonal
fault attack on APE which is a member of the PRIMATEs family of
authenticated encryption (AE) schemes. APE is the first nonce misuse-
resistant permutation based AE scheme and is one of the submissions to
the CAESAR competition. In this work we showcase how nonce reuse
can be *misused* in the context of differential fault analysis of on-line
authenticated encryption schemes like APE. Using the misuse, we finally
present a diagonal fault attack on APE-80 that is able to reduce the
key-search space from 2^{160} to 2^{25} using just two random uni-word (A
word in this context is a 5-bit vector.) diagonal faults. Increasing the
number of faults to 4 results in the unique identification of the key with
a high probability. We find that both the AES-like internal permutation
and the last round cipher-text output contribute to the reduction in key-
space. We also provide theoretical analysis on the average reduction in
the key-search space of the attack. To the best of our knowledge, this
work reports the first fault analysis of a Sponge based mode of operation
when used in the context of authenticated encryption.

Keywords: Fault analysis · Authenticated encryption · APE · Diagonal
fault attack · Sponge · Nonce misuse-resistance

1 Introduction

The idea of nonce-based encryption was formalized by Rogaway in [20]. The primary condition to be fulfilled is the uniqueness of the nonce in every instantiation
of the cipher. All security claims rely on this premise. An interesting consequence
of a unique nonce is the automatic protection from Differential Fault Analysis
(DFA). Fault analysis constitutes one of the most popular forms of side channel
attacks. The prospect of using faults to attack cryptographic hardware was first
explored by Boneh et. al. [8,9] in 1996. Biham and Shamir introduced Differential
fault analysis (DFA) in [6] which was later successfully applied on the Advanced
Encryption Standard (AES)[13,14,17–19]. One of the primary assumptions of
DFA is the ability to induce faults in the intermediate state of the cipher while
replaying the encryption with the same plaintext. Herein, comes the role of the

© Springer International Publishing Switzerland 2014
W. Meier and D. Mukhopadhyay (Eds.): INDOCRYPT 2014, LNCS 8885, pp. 197–216, 2014.
DOI: 10.1007/978-3-319-13039-2_12

unique nonce which prohibits the replaying criterion. Nonces have been used in Public-Key cryptography to thwart fault attacks. The famous Bellcore attack [9,15] on RSA-CRT signatures can be prevented if the message is padded with a random nonce which is recoverable only when verifying a correct signature. Though Coron et al. have shown in [10] that in some limited setting these nonces can be tackled, the techniques used rely on theoretical constructs which may not directly work in their private-key counterparts. Thus, nonce-based encryption as professed by Rogaway seem to be have an in-built protection against DFA. In practice, however, ensuring the uniqueness of the nonce incurs high overhead specifically in lightweight cryptographic applications where resources may be highly constrained. Hence, nonce reuse or misuse becomes an important issue that should be addressed while designing crypto primitives like authenticated encryption (AE). On the other hand this implies that DFA may again become relevant. In this work we explore this prospect by showcasing a technique that reinstates the replaying criterion of DFA on the authenticated cipher APE.

In FSE 2014, Andreeva et al. introduced a scheme named Authenticated Permutation-based Encryption (APE) [3] targeted for lightweight cryptography. It is the first permutation-based AE scheme that is resistant against nonce misuse. APE is basically a mode of operation which iterates a fixed permutation in a manner that is inspired from the Sponge [5] construction. In [3], the authors instantiated APE with permutations of lightweight hash functions like Spongent, Quark and Photon. However, in 2014, the authors reintroduced APE along with GIBBON and HANUMAN as part of the authenticated encryption family PRIMATEs [1] which is one of the submissions to the ongoing CAESAR competition. Unlike [3], in [1] the authors proposed an indigenous permutation called PRIMATE to serve as the underlying permutation for PRIMATEs. In terms of the internal state size the permutation has two variants : PRIMATE-80 and PRIMATE-120. The PRIMATE permutation family is inspired from FIDES [7] authenticated cipher and structurally follows the round function of Rijndael[12].

Finally, building upon the idea of misusing nonce reuse we present ESCAPE - a differential fault analysis of APE, the nonce misuse-resistant member of PRIMATEs family. It is interesting to note that, APE drops two important aspects of FIDES, firstly, the assumption of a nonce-respecting adversary and secondly, the final truncation operation. Our research reveals that both these changes result in efficient differential fault attacks. We first show how misuse resistance can be exploited to *repeat* the encryption on the same plaintext. Thus dropping the nonce constraint opens up the scheme to fault attacks, which require an attacker to be able to observe faulty ciphertexts by injecting faults while *repeating* the same encryption. We next mount diagonal fault attack exploiting the fact that PRIMATE closely follows the structure of AES round function. However, the inclusion of last round MixColumns transformation makes direct application of diagonal attack as described in [21] inefficient. In order to overcome this we come up with a tweak using the linearity of the MixColumns and ShiftRows operations. Finally, removal of the final truncation of FIDES in APE helps in improving the attack further by using the knowledge of the last block of the ciphertext. The results presented here are with reference PRIMATE-80. However, all the claims can be easily extended to PRIMATE-120. The contribution of this work is summarized below:

- Show how nonce reuse can make online AE schemes vulnerable to DFA.
- Present the first fault analysis that uses a Sponge based mode of operation.
- Efficient adaptation of the classical diagonal fault attack on APE
- Reduce average key-space from 2^{160} to 2^{25} using only two faults.
- Furnish theoretical arguments on the efficiency of the attack with respect to number of faults injected.

The rest of the paper is organized as follows: The notations used in this work along with a brief description of APE is given in section 2. The idea of misusing nonce reuse in the light of DFA and fault diffusion in the internal state of the PRIMATE permutation are discussed in section 3. Section 4 introduces the proposed diagonal attack while the theoretical analysis and experimental results are presented in section 5. The concluding remarks are furnished in section 6.

2   Preliminaries

2.1   The Design of PRIMATE

PRIMATE has two variants in terms of size : PRIMATE-80 (200-bit permutation) and PRIMATE-120 (280-bit) which operate on states of (5×8) and (7×8) 5-bit elements respectively. The family consists of four permutations p_1, p_2, p_3, p_4 which differ in the round constants used and the number of rounds. All notations introduced in this section are with reference to PRIMATEs-80 with the APE mode of operation.

Definition 1. *Let* $\mathbb{T} = \mathbb{F}[x]/(x^5 + x^2 + 1)$ *be the field* $\mathbb{F}_{2^5}$ *used in the PRIMATE MixColumn operation. Then a* **word** *is defined as an element of* $\mathbb{T}$.

Definition 2. *Let* $\mathbb{S} = (\mathbb{T}^5)^8$ *be the set of* (5×8)-word matrices. *Then the internal* **state** *of the PRIMATE-80 permutation family is defined as an element of* $\mathbb{S}$. *We denote a state* $s \in \mathbb{S}$ *with elements* $s_{i,j}$ *as* $[s_{i,j}]_{5,8}$.

$$s = [s_{i,j}]_{5,8}, \ where \begin{cases} s_{i,j} \in \mathbb{T} \\ 0 \leq i \leq 4, \ 0 \leq j \leq 7 \end{cases} \tag{1}$$

In the rest of the paper, for simplicity, we omit the dimensions in $[s_{i,j}]_{5,8}$ and use $[s_{i,j}]$ as the default notation for the 5×8 state. We denote a column of $[s_{i,j}]$ as $s_{*,j}$ and a row as $s_{i,*}$. We now describe in brief the design of PRIMATE permutation. APE instantiates p_1 which is a compositions of 12 round functions.

$$p_1 : \mathbb{S} \longrightarrow \mathbb{S}, \qquad p_1 = \mathcal{R}_{12} \circ \mathcal{R}_{11} \circ \cdots \circ \mathcal{R}_1$$
$$\mathcal{R}_r : \mathbb{S} \longrightarrow \mathbb{S}, \qquad \mathcal{R}_r = \alpha_r \circ \mu_r \circ \rho_r \circ \beta_r$$

where $\mathcal{R}_r$ is a composition of four bijective functions on $\mathbb{S}$. The index r denotes the rth round and may be dropped if the context is obvious. Here, the component function β represents the non-linear transformation SubBytes which constitutes word-wise substitution of the state according to predefined S-box.

$$\beta_r : \mathbb{S} \longrightarrow \mathbb{S}, \qquad s = [s_{i,j}] \longmapsto [S(s_{i,j})]$$

where $S : \mathbb{T} \longrightarrow \mathbb{T}$ is the S-box (Table 1). The transformation ρ corresponds to ShiftRows which cyclically shifts each row of the state based on a set of offsets.

$$\rho_r : \mathbb{S} \longrightarrow \mathbb{S}, \qquad s = [s_{i,j}] \longmapsto [s_{i,(j-\sigma(i)) \bmod 8}]$$

where, $\sigma = \{0, 1, 2, 4, 7\}$ is the ShiftRow offset vector and $\sigma(i)$ defines shift-offset for the ith row. The MixColumn operation, denoted by μ, operates on the state column-wise. μ is actually a left-multiplication by a 5×5 matrix (M_μ) over $\mathbb{T}$.

$$\mu_r : \mathbb{S} \longrightarrow \mathbb{S}, \qquad s = [s_{i,j}] \longmapsto s' = [s'_{i,j}], \qquad s'_{*,j} = M_\mu \times s_{*,j}$$

The last operation of the round function is α which corresponds to the round constant addition. The constants are the output $\{\mathcal{B}_1, \mathcal{B}_2, \cdots, \mathcal{B}_{12}\}$ of a 5-bit LFSR and are xored to the word $s_{1,1}$ of the state $[s_{i,j}]$. The APE mode of operation is depicted in Fig. 1. Here, $N[\cdot]$ represents a Nonce block while $A[\cdot]$ and $M[\cdot]$ denote blocks of associated data and message respectively. The IVs shown in Fig. 1 are predefined and vary according to the nature of the length of message and associated data.

$$\alpha_r : \mathbb{S} \longrightarrow \mathbb{S}, \qquad [s_{i,j}] \longmapsto [s'_{i,j}], \qquad s'_{i,j} = \begin{cases} s_{i,j} \oplus \mathcal{B}_r \text{ if } i, j = 1 \\ s_{i,j}, \text{ Otherwise} \end{cases}$$

2.2 Notations

Definition 3. *A **diagonal** of a state $(s = [s_{i,j}])$ is the set of words which map to the same column under the Shift-Row operation.*

$$d_k = \{s_{i,j} : \rho(s_{i,j}) \in s_{*,k}\}, \text{ where } \begin{cases} k = (j - \sigma(i)) \bmod 8 \\ \sigma = \{0, 1, 2, 4, 7\} \end{cases} \tag{2}$$

Definition 4. *A **differential state** is defined as the element-wise XOR between a state $[s_{i,j}]$ and the corresponding faulty state $[s'_{i,j}]$.*

$$s'_{i,j} = s_{i,j} \oplus \delta_{i,j}, \forall i, j \tag{3}$$

Table 1. The PRIMATE 5-bit S-box

x	0	1	2	3	4	5	6	7	8	9	10	11	12	13	14	15
$S(x)$	1	0	25	26	17	29	21	27	20	5	4	23	14	18	2	28
x	16	17	18	19	20	21	22	23	24	25	26	27	28	29	30	31
$S(x)$	15	8	6	3	13	7	24	16	30	9	31	10	22	12	11	19

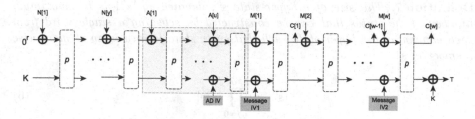

Fig. 1. The APE mode of operation (encryption)

δ fully captures the initial fault as well as the dispersion of the fault in the state. In this work we assume induction of random faults in some diagonal of a state. So, if the initial fault occurs in diagonal $d_k \in s$, the differential state is given by:

$$\delta_{i,j} = \begin{cases} f : f \xleftarrow{R} \mathbb{T} \setminus \{0\}, \text{ if } k = j - \sigma(i) \text{ for at least one } (i,j) \\ 0, \text{ Otherwise} \end{cases} \qquad (4)$$

If $\exists j : \delta_{i,j} = 0 \; \forall i$ then $\delta_{*,j}$ is called a *pure* column, otherwise $\delta_{*,j}$ is referred to as a *faulty* column.

Definition 5. *A **Hyper-state** of a state $s = [s_{i,j}]$, denoted by $s^h = [s^h_{i,j}]$, is a two-dimensional matrix, where each element $s^h_{i,j}$ is a non-empty subset of $\mathbb{T}$, such that s is an element-wise member of s^h.*

$$s^h = \begin{bmatrix} s^h_{00} & s^h_{01} & \cdots & s^h_{07} \\ s^h_{10} & s^h_{11} & \cdots & s^h_{17} \\ \vdots & \vdots & \ddots & \vdots \\ s^h_{40} & s^h_{41} & \cdots & s^h_{47} \end{bmatrix} \quad where \quad \begin{cases} s^h_{i,j} \subset \mathbb{T}, \; s^h_{i,j} \neq \varnothing \\ s_{i,j} \in s^h_{i,j} \; \forall i,j \end{cases} \qquad (5)$$

The significance of a hyper-state s^h is that the state s is in a way 'hidden' inside it. This means that if we create all possible states taking one word from each element of s^h, then one of them will be exactly equal to s. We now define a transformation ρ' on a hyper-state s^h which is analogous to ρ defined in the PRIMATE round functions.

Definition 6. *The **Hyper-state ShiftRow** transformation (ρ') corresponds to cyclically shifting each row of s^h based on the predefined set of offsets σ.*

$$\rho' : \bigtimes_{j=0}^{7} s^h_{i,j} \longrightarrow \bigtimes_{j=0}^{7} s^h_{i,(j-\sigma(i)) \bmod 8} \; \forall i$$

$$s^h_{i,*} \longmapsto s^h_{i,(*-\sigma(i)) \bmod 8} \; \forall i$$

It is interesting to note that, every word in the state $\rho(s)$ will be a member of the corresponding element of $\rho'(s^h)$, thereby implying that $\rho'(s^h) = (\rho(s))^h$.

Definition 7. *The size of a hyper-state s^h, denoted by $|s^h|$ is the maximum number of the states that can be constructed by selecting a single word from each element of the hyper-state such that every state formed is an element-wise member of s^h.*

$$|s^h| = \prod_{i,j=0}^{4,7} |s_{i,j}^h| \qquad (6)$$

The notion of **Hyper-state** will be used while we execute the INBOUND phase of ESCAPE described in subsection 4.1.

Definition 8. *If s^h is a hyper-state of s, then **Kernel** of a column $s_{*,j}^h \in s^h$, denoted by $\mathcal{K}^{s_{*,j}^h}$, is defined as the cross-product of $s_{0,j}^h, s_{1,j}^h, \cdots, s_{4,j}^h$.*

$$\mathcal{K}^{s_{*,j}^h} = \left\{ w_k : w_k^T \in \bigtimes_{i=0}^4 s_{i,j}^h, \ 1 \le k \le \prod_{i=0}^4 |s_{i,j}^h| \right\}$$

Subsequently, Kernel of the entire hyper-state is the set of the Kernels of all of its columns: $\mathcal{K}^{s^h} = \{\mathcal{K}^{s_{*,0}^h}, \mathcal{K}^{s_{*,1}^h}, \cdots, \mathcal{K}^{s_{*,7}^h}\}$.

Here, w_k^T represents the transpose of w_k, thereby implying that w_k is a column vector. One should note that $s_{*,j} \in \mathcal{K}^{s_{*,j}^h} \ \forall j$. Thus each column of s is contained in each element of $\mathcal{K}^{s^h}$. We now define an operation μ' over the Kernel of a hyper-state which is equivalent to μ that operates on a state.

Definition 9. *The **Kernel-MixColumn** transformation (μ') is the left multiplication of M_μ to each element of each $\mathcal{K}^{s_{*,j}^h} \in \mathcal{K}^{s^h}$.*

$$\mu'(\mathcal{K}^{s_{*,j}^h}) = \{M_\mu \times w_i, \ \forall w_i \in \mathcal{K}^{s_{*,j}^h}\}$$
$$\mu'(\mathcal{K}^{s^h}) = \{\mu'(\mathcal{K}^{s_{*,0}^h}), \mu'(\mathcal{K}^{s_{*,1}^h}), \cdots, \mu'(\mathcal{K}^{s_{*,7}^h})\}$$

An important implication is that $\mu'(\mathcal{K}^{s^h}) = \mathcal{K}^{\mu(s)^h}$. The notion of **Kernel** will be used in the OUTBOUND phase of ESCAPE detailed in subsection 4.2.

3   Misusing Nonce Reuse in Online AE

The fault model assumed in this work states that: The attacker can induce random uni-word faults in the internal state of APE permutation p_1. The primary aim here is to produce *faulty collisions* of the tag. A *faulty collision* is not a real collision. In a faulty collision, the attacker induces a fault in the state of the cipher so that two different plaintexts produce the same tag. We later show that the ability to collide the tag will result in highly efficient fault attacks on the overall scheme. It is interesting to note that the attacker is able to produce faulty collisions by exploiting two very desirable properties of authenticated ciphers :

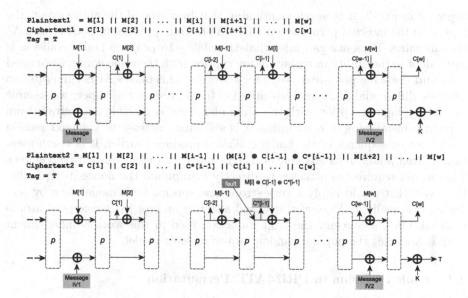

```
Plaintext1  = M[1] || M[2] || ... || M[i] || M[i+1] || ... || M[w]
Ciphertext1 = C[1] || C[2] || ... || C[i] || C[i+1] || ... || C[w]
Tag = T
```

```
Plaintext2  = M[1] || M[2] || ... || M[i-1] || (M[i] ⊕ C[i-1] ⊕ C*[i-1]) || M[i+2] || ... || M[w]
Ciphertext2 = C[1] || C[2] || ... || C*[i-1] || C[i] || ... || C[w]
Tag = T
```

Fig. 2. Exploiting Misuse-resistance to get faulty collisions in APE[1]

firstly, the nonce misuse-resistance and secondly, the online nature. In the next paragraph we describe the process of producing faulty collisions in APE.

The designers of APE in their revised[2] submission document [2] have emphasized that APE is misuse-resistant up to a common prefix. As per [4] this means that for an AE scheme, $E : K \times (\{0,1\}^n)^+ \to (\{0,1\}^n)^+$ if nonce and associated data are *same* for two plain-texts then the outputs are same for a common prefix. So, the first $|P|$ bits are same for $E_k(P||X)$ and $E_k(P||X')$ for any $P, X, X' \in (\{0,1\}^n)^+$. From, the point of view of the attacker this implies that the plain-texts must vary at least in one block. The attacker now arbitrarily chooses the first message $P||X = (x_0||x_1|| \cdots ||x_i|| \cdots ||x_w)$, $x_i \in \{0,1\}^n$ $\forall i$ and chooses the next $P||X' = (x_0||x_1|| \cdots ||x_i'|| \cdots ||x_w)$ in such a way that $\exists i$ $x_i' \neq x_i$. The attacker manipulates x_i' using the fact that APE is an online-cipher and using the assumed fault model. Let $E_k(P||X) = (y_0||y_1|| \cdots ||y_i|| \cdots ||y_w)$, $y_i \in \{0,1\}^n$ $\forall i$. Now, while processing the second message $(P||X')$, the attacker induces a fault at the output of APE in y_{i-1} and gets y_{i-1}'. He now prepares the ith block of the second message as $x_i' = y_{i-1} \oplus y_{i-1}' \oplus x_i$. This means that the input to the permutation is $x_i' \oplus y_{i-1}' = x_i \oplus y_{i-1}$. This ensures that $E_k(P||X) = E_k(P||X')$ for all $x_i \in X$ except the block where the fault is induced. This also implies that $TAG(P||X) = TAG(P||X')$ thereby producing a faulty collision. The process is

[1] According to APE specification, the 'Message IVs' differ depending on the length of the message. This attack is independent of the actual value of the 'Message IVs'.

[2] It is interesting to mention that in FSE 2014 paper [3] and in original submission [2] of PRIMATEs the authors mentioned that 'the use of nonce is optional'. In this scenario, the ability to produce faulty collisions would hold automatically.

depicted in Fig. 2. It is worth mentioning that by virtue of the above event, the inputs to the internal permutation PRIMATE while processing both X and X' remain same. Thus one can infer that the ability to produce faulty collision is equivalent to the ability to replay the encryption with the same nonce, associated data and the same plain-text. We later show that this makes APE vulnerable to efficient differential fault analysis attacks. In the rest of the paper, we assume that replaying encryption with the same plain-text is possible and refrain from explicitly mentioning it again unless it is absolutely necessary. The next section is a direct consequence of the faulty collisions produced earlier. It is worth mentioning that *faulty collisions may also be produced in a nonce-respecting scenario.* This would require the attacker to directly manipulate the nonce using faults. However, that would imply a very strong and specific fault model thereby rendering it unrealistic. Moreover, it would also rely on how the nonce-generator is realized. On the contrary the fault induction used in this work is independent of that and uses the popular random uni-word fault model.

3.1   Fault Diffusion in PRIMATE Permutation

In this section we describe the induction and diffusion of faults in the last iteration of APE. In the last section, we demonstrated the ability to replay the encryption with the same plain-text. We now build upon this and induce a secondary fault while replaying the encryption. In this way we are able to produce a faulty tag that is related to the original tag. In fact, our intention is to study the fault diffusion in the differential state of the PRIMATE permutation. Next we exploit the differential state to mount efficient diagonal fault attacks on APE. The secondary fault induction and subsequent differential state formation are illustrated in Fig 3. One can see that the fault is induced in the input of Round 10 of the PRIMATE permutation during the last iteration before tag generation. The logic behind this will be clear from the following important property of fault diffusion in the internal state of PRIMATE.

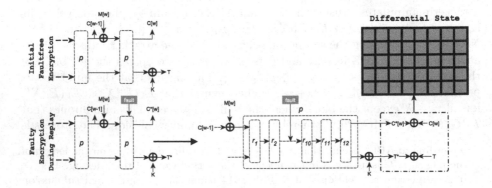

Fig. 3. Fault induction in APE and the differential state

Property 1. *If a single diagonal* (d_k) *is faulty at the start of* $\mathcal{R}_{r-2}$ *then there are exactly three pure columns after* β_r.

Analysis: This property is attributed to the non-square nature of the state matrix. To observe this we need to first look at how the diagonal fault diffuses in the state in the $(r-2)^{th}$ round. Let us denote the differential state at the input of $\mathcal{R}_{r-2}$ as $s = [s_{i,j}]$. Here, we are not concerned about the actual value of s, rather we track how the fault structurally disperses in it. At the beginning of $\mathcal{R}_{r-2}$ only diagonal d_k is faulty. In the following analysis we have omitted the operation α, since round-constant addition has no effect on the differential state.

- Fault diffusion in $\mathcal{R}_{r-2}$
  - β_{r-2} : No diffusion, fault limited to same diagonal.
  - ρ_{r-2} : Fault shifts from diagonal d_k to column $s_{*,k}$.
  - μ_{r-2} : Intra column diffusion. Fault diffuses within $s_{*,k}$.

$$d_k \xrightarrow{\beta_{r-2}} d_k \xrightarrow{\mu_{r-2}\circ\rho_{r-2}} s_{*,k} \tag{7}$$

- Fault diffusion in $\mathcal{R}_{r-1}$
  - β_{r-1} : No diffusion, fault limited to $s_{*,k}$.
  - ρ_{r-1} : Fault shifts from $s_{*,k}$ to five words $\{s_{i,(k-\sigma(i)) \bmod 8} : 0 \le i < |\sigma|\}$.
  - μ_{r-1} : Fault spreads to each column $s_{*,k-\sigma(i)}$.

$$c_k \xrightarrow{\beta_{r-1}} c_k \xrightarrow{\rho_{r-1}} \{s_{i,(k-\sigma(i)) \bmod 8}\} \xrightarrow{\mu_{r-1}} \{s_{*,(k-\sigma(i)) \bmod 8}\} \tag{8}$$

- Fault diffusion in $\mathcal{R}_r$
  - β_r : No diffusion, fault limited to same columns as after μ_{r-1}.

$$\{s_{*,(k-\sigma(i)) \bmod 8}\} \xrightarrow{\beta_r} \{s_{*,(k-\sigma(i)) \bmod 8}\} \tag{9}$$

From (7), (8) and (9) we have the following relation between the faulty diagonal d_k at the start $\mathcal{R}_{r-2}$ and the faulty columns after β_r.

$$d_k \xrightarrow{\beta_r \circ \mathcal{R}_{r-1} \circ \mathcal{R}_{r-2}} \{s_{*,(k-\sigma(i)) \bmod 8} : 0 \le i < |\sigma|\} \tag{10}$$

For PRIMATE-80, $\sigma = \{0,1,2,4,7\}$, implying that $|\sigma| = 5$. From (10), we have $|\{s_{*,(k-\sigma(i)) \bmod 8}\}| = |\sigma| = 5$. Thus a single faulty diagonal before $\mathcal{R}_{r-2}$ results in five faulty columns and respectively $8-5 = 3$ pure columns at the end of β_r. Fig. 4 shows an example of diffusion with the initial fault in diagonal d_1. ∎

In the next section we introduce EscApe, an adaptation of the classical diagonal fault attack [21] on AES. It is worth mentioning that straight forward application of the original diagonal attack on APE is inefficient. Arguments with respect to this is furnished in Appendix B for the interested reader.

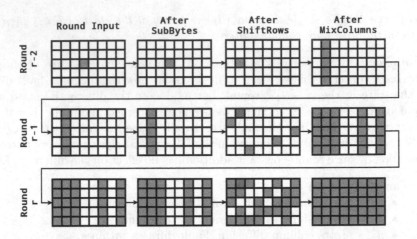

Fig. 4. 3-round fault diffusion with a uni-word fault in diagonal d_1

4   EscApe : An Efficient Diagonal Attack on APE

The EscApe attack proceeds in two phases namely, the INBOUND and OUT-BOUND phase. Both phases result in large-scale reduction in target key-search space. We first describe the INBOUND phase.

4.1   The INBOUND Phase

The first task in the INBOUND phase is to invert the differential output of APE up to β_{12}. While doing so we exploit the following property of APE.

Property 2. *If an attacker knows the correct and the faulty outputs[3] from the last iteration of APE, he can use the difference between them to find the differential state after β_r, r being the last round of PRIMATE permutation.*

Analysis: This property holds because the composition of the transformations α, μ and ρ is affine. Let us denote the affine map as $L_r = \alpha_r \circ \mu_r \circ \rho_r$. L_r is bijective and hence invertible. Let the correct and the faulty outputs of the last iteration of APE be x and x' respectively. Also let the outputs of the $(r-1)^{th}$ round in the respective cases be y and y'. We have the following derivation:

$$L_r^{-1}(x \oplus x') = L_r^{-1}((z \oplus k) \oplus (z' \oplus k))$$
$$= L_r^{-1}(z \oplus k) \oplus L_r^{-1}(z' \oplus k)$$
$$= \beta_r(y) \oplus \beta_r(y')[\text{Since, } \beta_r \circ L_r(y) = z]$$

It is evident from above that the output differential state leads an attacker to the differential state after β_r. However, as β is non-linear, the attacker can no longer deterministically penetrate further inside the last round.   ■

---

[3] This refers to the entire state where the rate part form the *last ciphertext block* and the capacity part forms the *tag* (Refer Fig. 1).

In case of p_1, $r = 12$. So using property (2) the attacker reaches the differential state after β_{12}. From this he tries to guess the source of the fault i.e., the faulty diagonal at the start of $\mathcal{R}_{10}$. For this he exploits an important property of PRIMATE that surfaces due to the non-square nature of the state matrix.

Property 3. *There exists a bijection between the position of faulty diagonal before $\mathcal{R}_{r-2}$ and the position of pure columns of the state at the end of β_r.*

Analysis: This is evident from (10). Since $\{(k - \sigma(i)) \bmod 8, 0 \leq i < |\sigma|\}$ is unique for each k, so the position of a faulty diagonal will correspond to a unique set of five faulty (respectively three pure) columns and vice-versa. This property is important for EscApe, since it aids in detecting the source of the fault just by observing the differential state at the output of APE. As there are a total of 8 diagonals, the diagonal detection reduces the attacking complexity by a factor of 8 and also implies the location independence of the induced fault.∎

In case of APE this implies that based on the diagonal which is faulty at the start of $\mathcal{R}_{10}$, one can predict the word inter-relations at the end of $\mathcal{R}_{11}$. Moreover, the diagonal principle as mentioned in [21] states that *multi-word faults which are confined to one diagonal before $\mathcal{R}_{10}$ are equivalent and result in the same word inter-relations at the end of $\mathcal{R}_{11}$.* For instance, if the faulty diagonal is d_0, then the corresponding relation matrix is given in Fig. 5. The relation matrix shows how words of the differential state are related at the end of $\mathcal{R}_{11}$. The empty columns in the relation matrix represent the columns which have been unaffected by the fault injected before $\mathcal{R}_{10}$. As shown in Fig. 5, by virtue of the diagonal principle all the faults in d_0 are equivalent in terms of the resulting word inter-relations. The relation matrices for other diagonals are depicted in Fig. 7 of Appendix A. We exploit these relations to reduce the state search-space.

Let the differential state after β_{12} computed using property (2) be $\delta = [\delta_{i,j}]$ and the corresponding correct and faulty states be s and s' respectively. Let

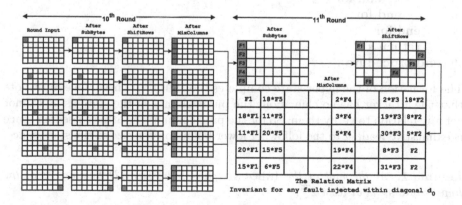

Fig. 5. Equivalence of uni-word faults in diagonal d_0 at the input of 10th round

the relation matrix be denoted by $\eta = [\eta_{i,j}]$. The attacker now guesses the actual values of s and uses δ to get the values of s'. He can guess any two words $(s_{i,j}, s_{k,j})$ from the same column $s_{*,j}$ and use the corresponding entries $(\eta_{i,j}, \eta_{k,j})$ from the relation matrix to form an equation of the form given in (11).

$$\eta_{i,j}^{-1} \times (\beta^{-1}(s_{i,j}) \oplus \beta^{-1}(s_{i,j} \oplus \delta_{i,j})) = \eta_{k,j}^{-1} \times (\beta^{-1}(s_{k,j}) \oplus \beta^{-1}(s_{k,j} \oplus \delta_{k,j})) \quad (11)$$

For the case of diagonal d_0, if the attacker chooses $(s_{1,0}, s_{2,0})$, he gets equation (12). It must be recalled that all multiplications and inverse operations are carried out in the finite field $\mathbb{T}$. Using these equations, the attacker verifies if the guessed words of s are correct. The guessed words for each $s_{i,j}$ that satisfy the equations form the candidate-vector for that word. All these candidate-vectors constitute the hyper-state, s^h (Definition 5) of the correct state s. The pseudo-code for the hyper-state generation in given below:

$$18^{-1} \times (\beta^{-1}(s_{1,0}) \oplus \beta^{-1}(s_{1,0} \oplus \delta_{1,0})) = 11^{-1} \times (\beta^{-1}(s_{2,0}) \oplus \beta^{-1}(s_{2,0} \oplus \delta_{2,0})) \quad (12)$$

```
1:  procedure GENHYPERSTATE(δ, η)          ▷ δ → Differential State after β₁₂
2:      Denote correct state after β₁₂ as s = [sᵢ,ⱼ].
3:      for all (i, j) do                            ▷ Initialize hyper-state
4:          sʰᵢ,ⱼ = {0, 1, ⋯ , 31}
5:      end for
6:      for all δ*,ⱼ ∈ δ : η*,ⱼ ≠ ∅ do               ▷ Process only faulty columns
7:          for all (δᵢ,ⱼ, δₖ,ⱼ) ∈ δ*,ⱼ do
8:              for all (a, b) ∈ sʰᵢ,ⱼ × sʰₖ,ⱼ do
9:                  Set (sᵢ,ⱼ, sₖ,ⱼ) = (a, b)
10:                 if ηᵢ,ⱼ⁻¹ × (β⁻¹(sᵢ,ⱼ) ⊕ β⁻¹(sᵢ,ⱼ ⊕ δᵢ,ⱼ)) ≠   ▷ Verify Equation (11)
11:                        ηₖ,ⱼ⁻¹ × (β⁻¹(sₖ,ⱼ) ⊕ β⁻¹(sₖ,ⱼ ⊕ δₖ,ⱼ)) then
12:                     sʰᵢ,ⱼ = sʰᵢ,ⱼ − {a}
13:                     sʰₖ,ⱼ = sʰₖ,ⱼ − {b}
14:                 end if
15:             end for
16:         end for
17:     end for
18:     return sʰ
19: end procedure
```

The formation of the hyper-state completes the INBOUND phase. One must note that the attacker has no clue about the pure columns of the state as he cannot get any relation to verify them. So for the pure columns the attacker search space is exhaustive leading to the following lower bound on the size of the state-space.

Lemma 1. *At the end of the* INBOUND *phase, the size (Definition 7) of the hyper-state s^h is at least 2^{75}.*

$$|s^h| \geq 2^{75}$$

Proof. This follows from property (1) which states that there will be exactly three pure-columns at the end of $\mathcal{R}_{11}$. For each $s_{i,j}^h \in s^h$ such that $s_{*,j}^h$ is a pure column $|s_{i,j}^h| = 2^5$. Each pure-column contributes five such vectors to s^h and there are three such columns. Thus we have 15 such $s_{i,j}^h$'s each contributing a factor of 2^5 to the size of the hyper-state.

$$|s^h| = \prod_{i,j=0}^{4,7} |s_{i,j}^h| = \prod_{\substack{i=0 \\ \forall j \in \mathcal{P}}}^{4} |s_{i,j}^h| \times \prod_{\substack{i=0 \\ \forall j \notin \mathcal{P}}}^{4} |s_{i,j}^h|$$

$$\geq \prod_{\substack{i=0 \\ \forall j \in \mathcal{P}}}^{4} |s_{i,j}^h| = (2^5)^{15} = 2^{75}$$

where $\mathcal{P}$ is the set of all pure-columns at the end of $\mathcal{R}_{11}$.   ∎

In the next subsection, we show that during the OUTBOUND phase the attacker can further reduce $|s^h|$ by reducing each $|s_{i,j}^h|, j \in \mathcal{P}$, thereby reducing the search space for the correct state (correspondingly, for the key). The following algorithm gives an overview of the INBOUND phase.

1: **procedure** INBOUND(δ)   ▷ $\delta \to$ Differential State
2:      $\delta' = \rho_{12}^{-1}(\mu_{12}^{-1}(\delta))$ ▷ Applying Property (2)
3:      $d_f \xleftarrow{\text{Property (3)}} \delta'$ ▷ Trace back faulty diagonal
4:      $\eta = [\eta_{i,j}] \xleftarrow{\text{Load relation matrix}} d_f.$ ▷ Refer Fig. 7
5:      $s^h \leftarrow$ GENHYPERSTATE(δ', η)
6:      **return** s^h
7: **end procedure**

4.2   The OUTBOUND Phase

This phase exploits the mode of operation of APE. Particularly, it exploits the knowledge of the last cipher-text block that forms the rate-part of the internal state of PRIMATE. The following property plays a pivotal role in this phase.

Property 4. *If the state before α_{12} is $t = [t_{i,j}]$, then the attacker knows the actual value of $t_{0,*}$.*

Analysis: This property is attributed to the APE mode of operation. According to APE, after processing the last message block the rate part (first row) of the internal state after $\mathcal{R}_{12}$ of PRIMATE is output in the clear as the last cipher-text block while the capacity part is xored with the key to form the tag. Also one can recall that in α the round constant is xored with word $(1,1)$ only while the rest of the state is unchanged. Now, let us denote the last output of APE by $a = [a_{i,j}]$ also let $k = [k_{i,j}]$ denote the key-state. Then we have:

$$a_{i,j} = \begin{cases} t_{i,j} & \text{if } (i = 0) \implies (k_{0,*} = \mathbf{0}) \\ a_{i,j} \oplus \mathcal{B}_{12} \oplus k_{i,j}, & \text{if } (i,j) = (1,1) \\ a_{i,j} \oplus k_{i,j}, & \text{otherwise} \end{cases} \qquad (13)$$

where $\mathcal{B}_{12}$ is the round constant for $\mathcal{R}_{12}$ in p_1. Since α_{12} has no effect on $t_{0,*}$ and since by Equation (13) $a_{0,*} = t_{0,*}$, so the last cipher-text block gives the attacker the actual value of $t_{0,*}$. This property plays a central role in the OUTBOUND phase and its use results in significant reduction of the key search space.   ■

The steps of the OUTBOUND phase are enlisted below. It can be recalled that the INBOUND phase gives us the hyper-state (s^h) of the state s after β_{12}.

1. The attacker starts the OUTBOUND phase by applying Hyper-state ShiftRow transformation (Definition 6) on the hyper-state from the INBOUND phase.

$$s^h \xrightarrow{\rho'} (\rho(s))^h$$

2. The next step is to compute the Kernel for $(\rho(s))^h : \mathcal{K}^{(\rho(s))^h}$. The concept of Kernel was introduced in Definition 8.

$$(\rho(s))^h \xrightarrow{\text{Compute Kernel}} \mathcal{K}^{(\rho(s))^h}$$

3. Then the attacker applies the Kernel-MixColumn transformation on the Kernel computed in the last step.

$$\mathcal{K}^{(\rho(s))^h} \xrightarrow{\mu'} \mathcal{K}^{(\mu(\rho(s)))^h}$$

4. Next comes the reduction step. In this step the attacker applies property (4). It can be noted that $\mathcal{K}^{\mu(\rho(s))^h}$ represents the kernel for the hyper-state of $\mu(\rho(s))$. i.e., the state just before the application of α_{12}. Now let $t = \mu(\rho(s))$. Then by property (4) the actual value of $t_{0,*}$ is known. This knowledge is used to reduce the size of each $\mathcal{K}^{t^h_{*,j}} \in \mathcal{K}^{t^h}$. The following pseudocode illustrates the reduction procedure.

```
1: procedure REDUCEKERNEL(K^{t^h}, t)
2:     for j = 0 : 7 do
3:         for all {e_0, e_1, e_2, e_3, e_4}^T ∈ K^{t^h_{*,j}} do
4:             if e_0 ≠ t_{0,j} then
5:                 K^{t^h_{*,j}} = K^{t^h_{*,j}} − {e_0, e_1, e_2, e_3, e_4}^T
6:             end if
7:         end for
8:     end for
9:     K^{t^h}_{red} = K^{t^h}
10:    return K^{t^h}_{red}
11: end procedure
```

The cross-product of $\mathcal{K}^{t^h}_{red}$ gives the final reduced state-space for t. The final key-space can be expressed as follows :

$$\mathbb{K} = \left\{ k : k = a \oplus \alpha_{12}(w), \; \forall w \in \bigtimes_{j=0}^{7} \mathcal{K}^{t^h_{*,j}}_{red} \right\}$$

The following algorithm summarizes the OUTBOUND phase. In the next subsection we outline the complete attack.

```
1: procedure OUTBOUND(s^h, t)
2:     (ρ(s))^h ← ρ'(s^h)
3:     𝒦^(ρ(s))^h ← (ρ(s))^h
4:     𝒦^{t^h} ← μ'(𝒦^{ρ(s)^h})
5:     𝒦^{t^h}_{red} ← REDUCEKERNEL(𝒦^{t^h}, t)
6:     return 𝒦^{t^h}_{red}
7: end procedure
```

4.3   The Complete Attack

The EscApe attack in its complete form takes into account multiple faults and further reduces the key-space. When dealing with multiple faulty cipher-texts, the INBOUND phase is repeated to get a hyper-state for each faulty cipher-text and finally an element-wise intersection is taken over all the hyper-states. This intersection largely reduces the size of the final hyper-state. This reduced-size hyper-state is given as output of the INBOUND phase. The OUTBOUND phase proceeds as described above. The following algorithm presents the complete EscApe attack while the pictorial description is given in Fig. 6. It is implied that the number of faults is inversely proportional to the size of the final key-search space. A discussion on this is furnished in the next section.

```
 1: procedure EscApe(c, {c'_1, c'_2, ⋯, c'_n})              ▷ n → # of faulty outputs
 2:     for f = 1 : n do
 3:         s^{h_f} ← INBOUND(c ⊕ c'_f)                    ▷ Compute hyper-state
 4:     end for
 5:     s^h = ⋂_{f=1}^{n} s^{h_f}                          ▷ Intersect all hyper-states
 6:     𝒦^{t^h}_{red} ← OUTBOUND(s^h, c)
 7:     𝕂 = ∅
 8:     for all w ∈ (×_{j=0}^{7} 𝒦^{t^h,j}_{red}) do
 9:         k = α_{12}(w) ⊕ c
10:         𝕂 = 𝕂 ∪ {k}
11:     end for
12:     return 𝕂
13: end procedure
```

5   Results

While running computer simulations, we have observed that there exists a relation between the distribution of the induced fault and the average reduction in the key-space. In this section, we analyze this observation in detail.

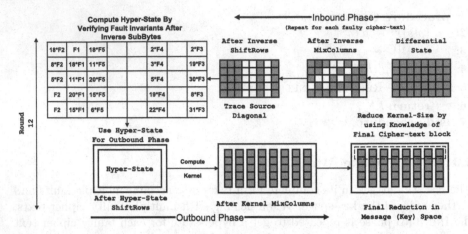

Fig. 6. The EscApe attack using a diagonal fault in d_1

5.1   A Discussion on the Size of $\mathbb{K}$

The efficiency of the attack when dealing with multiple faulty outputs relies on the location of the individual faults. The basic implication is that if the same diagonal is faulty across different faults then the average reduction in $|\mathbb{K}|$ is less than the case where different diagonals are affected. In order to study the likelihood of these events we give the following lemma.

Lemma 2. *If there are $\mathcal{N} = n$ faults then the following expression gives the probability that the number of affected diagonals $\mathcal{D}$ is at least d.*

$$\Pr(\mathcal{N} = n, \mathcal{D} \geq d) = 1 - \sum_{i=1}^{d-1} \frac{\binom{8}{i} \times \binom{n-1}{i-1}}{\binom{n+7}{7}}$$

Proof. We prove this by finding the probability of the complement event and then employing the law of total probability along with a result from combinatorics. The complement event is when at most $(d-1)$ diagonals are faulty. We divide this event into sub-cases and handle each case separately starting from *exactly* 1 diagonal being faulty, followed by 2 and going up to the event when exactly $(d-1)$ diagonals are affected. This leads us to the following relation:

$$\begin{aligned} \Pr(\mathcal{N} = n, \mathcal{D} \geq d) &= 1 - \Pr(\mathcal{N} = n, \mathcal{D} < d) \\ &= 1 - \Pr(\mathcal{N} = n, \mathcal{D} = 1) - \Pr(\mathcal{N} = n, \mathcal{D} = 2) - \cdots \\ &\quad - \Pr(\mathcal{N} = n, \mathcal{D} = i) - \cdots - \Pr(\mathcal{N} = n, \mathcal{D} = d - 1) \\ &= 1 - \sum_{i=1}^{d-1} \Pr(\mathcal{N} = n, \mathcal{D} = i) \end{aligned} \tag{14}$$

Here the term, $\Pr(\mathcal{N} = n, \mathcal{D} = i)$ denotes the probability that exactly i diagonals are faulty after n faults. Here, the ordering among the faults as well as the diagonals is irrelevant. Now, the complete sample space is given by the number of ways n faults can be distributed among 8 diagonals which is given by $\binom{n+8-1}{8-1} = \binom{n+7}{7}$. To find $\Pr(\mathcal{N} = n, \mathcal{D} = i)$, we need to find the number of ways n faults can affect *exactly* i diagonals. This is given by $\binom{8}{i} \times \binom{n-1}{i-1}$. Here, the first term comes the number of ways of choosing i diagonals from the 8 diagonals. For each such case n faults can be distributed among exactly i diagonals in $\binom{n-1}{i-1}$ ways. Now the probability of this event is $\Pr(\mathcal{N} = n, \mathcal{D} = i) = \frac{\binom{8}{i} \times \binom{n-1}{i-1}}{\binom{n+7}{7}}$. Substituting this in Equation (14) we get:

$$\Pr(\mathcal{N} = n, \mathcal{D} \geq d) = 1 - \sum_{i=1}^{d-1} \frac{\binom{8}{i} \times \binom{n-1}{i-1}}{\binom{n+7}{7}}$$

∎

Now, from Lemma 2 we get an idea that for lower values for d, $\Pr(\mathcal{N} = n, \mathcal{D} \geq d) \to 1$ implying that the event is highly probable. Another observation is that the probability of only a single faulty diagonal across all faults is $\Pr(\mathcal{N} = n, \mathcal{D} = 1) \ll \Pr(\mathcal{N} = n, \mathcal{D} \geq d)$ for lower values of d. From the point of view of an attacker, n should be as low as possible at the same time d should be such that $\Pr(\mathcal{N} = n, \mathcal{D} \geq d)$ is high while reaching an optimally low bound for $|\mathbb{K}|$. We performed our experimentation using random diagonal faults in APE instantiated with PRIMATE permutation p_1 for various values of n, d and found the most optimal values to be $n = 2, d = 2$. The results of the experiments are furnished in Table 2. As one can see $|\mathbb{K}|$ reduces to practical limits $\forall n > 1$. However, the least value of $\mathcal{N}$ for which $|\mathbb{K}|$ is optimally low is 2 and this happens when $\mathcal{D} \geq 2$. The probability of this event ($\Pr(\mathcal{N} = 2, \mathcal{D} \geq 2)$) is as high as 0.778.

Table 2. Avg. key-search space $|\mathbb{K}|$ for various values of n, d along with their probabilities ($\mathcal{D} \leq min\{n, 8\}$)

n	d	$\Pr(\mathcal{N} = n, \mathcal{D} = d)$	$\|\mathbb{K}\|$(avg.)	$\Pr(\mathcal{N} = n, \mathcal{D} \geq d)$	$\|\mathbb{K}\|$(avg.)
1	1	1	2^{80}		
2	1	0.222	2^{45}	1	2^{27}
	2	**0.778**	2^{25}		
3	2	0.4667	2^{13}	0.9333	2^{5}
	3	0.4667	2^{25}		
4	2	0.25454	2^{7}	0.97575	1
	3	0.50909	2^{1}	0.72121	1

6   Conclusion

We present arguments stating that the otherwise desirable feature of nonce misuse-resistance can become the gateway for mounting differential faults attacks in online AE schemes. Building upon this idea we develop an efficient diagonal fault attack on the APE authenticated cipher. APE is the first permutation based AE scheme which uses the Sponge based mode of operation. Hence the result presented here can be interpreted as the first fault analysis of Sponge when used in the context of authenticated encryption. The number of faulty cipher-texts required to reduce the key-search space to a practical limit is only 2 while 4 faulty outputs can uniquely identify the key with a very high probability. Finally, the results presented in this work highlight that the design of the underlying permutation and the choice of the mode of operation can both contribute to the susceptibility of the AE scheme to fault analysis attacks.

A   The Relation Matrix

The relation matrix represents the relations between individual words of a differential state. The relations form at the end of $\mathcal{R}_{11}$ due to the propagation of the fault injected in a diagonal before $\mathcal{R}_{10}$. These relations are invariant for all faults confined within a diagonal and are also unique for each diagonal. The relation matrices for all diagonals of PRIMATE-80 permutation are furnished in Fig. 7.

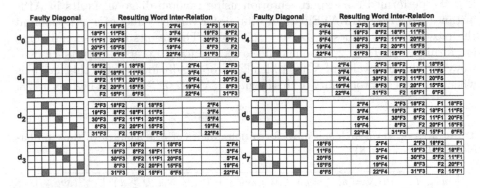

Fig. 7. Word inter-relations at the start of $\mathcal{R}_r$ due to corresponding faulty diagonal at the start of $\mathcal{R}_{r-2}$ (Empty relations refer to the fault-free columns)

B   Inefficiency of Classical Diagonal Attack on APE

As stated earlier, the diagonal attack [21] has been stated in literature [16] as one of the most effective differential fault analysis attacks on AES. As the design of PRIMATE permutation family is inspired from AES, so this work concentrates on the application of Diagonal attack on APE. However, it is interesting to note that the classical diagonal attack cannot efficiently be applied to the PRIMATE permutations. The structure of PRIMATE, though similar to AES, differs from AES in the last round. All rounds of AES consist of the same round function except the last round where the MixColumn transformation is dropped. On the contrary in PRIMATE all the rounds are uniform. This imposes a problem in the straight-forward application of the diagonal attack.

The diagonal attack requires an attacker to invert both the correct and faulty outputs of the cipher up to the last round SubBytes operation by guessing the key-bytes. Then by using the byte inter-relations that hold due to the fault, the wrong guesses are discarded thereby reducing the key-search space. In case of AES, the last round inversion only involves the InverseShiftRows operation (as MixColumn is missing). Thus while verifying the byte inter-relations in case of a single diagonal being faulty each call of the inverse-sbox involves just one key-byte. Unfortunately, the inclusion of μ in the last round of PRIMATE makes four key-words to be involved in each each call of β^{-1}.

Let the last output of APE be a and the key be k. As the key is not xored with the first row, so $k_{0,*} = \mathbf{0}$. Let the output of PRIMATE be b. Each word of b is of the form $b_{i,j} = a_{i,j} \oplus k_{i,j}$. Let the inverse MixColumn matrix be denoted by $M_{\mu}^{-1} = [m_{i,j}]_{5,5}$. If the $d = \mu^{-1}(b)$, then each word of $d_{i,j}$ will be of the form

$$d_{i,j} = \sum_{l=0}^{4} m_{i,l} * b_{l,j} = \sum_{l=0}^{4} m_{i,l} * (a_{l,j} \oplus k_{l,j})$$

$$= m_{i,0} * a_{0,j} + \sum_{l=1}^{4} m_{i,l} * (a_{l,j} \oplus k_{l,j}) [\text{Since, } k_{0,*} = \mathbf{0}]$$

The above expression shows that each word involves four key-words. Now ρ^{-1} will only rearrange the words without affecting the number of key-words involved per word of the state. The relation verification requires guessing two such words. This in turn implies that the total number of key guesses to verify the relation will be $2^{20 \times 2}$. Thus total number of key-guesses considering the entire state will be of the order of 2^{40}. This makes the straight forward application of the diagonal attack inefficient. One can note that this observation is inline with the claims made by Daemen et al. in [11] where it is shown that key diffusion results in increased resistance to differential attacks. However, in section 4 we show that we can exploit the differential state of the last iteration of APE to tweak the diagonal attack by guessing the state instead of the key resulting in a very efficient attack on APE.

References

1. Andreeva, E., Bilgin, B., Bogdanov, A., Luykx, A., Mennink, B., Mouha, N., Wang, Q., Yasuda, K.: PRIMATEs v1 (2014). http://competitions.cr.yp.to/round1/primatesv1.pdf
2. Andreeva, E., Bilgin, B., Bogdanov, A., Luykx, A., Mennink, B., Mouha, N., Wang, Q., Yasuda, K.: PRIMATEs v1.01 (2014). http://primates.ae/wp-content/uploads/primatesv1.01.pdf

3. Andreeva, E., Bilgin, B., Bogdanov, A., Luykx, A., Mennink, B., Mouha, N., Yasuda, K.: APE: Authenticated Permutation-Based Encryption for Lightweight Cryptography. In: FSE. LNCS. Springer (2014). https://lirias.kuleuven.be/handle/123456789/450105
4. Andreeva, E., Bogdanov, A., Luykx, A., Mennink, B., Tischhauser, E., Yasuda, K.: Parallelizable and Authenticated Online Ciphers. In: Sako, K., Sarkar, P. (eds.) ASIACRYPT 2013, Part I. LNCS, vol. 8269, pp. 424–443. Springer, Heidelberg (2013)
5. Bertoni, G., Daemen, J., Peeters, M., Assche, G.V.: Cryptographic sponge functions. http://sponge.noekeon.org/CSF-0.1.pdf
6. Biham, E., Shamir, A.: Differential Fault Analysis of Secret Key Cryptosystems. In: Kaliski Jr, B.S. (ed.) CRYPTO 1997. LNCS, vol. 1294, pp. 513–525. Springer, Heidelberg (1997)
7. Bilgin, B., Bogdanov, A., Knežević, M., Mendel, F., Wang, Q.: FIDES: Lightweight Authenticated Cipher with Side-Channel Resistance for Constrained Hardware. In: Bertoni, G., Coron, J.-S. (eds.) CHES 2013. LNCS, vol. 8086, pp. 142–158. Springer, Heidelberg (2013)
8. Boneh, D., DeMillo, R.A., Lipton, R.J.: On the Importance of Checking Cryptographic Protocols for Faults. In: Fumy, W. (ed.) EUROCRYPT 1997. LNCS, vol. 1233, pp. 37–51. Springer, Heidelberg (1997)
9. Boneh, D., DeMillo, R.A., Lipton, R.J.: On the Importance of Eliminating Errors in Cryptographic Computations. J. Cryptology 14(2), 101–119 (2001)
10. Coron, J.-S., Joux, A., Kizhvatov, I., Naccache, D., Paillier, P.: Fault Attacks on RSA Signatures with Partially Unknown Messages. In: Clavier, C., Gaj, K. (eds.) CHES 2009. LNCS, vol. 5747, pp. 444–456. Springer, Heidelberg (2009)
11. Daemen, J., Knudsen, L.R., Rijmen, V.: The Block Cipher SQUARE. In: Biham, E. (ed.) FSE 1997. LNCS, vol. 1267, pp. 149–165. Springer, Heidelberg (1997)
12. Daemen, J., Rijmen, V.: The Design of Rijndael: AES - The Advanced Encryption Standard. Information Security and Cryptography. Springer (2002)
13. Dusart, P., Letourneux, G., Vivolo, O.: Differential Fault Analysis on A.E.S. IACR Cryptology ePrint Archive 2003, 10 (2003). http://eprint.iacr.org/2003/010
14. Giraud, C.: DFA on AES. In: Dobbertin, H., Rijmen, V., Sowa, A. (eds.) AES 2005. LNCS, vol. 3373, pp. 27–41. Springer, Heidelberg (2005)
15. Joye, M., Lenstra, A.K., Jacques Quisquater, J.: Chinese Remaindering Based Cryptosystems in the Presence of Faults. Journal of Cryptology 12, 241–245 (1999)
16. Joye, M., Tunstall, M. (eds.): Fault Analysis in Cryptography. Information Security and Cryptography. Springer (2012)
17. Moradi, A., Shalmani, M.T.M., Salmasizadeh, M.: A Generalized Method of Differential Fault Attack Against AES Cryptosystem. In: Goubin, L., Matsui, M. (eds.) CHES 2006. LNCS, vol. 4249, pp. 91–100. Springer, Heidelberg (2006)
18. Mukhopadhyay, D.: An Improved Fault Based Attack of the Advanced Encryption Standard. In: Preneel, B. (ed.) AFRICACRYPT 2009. LNCS, vol. 5580, pp. 421–434. Springer, Heidelberg (2009)
19. Piret, G., Quisquater, J.-J.: A Differential Fault Attack Technique against SPN Structures, with Application to the AES and KHAZAD. In: Walter, C.D., Koç, Ç.K., Paar, C. (eds.) CHES 2003. LNCS, vol. 2779, pp. 77–88. Springer, Heidelberg (2003)
20. Rogaway, P.: Nonce-Based Symmetric Encryption. In: Roy, B., Meier, W. (eds.) FSE 2004. LNCS, vol. 3017, pp. 348–359. Springer, Heidelberg (2004)
21. Saha, D., Mukhopadhyay, D., RoyChowdhury, D.: A Diagonal Fault Attack on the Advanced Encryption Standard. Cryptology ePrint Archive, Report 2009/581 (2009). http://eprint.iacr.org/

Cryptanalysis

Using Random Error Correcting Codes in Near-Collision Attacks on Generic Hash-Functions

Inna Polak[✉] and Adi Shamir

Department of Computer Science and Applied Mathematics,
Weizmann Institute of Science, 76100 Rehovot, Israel
innapolak@gmail.com, adi.shamir@weizmann.ac.il

Abstract. In this paper we consider the problem of finding near-collisions with Hamming distance bounded by r in generic n-bit hash functions. In 2011, Lamberger and Rijmen proposed a modified version of Pollard's rho method, and in 2012 Leurent improved this memoryless algorithm by using any available memory to store chain endpoints. Both algorithms use a perfect error correcting code to change near-collisions into full-collisions, but such codes are rare and have very small distance. In this paper we propose using randomly chosen linear codes, whose decoding can be made efficient by using some of the available memory to store error-correction tables. Compared to Leurent's algorithm, we experimentally verified an improvement ratio of about 3 in a small example with $n = 160$ and $r = 33$ which we implemented on a single PC, and mathematically predicted a significant improvement ratio of about 730 in a larger example with $n = 1024$ and $r = 100$, using 2^{40} memory.

Keywords: Hash function · Near-collision · Random-code · Time-memory trade-off · Generic attack

1 Introduction

A *hash function* h maps an arbitrarily long input into an n-bit digest. Cryptographically strong hash functions should be indistinguishable from random functions, and in particular it should be difficult to find collisions (defined as pairs (m_1, m_2) s.t. $m_1 \neq m_2$ and $h(m_1) = h(m_2)$) in fewer than $2^{n/2}$ evaluations of h.

In this paper we consider a weaker notion of collision called *r-near-collision*, in which up to r bits in $h(m_1)$ and $h(m_2)$ are allowed to be different. There are several reasons why we may want to study such near-collisions. First of all, in many applications such as the generation of cryptographic keys or MAC's, the standard output of a hash function is too long, and we use only a subset of its bits. In this case, a near-collision can become a real collision if the differing bits are thrown away. In addition, finding near-collisions is often a useful first step when we try to find a multi-block collision in a hash function, as demonstrated

© Springer International Publishing Switzerland 2014
W. Meier and D. Mukhopadhyay (Eds.): INDOCRYPT 2014, LNCS 8885, pp. 219–236, 2014.
DOI: 10.1007/978-3-319-13039-2_13

in [13, 19, 21]). By studying the complexity of near-collision attacks on generic hash functions (which are modeled as random functions), we can get upper-bounds on the near-collision resistance of any concrete hash-function, but in some cases we can do much better (see [1, 3, 16]). Even when we cannot turn such a near-collision attack into a full collision attack, the mere existence of a better than expected near-collision attack may suffice to disqualify a new hash function proposal. Finally, the task of finding a near-collision can be used as a more flexible and accurate type of *proof-of-work* [6, 14] than finding a full collision since there are more parameters that we can specify in defining the computational task.

Let us now introduce our notation. The vector space whose elements are n-bit words is denoted by $\mathbb{F}_2^n$. We use the notation of d_H to describe the Hamming distance function, and $\|\|_H$ for the Hamming-weight. We call $x, y \in \mathbb{F}_2^n$ *R-close* vectors if $d_H(x, y) \leq R$. The ball of radius R around x in $\mathbb{F}_2^n$, defined as $\{y | d_H(y, x) \leq R\}$, is denoted by $B_R^n(x)$. Its volume is denoted by V_R^n, and is defined as

$$V_R^n = \sum_{i=0}^{R} \binom{n}{i}$$

Any r_1-near-collision is in particular also an r_2-near-collision for every $r_1 \leq r_2$ (when $r = 0$ it is simply a full-collision). Therefore, the difficulty of finding r-near-collision decreases as r increases. However, detecting such a collision as soon as it occurs becomes algorithmically harder as r increases. The probability of a random pair of points in $\mathbb{F}_2^n$ to be r-close is $q_r^n := V_r^n \cdot 2^{-n}$. By the birthday paradox, the first r-near-collision is thus expected to be seen after about $1/\sqrt{q_r^n} = 2^{n/2}/\sqrt{V_r^n}$ hash evaluations. These evaluations require less than $2^{n/2}$ time, but actually finding the unique near-collision among them requires more than $2^{n/2}$ time (since the property of being r-close for $r \geq 1$ is not transitive, there is no sorting order which will always place the nearly colliding values next to each other, see Appendix A). Note that if we continue to evaluate h on $O(2^{n/2})$ additional inputs, we expect to have one full-collision which is very easy to find in $O(2^{n/2})$ time (see [22]), and by definition it will also be an r-near-collision for any r. If we consider both the evaluation of h and the search step as unit time operations and try to minimize the total time complexity, our goal is to evaluate h on more points than absolutely necessary in order to make the search part faster, keeping each one of these complexities below the trivial bound of $2^{n/2}$.

When we consider the issue of memory complexity, there are many known algorithms ([2, 7, 15, 17, 18]) which use only constant or logarithmic amount of memory in order to find a full collision shortly after it is first created. Most of these algorithms are based on Pollard's rho method, which evaluates a chain of values of h, and uses the fact that any equality between two values on the chain implies an equality between their successors. Unfortunately, we cannot directly use this technique to find near-collisions, since the h-successors of two values which nearly collide can be arbitrarily far apart.

In 2011, Lamberger and Rijmen [9] suggested using an error correction code in order to turn some near-collisions into full collisions, and further studied such constructions in [8,10]. His proposed algorithm (described in greater detail in Section 2.1) uses a variant of Pollard's rho method which alternately applies the hash function h and the error correction operation e to some random initial value x_0 until it loops. He then hopes that the two colliding values after some e will be nearly colliding values after the previous h. In 2012, Leurent [11] extended the algorithm of Lamberger and Rijmen by using Van Oorschot and Wiener's technique [20] to find many simultaneous collisions with an algorithm which can be parallelized. Leurent's algorithm is not memoryless, and suggests a time-memory tradeoff for the problem of finding near-collisions. However, it used the same type of error correction codes that Lamberger and Rijmen used, along with some bit truncation (which can be viewed as a primitive type of error correction code whose code-words have zeroes in all the truncated positions).

Note that in standard error correction applications, we only have to correct bit strings which are in small balls surrounding each code-word, but the union of all these balls can be a tiny fraction of the whole space and thus we may be unable to change the vast majority of bit strings into code-words. In our application, we have to efficiently correct *any* bit string provided by h into a nearby codeword, and thus we want to use a *covering* code in which the union of all the sets of vectors which are corrected to each code-word exactly covers the space. A good code should minimize the following two types of errors in the algorithm: Two outputs of h may be very close to each other but will be missed by the algorithm if they are corrected by e into two different code-words, or they may be more than r apart but still mapped by e to the same code-word. The first type of error is very common since in a high dimensional space a random vector x is expected to be at maximal distance from its associated code-word, and thus even when we change x into a neighboring x' by a single bit flip, it is likely to move further away from the code-word and thus into the region surrounding a different code-word. The second type of error (which we call a "false alarm") is likely to occur when the error correction region around each code-word is a highly elongated ellipsoid rather than a sphere, which allows some pairs of vectors (e.g., at the opposite ends of the ellipsoid) to be mapped into the same code-word even though they are very far apart.

The ideal codes in our application are thus codes which partition the whole space into the disjoint union of spheres of the same radius. Such codes (called perfect codes) are extremely rare, and their known constructions can correct only one or three errors, which is too small for our application. To overcome this problem, Lamberger and Rijmen and Leurent proposed using a concatenation of several Hamming codes (which are perfect codes capable of correcting a single error), where each code is applied to a different set of bits. Unfortunately, this severely distorts the error correction regions surrounding each code-word. For example, if we divide $n = 240$ bits into 16 substrings and use a concatenation of 16 Hamming codes which can correct a single error among the 15 bits in each substring, then we can correct some combinations of 16 errors (if each error occurs in a different substring), but we cannot correct some combinations with just two errors (if they occur in the same substring). In addition, random patterns of 4 or more errors are likely to have such repetitions by the birthday paradox, and thus the average error correction capability of such concatenated codes is much smaller than the number of codes.

In this paper we propose to replace the Hamming codes by random linear error correcting codes denoted by $[n, k]$. Their code-words (which we hope to be uniformly distributed in the whole space) form a k-dimensional linear subset of vectors of length n over a finite field. The whole space can be partitioned into 2^k regions, where each region contains all the vectors which are closest to a particular code-word. A code has *covering radius* R if R is the smallest integer such that each region is contained in the R-ball around the code-word (and thus the union of these balls covers the entire space). A $[n, k]$ linear code with covering radius R is denoted by $[n, k] R$. In this case, when we find a collision in $f = e \circ h$, it is guaranteed to be a $2R$-near-collision in h by the triangle inequality.

Randomly selected linear codes do not have efficient error correction algorithms. We overcome this problem by devoting some of the available memory to an array, which can be prepared in advance (but not for free - to have a fair comparison with previous algorithms we take this preprocessing time into account). We show that the performance of such codes differs from the theoretically best possible covering codes only by small constant factors, and is considerably better than the concatenation of Hamming codes proposed by Lamberger and Rijmen and Leurent. In fact, the gap between the codes is already practically significant for small values of n, and grows in an unbounded way as we increase the parameters of the problem. In particular, we present experimental evidence that by using random codes we can improve Leurent's algorithm by a factor of at least 3 when trying to solve a simple problem such as finding a 33-near-collisions in the SHA-1 hash function, and present theoretical analysis which shows that finding 100-near-collisions in a hash function with $n = 1024$ can be improved by about three orders of magnitude for practical amounts of memory (see Table 1).

The paper is organized as follows. In Section 2 we describe previous algorithms for finding near-collisions. In Section 3 we analyze the properties of error correcting codes in the context of finding near-collisions. In Section 4 we consider random codes and linear random codes, and show how to construct them and how to implement their decoding function in constant time using a sufficiently large preprocessed array. In Section 5 we show how previous memoryless algorithms for finding near-collisions can be improved by using random codes. In Section 6 we analyze the time-memory tradeoffs of the new algorithm and demonstrate that it improves Leurent's algorithm. We conclude in Section 7.

2   Previous Work

2.1   Lamberger and Rijmen's Construction of a Linear Covering-Code

Lamberger and Rijmen [9] try to find codes which have an efficient error correction function e, have the desired covering radius R, and have a minimum number of code-words K. Their algorithm then uses Pollard's Rho-method to find a collision in $e \circ h$ in $O(\sqrt{K})$ time. By the triangle inequality, the distance between two vectors decoded into the same code-word is at most $2R$, and thus they find a $r = 2R$-near-collision in h.

Error correction codes which can correct any received message with errors in up to R coordinates are also covering-codes of radius R only if they are *perfect codes of radius R* [12]. Unfortunately, the only non-trivial known perfect codes in $\mathbb{F}_2^n$ are the $[23, 12]$ 3 Golay code and the $[2^i - 1, 2^i - i - 1]$ 1 Hamming codes $\mathcal{H}_i$ for $i \geq 1$. To efficiently handle more errors, Lamberger and Rijmen's approach is to use direct sum of several Hamming codes. Leurent [11] represents the direct sum as a concatenation of their input and output bits and we continue with Leurent's notation. The concatenation of m linear codes $[n_i, k_i]$ R_i for $1 \leq i \leq m$, results in $[n, k]$ R code where $n = \sum_{i=1}^{m} n_i$, $k = \sum_{i=1}^{m} k_i$ and $R = \sum_{i=1}^{m} R_i$. The nearest-neighbor of a vector in $\mathbb{F}_2^n$ in such code is the nearest-neighbor in each one of the substrings separately. We denote such a concatenation of codes by the operator $\oplus$, and $t \times C$ will stand for $\bigoplus_{i=1}^{t} C$ (concatenation of the code C t times).

Hamming codes exist only for $n = 2^k - 1$. For arbitrary values of n, Lamberger and Rijmen suggest using an $[n, k]R$ code which is the concatenation of several Hamming codes of two consecutive sizes, along with the trivial projection code $[i, i]0$ in $\mathbb{F}_2^i$, i.e.,

$$\mathcal{H}_R^n = s \times \mathcal{H}_{l+1} \oplus (R - s) \times \mathcal{H}_l \oplus \mathbb{F}_2^x \tag{1}$$

where $l := \lfloor \log_2 \left(\frac{n}{R} + 1 \right) \rfloor$, $s := \left\lfloor \frac{n - R(2^l - 1)}{2^l} \right\rfloor$ and $x := s \cdot \left(2^{l+1} - 1 \right) + (R - s) \left(2^l - 1 \right)$. The dimension of the code is:

$$k = n - R \cdot l - s \tag{2}$$

The size of the code is $K = 2^k$ and therefore his algorithm's complexity is $2^{k/2}$.

Lamberger and Rijmen prove that this method gives lower complexity than what can be achieved by projection alone (formally defined as the $[n, k]$ code $\mathcal{P}_k^n$ whose function $e = \pi_k^n$ sets certain $n - k$ coordinates to zero, which can also be viewed as the truncation of $n - k$ bits). Related analysis was carried out by Gordon et al. [4], who compared the minimal-hamming-weight decoding functions of projection and random codes from the viewpoint of *locally sensitive hash-functions*, and proved that random codes are asymptotically better than projections in minimizing the distance between random points in the space and their corresponding code-words for given ratios of k/n when $n \to \infty$.[1] Using $\pi_{n-r}^n \circ h$ in Pollard's algorithm, Lamberger and Rijmen find an r-near-collision after about $2^{(n-r)/2}$ hash computations. It can be improved by truncating $2r + 1$ bits, and using π_{n-2r-1}^n, for which a single trial of the Rho-method finds an r-near-collision with probability $1/2$. This gives an overall complexity of $2 \cdot 2^{(n-2r-1)/2} = 2^{(n+1)/2-r}$.

---

[1] This is the special case with $p = 1/2$ of his claim, in which the distribution of the points in the n-dimensional space is uniform, and there are no assumptions about them being close to the code-words.

2.2   Time-Memory Trade-Offs for Near-Collisions

Leurent in [11] provides a near-collision attack with a time-memory tradeoff. His main idea is to use the algorithm of Van Oorschot and Wiener [20] for parallel collision search of many collisions when having some memory available.

Let $\pi_{n'}^n$ be the projection function that truncates $\tau = n - n'$, and let $\psi_R^{n'}$ be the decoding function of $\mathcal{H}_R^{n'}$ for a given R. Then the function used in Pollard's algorithm is $\psi_R^{n'} \circ \pi_{n'}^n \circ h$, and it finds as many full collisions in its domain as needed, until one of them happens to be an r-near-collision in the original hash function h. An equivalent representation of the code is:

$$\mathcal{Y}_{R,\tau}^n := \mathcal{H}_R^{n'} \oplus \mathcal{P}_0^\tau \tag{3}$$

Let $p_{\tau,R}$ denote the probability that a detected near-collision in the algorithm is an r-near-collision, which can be calculated for given τ and R. Then $l_{\tau,R} := 1/p_{\tau,R}$ is the expected number of collisions that have to be considered until an r-near-collision in h is found. The dimensions of the codes $\mathcal{Y}_{R,\tau}^n$, which are the lengths of $\mathcal{H}_R^{n'}$, are denoted by $k_{\tau,R}$, and are given by Formula (2). Having M memory units, the time-complexity of the algorithm is bounded by:

$$\left(\sqrt{\frac{\pi}{2}} + \frac{5\sqrt{l_{\tau,R}}}{\sqrt{M}} \right) \cdot \sqrt{l_{\tau,R}} \cdot 2^{k_{\tau,R}/2} \tag{4}$$

Leurent provides a script that calculates the complexity for every possible R, τ in the ranges $0 \leq \tau \leq n$ and $0 \leq R \leq 2r$, and returns the estimated optimal parameters which gives the lowest complexity bound.

3   Properties of Code-Systems for Finding Near-Collisions

We analyze algorithms for finding r-near-collisions $(r > 0)$ using maps applied on the hash-values that increase the chance of colliding after the map for nearby hash values. We notice that error-correction codes are designed for different applications. Due to some analogous properties, we still use the term "decoding function" to describe the map, and the term "code" to describe the domain set of a map. However, the decoding function doesn't have to be the nearest-neighbor function. We will use the term *"code-system"* to describe the code together with its related decoding-function e. For convenience, we will use the same letter to describe the code-set and the code-system. The *radius of a code-system* in the generalized meaning is the maximum number of bit-flips made by the decoding function e.

The only near-collisions in h detectable by these algorithms are (m_1, m_2) such that both $h(m_1)$ and $h(m_1)$ are decoded to the same code-word ($e(h(m_1)) = e(h(m_2))$). This implies theoretical lower bounds for these methods.

From now on, we will use the following notations:

- β_C will be the probability that a random pair x, y is decoded to the same code-word in C.

- $\rho_C (R)$ will be the *probability mass function* (PMF)[2] of the distance between pairs decoded to the same code-word in C.
- $\mathcal{R}_C (R)$ will be the *cumulative distribution function* (CDF) of ρ_C , which describes the probability for a pair decoded to the same code-word to be R-close.
- $\varphi_C (R)$ is the chance that a random R-close pair is decoded to the same code-word in C.

In this section we denote:

A_r := the event that x, y are r-close. When r is fixed, we will simply use A.

B_C := the event that a random pair of vectors x, y are decoded to the same code-word in C. When C is fixed, we will simply use B.

It is easy to verify that $\Pr[A] = q_r^n$ which is a constant for given n and r, and $\Pr[B_C] = \beta_C$.

The probability that a random pair of messages is a detectable r-near-collision is $\Pr[A \wedge B]$, and by the birthday paradox, $1/\sqrt{\Pr[A \wedge B]}$ messages are required, while the lower bound without the code is $1/\sqrt{\Pr[A]}$.

By the conditional probability formula:

$$\Pr[A \wedge B] = \Pr[B|A] \cdot \Pr[A] = \Pr[A|B] \cdot \Pr[B] \tag{5}$$

Therefore, using the code increases the lower bound for hash computations by a factor of $1/\sqrt{\Pr[B|A]}$, which is exactly:

$$1/\sqrt{\varphi_C (r)} \tag{6}$$

The overall bound is:

$$\frac{1}{\sqrt{q_r^n} \cdot \sqrt{\varphi_C (r)}} \tag{7}$$

Hence, a higher $\varphi_C (r)$ is an indication of a potentially better code in terms of the number of hash calculations required.

We can also specifically look at an algorithm that ignores random collisions in the code-space until an r-near-collision is detected. The chance of a collision to be an r-near-collision is $\Pr[A|B]$ and therefore $1/\Pr[A|B]$ collisions are expected to be examined. As $\Pr[A|B] = \mathcal{R}_C (r)$, the time complexity of the algorithm is at least:

$$1/\mathcal{R}_C (r) \tag{8}$$

Hence, a higher $\mathcal{R}_C (r)$ is also an indication of a potentially better code in terms of the time complexity of the algorithm.

When we construct a code-system, there is a tradeoff between getting a higher φ_C and getting a higher $\mathcal{R}_C$. For example, the only code-system which satisfies $\varphi_C (r) = 1$ for $r \geq 1$ has a single code-word, and thus has the property $\Pr[A|B] = \Pr[A]$. So $1/\Pr[A] = 1/q_r^n$ computations are required, which is the square of the

---

[2] In statistics, a probability mass function f of a discrete random variable X is defined as: $f(k) = \Pr[X = k]$.

lower bound on hash-computations when the code is not used. However, code-systems which have equal β_C are comparable. By Equation (5):

$$\varphi_C(r) \cdot q_r^n = \mathcal{R}_C(r) \cdot \beta_C \tag{9}$$

Thus, $\varphi_C(r)$ and $\mathcal{R}_C(r)$ are proportional to each other with ratio β_C / q_r^n. In particular, when the codes are of the same size and the pre-images of all the code-words are of uniform size, $\beta_C = 1/|C|$, and then:

$$\mathcal{R}_C(r) = (q_r^n \cdot |C|) \cdot \varphi_C(r) \tag{10}$$

This holds for collections of $[n, k]$ linear codes, for which $|C| = 2^k$.

3.1 Concatenation of Several Code-Systems

Lamberger and Rijmen's algorithm described in Section 2.1 uses covering codes which are a concatenation of error-correction codes. We can similarly combine several code-systems with codes $C_1, C_2, \ldots, C_t$ of lengths $n_1, n_2, \ldots, n_t$ by concatenating the codes into a code $C = \bigoplus_{i=1}^{t} C_i$ of length $n = \sum_{i=1}^{t} n_i$. The decoding function of the resulting system is the composition of all the decoding functions applied to their appropriate substrings. If the systems are of radii $R_1, R_2, \ldots, R_t$, the resulting system is of radius $R = \sum_{i=1}^{t} R_i$.

As the bits of random vectors are unrelated:

$$\beta_C = \Pr[B_C] = \Pr\left[\bigwedge_{i=1}^{t} B_{C_i}\right] = \prod_{i=1}^{t} \Pr[B_{C_i}] = \prod_{i=1}^{t} \beta_{C_i} \tag{11}$$

By the definition of the distribution ρ_C, it will be the convolution of all ρ_{C_i}:

$$\rho_C(R) = (\rho_{C_1} * \rho_{C_2} * \ldots * \rho_{C_t})[R] \tag{12}$$

The $\mathcal{R}_C(R)$ is the CDF of $\rho_C(R)$, and $\varphi_C(r)$ is given by Equation (9).

Let's denote:

$$V_C := \prod_{i=1}^{t} V_{R_i}^{n_i} \tag{13}$$

It describes the actual volume of a single code-word's pre-image if the concatenated codes are prefect, or an upper bound on the volume otherwise.

We have $\mathcal{R}_C(2R) = 1$ and we can calculate $\varphi_C(2R)$:

$$\beta_C = \prod_{i=1}^{t} \beta_{C_i} \leq \prod_{i=1}^{t} q_{R_i}^{n_i} = \prod_{i=1}^{t} \frac{V_{R_i}^{n_i}}{2^{n_i}} = \frac{\prod_{i=1}^{t} V_{R_i}^{n_i}}{2^n} = \frac{V_C}{2^n} \tag{14}$$

So:

$$\varphi_C(2R) = \frac{\beta_C}{q_{2R}^n} \leq \frac{V_C}{V_{2R}^n} \tag{15}$$

For perfect codes the inequalities becomes equalities. It is easy to see that V_C is strictly maximal when $t = 1$, and is V_R^n in that case. No other case reaches the bound of the perfect code of length n and radius R. The bound decreases when the vector is partitioned into more parts.

4 Random Codes

Suppose we could randomly sample a function $f : \{0,1\}^n \to \{0,1\}$ that returns 1 on $2^n/p$ of the inputs (i.e, it returns 1 with probability $p = 2^{-\mu}$ on a random point). We can use it as code-word indicator to define the code:

$$C_f = \{x \in \{0,1\}^n \,|\, f(x) = 1\}$$

Each element is decoded to its nearest-neighbor. To make it well defined in our metric, we have to break ties by defining a secondary order. For example, we can decide that among two vectors with the same Hamming distance to x, one is closer if the XOR between it and x has a larger numeric value in binary representation.

4.1 Limited Radius Version

We may force the above system into a system of some bounded radius R by changing the decoding function so that if there is no code-word in $B_R^n(x)$ then x is not corrected into any member of C_f, and remains unchanged. This applies to vectors in:

$$\widehat{C_f} := \overline{\bigcup_{c \in C_f} B_R^n(c)}$$

Let's denote the probability that x is decoded into C_f by α. Given the probability p, α is the complement of the probability of not succeeding in V_R^n trials:

$$\alpha = \alpha(n, R, p) = 1 - (1-p)^{V_R^n} \tag{16}$$

Notice that if $p = 1/V_R^n$ then:

$$\alpha = 1 - \left(1 - \frac{1}{V_R^n}\right)^{V_R^n} \approx 1 - \frac{1}{e}$$

Compared to a theoretical perfect code of radius R, β_C is smaller by α^2. However, due to the search order for code-words within $B_R^n(x)$, the function $\mathcal{R}(\cdot)$ increases and so does the bound in Equation (8). Therefore $\varphi(\cdot)$ decreases by less than α^2 and the lower bound in Equation (7) increases by at most $1/\alpha \approx 1.58$ which is a small constant.

4.2 Estimating the Distribution Functions

The size of the code is $|C| = 2^n \cdot p$ and the distribution of the code-words in the full space is expected to be close to uniform, so the pre-images of the code-words are of similar volumes and therefore:

$$\beta_C \approx 1/|C| = \frac{2^{-n}}{p} \tag{17}$$

So $\mathcal{R}_C$ and φ_C are related by Equation (10). By definition, $\mathcal{R}_C$ can be calculated as CDF of ρ_C, and we describe how to calculate the latter in Appendix B.

4.3   Linear Random Code

If f can be easily sampled or calculated, the decoding of a single x can be conducted by an average of $1/p$ evaluations of f. Even though the decoding does not require any hash-evaluations, this may be a high complexity operation for a small p. For a truly random function a full description of the decoding process, for example using a full decoding table, would be impractical in terms of memory. However, as described below the situation is much better for linear codes.

A random $[n, n - \mu]$ code can be defined as the kernel of a randomly chosen matrix $H \in \{0, 1\}^{\mu \times n}$ of maximal rank, which is called the *parity-check* matrix of the code. We denote such a code by $\mathcal{C}_\mu^n$. The code-word indicator function $f(x)$ returns 1 when $Hx = 0$. In a neighborhood of a randomly chosen x it behaves similarly to a random function that returns 1 with probability $p = 2^{-\mu}$. As shown in Appendix C, an array of length $K = 2^\mu$ that describes the decoding operation in the vicinity of the zero code-word makes it possible to find the nearest neighbor in the entire space in constant time via simple shifts.

Even though we count the memory in units which are array elements, we would like to point out that due to the fact that we store only vectors of very low hamming-weight, the content of the array can be compressed and stored much more efficiently.

Read-Only-Memory vs. Random-Access-Memory. Notice that the array we construct does not depend on the hash-function, but only on the chosen linear code. For a constant parity-check matrix H which is randomly chosen in advance, the array can be constructed and hard-coded on a Read-Only-Memory (ROM), which is much cheaper than Random-Access-Memory (RAM), and thus a special-purpose computing machine may have more of it. Therefore, in scenarios which have almost no RAM, where we would normally use the Rho-method to find near-collisions, we may still use a large array if we have enough ROM. Such a array could also be stored on an external device or some common server which can be queried.

5   Rho-Method Algorithm Using Random-Codes

In Section 2.1 we described Lamberger and Rijmen's algorithm for finding $2R$-near-collisions which is based on a single run of the Rho-method algorithm. For a general code-system C with the restriction of $\mathcal{R}_C = 1$, the complexity of this algorithm is:

$$1/\sqrt{\beta_C}$$

For a random-code with limited radius R, the complexity is larger by a factor of about $1/\alpha$ compared to a theoretical perfect code (see Section 4). The gap between the β of Lamberger and Rijmen's construction and a perfect code grows as R grows, due to the partition of the vector into more parts, as can be seen in Formula (14).

For example when looking for a 24-near-collision in $\mathbb{F}_2^{128}$, a random-code could improve Lamberger and Rijmen's algorithm by a significant factor of 69.3, ignoring the cost of the decoding. However, a linear-random-code of radius R requires an array of about V_R^n length, which is impractical for V_{12}^{128}.

Using a random-code-system $2 \times \mathcal{C}_{26}^{64}$, when $\mathcal{C}_{26}^{64}$ has limited radius 6 and $\alpha \approx 0.7$, requires an array of practical length 2^{26}. We get $\beta_C = \left(\beta_{\mathcal{C}_{26}^{64}}\right)^2$ and $\beta_{\mathcal{C}_{26}^{64}} = \alpha^2 \cdot 2^{-64} \cdot V_6^{64} \approx 2^{-38.67}$. The total expected number of hash evaluations is $1/\sqrt{\beta_C} = 1/\beta_{\mathcal{C}_{26}^{64}} = 2^{38.67}$ (multiplied by a small constant that depends on the Rho-method's implementation, and can be ignored since it affects the two algorithms we compare similarly). Lamberger and Rijmen's algorithm uses a $[128, 87]12$ code and makes an expected number of $2^{43.5}$ hash evaluations. Therefore our code improves Lamberger and Rijmen's algorithm's time-complexity in this case by a factor of 28.4.

When we do not restrict the code to have $\mathcal{R}_C(2R) = 1$, the Rho-method has to be repeated $l_C := 1/\mathcal{R}_C(2R)$ times on average and the complexity increases to:

$$l_C/\sqrt{\beta_C}$$

Due to the large number of available parameters and the fact that we do not want to ignore constant factors, we estimated the optimal parameters for a random-code and for Lamberger and Rijmen's construction using a script which exhaustively searches over all their possible choices rather than via some asymptotic formula. For Lamberger and Rijmen's construction, we got that the optimal radius for finding a 24-near-collision is 28, and the number of expected hash-computations is about $2^{39.1}$, where $l_C \approx 8.6$. Using a linear random-code $2 \times \mathcal{C}_{34}^{128}$ our algorithm finds 24-near-collision after an expected number of $2^{35.3}$ hash computations, when $l_C \approx 39.1$. This is an improvement by a factor of 14.2.

6 Time-Memory Trade-Off Using Random-Codes

In Section 2.2 we described Leurent's algorithm [11] for finding near-collisions, using a table of size $M = 2^m$ to store the endpoints of chains in Van Oorschot and Wiener's algorithm. Our goal is to improve the algorithm by using random-codes instead of the concatenation of Hamming-codes.

Although we can sometimes distinguish between the memory that is used for the table and the memory that is used for the arrays (as described in Section 4.3), in this section we consider the harder case in which we have only one type of memory of size M. Therefore, we are limited to use not more than M memory units *including* the array size. Considering the fact that we can compress the array (see the remark in Section 4.3) and the fact that when such codes are concatenated more than once we still store the array only once, we can assume for the sake of simplicity that we can use a random-code with up to m equations, and still consume only a small fraction of the available memory for its associated array.

The code-systems we use are of the form:

$$\mathcal{Z}_{\mu,j,\tau}^n := j \times \mathcal{C}_{\mu}^{n'} \oplus \mathbb{F}_2^x \oplus \mathcal{P}_0^\tau \qquad (18)$$

where $n' = \lfloor (n - \tau)/j \rfloor$ and $x = (n - \tau) \mod j$.

When using the linear version of $\mathcal{C}_{\mu}^{n'}$, $\mathcal{Z}_{\mu,j,\tau}^n$ is a $[n, n - j \cdot \mu - \tau]$ linear code. The projection $\mathcal{P}_0^\tau$, for any value of τ, is a code-system that decodes all the vectors in $\mathbb{F}_2^x$ into the a single code-word and therefore has the properties: $\beta_{\mathcal{P}_0^\tau} = 1$, $\varphi_{\mathcal{P}_0^\tau} \equiv 1$ and $\mathcal{R}_{\mathcal{P}_0^\tau}$ distributes according to the binomial distribution $N\left(\tau, \frac{1}{2}\right)$. We calculate the distribution $\mathcal{R}_{\mathcal{Z}_{\mu,j,\tau}^n}(R)$ as a convolution between $N\left(\tau, \frac{1}{2}\right)$ and j times $\mathcal{R}_{\mathcal{C}_{\mu}^{n'}}(R)$. Then, the expected number of distinct collisions we have to find is:

$$l_{\mu,j,\tau} := 1/\mathcal{R}_{\mathcal{Z}_{\mu,j,\tau}^n}(r) \qquad (19)$$

In order to optimize the algorithm, we choose τ and μ that minimize the following upper bound on the time complexity:

$$T(\mu, j, \tau) = \left(\sqrt{\frac{\pi}{2}} + \frac{5\sqrt{l_{\mu,j,\tau}}}{\sqrt{M}} \right) \cdot \sqrt{l_{\mu,j,\tau}} \cdot 2^{(n-j\cdot\mu-\tau)/2} \qquad (20)$$

This can be done using brute-force computations of at most one value of $T(\mu, j, \tau)$ for every possible $\tau, j \cdot \mu \leq n$ (in fact, most of the parameters within the range can be easily ruled-out by simple estimations). We used a script to compute the exact values of the optimal parameters. Due to space limitations, we cannot include the script in this paper and refer the reader to the *eprint* version of the paper [5, Apendix D].

We can generalize the formula for code-systems which are not necessarily linear.[3] Given a code-system C and the parameter r of the problem, if we calculate β_C and $\mathcal{R}_C(r)$, we can get the following complexity upper bound:

$$T = \left(\sqrt{\frac{\pi}{2}} + \frac{5}{\sqrt{M \cdot \mathcal{R}_C(r)}} \right) \cdot \sqrt{\frac{1}{\mathcal{R}_C(r) \cdot \beta_C}} \qquad (21)$$

We would like to remind the reader that $\mathcal{R}_C(r) \cdot \beta_C = \varphi_C(r) \cdot q_r^n$ by Equation (9) which describes the probability of a random pair of vectors to be a detectable r-near-collision. This is another way to see that the significance of $\varphi_C(r)$ over $\mathcal{R}_C(r)$ grows when we have more memory. This form also emphasizes how Formula (20) should be adapted when limiting the radius.

---

[3] Van Oorschot and Wiener analyzed his algorithm for hash-functions that induce random-graphs on their output domains. We assume that the special properties of the graph induced by $e \circ h$ do not affect much the performance of the algorithm. However, such effect may exist and could be further analyzed.

In some application we may want to find a large number i of r-near-collisions. In this case we will have to find $i/\mathcal{R}_C(r)$ collisions and the formula becomes:

$$T = \left(\sqrt{\frac{\pi}{2}} + \frac{5 \cdot i}{\sqrt{M \cdot \mathcal{R}_C(r)}} \right) \cdot \sqrt{\frac{i}{\mathcal{R}_C(r) \cdot \beta_C}} \qquad (22)$$

6.1 Complexity Analysis

For relevant parameters, a single optimal random-code is better than 3 or more concatenated Hamming-Codes and this advantage grows when the number of concatenated codes increases, because it is particularly true for random-codes of similar dimension (which are comparable). However, the memory requirements for these random-codes also grows. The truncation-code $\mathcal{P}_0^\tau$ may have low $\mathcal{R}_{\mathcal{P}_0^\tau}$ for large τ, but at the same time it has the property $\varphi_{\mathcal{P}_0^\tau} \equiv 1$, which cannot be achieved by any other code-system. As was shown before, the importance of the φ distribution over $\mathcal{R}$ grows as M gets larger. Thus, generally speaking, the optimal τ is higher for larger M values, and the optimal code on the remaining part makes fewer bit-flips on random inputs. This can also be seen in [11, Table 2]. Thus, on one hand, when we have little memory we may not be able to construct the optimal random-code, and on the other hand, when we have a lot of memory the random-code does not improve much if at all.

Table 1. Comparison of number of the hash-calculations using Leurent's algorithm that uses Lamberger and Rijmen's construction versus our algorithm that uses random-codes in two variants: with and without limiting the radius. The upper entry is based on experimental results. The lower 3 entries are predictions based on calculations made using a script. T values are in logarithmic scale.

n, r	m	Leurent's algorithm [(3),(4)] $T(R,\tau)$	Our algorithm [(18),(20)] $T(\mu,j,\tau)$	Improvement ratio
160, 33	16	above 35.06 (2, 106)[a]	33.46 (15, 1, 98) - limited radius $R = 3$	above 3
			34.91 (15, 1, 98)	minor
1024, 80	38	389.7 (14, 206)	381.3 (38, 6, 70)	354.5
1024, 100	40	363.6 (13, 269)	354.1 (40, 7, 79)	728.8
1024, 100	52	358.6 (12, 294)	348.1 (52, 7, 16)	1482

[a] In 10 executions of Leurent's algorithm we got one unusually bad result ($2^{38.4}$ hash-computations) and two mildly bad ones. It could be a result of unoptimized elements of the algorithm. Thus, we reduced the time-complexity of Leurent's algorithm by artificially terminating it after a certain number of steps, which is the number of computations as in the 2nd worst result, and took into consideration that our effort resulted in 9 successful experiments instead of 10. Van Oorschot and Wiener suggest to restart the process after $10M = 10 \cdot 2^{16}$ collisions, but then it would not be optimal. However, we did not modify the experimental results of our algorithm in any way, and thus, the comparison in this table is actually biased in favor of Leurent's algorithm. In spite of this, our algorithm is about 3 times better in this small example in the limited radius version.

Although the concatenation of random-codes is never optimal when we ignore memory considerations, the actual memory requirements for codes of smaller dimensions are much smaller. Moreover, a concatenation of a competitive random-code several times requires only a single common array, and by the relations described in Section 3.1, the improvement factors over Lamberger and Rijmen's construction of similar dimensions are roughly multiplied. Therefore, the advantage of random-codes grows with n, since we can truncate some of the bits and still have enough bits to partition into large-enough bit-ranges for which practical sized random-codes can be constructed. In Table 1 we show concrete sets of parameters for which our algorithm improves Leurent's algorithm by several orders of magnitude.

Although the predicted upper bounds for the case we tested experimentally were similar in Leurent's algorithm and in the two versions of our algorithm, in practice our code with limited radius improved Leurent's algorithm for finding a 33-near-collision in SHA-1 hash function using 2^{16} memory by a factor of at least 3. This indicates that the improvement factors we obtain may be even higher than those we predict from our script. When we consider larger values of n, the improvement factors become much larger. For example, even our pessimistic estimates indicate that our algorithm is expected to improve Leurent's algorithm by a factor of about 730, which is almost three orders of magnitude, when looking for a 100-near-collision in a 1024-dimensional space using 2^{40} memory.

7   Conclusions and Further Work

In this paper we analyzed the two major statistical properties that make certain codes better for finding near-collisions, which are $\varphi(r)$ and $\mathcal{R}(r)$. We showed how to choose the optimal parameters of these random codes, described how to use arrays in order to decode an arbitrary vector into its related code-words in constant time, and discussed their advantages and disadvantages. We saw that random-codes have better properties than the concatenation of Hamming-codes of radius 1 for overall radii larger than 3, and that the gaps grow when the radius grows. We re-analyzed the time-memory trade-off of Leurent's algorithm after replacing the Hamming-codes with random-codes.

If we are allowed to use an unbounded amount of cheap ROM to store the fixed array used to decode vectors, we can achieve even larger improvements in many ranges of the parameters. For example, we can improve Lamberger and Rijmen's construction which uses the rho method by a factor of 69.3 in the settings described in Section 5. Without this assumption, i.e, when we had to use the available memory both for the array and the chain-endpoints, we still showed experimentally that for a small example there is a reduction of the number of hash evaluations by a factor of at least 3. The improvement ratio increases with n, and in Table 1 we showed concrete examples in which the improvement ratio is several orders of magnitude.

7.1   Further Work

We tried to analyze a *multi-code* variant of a random-code with limited radius R. Instead of leaving $1 - \alpha$ fraction of the hash-values unchanged by the decoding function, we serially try a sequence of decoding-functions with various linear shifts of the code-words, until one of them succeeds. In other words, we used the same random matrix A to define a series of codes as the solutions of $Ax = v_i$. We chose $v_0 = 0$ and then each v_i was chosen from among the image-points whose pre-images cannot be decoded into the previous codes. This way, the distances between different code-components are at least R. This variation has negligible effect on $\mathcal{R}_C$ but increases the β_C and φ_C to about $\frac{\alpha}{2-\alpha}$ of the linear-version instead of α^2. However, in our practical experiments we did not get conclusive results about the effect on the time-complexity when using Van Oorschot and Wiener's algorithm. This is possibly due to the effect of both variations on the statistical properties of the graph induced by $e \circ h$, which influences the probabilities of having new collisions overviewed along the run of the algorithm. This effect should be further analyzed.

We also suggest to consider other models in which the distance function is not the Hamming-distance. For example, we can consider a *weighted*-Hamming-distance in which bits have weights that correspond to the probability that they will be discarded when we extract a smaller number of random bits from the large output of the hash function.

A   A Full-Memory Approach to Near-Collisions

Yuval's algorithm for finding a collision [22] computes distinct hash values and stores them in a hash-table until a collision between a new value and previous values is detected. It is based on the birthday paradox and has the time and memory complexity of $2^{n/2}$.

As mentioned in [9], its generalization for near-collisions requires checking the presence of all the values in $B_r^n(x)$ for each new hash value x. It reaches the lower bound in hash-computations but has time-complexity of $O\left(2^{n/2} \cdot \sqrt{V_r^n}\right)$ and requires $O\left(2^{n/2}/\sqrt{V_r^n}\right)$ memory. We can reduce the time-complexity by storing for every computed x all the pairs (y, m) such that $y \in B_R^n(x)$, for some $R \leq r/2$, and beforehand check for collisions with previous points by querying the points in $B_{r-R}^n(x)$. It is easy to see that the first r-collision is detected by the algorithm. The time complexity is reduced to $O\left(2^{n/2} \cdot V_{r-R}^n/\sqrt{V_r^n}\right)$, which is still above $O\left(2^{n/2}\right)$, but the memory demand is increased to $O\left(2^{n/2} \cdot V_R^n/\sqrt{V_r^n}\right)$.

B   Calculation of ρ Distribution of Random-Codes

First we calculate the distribution of the number of bits flipped by the decoding function on a random value x, that we denote by χ:

$$\chi(R) = \Pr\left[d_H(x, Dec(x)) = R\right] =$$
$$\Pr\left[d_H(x, Dec(x)) \leq R\right] - \Pr\left[d_H(x, Dec(x)) \leq R - 1\right] \approx$$
$$\left(1 - e^{-pV_R^n}\right) - \left(1 - e^{-pV_{R-1}^n}\right) = e^{-pV_{R-1}^n} - e^{-pV_R^n} \quad (23)$$

Then we estimate the distribution over the triples $(dist1, dist2, \#overlaps)$, when $dist1 = d_H(c, x_1)$, $dist2 = d_H(c, x_1)$ and $\#overlaps$ stands for the number of bits that are flipped by the Dec function for both x_1 and x_2.

$dist1$ and $dist2$ are independent and distributed according to χ. Therefore:

$$\Pr\left[(dist1, dist2, \#overlaps) = (b_1, b_2, s)\right] =$$
$$\chi(b_1) \cdot \chi(b_2) \cdot \Pr\left[\#overlaps = s | dist1 = b_1 \wedge dist2 = b_2\right] \quad (24)$$

For any given pair, if the chances of every bit to be flipped on any message is the same, given $dist1$ and $dist2$, the number of overlaps between $dist1$ and $dist2$ bits distributes according to a hyper-geometric distribution. Combinatorially this is the number of "special items" that are picked when choosing $dist2$ distinct items from a pool of n items, of which $dist1$ are "special":

$$\Pr\left[\#overlaps = s | dist1 = b_1 \wedge dist2 = b_2\right] = \frac{\binom{b_1}{s} \cdot \binom{n-b_1}{b_2-s}}{\binom{n}{b_2}} \quad (25)$$

However, this estimate does not take into account that the secondary order we use works for our benefit. Generally speaking, the lower bits have larger probability to be flipped. For example, if not more than b bits are flipped in the majority of the vectors, the probability of finding a code-word within the first V_b^n trials is some $\alpha > 1/2$. If a certain vector is not decoded within these trials, there are V_b^{n-1} next trials to find a code-word that differs in $b+1$ bits, one of which is the first bit. Since b is much smaller than n, the ratio between V_b^n and V_b^{n-1} is close to 1. Therefore, in probability of almost α, if more than b bits are flipped, one of the flipped bits is going to be the first.

The actual distance between the hash values of pairs that correspond to a certain triple is $dist = dist1 + dist2 - 2 \cdot \#overlaps$.

$$\rho(R) := \sum_{b_1+b_2-2s=R} \Pr\left[(dist1, dist2, \#overlaps) = (b_1, b_2, s)\right] \quad (26)$$

When we limit the radius to R, for every $b > R$ the value of $\chi(b)$ should be set to 0, and for every $b \leq R$ it should be divided by $\alpha = \sum_{i \leq R} \chi(i)$.

C   Efficient Decoding with Precomputed Arrays

For a linear code defined in Section 4.3, the nearest-neighbor of a given x is $c = x + \Delta$ when Δ is the minimal-weight vector such that $H(x \oplus \Delta) = 0$ (defined primarily by the Hamming distance and secondly by the secondary order described in Section 4), or equivalently $Hx = H\Delta$. In coding theory, such Δ is called a *coset-leader* of $x + C$ and corresponds to the syndrome $y = Hx$ [12, Chapter 1].[4]

---

[4] For regular *maximum-likelihood* decoding, the coset-leader can be chosen arbitrarily among the minimum Hamming weight vectors.

We construct an array, which is called a *standard-array* of the code, that stores the coset-leaders for any syndrome $y \in \mathbb{F}_2^\mu$ in index y. Then the decoding process for a given x takes a constant time: calculate the value $Hx \in \mathbb{F}_2^\mu$, find its corresponding coset-leader from the array, and get the code-word $c = x \oplus \Delta$.

The array can be initialized efficiently by generating the smallest vectors in an increasing order starting from $\Delta = 0^n$, in a sort of a spiral (according to the primary and the secondary order). In each iteration we calculate $y = H\Delta$, and store the Δ at the y-th entry of the array if it is empty. We stop when the entries of the array are filled. The length of the array is $K = 2^\mu$. By the coupon collector argument, the expected number of vectors being overviewed is $K \cdot \log K$, which is almost linear in K.

The distance between two vectors x_1 and x_2 that are encoded to the same code-word, by addition of Δ_1 and Δ_2 respectively, is:

$$d_H(x_1, x_2) = \|x_1 \oplus x_2\|_H = \|(c \oplus \Delta_1) \oplus (c \oplus \Delta_2)\|_H = \|\Delta_1 \oplus \Delta_2\|_H$$

Due to the linear independence of the rows of H, which we may assume, both Δ_1 and Δ_2 could be seen as randomly taken from the values in the array, independently of the code-word. Having the array set, the distribution $\rho(r)$ is exactly the distribution of distances between pairs of entries in the array. It can be calculated in about $K^2/2$ steps or approximated experimentally by pair sampling.

References

1. Biham, E., Chen, R.: Near-collisions of sha-0. In: Franklin, M. (ed.) CRYPTO 2004. LNCS, vol. 3152, pp. 290–305. Springer, Heidelberg (2004)
2. Brent, R.P.: An improved monte carlo factorization algorithm. BIT Numerical Mathematics **20**(2), 176–184 (1980)
3. Chabaud, F., Joux, A.: Differential collisions in sha-0. In: Krawczyk, H. (ed.) CRYPTO 1998. LNCS, vol. 1462, pp. 56–71. Springer, Heidelberg (1998)
4. Gordon, D.M., Miller, V.S., Ostapenko, P.: Optimal hash functions for approximate matches on the-cube. IEEE Transactions on Information Theory **56**(3), 984–991 (2010)
5. Shamir, A., Polak, I.: Using random error correcting codes in near-collision attacks on generic hash-functions. Cryptology ePrint Archive, Report 2014/417 (2014). http://eprint.iacr.org/2014/417.pdf
6. Jakobsson, M., Juels, A.: Proofs of work and bread pudding protocols. In: Preneel, B. (ed.) Secure Information Networks. IFIP, vol. 23, pp. 258–272. Springer, Boston (1999)
7. Knuth, D.E.: Seminumerical algorithm (arithmetic) the art of computer programming, vol. 2 (1981)
8. Lamberger, M., Mendel, F., Rijmen, V., Simoens, K.: Memoryless near-collisions via coding theory. Designs, Codes and Cryptography **62**(1), 1–18 (2012)
9. Lamberger, M., Rijmen, V.: Optimal covering codes for finding near-collisions. In: Biryukov, A., Gong, G., Stinson, D.R. (eds.) SAC 2010. LNCS, vol. 6544, pp. 187–197. Springer, Heidelberg (2011)
10. Lamberger, M., Teufl, E.: Memoryless near-collisions, revisited. Information Processing Letters **113**(3), 60–66 (2013)

11. Leurent, G.: Time-memory trade-offs for near-collisions. In: Moriai, S. (ed.) FSE 2013. LNCS, vol. 8424, pp. 205–218. Springer, Heidelberg (2014)
12. MacWilliams, F.J., Sloane, N.J.A.: The theory of error-correcting codes, vol. 16. Elsevier (1977)
13. Mendel, F., Schläffer, M.: On free-start collisions and collisions for TIB3. In: Samarati, P., Yung, M., Martinelli, F., Ardagna, C.A. (eds.) ISC 2009. LNCS, vol. 5735, pp. 95–106. Springer, Heidelberg (2009)
14. Nakamoto, S.: Bitcoin: A peer-to-peer electronic cash system. Consulted 1, 2012 (2008)
15. Nivasch, G.: Cycle detection using a stack. Information Processing Letters 90(3), 135–140 (2004)
16. Pramstaller, N., Rechberger, C., Rijmen, V.: Exploiting coding theory for collision attacks on sha-1. In: Smart, N.P. (ed.) Cryptography and Coding 2005. LNCS, vol. 3796, pp. 78–95. Springer, Heidelberg (2005)
17. Quisquater, J.-J., Delescaille, J.-P.: How easy is collision search. new results and applications to des. In: Brassard, G. (ed.) CRYPTO 1989. LNCS, vol. 435, pp. 408–413. Springer, Heidelberg (1990)
18. Sedgewick, R., Szymanski, T.G., Yao, A.C.: The complexity of finding cycles in periodic functions. SIAM Journal on Computing 11(2), 376–390 (1982)
19. Stevens, M., Sotirov, A., Appelbaum, J., Lenstra, A., Molnar, D., Osvik, D.A., de Weger, B.: Short chosen-prefix collisions for md5 and the creation of a rogue CA certificate. In: Halevi, S. (ed.) CRYPTO 2009. LNCS, vol. 5677, pp. 55–69. Springer, Heidelberg (2009)
20. Van Oorschot, P.C., Wiener, M.J.: Parallel collision search with cryptanalytic applications. Journal of Cryptology 12(1), 1–28 (1999)
21. Wang, X., Yu, H.: How to break MD5 and other hash functions. In: Cramer, R. (ed.) EUROCRYPT 2005. LNCS, vol. 3494, pp. 19–35. Springer, Heidelberg (2005)
22. Yuval, G.: How to swindle rabin. Cryptologia 3(3), 187–191 (1979)

Linear Cryptanalysis of FASER128/256 and TriviA-ck

Chao Xu[1]([✉]), Bin Zhang[1,2], and Dengguo Feng[1]

[1] TCA, Institute of Software, Chinese Academy of Sciences, Beijing, China
{zhangbin,xuchao}@tca.iscas.ac.cn
[2] State Key Laboratory of Computer Science, Institute of Software,
Chinese Academy of Sciences, Beijing, China

Abstract. In this paper, we evaluate the security of FASER and TriviA-ck, two authenticated encryption schemes submitted to the CAESAR competition, by linear cryptanalysis method. It is pointed out that the most serious weakness of FASER is that the linear FSRs and nonlinear FSRs do not interact with each other. Thus by linear approximation of the MAJ function, it is possible to derive linear approximations involving the keystream words and the linear FSR initial states only. We found some such equations with correlation coefficient 2^{-1} for FASER128 and FASER256, which lead to the initial state recovery of the linear FSRs with an off-line time complexity of 2^{36} to compute a low weight multiple polynomial, and a negligible online time complexity, which is the polynomial time of the total length of linear FSRs, given 2^{36} keystream words. Moreover, we construct some distinguishers involving two consecutive steps of keystream words with a correlation coefficient of 2^{-2} for FASER128 and FASER256. Thus we only need 16 keystream words for FASER128 and FASER256 to distinguish the corresponding keystream from random sequence, respectively. These distinguishers do not rely on any weakness of the MIX operation, so the distinguishing attack will still work even when the FASER designers modify the MIX function. Finally, we use the linear sequential circuit approximation (LSCA) method to analyze TriviA-ck, a stream cipher similar to Trivium, and derive a linear function of consecutive keystream bits with a correlation coefficient of 2^{-76}. This shows that TriviA-ck has much more weaker immunity against linear cryptanalysis than Trivium.

Keywords: Stream ciphers · CAESAR · FASER128/256 · TriviA-ck · Correlation attack · Distinguisher

1 Introduction

FASER [5] and TriviA-ck [4] are two authenticated encryption schemes submitted to the ongoing CAESAR competition [1]. FASER uses two state registers,

This work was supported by the National Grand Fundamental Research 973 Program of China (Grant No. 2013CB338002) and the programs of the National Natural Science Foundation of China (Grant No. 60833008, 60603018, 61173134, 91118006, 61272476).

W. Meier and D. Mukhopadhyay (Eds.): INDOCRYPT 2014, LNCS 8885, pp. 237–254, 2014.
DOI: 10.1007/978-3-319-13039-2_14

identical in size, one for encryption and one for authentication. The state registers used in FASER are updated using a Feedback Shift Register(FSR). And FASER has been withdraw from the CAESAR competition, since the guess and determine attack proposed in [8]. TriviA-ck uses a stream cipher Trivia-SC and VPV-Hash as the mathematical components. Trivia-SC is an extended version of Trivium [7], which is one of the eStream finalists and can be efficiently implemented both in software and hardware. In this paper, we apply the linear attack to these two stream ciphers.

Linear cryptanalysis has been widely used to analyse stream cipher [10], especially the correlation attack [12,13] and distinguishing attack [6]. Correlation attack is a classical method in the cryptanalysis of stream ciphers, which exploits some statistically correlation between the produced keystream and the output of certain underlying linear FSR sequences. Intuitively, if the correlation is large it should be possible to use it to recover the initial state of the linear FSR. Distinguishing attack uses some statistical bias in the keystream to distinguish the sequence whether it is generated by the cipher or not.

In this paper, we first propose a linear attack against FASER128/256. The proposed attack exploits linear approximations of the output function, called NLF in FASER. Our attack mainly uses three weaknesses in the design of FASER. First, the sub-FSRs update different regions of the state independently. So the linear FSRs and the non-linear FSRs do not interact with each other. Therefore, we can find some linear approximations that only involve the linear FSR state bits and the keystream bits with large correlation coefficients. Then we apply the correlation attack to recover the initial state of the linear FSRs. Second, even though the MIX function, which combines information from across the state register, has a good diffusion property, there are still some bits in the output of MIX only containing the state bits coming from linear FSRs. Third, the output function is the bitwise majority function, called MAJ. It is too simple and has some linear approximations with large correlation coefficients. Combining these weaknesses, we find 2 and 27 linear approximations for FASER128 and FASER256, respectively, which are only containing the linear FSR state bits and the keystream bits, with correlation coefficients $1/2$. Then, with the fast correlation attack algorithm [12], we recover the initial state of the linear FSRs. After that, we give a distinguishing attack of FASER128/256. This attack does not use the design flaw of MIX. Our attack only need 16 keystream words for FASER128 and FASER256, respectively. We also do some experiments to verify our analysis. Golić [9] developed an effective method for the linear model determination based on Linear Sequential Circuit Approximation (LSCA) of autonomous finite state machines. By this method, we could find a unbalanced linear function of consecutive keystream bits and then applying the standard chi-square statistical test. Using Golić's method, we extract a linear distinguisher of the TriviA-ck with correlation coefficient 2^{-76}. For Trivium, the correlation coefficient is about 2^{-72} [11]. But Trivium only has 80-bit key and 288-bit state, while TriviA-ck has 128-bit key and 384-bit state. Furthermore, the output of Trivium is a linear function but TriviA-ck is not. If we remove the non-linear term of it, the correlation coefficient becomes 2^{-60}, the cipher will not be secure against linear cryptanalysis. All these facts show that as an extended version of Trivium, TriviA-ck has much more weaker immunity against linear cryptanalysis than Trivium.

This paper is organized as follows. FASER and TriviA-ck stream cipher are introduced in Section 2. The linear cryptanalysis of FASER is presented in Section 3. And the LSCA of TriviA-ck is shown in Section 4. In Section 5, we conclude this paper.

2   Description of FASER and TriviA-ck

2.1   The Stream Cipher FASER

FASER is a family of authenticated ciphers. The family consists of two parent ciphers: FASER128 and FASER256. Here we only consider the encryption process. Denote the register by X, and X_i denotes the i-th 64-bit word of register X. The state register in FASER128 can be represented as $X = (X_3, X_2, X_1, X_0)$ and similarly, in FASER256 the state is $X = (X_5, X_4, X_3, X_2, X_1, X_0)$. The state register is updated using a FSR. The FSR is made up of n sub-FSRs, where n is 8 and 12 for FASER128 and FASER256, respectively. The sub-FSRs are divided into two kinds: linear FSRs and non-linear FSRs. Each 64-bit word in the register X is divided into 2 sub-FSRs. Here we denote the i most significant bits and i least significant bits of a 64-bit word x by $H_i(x)$ and $L_i(x)$, respectively. In FASER128, the FSR is

$$FSR(X) = (FSR_3(X_3), FSR_2(X_2), FSR_1(X_1), FSR_0(X_0)),$$

where

$$FSR_0(X_0) = FSR33(H_{33}(X_0)) \parallel FSR31(L_{31}(X_0)),$$
$$FSR_1(X_1) = FSR35(H_{35}(X_1)) \parallel FSR29(L_{29}(X_1)),$$
$$FSR_2(X_2) = FSR41(H_{41}(X_2)) \parallel FSR23(L_{23}(X_2)),$$
$$FSR_3(X_3) = FSR47(H_{47}(X_3)) \parallel FSR17(L_{17}(X_3)).$$

In FASER256, the FSR is

$$FSR(X) = (FSR_5(X_5), FSR_4(X_4), FSR_3(X_3), FSR_2(X_2), FSR_1(X_1), FSR_0(X_0)),$$

where

$$FSR_0(X_0) = FSR33(H_{33}(X_0)) \parallel FSR31(L_{31}(X_0)),$$
$$FSR_1(X_1) = FSR35(H_{35}(X_1)) \parallel FSR29(L_{29}(X_1)),$$
$$FSR_2(X_2) = FSR37(H_{37}(X_2)) \parallel FSR27(L_{27}(X_2)),$$
$$FSR_3(X_3) = FSR41(H_{41}(X_3)) \parallel FSR23(L_{23}(X_3)),$$
$$FSR_4(X_4) = FSR43(H_{43}(X_4)) \parallel FSR21(L_{21}(X_4)),$$
$$FSR_5(X_5) = FSR47(H_{47}(X_5)) \parallel FSR17(L_{17}(X_5)).$$

Here, $\|$ denotes concatenation. For each X_i, there is one non-linear FSR and one linear FSR. The notation x_i denotes the i-th bit of x. And all the sub-FSRs are denoted as follow.

$$FSR17(x) : y \leftarrow x_{16} + x_{15} + x_{14} \cdot x_{13}, \qquad (x_{16}, \cdots, x_1, x_0) \leftarrow (x_{15}, \cdots, x_0, y)$$
$$FSR21(x) : y \leftarrow x_{20} + x_{19} + x_{17} \cdot x_{15}, \qquad (x_{20}, \cdots, x_1, x_0) \leftarrow (x_{19}, \cdots, x_0, y)$$
$$FSR23(x) : y \leftarrow x_{22} + x_{21} + x_{12} \cdot x_{11}, \qquad (x_{22}, \cdots, x_1, x_0) \leftarrow (x_{21}, \cdots, x_0, y)$$
$$FSR27(x) : y \leftarrow x_{26} + x_{24} + x_{21} \cdot x_{16} + 1, \quad (x_{26}, \cdots, x_1, x_0) \leftarrow (x_{25}, \cdots, x_0, y)$$
$$FSR29(x) : y \leftarrow x_{28} + x_{27} + x_{19} \cdot x_{12}, \qquad (x_{28}, \cdots, x_1, x_0) \leftarrow (x_{27}, \cdots, x_0, y)$$
$$FSR31(x) : y \leftarrow x_{30} + x_{11} + x_{21} \cdot x_{13}, \qquad (x_{30}, \cdots, x_1, x_0) \leftarrow (x_{29}, \cdots, x_0, y)$$
$$FSR33(x) : y \leftarrow x_{32} + x_{19}, \qquad\qquad\qquad (x_{32}, \cdots, x_1, x_0) \leftarrow (x_{31}, \cdots, x_0, y)$$
$$FSR35(x) : y \leftarrow x_{34} + x_{32}, \qquad\qquad\qquad (x_{34}, \cdots, x_1, x_0) \leftarrow (x_{33}, \cdots, x_0, y)$$
$$FSR37(x) : y \leftarrow x_{36} + x_{35} + x_{32} + x_{30}, \qquad (x_{36}, \cdots, x_1, x_0) \leftarrow (x_{35}, \cdots, x_0, y)$$
$$FSR41(x) : y \leftarrow x_{40} + x_{37}, \qquad\qquad\qquad (x_{40}, \cdots, x_1, x_0) \leftarrow (x_{39}, \cdots, x_0, y)$$
$$FSR43(x) : y \leftarrow x_{42} + x_{41} + x_{37} + x_{36}, \qquad (x_{42}, \cdots, x_1, x_0) \leftarrow (x_{41}, \cdots, x_0, y)$$
$$FSR47(x) : y \leftarrow x_{46} + x_{41}, \qquad\qquad\qquad (x_{46}, \cdots, x_1, x_0) \leftarrow (x_{45}, \cdots, x_0, y)$$

Here $+$ and $\cdot$ denote addition and multiplication over $GF(2)$, respectively. After updating the state, the MIX function combines information from across the state register. In FASER128, the MIX function input is the entire state (X_3, X_2, X_1, X_0) and the output is three 64-bit words such that:

$$Y_0 = (X_0 \lll 3) \oplus (X_1 \lll 12) \oplus (X_2 \lll 43) \oplus (X_3 \lll 27),$$
$$Y_1 = (X_0 \lll 22) \oplus (X_1 \lll 54) \oplus (X_2 \lll 5) \oplus (X_3 \lll 30), \qquad (1)$$
$$Y_2 = (X_0 \lll 50) \oplus (X_1 \lll 35) \oplus (X_2 \lll 14) \oplus (X_3 \lll 60).$$

In FASER256, the MIX function operates on the state $(X_5, X_4, X_3, X_2, X_1, X_0)$ and the output is also three 64-bit words such that:

$$Y_0 = (X_0 \lll 3) \oplus (X_1 \lll 11) \oplus (X_4 \lll 26) \oplus (X_5 \lll 36),$$
$$Y_1 = (X_1 \lll 49) \oplus (X_2 \lll 16) \oplus (X_3 \lll 24) \oplus (X_4 \lll 59), \qquad (2)$$
$$Y_2 = (X_0 \lll 30) \oplus (X_2 \lll 41) \oplus (X_3 \lll 51) \oplus (X_5 \lll 2).$$

The notation $\lll$ denotes bitwise rotation to the left. The output function of FASER is the bitwise majority function, called MAJ function. The MAJ function operates on 64-bit words and is denoted by $Z = (Y_0 \wedge Y_1) \vee (Y_0 \wedge Y_2) \vee (Y_1 \wedge Y_2)$. The pseudo-code for the updating process in encryption is

---
UPDATE(X^{t-1}):
  $X^t := \mathrm{FSR}^8(X^{t-1})$;
  $(Y_2^t, Y_1^t, Y_0^t) := \mathrm{MIX}(X^t)$;
  $Z^t := \mathrm{MAJ}(Y_2^t, Y_1^t, Y_0^t)$;

---

where FSR^8 represents clock the FSR 8 times and the superscript t denotes the time instant.

2.2   The Stream Cipher Trivia-SC

TrivaA-ck uses a stream cipher Trivia-SC and an efficient universal hash function VPV-Hash. Here, we only consider the encryption process, i.e., Trivia-SC. It is a stream cipher which is a modified version of Trivium [7]. TriviaSC is loaded with 128-bit key and 128-bit IV and generates keystream bit by bit. It is consist of three non-linear feedback shift registrars (NFSR) A, B and C of size 132 bits, 105 bits and 147 bits respectively. The internal state can be represent by $S_1, S_2, \cdots, S_{384}$, where $A = (S_1, S_2, \cdots, S_{132}), B = (S_{133}, S_{134}, \cdots, S_{237})$ and $C = (S_{238}, S_{239}, \cdots, S_{384})$.

The key stream generation consists of an iterative process which extracts the values of 17 specific state bits and used them both to update 3 bits of the state and to compute 1 bit of keysteam z_i. The state bits are then rotated and the process repeats itself until the requested N bits of the keystream have been generated. The update function and the output function is denote as follows. $\wedge$ represents "and" between two bits.

---

for $i = 1$ **to** N **do**:

$z_i = S_{66} \oplus S_{132} \oplus S_{201} \oplus S_{237} \oplus S_{303} \oplus S_{384} \oplus (S_{102} \wedge S_{198})$;

$t_1 \leftarrow S_{66} \oplus S_{132} \oplus (S_{130} \wedge S_{131}) \oplus S_{228}$;

$t_2 \leftarrow S_{201} \oplus S_{237} \oplus (S_{235} \wedge S_{236}) \oplus S_{357}$;

$t_3 \leftarrow S_{303} \oplus S_{384} \oplus (S_{382} \wedge S_{383}) \oplus S_{75}$;

$(S_1, S_2, \cdots, S_{132}) \leftarrow (t_3, S_1, S_2, \cdots, S_{131})$;

$(S_{133}, S_{134}, \cdots, S_{237}) \leftarrow (t_1, S_{133}, S_{134}, \cdots, S_{236})$:

$(S_{238}, S_{239}, \cdots, S_{384}) \leftarrow (t_2, S_{238}, S_{239}, \cdots, S_{383})$;

end for

---

3   Linear Cryptanalysis of FASER128/256

In this section, we first give some design flaws of the FASER. Then, we evaluate the security of FASER by linear cryptanalysis method.

3.1   Three Design Flaws of FASER

According to the Subsection 2.1, each state word X_i can be divided into two parts: the linear part and the non-linear part. And these two parts are updated independently. Therefore, FASER can be viewed as in Fig.1. This structure of stream cipher is not a strong design, since the security of cipher is mainly dependent on the properties of the filter function. If the filter function are not very strong, some attacks can be applied.

Furthermore, FASER uses a MIX function to achieve the diffusion within each sub-FSR state. The authors claim that each output bit of MIX is affected by at least one input bit from a non-linear FSR. But by Eq.(1) and Eq.(2), we can easily find 2 and 27 bits in the output of $(Y_2, Y_1, Y_0) = \text{MIX}(X)$ that only contain state bits coming from linear FSRs for FASER128 and FASER256,

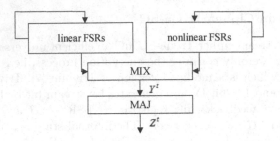

Fig. 1. The equivalent structure of FASER

respectively. Hereafter in this section, for a 64-bit state word X_i, we use the superscript t, i.e., X_i^t, to denote the time instant and the second subscript j, i.e., $X_{i,j}$, to denote the j-th significant bit of X_i. Then for FASER128 these two output bits are given as follows.

$$Y_{0,2}^t = X_{0,63}^t \oplus X_{1,54}^t \oplus X_{2,23}^t \oplus X_{3,39}^t,$$
$$Y_{1,53}^t = X_{0,31}^t \oplus X_{1,63}^t \oplus X_{2,48}^t \oplus X_{3,23}^t.$$

And for FASER256, the 27 bits are shown in the Appendix A. Note that the variables on the right side of the above equations are all coming from the linear FSRs. In the following subsections, we will show that even when assuming that each output bit of MIX is affected by at least one input bit from a non-linear FSR, we can still apply the distinguishing attack and the state recovering attack.

FASER chooses a bitwise majority function (MAJ) that operates on 64-bit words to generate keystream. Denote the keystream word $Z = (Z_0, Z_1, \cdots, Z_{63})$, where $Z_i \in GF(2)$. Then the MAJ function can be expressed in bitwise as follows.

$$(Z_0, \cdots, Z_{63}) = (\text{maj}(Y_{2,0}, Y_{1,0}, Y_{0,0}), \cdots, \text{maj}(Y_{2,63}, Y_{1,63}, Y_{0,63})), \qquad (3)$$

where maj is the Boolean majority function. So each bit in the keystream Z is only dependent on 3 bits of (Y_2, Y_1, Y_0). Further, the MAJ function is known to have large correlations. We can derive some linear approximations only containing the linear FSRs' bits.

3.2   Deriving Linear Approximations

The linear space of n-dimensional binary vectors is denoted by $GF(2)^n$. The inner product for $a = (a_1, \cdots, a_n), b = (b_1, \cdots, b_n) \in GF(2)^n$ is defined as $a \cdot b = a_1 b_1 \oplus \cdots \oplus a_n b_n$. The correlation coefficient of a random Boolean variable X is denoted by $c(X) = Pr(X = 0) - Pr(X = 1)$. Assume a 64-bit linear mask $\Gamma = (\gamma_0, \cdots, \gamma_{63})$. Then we have

$$\Gamma \cdot Z^t = \Gamma \cdot \text{MAJ}(Y_2^t, Y_1^t, Y_0^t) = \sum_{i=0}^{63} \gamma_i \text{maj}(Y_{2,i}^t, Y_{1,i}^t, Y_{0,i}^t).$$

and we choose two linear masks as follows.

$$\Gamma_1 = 0x0000000000000004, \Gamma_2 = 0x0020000000000000.$$

So we have $\Gamma_1 \cdot Z^t = \text{maj}(Y_{2,2}^t, Y_{1,2}^t, Y_{0,2}^t)$ and $\Gamma_2 \cdot Z^t = \text{maj}(Y_{2,53}^t, Y_{1,53}^t, Y_{0,53}^t)$. Since $\text{maj}(x, y, z)$ is correlated to each input x, y or z with correlation coefficient $1/2$, we have two linear approximations with correlation coefficient $1/2$ as follows.

$$\Gamma_1 \cdot Z^t = Z_2^t = Y_{0,2}^t = X_{0,63}^t \oplus X_{1,54}^t \oplus X_{2,23}^t \oplus X_{3,39}^t,$$
$$\Gamma_2 \cdot Z^t = Z_{53}^t = Y_{1,53}^t = X_{0,31}^t \oplus X_{1,63}^t \oplus X_{2,48}^t \oplus X_{3,23}^t.$$

Note that these two linear approximations only involve the keystream bits and the state bits of linear FSRs. The same techniques can be easily applied to FASER256, shown in Appendix A. Next, with these linear approximations, we apply the fast correlation attack [12] to recover the initial state of the linear FSRs.

3.3   Fast Correlation Attack on FASER128

Now, we use the linear approximation $Z_2^t = X_{0,63}^t \oplus X_{1,54}^t \oplus X_{2,23}^t \oplus X_{3,39}^t$ of FASER128 with correlation coefficient $1/2$ as an example to recover the initial state of linear FSRs. This attack process can be regarded as the model shown in Fig. 2. The cryptanalysis's problem can be formulated as follows. Given a length N received sequence $Z^1, Z^2, \cdots, Z^N$, recover the initial state of the four FSRs. We can use the divide and conquer method to solve this problem. Assume that we first want to recover the state of FSR47. So we need to find a low-weight parity check of the minimal common multiple polynomial of the other three FSRs's feedback polynomial, i.e., $f = (x^{33} + x^{20} + 1)(x^{35} + x^{33} + 1)(x^{41} + x^{38} + 1)$. In the precomputation phase, we apply the generalized birthday algorithm [2,14] to find a multiple polynomial with weight 5 of f, which need 2^{36} time complexity and the degree is 2^{36}. Using this low-weight polynomial, we can eliminate the state

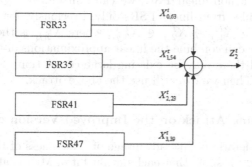

Fig. 2. The model of correlation attack

variables of the other three FSRs. Now the linear approximation is transformed into

$$Z_2^{t_1} \oplus Z_2^{t_2} \oplus Z_2^{t_3} \oplus Z_2^{t_4} \oplus Z_2^{t_5} = X_{1,54}^{t_1} \oplus X_{1,54}^{t_2} \oplus X_{1,54}^{t_3} \oplus X_{1,54}^{t_4} \oplus X_{1,54}^{t_5}.$$

with correlation coefficient 2^{-5}. For brevity, let $z_t = Z_2^{t_1} \oplus Z_2^{t_2} \oplus Z_2^{t_3} \oplus Z_2^{t_4} \oplus Z_2^{t_5}$ and $u_t = X_{1,54}^{t_1} \oplus X_{1,54}^{t_2} \oplus X_{1,54}^{t_3} \oplus X_{1,54}^{t_4} \oplus X_{1,54}^{t_5}$. Therefore, the problem of recovering the state of FSR47 can be modeled as a binary symmetric channel (BSC), with the crossover probability $p = 0.5 - 2^{-5}$ and $Pr[u_i = z_i] = 1 - p = 0.53125$.

Meier and Staffelbach [12] presented two different algorithms for fast correlation attacks, where they use a correlation between the keystream and the output stream of LFSR. The basic idea in our attack is similar to the fast correlation attack Algorithm B in [12]. Each linear FSR is clocked 8 times and then outputs a bit. Thus the output of linear FSR is a 8 sampling sequence of the original one. Since α^8 is a primitive element, where α is a root of feedback polynomial, this output sequence has the same feedback polynomial with the original one. Therefore, these four linear FSRs, i.e., FSR33,FSR35,FSR41,FSR47, are all have taps $t = 2$. As stated in [12] for $t = 2$ Algorithm B is very effective, we can directly apply the Meier's Algorithm B to recover the state of FSR47. The detail of Algorithm B is given in the Appendix C.

Here we analyse the complexities. Suppose N outputs are available. N is lower bounded by roughly $N \geq l/(1 - H(p)) = 2^{12}$, where $H(\cdot)$ is the binary entropy function and $l = 47$ is the length of FSR47. The degree of the multiple polynomial is 2^{36}, which means that we need about $O(2^{36})$ keystream words. So we set $N = 2^{36}$. The time complexity of Algorithm B is a polynomial time attack of the length of FSR47. So the time complexity of recovering the state of FSR47 is $O(47)$ and the number of keystream is $O(2^{36})$. After recovering the FSR47, then we can use the same technique to recovery the remaining linear FSRs. In total, the time complexity is $O(109)$ and the keystream are not exceed 2^{36}.

Furthermore, we assume that each output bit of MIX is affected by at least one input bit from a non-linear FSR, we can still find a linear approximation involving the variables from linear FSRs only. Preciously, for a linear approximation $Z_j^t = X_{0,j_1}^t \oplus X_{1,j_2}^t \oplus X_{2,j_3}^t \oplus X_{3,j_4}^t$, where X_{3,j_4}^t is the variable from a non-linear FSR. We can combine the linear approximations of the above form at different time. And there may be existing large correlation of $\sum_{k=1}^{a} X_{3,j_4}^{t_k}$, this idea is used in [3]. Then we can still use the above attack.

3.4   Distinguishing Attack on the Improved Version of FASER

The previous considerations take advantage of a weakness of the MIX function. In this subsection, we assume that each output bit of MIX is affected by at least one input bit from a non-linear FSR.

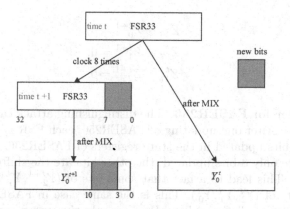

Fig. 3. The part state of Y_0^t and Y_0^{t+1}

Distinguisher for FASER128. Note that in one update of FASER128, the FSR is clocked 8 times, where each clock of the FSR results in each sub-FSR being clocked. Consequently, there are 64-bits updated in the state registers of FASER128. Since the sub-FSRs are updated indecently, after one clock of FASER128 there are still some state bits that are not updated, but shifted. For example, the FSR33 updating process is shown in Fig.3. We find that the output $(Y_2^{t+1}, Y_1^{t+1}, Y_0^{t+1})$ of MIX at time $t+1$ may have some bits are just the shifted version of (Y_2^t, Y_1^t, Y_0^t). For example, assume that the state variables at time t are denoted by $X_i^t = (X_{i,0}^t, X_{i,1}^t, \cdots, X_{i,63}^t)$. So we have

$$Y_{0,3}^t = X_{0,0}^t \oplus X_{1,55}^t \oplus X_{2,24}^t \oplus X_{3,40}^t,$$
$$Y_{2,3}^t = X_{0,17}^t \oplus X_{1,32}^t \oplus X_{2,53}^t \oplus X_{3,7}^t;$$
$$Y_{0,3+8}^{t+1} = X_{0,0}^t \oplus X_{1,55}^t \oplus X_{2,24}^t \oplus X_{3,40}^t,$$
$$Y_{2,3+8}^{t+1} = X_{0,17}^t \oplus X_{1,32}^t \oplus X_{2,53}^t \oplus X_{3,7}^t.$$

Then combining with the MAJ function, we have

$$Z_3^t = \mathrm{maj}(Y_{2,3}^t, Y_{1,3}^t, Y_{0,3}^t),$$
$$Z_{3+8}^{t+1} = \mathrm{maj}(Y_{2,3+8}^{t+1}, Y_{1,3+8}^{t+1}, Y_{0,3+8}^{t+1}).$$

From the above equation, we note that $Y_{0,3}^t = Y_{0,3+8}^{t+1}$ and $Y_{2,3}^t = Y_{2,3+8}^{t+1}$. Then we approximate the maj function by $Z_3^t = Y_{0,3}^t$ and $Z_{3+8}^{t+1} = Y_{0,3+8}^{t+1}$, we get a distinguisher

$$Z_3^t \oplus Z_{3+8}^{t+1} = 0,$$

with correlation coefficient 2^{-2}. Note that here we do not use the weakness of MIX function and the output of MIX can also involve the non-linear FSRs state bits. This is only one example of distinguishers, there also exist many other distinguishers with correlation coefficient 2^{-2}. We give some of them as follows.

$$Z_{56}^t \oplus Z_0^{t+1} = 0,$$
$$Z_{57}^t \oplus Z_1^{t+1} = 0,$$
$$Z_{17}^t \oplus Z_{25}^{t+1} = 0.$$

Distinguisher for FASER256. The distinguishing attack can also be applied to FASER256. After one updating of FASER256, each FSR is clock 8 times. So there are 96 bits updated in the state registers of FASER256. But for each sub FSR there are only 8 bits updated, the other bits are the shifted version of last time instant. This lead to a fact that some bits of $(Y_2^{t+1}, Y_1^{t+1}, Y_0^{t+1})$ are just shifting 8 bits of (Y_2^t, Y_1^t, Y_0^t). This is the same case in FASER128. There are also many different distinguishers. Here we only give one example.

$$Y_{0,56}^t = X_{0,53}^t \oplus X_{1,45}^t \oplus X_{4,30}^t \oplus X_{5,20}^t,$$
$$Y_{0,0}^{t+1} = X_{0,53}^t \oplus X_{1,45}^t \oplus X_{4,30}^t \oplus X_{5,20}^t.$$

According to the MAJ function, we have

$$Z_{56}^t = \mathrm{maj}(Y_{2,56}^t, Y_{1,56}^t, Y_{0,56}^t),$$
$$Z_0^{t+1} = \mathrm{maj}(Y_{2,0}^{t+1}, Y_{1,0}^{t+1}, Y_{0,0}^{t+1}).$$

Then approximating the maj function with $Z_{56}^t = Y_{0,56}^t$ and $Z_0^{t+1} = Y_{0,0}^{t+1}$, we get a distinguisher $Z_0^{t+1} \oplus Z_{56}^t = 0$ with 2^{-2}.

We need 16 keystream words to distinguish the keystream of FASER from the random sequence for FASER128 and FASER256, respectively. These distinguishers tell us that the MIX and MAJ functions are not designed very well to against the linear cryptanalysis.

Experiment In our experiment, we test 2^{20} random keys and IVs. For each pair of key and IV, we generate 2^{25} keystream words, and obtain the correlation coefficient of the above distinguishers for FASER128 and FASER256 are near 2^{-2}. The experimental results confirm our analysis.

4   Linear Sequential Circuit Approximation of the Trivia-SC

In this section, we construct a distinguisher of the Trivia-SC stream cipher, which is an extended version of Trivium, with the Linear Sequential Circuit Approximation (LSCA) proposed by Golić [9]. And then we point out that the immunity against linear cryptanalysis of Trivia-SC is much weaker than Trivium. Here we use the first subscript to denote the time and the second one represents the bit position in the state vector, i.e., $S_{t,i}$.

Keystream generators of stream cipher can be regarded as autonomous finite-state machines (FSM) whose initial state are dependent on a secret key. A binary autonomous FSM is defined by $\mathbf{S}_t = F(\mathbf{S}_{t-1})$ and $z_t = f(\mathbf{S}_t)$, where

$F : GF(2)^M \to GF(2)^M$ is the next-state function, $f : GF(2)^M \to GF(2)$ is the output function, $\mathbf{S}_t$ is the state vector at time t, and z_t is the keystream bit.

Golić [9] has proved that for a binary keystream generator with M bits of memory whose initial state is chosen uniformly at random, there exists a linear function of at most $M + 1$ consecutive output bits, i.e., $L(z_t, z_{t+1}, \cdots, z_{t+M})$, which is an unbalanced function of the initial state variables. With this function, the adversary can apply the standard chi-square frequency statistical test to make a distinguishing attack. The test is successful if and only if the length of the keystream is $1/c^2$, where c is the correlation coefficient of this linear function. Next, we use Golić's method to extract the LSCA of Trivia-SC step by step.

For Trivia-SC we have $M = 384$ and $\mathbf{S}_t = (S_{t,1}, S_{t,2}, \cdots, S_{t,384})^T$ be the state vector at time t. Now Trivia-SC can be regarded as an autonomous FSM denoted above. We first decompose the output function f and the component functions in the next-state function F of the keystream generator into the sum of linear functions and an unbalanced function. Thus we have

$$\mathbf{S}_t = A\mathbf{S}_{t-1} \oplus \Delta(\mathbf{S}_{t-1}), \ t \geq 1 \tag{4}$$

$$z_t = B\mathbf{S}_t \oplus \gamma(\mathbf{S}_t), \qquad t \geq 1 \tag{5}$$

where $\mathbf{S}_t$ is a 384×1 column vector, A is an 384×384 matrix and B is a 1×384 row vector, Δ is a 384×1 noise vector and $\gamma(\mathbf{S}_t)$ is a noise Boolean function. In order to acquire the parameters of the above the equations, we first study the update function and output function in details. There are only four non-linear Boolean functions in each round of Trivia-SC. They are given as follows.

$$S_{t+1,1} = S_{t,66} \oplus S_{t,132} \oplus (S_{t,130} \wedge S_{t,131}) \oplus S_{t,228},$$

$$S_{t+1,133} = S_{t,201} \oplus S_{t,237} \oplus (S_{t,235} \wedge S_{t,236}) \oplus S_{t,357},$$

$$S_{t+1,238} = S_{t,303} \oplus S_{t,384} \oplus (S_{t,382} \wedge S_{t,383}) \oplus S_{t,75},$$

$$z_i = S_{t,66} \oplus S_{t,132} \oplus S_{t,201} \oplus S_{t,237} \oplus S_{t,303} \oplus S_{t,384} \oplus (S_{t,102} \wedge S_{t,198}).$$

We note that the only non-linear term is $x_1 \wedge x_2$, which has the best linear approximation $x_1 \wedge x_2 = 0$ with correlation coefficient 0.5. Since only three new bits are introduced to $\mathbf{S}_t$ in each round of TriviaSC. Thus, we can rewrite the Eq.(4) and Eq.(5) as follows.

$$\mathbf{S}_t = A\mathbf{S}_{t-1} \oplus H\Delta'_t, \ t \geq 1 \tag{6}$$

$$z_t = B\mathbf{S}_t \oplus \gamma_t, \qquad t \geq 1 \tag{7}$$

where $H = [h_{i,j}]$ is a 384×3 matrix whose all entries are zero, except $h_{1,1}, h_{133,2}, h_{238,3}$ are one, $\Delta'_t = [\delta^1_t, \delta^2_t, \delta^3_t]^T$ and γ_t are the noise terms corresponding to the linear approximations of four non-linear functions as shown above. According to the updating and output functions of TriviaSC, the matrices A and B are as follows[1].

---

[1] e_i denotes the i-th row of 384×384 identity matrix.

$$A = \begin{bmatrix} e_{303} \oplus e_{384} \oplus e_{75} \\ e_1 \\ \cdots \\ e_{131} \\ e_{66} \oplus e_{132} \oplus e_{228} \\ e_{133} \\ \cdots \\ e_{236} \\ e_{201} \oplus e_{237} \oplus e_{357} \\ e_{238} \\ \cdots \\ e_{383} \end{bmatrix},$$

$$B = e_{66} \oplus e_{132} \oplus e_{201} \oplus e_{237} \oplus e_{303} \oplus e_{384}.$$

By Eq.(6) we can get the relation between $\mathbf{S}_t$ and $\mathbf{S}_0$ as follows[2].

$$\mathbf{S}_t = A^t \mathbf{S}_0 \oplus \sum_{l=0}^{t-1} A^l H \Delta'_{t-l}, \ t \geq 1$$

The degree of the minimal polynomial of A is $m = 375$, which is denoted by $\varphi(x) = \sum_{k=0}^{m} \varphi_k x^k$ in Appendix B. Since $\sum_{k=0}^{m} \varphi_k A^k = 0$, it follows that

$$\sum_{k=0}^{m} \varphi_k \mathbf{S}_{t+k} = \sum_{k=0}^{m} \varphi_k \left(A^{t+k} \mathbf{S}_0 \oplus \sum_{l=0}^{t+k-1} A^l H \Delta'_{t+k-l} \right)$$

$$= \sum_{k=0}^{m} \varphi_k \sum_{l=0}^{t+k-1} A^l H \Delta'_{t+k-l}$$

$$= \sum_{\tau=0}^{m} \sum_{r=0}^{m-\tau} \varphi_{r+\tau} A^r H \Delta'_{t+\tau}$$

And according to Eq.(7), we have

$$\sum_{k=0}^{m} \varphi_k z_{t+k} = \sum_{k=0}^{m} (\varphi_k B \mathbf{S}_{t+k} \oplus \varphi_k \gamma_{t+k}) = \sum_{\tau=0}^{m} \sum_{r=0}^{m-\tau} \varphi_{\tau+r} B A^r H \Delta'_{t+\tau} \oplus \sum_{k=0}^{m} \varphi_k \gamma_{t+k}$$

For brevity, let $C_\tau = \sum_{r=1}^{m-\tau} \varphi_{r+\tau} B A^r H$ be a 1×3 vector for $\tau = 0, \cdots, m$ and $\Delta_t^{(i)} = (\delta_t^i, \delta_{t+1}^i, \cdots, \delta_{t+m}^i)$ be a $1 \times (m+1)$ vector for $i = 1, 2, 3$, that is

$$\begin{bmatrix} \Delta_t^{(1)} \\ \Delta_t^{(2)} \\ \Delta_t^{(3)} \end{bmatrix} = [\Delta'_t, \Delta'_{t+1}, \cdots, \Delta'_{t+m}]$$

---

[2] For simplicity, we use $\sum$ to represent the addition over $GF(2)$.

Therefore, we have

$$\sum_{k=0}^{m} \varphi_k z_{t+k} = \sum_{\tau=0}^{m} C_\tau \Delta'_{t+\tau} \oplus \sum_{k=0}^{m} \varphi_k \gamma_{t+k}$$

$$= \Delta_t^{(1)} C^{(1)} \oplus \Delta_t^{(2)} C^{(2)} \oplus \Delta_t^{(3)} C^{(3)} \oplus \sum_{k=0}^{m} \varphi_k \gamma_{t+k} \qquad (8)$$

where $C^{(i)}$ is the i-th column of the $(m+1) \times 3$ matrix

$$\begin{bmatrix} C_0 \\ C_1 \\ \cdots \\ C_m \end{bmatrix} = \begin{bmatrix} C^{(1)}, C^{(2)}, C^{(3)} \end{bmatrix}$$

Now by Eq.(8), $\sum_{k=0}^{m} \varphi_k z_{t+k}$ is unbalanced and we can apply the standard chi-square statistical test to distinguish the keystream sequence from a purely random binary sequence. And the length of sequence is $1/c^2$, where c is the correlation of $\sum_{k=0}^{m} \varphi_k z_{t+k}$.

$C^{(1)}, C^{(2)}$ and $C^{(3)}$ can be expressed in the polynomial forms. They are listed in Appendix B. Since every component of $\Delta_t^{(1)}, \Delta_t^{(2)}, \Delta_t^{(3)}$ and δ_t is product of two binary terms, therefore it has correlation coefficient $1/2$. Then, the total correlation coefficient of $\sum_{k=0}^{m} \varphi_k z_{t+k}$ is

$$c = 2^{-(wt(C^{(1)}(x))+wt(C^{(2)}(x))+wt(C^{(3)}(x))+wt(\varphi(x)))}, \qquad (9)$$

where $wt(\cdot)$ denote the number of terms in a polynomial. Therefore, according to the Appendix B the correlation coefficient of Eq.(8) is 2^{-228}.

From Eq.(8) and Eq.(9) we find that linear equations with greater correlations may be found by searching the low weight multiple polynomials of $\varphi(x), C^{(1)}(x), C^{(2)}(x)$ and $C^{(3)}(x)$, i.e., multiplying both the most right and the most left sides of Eq.(8) by a polynomial $g(x)$. Then we compute

$$wt(g(x)\varphi(x)) + wt(g(x)C^{(1)}(x)) + wt(g(x)C^{(2)}(x)) + wt(g(x)C^{(3)}(x)).$$

We have made a thorough search over all the polynomials $g(x)$ with the degree up to 29. When $g(x) = x^9 + x^6 + x^3 + 1$, we get the lowest weight $wt(g(x)\varphi(x)) + wt(g(x)C^{(1)}(x)) + wt(g(x)C^{(2)}(x)) + wt(g(x)C^{(3)}(x)) = 76$. Therefore we get a linear relation of keystream bit determined by $g(x)\varphi(x)$ with the correlation coefficient 2^{-76}. It means that the time complexity for distinguishing the output sequence of the Trivia-SC from a truly random sequence is $O(2^{152})$. The complexity is higher than the security of Trivia-SC, i.e., $O(2^{128})$.

The LSCA evaluation result of Trivia-SC reveals that compared to the Trivium stream cipher, Trivia-SC has the following shortcomings against linear cryptanalysis. First remind that though the output function of Trivium has no non-linear term, the linear cryptanalysis of it has a complexity of 2^{144}. However, if we remove the non-linear term of the output function of Trivia-SC, we have $wt(g(x)C^{(1)}(x)) + wt(g(x)C^{(2)}(x)) + wt(g(x)C^{(3)}(x)) = 60$, then the correlation coefficient becomes 2^{-60} and the complexity of linear cryptanalysis becomes 2^{120}, which is lower then the security level of Trivia-SC. This shows that the tap positions of Trivia-SC are not carefully chosen. Second, Trivia-SC has a 384-bit state, which is much larger than Trivium's 288-bit state, but the immunity against linear cryptanalysis is not improved accordingly to a desirable level. Since Trivium only has 80-bit key, while Trivia-SC has 128-bit key, the complexity of linear cryptanalysis of Trivium is 2^{64} higher than the required security and for Trivia-SC is only 2^{24}.

5   Conclusion

In this paper, we have evaluated the security of FASER128/256 and Trivia-SC by linear cryptanalysis method. Our results have clearly shown that the design of FASER128/256 are totally insecure. The MIX function has some serious flaws, and even improving the MIX function to such a function that each output bit is affected by at least one input bit from a nonlinear FSR, we can still find a linear approximation involving the variables from linear FSRs only. We use the LSCA method to analyse the security of Trivia-SC, and compare the results with linear cryptanalysis of Trivium. We point out that Trivia-SC has some shortcomings against linear cryptanalysis. Trivia-SC has much more weaker immunity against linear cryptanalysis than Trivium.

A   The Linear Approximations of FASER256

The 27 bits in the output of $(Y_2, Y_1, Y_0) = MIX(X)$ that only contain state bits coming from linear FSRs are as follows.

$$Y_{0,0}^t = X_{0,61}^t \oplus X_{1,53}^t \oplus X_{4,38}^t \oplus X_{5,28}^t,$$
$$Y_{0,1}^t = X_{0,62}^t \oplus X_{1,54}^t \oplus X_{4,39}^t \oplus X_{5,29}^t,$$
$$Y_{0,2}^t = X_{0,63}^t \oplus X_{1,55}^t \oplus X_{4,40}^t \oplus X_{5,30}^t,$$
$$Y_{0,53}^t = X_{0,50}^t \oplus X_{1,42}^t \oplus X_{4,27}^t \oplus X_{5,17}^t,$$
$$Y_{0,54}^t = X_{0,51}^t \oplus X_{1,43}^t \oplus X_{4,28}^t \oplus X_{5,18}^t,$$
$$Y_{0,55}^t = X_{0,52}^t \oplus X_{1,44}^t \oplus X_{4,29}^t \oplus X_{5,19}^t,$$
$$Y_{0,56}^t = X_{0,53}^t \oplus X_{1,45}^t \oplus X_{4,30}^t \oplus X_{5,20}^t,$$
$$Y_{0,57}^t = X_{0,54}^t \oplus X_{1,46}^t \oplus X_{4,31}^t \oplus X_{5,21}^t,$$
$$Y_{0,58}^t = X_{0,55}^t \oplus X_{1,47}^t \oplus X_{4,32}^t \oplus X_{5,22}^t,$$
$$Y_{0,59}^t = X_{0,56}^t \oplus X_{1,48}^t \oplus X_{4,33}^t \oplus X_{5,23}^t,$$
$$Y_{0,60}^t = X_{0,57}^t \oplus X_{1,49}^t \oplus X_{4,34}^t \oplus X_{5,24}^t,$$
$$Y_{0,61}^t = X_{0,58}^t \oplus X_{1,50}^t \oplus X_{4,35}^t \oplus X_{5,25}^t,$$
$$Y_{0,62}^t = X_{0,59}^t \oplus X_{1,51}^t \oplus X_{4,36}^t \oplus X_{5,26}^t,$$
$$Y_{0,63}^t = X_{0,60}^t \oplus X_{1,52}^t \oplus X_{4,37}^t \oplus X_{5,27}^t,$$

$$Y_{1,47}^t = X_{1,62}^t \oplus X_{2,31}^t \oplus X_{3,23}^t \oplus X_{4,52}^t,$$

$$Y_{1,48}^t = X_{1,63}^t \oplus X_{2,32}^t \oplus X_{3,24}^t \oplus X_{4,53}^t,$$

$$Y_{2,19}^t = X_{0,53}^t \oplus X_{2,42}^t \oplus X_{3,32}^t \oplus X_{5,17}^t,$$

$$Y_{2,20}^t = X_{0,54}^t \oplus X_{2,43}^t \oplus X_{3,33}^t \oplus X_{5,18}^t,$$

$$Y_{2,21}^t = X_{0,55}^t \oplus X_{2,44}^t \oplus X_{3,34}^t \oplus X_{5,19}^t,$$

$$Y_{2,22}^t = X_{0,56}^t \oplus X_{2,45}^t \oplus X_{3,35}^t \oplus X_{5,20}^t,$$

$$Y_{2,23}^t = X_{0,57}^t \oplus X_{2,46}^t \oplus X_{3,36}^t \oplus X_{5,21}^t,$$

$$Y_{2,24}^t = X_{0,58}^t \oplus X_{2,47}^t \oplus X_{3,37}^t \oplus X_{5,22}^t,$$

$$Y_{2,25}^t = X_{0,59}^t \oplus X_{2,48}^t \oplus X_{3,38}^t \oplus X_{5,23}^t,$$

$$Y_{2,26}^t = X_{0,60}^t \oplus X_{2,49}^t \oplus X_{3,39}^t \oplus X_{5,24}^t,$$

$$Y_{2,27}^t = X_{0,61}^t \oplus X_{2,50}^t \oplus X_{3,40}^t \oplus X_{5,25}^t,$$

$$Y_{2,28}^t = X_{0,62}^t \oplus X_{2,51}^t \oplus X_{3,41}^t \oplus X_{5,26}^t,$$

$$Y_{2,29}^t = X_{0,63}^t \oplus X_{2,52}^t \oplus X_{3,42}^t \oplus X_{5,27}^t.$$

Here we only give one example using the first equation. Let the linear mask be $\Gamma_1 = 0x0000000000000001$, then we have $\Gamma \cdot Z^t = \mathrm{maj}(Y_{2,0}^t, Y_{1,0}^t, Y_{0,0}^t)$. With approximating the maj function, we can acquire the following relation.

$$\Gamma \cdot Z^t = Z_0^t = Y_{0,0}^t = X_{0,61}^t \oplus X_{1,53}^t \oplus X_{4,38}^t \oplus X_{5,28}^t,$$

with correlation coefficient $\frac{1}{2}$.

B   The Expression of $\varphi(x), C^{(1)}(x), C^{(2)}(x), C^{(3)}(x)$

$$\begin{aligned}
\varphi(x) = {} & x^{375} + x^{372} + x^{363} + x^{360} + x^{351} + x^{348} + x^{339} + x^{336} + x^{327} + \\
& x^{324} + x^{315} + x^{312} + x^{303} + x^{297} + x^{291} + x^{285} + x^{276} + x^{273} + \\
& x^{264} + x^{261} + x^{255} + x^{249} + x^{243} + x^{237} + x^{231} + x^{225} + x^{219} + \\
& x^{213} + x^{207} + x^{204} + x^{195} + x^{192} + x^{183} + x^{177} + x^{174} + x^{165} + \\
& x^{162} + x^{159} + x^{156} + x^{153} + x^{150} + x^{147} + x^{144} + x^{141} + x^{132} + \\
& x^{129} + x^{120} + x^{117} + x^{93} + x^{90} + x^{84} + x^{78} + x^{69} + x^{66} + \\
& x^{27} + x^{24} + x^{15} + x^{12} + x^3 + 1, \\
C^{(1)}(x) = {} & x + x^4 + x^{13} + x^{16} + x^{25} + x^{31} + x^{37} + x^{43} + x^{49} + x^{55} + \\
& x^{61} + x^{64} + x^{67} + x^{70} + x^{73} + x^{76} + x^{79} + x^{88} + x^{91} + x^{94} + \\
& x^{97} + x^{100} + x^{103} + x^{106} + x^{109} + x^{112} + x^{115} + x^{124} + x^{127} + \\
& x^{130} + x^{139} + x^{142} + x^{151} + x^{154} + x^{157} + x^{160} + x^{163} + x^{166} + \\
& x^{169} + x^{172} + x^{175} + x^{178} + x^{181} + x^{184} + x^{187} + x^{190} + x^{193} +
\end{aligned}$$

$$x^{196} + x^{205} + x^{208} + x^{214} + x^{220} + x^{223} + x^{232} + x^{235} + x^{244} +$$
$$x^{247} + x^{250} + x^{259} + x^{262} + x^{271} + x^{274} + x^{283} + x^{286} + x^{295} +$$
$$x^{298} + x^{307} + x^{310},$$

$$C^{(2)}(x) = x + x^4 + x^{13} + x^{16} + x^{25} + x^{28} + x^{58} + x^{61} + x^{67} + x^{73} +$$
$$x^{79} + x^{82} + x^{85} + x^{88} + x^{91} + x^{100} + x^{106} + x^{112} + x^{133} +$$
$$x^{136} + x^{139} + x^{142} + x^{145} + x^{154} + x^{157} + x^{160} + x^{163} + x^{166} +$$
$$x^{184} + x^{187} + x^{205} + x^{208} + x^{214} + x^{220} + x^{226} + x^{232} + x^{238} +$$
$$x^{241} + x^{280} + x^{283} + x^{292} + x^{295} + x^{304} + x^{307},$$

$$C^{(3)}(x) = x + x^4 + x^{10} + x^{16} + x^{22} + x^{28} + x^{34} + x^{37} + x^{46} + x^{49} + x^{58} +$$
$$x^{61} + x^{70} + x^{73} + x^{76} + x^{79} + x^{88} + x^{94} + x^{100} + x^{103} + x^{106} +$$
$$x^{109} + x^{112} + x^{115} + x^{124} + x^{127} + x^{136} + x^{139} + x^{142} + x^{145} +$$
$$x^{148} + x^{151} + x^{154} + x^{172} + x^{175} + x^{187} + x^{190} + x^{199} + x^{202} +$$
$$x^{211} + x^{214} + x^{238} + x^{241} + x^{244} + x^{247} + x^{250} + x^{259} + x^{262} +$$
$$x^{271} + x^{274} + x^{283} + x^{286} + x^{295} + x^{298} + x^{307} + x^{310}.$$

C   Fast Correlation Attack

Algorithm 1. Meier and Staffelbach's Algorithm B

Input: $\mathbf{z} = z_1, z_2, \cdots, z_N$ and $\mathbf{u} = u_1, \cdots, u_N$
$Pr\{z_i = u_i\} = 1 - p$. m different parity checks of weight $t + 1$.
Processing:
1: compute the two thresholds p_{thr} and N_{thr}.
2: for round $r = 1, 2, \cdots$
3:    for iteration $r = 1, 2, \cdots, \alpha$
4:      for $n = 0, 1 \cdots$ calculate the value $p*$ and assign $p_n = p*$.
5:      find $N_\omega = |\{n | p_n > p_{thr}\}|$.
6:      if $N_\omega \geq N_{thr}$, break.
7:    complement the bits of $\mathbf{z}$ with $p_n > p_{thr}$. Reset all probabilities to p.
8:    if all bits z_n fulfil the parity check defining the LFSR, break.
9: Terminate with $\mathbf{u} = \mathbf{z}$.

According to [12], the average amount of relations obtained for each LFSR output u_n can be computed as $m = (t+1)\log(\frac{N}{2l}) = 2^{29}$. These $m = 2^{29}$ equations can be represented as

$$u_n + b_n^{(1)} = 0,$$
$$u_n + b_n^{(2)} = 0,$$
$$\cdots$$
$$u_n + b_n^{(m)} = 0,$$

where $b_n^{(j)} = u_{n+i_1} + u_{n+i_2}$. These variables can not be observed directly, we define L_j as follows.

$$z_n + y_n^{(1)} = L_n^{(1)},$$
$$z_n + y_n^{(2)} = L_n^{(2)},$$
$$\cdots$$
$$z_n + y_n^{(m)} = L_n^{(m)},$$

where $y_n^{(j)}$ is the sum of z_n's corresponding to the u_n used in $b_n^{(j)}$ above. Recall $Pr[u_n = z_n] = 1 - p$, and the bias $\epsilon = \frac{1}{2} - p$. Define $s = Pr[y = b] = \frac{1}{2} + 2^{t-1}\epsilon^t$. When the L are observed, some number h of them are 0 and $m - h$ are 1. So the conditional probability is

$$p* = Pr[z_n \neq u_n | h \text{ equations hold}] = \frac{p(1-s)^h s^{m-h}}{p(1-s)^h s^{m-h} + (1-p)(1-s)^{m-h} s^h}.$$

The general idea is to make several passes over the sequence, and each time the probabilities $p*$ is updated. The values $u_n + z_n$ ia converged toward to $0(1)$ when u_n is (not) equal to z_n. In such iterated decoding, all $p*$ probabilities for sequence symbols start at $p* = p$ but can then diverge differently, as the $s = Pr[y = b]$ varies with j and n. And for each u_n stored its own $p*$ by p_n, $n = 1, 2, \cdots, N$. These p_n values are used to compute new $p*$ values as follows.

$$s(p_1, \cdots, p_t, t) = p_t s(p_1, \cdots, p_{t-1}, t-1) + (1-p_t)(1 - s(p_1, \cdots, p_{t-1}, t-1)),$$
$$s(p_1, 1) = p,$$

where $Pr[b^{(j)} = 0] = s(p_1, \cdots, p_t, t)$.

For each u_n, put all equation indices j in M, and those for which the equation $y = b$ holds in H. Then

$$p* = Pr[z_n \neq u_n | \text{equation } j \text{ holds for } j \in H \text{ but not for } j \in M \setminus H]$$
$$= \frac{p \prod_{j \in H}(1 - s_j) \prod_{j \in M \setminus H} s_j}{p \prod_{j \in H}(1 - s_j) \prod_{j \in M \setminus H} s_j + (1-p) \prod_{j \in M \setminus H}(1 - s_j) \prod_{j \in H} s_j}.$$

The algorithm use two different thresholds, p_{thr} and N_{thr}. p_{thr} is used to decide whether $p*$ is so high that the corresponding z_j should likely be changed. Those $p*$ are counted and the count N_w is compared to N_{thr}: if enough bits have been identified that ought to be changed, those bits are complemented, otherwise the calculation of $p*$ is iterated.

References

1. Caesar: Competition for authenticated encryption: Security, applicability, and robustness. http://competitions.cr.yp.to/caesar.html

2. Ågren, M., Löndahl, C., Hell, M., Johansson, T.: A survey on fast correlation attacks. Cryptography and Communications **4**(3–4), 173–202 (2012)
3. Berbain, C., Gilbert, H., Maximov, A.: Cryptanalysis of grain. In: Robshaw, M. (ed.) FSE 2006. LNCS, vol. 4047, pp. 15–29. Springer, Heidelberg (2006)
4. Chakraborti, A., Nandi, M.: Trivia-ck v1. CAESAR (2014)
5. Chaza, F., McDonald, C., Avanzi, R.: Faser v1: Authenticated encryption in a feedback shift register. CAESAR (2014)
6. Coppersmith, D., Halevi, S., Jutla, C.: Cryptanalysis of stream ciphers with linear masking. In: Yung, M. (ed.) CRYPTO 2002. LNCS, vol. 2442, pp. 515–532. Springer, Heidelberg (2002)
7. De Cannière, C.: TRIVIUM: A stream cipher construction inspired by block cipher design principles. In: Katsikas, S.K., López, J., Backes, M., Gritzalis, S., Preneel, B. (eds.) ISC 2006. LNCS, vol. 4176, pp. 171–186. Springer, Heidelberg (2006)
8. Feng, X., Zhang, F.: A realtime key recovery attack on the authenticated cipher faser128. Cryptology ePrint Archive, Report 2014/258 (2014). http://eprint.iacr.org/
9. Golić, J.D.: Intrinsic statistical weakness of keystream generators. In: Safavi-Naini, R., Pieprzyk, J.P. (eds.) ASIACRYPT 1994. LNCS, vol. 917, pp. 91–103. Springer, Heidelberg (1995)
10. Hell, M., Johansson, T.: Linear attacks on stream ciphers. Advanced Linear Cryptanalysis of Block and Stream Ciphers/Cryptology and Information Security Series, pp. 55–85 (2011)
11. Khazaei, S., Hassanzadeh, M.: Linear sequential circuit approximation of the trivium stream cipher. eSTREAM, ECRYPT Stream Cipher Project (2005)
12. Meier, W., Staffelbach, O.: Fast correlation attacks on certain stream ciphers. Journal of Cryptology **1**, 159–176 (1989)
13. Siegenthaler, T.: Decrypting a class of stream ciphers using ciphertext only. IEEE Transactions on Computers **C-34**, 81–85 (1985)
14. Wagner, D.: A generalized birthday problem. In: Yung, M. (ed.) CRYPTO 2002. LNCS, vol. 2442, pp. 288–304. Springer, Heidelberg (2002)

Partial Key Exposure Attack on CRT-RSA

Santanu Sarkar[1]([⊠]) and Ayineedi Venkateswarlu[2]

[1] Chennai Mathematical Institute, H1, SIPCOT IT Park, Siruseri, Kelambakkam,
Chennai 603103, India
sarkar.santanu.bir@gmail.com
[2] Computer Science Unit, Indian Statistical Institute - Chennai Centre,
MGR Knowledge City Road, Taramani, Chennai 600113, India
venku@isichennai.res.in

Abstract. In Eurocrypt 2005, Ernst et al. proposed an attack on RSA
allowing to recover the secret key when the most or least significant
bits of the decryption exponent d are known. In Indocrypt 2011, Sarkar
generalized this by considering the number of unexposed blocks in the
decryption exponent is more than one. In this paper, for the first time,
we study this situation for CRT-RSA. Further, we consider the case when
random bits of one decryption exponent are exposed in this model. These
results have implications in side channel attacks.

Keywords: CRT-RSA · Factorization · Partial key exposure · Coppersmith's techniques · Lattice · Multi-point evaluation of a polynomial

1 Introduction

RSA cryptosystem, introduced in 1977, is the most popular Public Key Cryptosystem till date. Let us recall the basic RSA scheme: an RSA public modulus
$N = pq$, where p and q are large secret primes. A public exponent e and the
corresponding private exponent d are chosen such that $ed \equiv 1 \bmod \phi(N)$, where
$\phi(N)$ is Euler's totient function. To encrypt a message, we compute its e-th
power modulo N, while decrypting a ciphertext requires computing its d-th
power modulo N.

Consider that one likes to make the decryption process faster. Then the secret
decryption exponent d has to be made small. However, Wiener [29] proved that
when $d < \frac{1}{3}N^{0.25}$, one can factor N efficiently. Boneh and Durfee [5] extended
this to $d < N^{0.292}$ by reducing the recovery of d to finding small roots of polynomial equations, using the seminal algorithms introduced by Coppersmith [8].

CRT-RSA, the faster variant of RSA proposed by Quisquater and Couvreur [23], is the most widely accepted version of RSA in practice. The encryption technique is similar to the standard RSA, but the decryption process is
little different. Instead of one decryption exponent as in the standard RSA,
here two decryption exponents (d_p, d_q) are used, where $d_p \equiv d \bmod (p-1)$ and
$d_q \equiv d \bmod (q-1)$. To decrypt a ciphertext C, one needs to calculate both

© Springer International Publishing Switzerland 2014
W. Meier and D. Mukhopadhyay (Eds.): INDOCRYPT 2014, LNCS 8885, pp. 255–264, 2014.
DOI: 10.1007/978-3-319-13039-2_15

$C_p \equiv C^{d_p} \bmod p$ and $C_q \equiv C^{d_q} \bmod q$. From C_p and C_q, one can get the plaintext M by an application of Chinese Remainder Theorem (CRT). The CRT-RSA decryption is 4 times faster than the standard RSA on an average if the schoolbook multiplication is used in exponentiation. However, it is a common practice to use Karatsuba multiplication [21] technique for RSA with large primes, and in such a case the CRT-RSA decryption is around 3 times faster than the standard RSA.

May [20] described two weaknesses in CRT-RSA that works when the smallest prime factor is less than $N^{0.382}$. Bleichenbacher and May [2] improved the idea of [20] even when the smallest prime factor is less than $N^{0.468}$. In [11], an attack on CRT-RSA has been presented for small e when the primes are of same bit size. In [17], it is shown that CRT-RSA can be attacked when the encryption exponents are of the order of N, and d_p and d_q are smaller than $N^{0.073}$.

Kocher [18] proposed a new attack on RSA to obtain the private exponent d. He showed that an attacker can get a few bits of d by timing characteristic of an RSA implementing device. In [4], it has been studied how many bits of d need to be known to mount an attack on RSA. The constraint in the work of [4] was that the public exponent e is bounded above by $\sqrt{N}$. The study attracted interest and the idea of [4] has been improved in [3] where the upper bound of e was increased up to $N^{0.725}$. Then the work of [10] improved the result for full size public exponent e. These attacks require knowledge of contiguous block of bits of the RSA secret keys. Later in [26], partial key exposure attacks on RSA are proposed for the case where the number of unexposed blocks in the decryption exponent is more than one.

CRT-RSA is specially interesting for time critical applications. Therefore it is frequently used on smart cards. On the other hand, it is well known that smart cards are highly vulnerable to different kinds of side channel attacks. So, it is very important to study partial key exposure attacks on CRT-RSA. Partial key exposure attack on CRT-RSA for small e was first studied in [3]. Then in [25], attacks were proposed for the case where e is of full bit size. Another striking example of recovering key bits is the cold boot attack [14], where random bits of RAM can be forced to remain accessible.

Our Contribution: In all partial key exposure attacks on CRT-RSA, it is assumed that the most significant or the least significant part of the decryption exponent is exposed. However, it is more realistic to assume that the attacker knows some bits at random positions. Indeed, side channel attacks on embedded devices highlight the fact that blocks of data can leak information when being loaded from non-volatile memory onto the random access memory (RAM). In [13], it is shown that random bits of the RAM can be forced to remain accessible. In the case of RSA, this could be a secret element manipulated during computation. Still specific to the RSA, advanced simple power analysis techniques can enable the attacker to guess some of the operations-thus of d-during regular exponentiation computations; this includes the square and multiply algorithm and Montgomery ladder algorithm, faced to pattern-matching techniques or template attacks [6]. These techniques can also be applied on CRT-RSA, thus leaking information of the secret exponents d_p or d_q.

In [26], author considered the situation where a few contiguous blocks of the RSA secret exponent d are unknown. In this paper, for the first time, we analyse CRT-RSA under this partial key exposure attack model in Section 2. We propose attacks using two different approaches: lattice-based and multi-point evaluation of a polynomial. We have implemented all programs in SAGE 6.1.1 over Linux Mint 17 (64-bit) on a desktop with Intel Core i7-2600 CPU 3.4 GHz, 16 GB RAM and 8 MB Cache.

2   Partial Key Exposure Attack on CRT-RSA

In this section we propose attacks using two different approaches. The first one uses the LLL algorithm.

2.1   Lattice Based Approach

In this section, we consider the situation where the number of unexposed blocks in one decryption exponent of CRT-RSA is more than one. In [3], authors considered the situation where the number of unexposed blocks in one decryption exponent is one.

We will need the following two results. The first one is due to Howgrave-Graham [16].

Lemma 1. Let $h(x_1, \cdots, x_n) \in \mathbb{Z}[x_1, \cdots, x_n]$ be the sum of at most ω monomials. Suppose that $h(x_1^{(0)}, \cdots, x_n^{(0)}) \equiv 0 \bmod N^m$ where $|x_i^{(0)}| \leq X_i$ for $1 \leq i \leq n$ and

$$||h(x_1 X_1, \ldots, x_n X_n)|| < \frac{N^m}{\sqrt{\omega}}.$$

Then $h(x_1^{(0)}, \cdots, x_n^{(0)}) = 0$.

Also note that the basis vectors of an LLL-reduced basis fulfil the following property [19].

Lemma 2. Let L be an integer lattice of dimension ω. The LLL algorithm applied to L outputs a reduced basis of L spanned by $\{v_1, \ldots, v_\omega\}$ with

$$||v_1|| \leq ||v_2|| \leq \cdots \leq ||v_i|| \leq 2^{\frac{\omega(\omega-1)}{4(\omega+1-i)}} \det(L)^{\frac{1}{\omega+1-i}}, \; for \; i = 1, \ldots, \omega,$$

in polynomial time of dimension ω and the bit size of the entries of L.

Our technique works in practice as noted from the experiments we perform, but this may not always happen theoretically as we are working with heuristics only. Thus our approach relies on the following heuristic assumption like many other lattice based approaches [10,15]:

Assumption 1. *In this work, the lattice based constructions yields a polynomial and the root can be computed efficiently.*

It is clear that if the attacker knows any one of the decryption exponents say d_p then he/she can factor N with high probability in polynomial time by calculating the $\gcd(w^{ed_p-1} - 1, N)$ for some random integer w. Now we consider the case that all the bits of d_p are known except n unknown blocks of size $\delta_1 \log N, \ldots, \delta_n \log N$. In this situation, we have the following result. Our approach is similar to [15].

Theorem 1. *Let $e = N^\alpha$. Suppose the bits of d_p are exposed except n many blocks, with sizes $\delta_i \log N$ bits for $1 \leq i \leq n$. Then under Assumption 1, one can factor N in time polynomial in $\log N$ and exponential in n if*

$$\sum_{i=1}^{n} \delta_i + \alpha < 1 - \frac{1}{2^{\frac{n+2}{n+1}}} - \frac{n+2}{2} \left(1 - \frac{1}{2^{\frac{1}{n+1}}}\right).$$

Proof. The unknown portion of d_p is n many contiguous blocks, so we can write $d_p = a_0 + a_1 y_1 + \cdots + a_n y_n$, where $y_1, y_2, \ldots, y_n$ are the unknown blocks. We have $ed_p = 1 + k_p(p - 1)$ for some integer k_p which is also not known, and so

$$(ea_0 - 1) + ea_1 y_1 + \cdots + ea_n y_n + k_p \equiv 0 \bmod p.$$

Then the unknown values $(y_1, \ldots, y_n, k_p)$ is a root of the polynomial

$$f(x_1, \ldots, x_n, x_{n+1}) = (ea_0 - 1) + ea_1 x_1 + \cdots + ea_n x_n + x_{n+1}$$

in $\mathbb{Z}_p$. Our aim is to find a small root of the polynomial f given the bounds on the unknown blocks. We expect that the small root obtained is equal to $(y_1, \cdots, y_n, k_p)$. For this purpose, we generate a collection of polynomials for constructing a lattice to work on. Let $X_i = N^{\delta_i}$ for $1 \leq i \leq n, X_{n+1} = N^\alpha$. Clearly, $X_1, \ldots, X_{n+1}$ are the upper bounds of absolute values of $y_1, \ldots, y_n, k_p$, respectively. Now we define a collection of shift polynomials

$$g_{i_1, i_2, \cdots, i_n, l} = x_1^{i_1} x_2^{i_2} \cdots x_n^{i_n} f^l N^{\max\{t-l, 0\}}, \tag{1}$$

for $0 \leq i_1, \cdots, i_n \leq m$ such that $i_1 + \cdots + i_n \leq m - l$ and $t \geq 0$ with $0 \leq l \leq m$.

It is clear that $g(y_1, \cdots, y_n, k_p) \equiv 0 \bmod p^t$ for any shift polynomial described above. We construct a lattice L from the coefficient vectors of the shift polynomials $g_{i_1, \cdots, i_n, l}(x_1 X_1, \cdots, x_n X_n, x_{n+1} X_{n+1})$. The dimension ω of L is $\omega = \binom{m+n+1}{m}$. The determinant of L is

$$\det(L) = \prod_{i=1}^{n+1} X_i^{s_1} N^{s_2},$$

where $s_1 = \binom{m+n+1}{m-1}$ and $s_2 = \omega t - \binom{m+n+1}{m-1} + \binom{m+n+1-t}{m-1-t}$.

From Lemma 1 and Lemma 2, we know that the small root of f can be found if

$$2^{\frac{\omega(\omega-1)}{4(\omega-n)}} \det(L)^{\frac{1}{\omega-n}} < \frac{p^t}{\sqrt{\omega}}.$$

Now putting the values of $\det(L)$ and ω in the above condition, using the fact that $p \approx N^{\frac{1}{2}}$ and taking $t = \tau m$ with $\tau = 1 - \frac{1}{2^{\frac{1}{n+1}}}$ (obtained by a similar calculation of [15]), we get the following condition

$$\sum_{i=1}^{n} \delta_i + \alpha < 1 - \frac{1}{2^{\frac{n+2}{n+1}}} - \frac{n+2}{2}\left(1 - \frac{1}{2^{\frac{1}{n+1}}}\right).$$

The running time of our algorithm is dominated by the LLL algorithm, which is polynomial in the dimension of the lattice and in the bitsize of the entries. Since the lattice dimension in our case is exponential in n, so the running time is polynomial in $\log N$ but exponential in n. □

Now we consider the case when the number of unknown blocks n approaches to infinity. Since

$$\lim_{n\to\infty} 1 - \frac{1}{2^{\frac{n+2}{n+1}}} - \frac{n+2}{2}\left(1 - \frac{1}{2^{\frac{1}{n+1}}}\right) = \frac{1}{2}\left(1 - \ln(2)\right) = 0.153,$$

asymptotically our approach works when $\sum_{i=1}^{\infty} \delta_i + \alpha < 0.153$. When size of X_i are different, upper bound can be improved by giving different weight over the shift polynomials using the idea of [27].

In Table 1 we present a few experimental results.

Table 1. Experimental results when few bits of d_p are known

n	$\log_2 N$	$\log_2 e$	$\sum_{i=1}^{n} \delta_i$	(m,t)	LD	Time (Sec)
1	1024	60	0.09	(9,4)	55	5.56
1	2048	60	0.09	(9,4)	55	47.31
2	1024	40	0.04	(6,2)	84	67.26
2	2048	50	0.04	(6,2)	84	205.71
3	2048	50	0.03	(5,3)	126	4700.23

From the first row of Table 1, note that in the case e is of bit size 60, we can reconstruct d_p within few seconds if the number of unknown blocks is $n = 1$ and it is of size at most $0.09 \times 1024 = 92$ and $0.09 \times 2048 = 184$ for 1024 and 2048 bit RSA modulus respectively. If this unknown block corresponds to the MSB side of d_p, the time complexity of the existing result [4] is of the order of $e \cdot \text{poly}(\log N)$. Hence in this situation, the existing approach is not efficient, but our proposed method is. So our method gives a better result from a practical point of view in such cases.

2.2   Approach Using Multi-point Evaluation of a Polynomial

In this section we will study partial key exposure attack on CRT-RSA by using simultaneous evaluation of a univariate polynomial at different points. Coron et al. in PKC 2011 [9] use this idea to cryptanalyse RSA subgroup assumption [12]. In Eurocrypt 2012, Chen and Nguyen [7] presented cryptanalysis of Fully Homomorphic Encryption over Integers [28] using the same.

Here we assume that the missing bits are randomly distributed over the decryption exponent d_p. Lattice-based method that we discussed earlier may not be efficient in this situation as it is exponential in the number of unknown blocks in d_p.

Let the bit size of d_p be ℓ. So we can write $d_p = \sum_{i=0}^{\ell-1} d_p^{(i)} 2^i$, where $d_p^{(i)}$ denotes the i-th bit in the binary representation of d_p. Now assume that we know all the bits of d_p except at k positions, say $i_1, \ldots, i_k$. Let us set

$$d_p' = \sum_{i=0,\ i \neq i_1,\ldots,i_k}^{\ell-1} d_p^{(i)} 2^i \quad \text{and} \quad \hat{d}_p = \sum_{j=1}^{k} d_p^{(i_j)} 2^{i_j}.$$

We have $d_p = d_p' + \hat{d}_p$. We know the value of d_p', and there are 2^k possibilities for the unknown $\hat{d}_p$ given by

$$\mathcal{D} = \left\{ \sum_{j=1}^{k} t^{(j-1)} 2^{i_j} : 0 \leq t \leq 2^k - 1 \right\},$$

where $t^{(j-1)}$ denotes the $(j-1)$-th bit in the binary representation of t.

Consider the situation where e is small to reduce the cost of encryption. It is very common practice to use $e = 2^{16} + 1$. UCSD TLS Corpus use $e = 2^{16} + 1$ in 99.5% cases [30]. We may assume that the attacker knows k_p for small e as $k_p < e$. We first consider this case and discuss a method to factor N.

Recall that $\gcd(ed_p - 1 + k_p, N) = p$. Now one can check that

$$p = \gcd\left(N, \prod_{c \in \mathcal{D}} ((ed_p' + ec - 1 + k_p) \bmod N) \right) \qquad (2)$$

is satisfied with overwhelming probability. We can rewrite the above identity as follows.

$$p = \gcd\left(N, e \prod_{c \in \mathcal{D}} (c - \theta_1) \bmod N \right),$$

where $\theta_1 = e^{-1}(1 - k_p) - d_p' \pmod{N}$. One can also see that p can be obtained by using 2^k modular multiplications and with a single gcd calculation. We improve upon this naive idea.

Let $l = \lfloor \frac{k}{2} \rfloor$. Define the the following two sets

$$\mathcal{D}_1 = \left\{ a \in \mathcal{D} : a < 2^{i_l} \right\} \quad \text{and} \quad \mathcal{D}_2 = \left\{ a \in \mathcal{D} : a \geq 2^{i_l} \right\}.$$

Note that for any $c \in \mathcal{D}$ there exists $a \in \mathcal{D}_1$ and $b \in \mathcal{D}_2$ uniquely such that $c = a + b$.

We now present a square root time algorithm for factorization in Algorithm 1. The main idea is to compute the product term efficiently in a two-step process.

Algorithm 1. Factorization with partial knowledge of d_p

    Input: N, e, k_p, d'_p and the k unknown bit positions $i_1, \ldots, i_k$.

1   Compute the polynomial $f(x) = \displaystyle\prod_{b \in \mathcal{D}_2} (x + b - \theta_1)$;

2   Evaluate $f(x)$ at values $a \in \mathcal{D}_1$;

3   Report $\gcd \left(N, \displaystyle\prod_{a \in \mathcal{D}_1} f(a) \bmod N \right)$;

We know from [1] that the computation and evaluation of the polynomial $f(x)$ in steps 1 and 2 of the Algorithm 1 can be done in $O(l2^l)$ many operations, which is essentially square root of the naive idea.

We set $e = 2^{16} + 1$ and the bit size of d_p is 512. We present our experimental results in Table 2.

Table 2. Experiments for 1024 bit RSA

Number of unknown bits (k)	Time (Seconds)
30	27.17
32	60.21
36	292.48
40	1351.27
44	7362.43

From Table 2, it seems that the parameter $k = 60$ is achievable with more resources. Using the approach of [14], one can factor N from the knowledge of 58% bits both from d_p and d_q. So for 1024 bit N, one needs a total of $1024 \times 0.58 = 594$ bits. In our approach, we need a total of 512-60=452 bits.

However, the approach of [14] is polynomial time whereas our method is exponential. Also it is difficult to deal with larger k as the attack is heavily memory bound: the $2^{\frac{k}{2}}$ memory requirement is a serious problem. It seems parameter k larger than 60 can be attacked if the storage can be efficiently distributed.

Several issues may be optimized in the experiment. We have written code in SAGE just for the proof of concept. Dedicated hardware implementation would give much better running time.

We now consider the case where k_p is not known. In this case the above technique can be modified to include the unknown bits of k_p. Then the time complexity increases by $size(e)\sqrt{e}/2$ times, which makes the attack less practical for larger e. We now present a different technique to factor N which does not require the information of k_p.

We know that p divides $(2^{ed_p}-2) \bmod N$, and $p = \gcd\left(2^{ed_p}-2 \pmod{N}, N\right)$

$= \gcd\left(2^{ed'_p + e\hat{d}_p} - 2 \pmod{N}, N\right)$ holds with overwhelming probability. Therefore we can obtain p by computing

$$\gcd\left(\prod_{c \in \mathcal{D}}(2^{ed'_p + ec} - 2) \bmod N, N\right) = \gcd\left(2^{ed'_p}\prod_{c \in \mathcal{D}}((2^e)^c - \theta_2) \bmod N, N\right),$$

where $\theta_2 = 2^{1 - ed'_p} \pmod{N}$. But this would require 2^k exponentiations and multiplications modulo N. Improving upon this naive idea, we now present a square-root time algorithm which uses the efficient simultaneous evaluation of a polynomial at several values.

Algorithm 2. Factorization with partial knowledge of d_p

    Input: N, e, d'_p and the k unknown bit positions $i_1, \ldots, i_k$.

1   Compute the polynomial $f(x) = \prod_{b \in \mathcal{D}_2}((2^e)^b x - \theta_2)$;

2   Evaluate $f(x)$ at values $(2^e)^a$ for $a \in \mathcal{D}_1$;

3   Report $\gcd\left(N, \prod_{a \in \mathcal{D}_1} f((2^e)^a) \bmod N\right)$;

One can see that the complexity of the above algorithm is $O(2^l)$ modular exponentiations. We set $e = 2^{16} + 1$ and the bit size of d_p is 512. We present our experimental results in Table 3.

Table 3. Experiments for 1024 bit RSA

Number of unknown bits (k)	Time (Seconds)
30	38.82
32	82.77
34	184.80
36	393.46
38	838.76

3   Conclusion

All the existing partial key exposure attacks on CRT-RSA assume that the bits of the secret exponent are exposed block-wise. In this paper we first proposed an attack on CRT-RSA when the number of unexposed blocks of the decryption exponent d_p is more than one. Then, we considered the case when some bits of d_p at random positions are unknown. We have proposed two methods to factor N using the efficient simultaneous evaluation of a univariate polynomial at several values. Our methods can perform better than the existing results for certain parameter ranges.

References

1. Bernstein, D.J.: Fast multiplication and its applications. Algorithmic number theory: lattices, number fields, curves and cryptography, vol. 44, pp. 325–384. Cambridge University Press, MSRI Publications (2008)
2. Bleichenbacher, D., May, A.: New Attacks on RSA with Small Secret CRT-Exponents. In: Yung, M., Dodis, Y., Kiayias, A., Malkin, T. (eds.) PKC 2006. LNCS, vol. 3958, pp. 1–13. Springer, Heidelberg (2006)
3. Blömer, J., May, A.: New Partial Key Exposure Attacks on RSA. In: Boneh, D. (ed.) CRYPTO 2003. LNCS, vol. 2729, pp. 27–43. Springer, Heidelberg (2003)
4. Boneh, D., Durfee, G., Frankel, Y.: Exposing an RSA Private Key Given a Small Fraction of its Bits. In: Ohta, K., Pei, D. (eds.) ASIACRYPT 1998. LNCS, vol. 1514, pp. 25–34. Springer, Heidelberg (1998)
5. Boneh, D., Durfee, G.: Cryptanalysis of RSA with Private Key d Less Than $N^{0.292}$. IEEE Transactions on Information Theory 46(4), 1339–1349 (2000)
6. Chari, S., Rao, J.R., Rohatgi, P.: Template Attacks. In: Kaliski, B.S., Koç, Ç.K., Paar, C. (eds.) CHES 2002. LNCS, vol. 2523, pp. 13–28. Springer, Heidelberg (2002)
7. Chen, Y., Nguyen, P.Q.: Faster Algorithms for Approximate Common Divisors: Breaking Fully-Homomorphic-Encryption Challenges over the Integers. In: Pointcheval, D., Johansson, T. (eds.) EUROCRYPT 2012. LNCS, vol. 7237, pp. 502–519. Springer, Heidelberg (2012)
8. Coppersmith, D.: Small solutions to polynomial equations and low exponent vulnerabilities. Journal of Cryptology 10(4), 223–260 (1997)
9. Coron, J.-S., Joux, A., Mandal, A., Naccache, D., Tibouchi, M.: Cryptanalysis of the RSA Subgroup Assumption from TCC 2005. In: Catalano, D., Fazio, N., Gennaro, R., Nicolosi, A. (eds.) PKC 2011. LNCS, vol. 6571, pp. 147–155. Springer, Heidelberg (2011)
10. Ernst, M., Jochemsz, E., May, A., de Weger, B.: Partial key exposure attacks on rsa up to full size exponents. In: Cramer, R. (ed.) EUROCRYPT 2005. LNCS, vol. 3494, pp. 371–386. Springer, Heidelberg (2005)
11. Galbraith, S.D., Heneghan, C., McKee, J.F.: Tunable Balancing of RSA. In: Boyd, C., González Nieto, J.M. (eds.) ACISP 2005. LNCS, vol. 3574, pp. 280–292. Springer, Heidelberg (2005)
12. Groth, J.: Cryptography in Subgroups of Z_n^*. In: Kilian, J. (ed.) TCC 2005. LNCS, vol. 3378, pp. 50–65. Springer, Heidelberg (2005)
13. Halderman, A., Schoen, S.D., Heninger, N., Clarkson, W., Paul, W., Calandrino, J.A., Feldman, A.J., Appelbaum, J., Felten, E.W.: Lest We Remember: Cold Boot Attacks on Encryption Keys. In: USENIX Security Symposium (February 2008)
14. Heninger, N., Shacham, H.: Reconstructing RSA Private Keys from Random Key Bits. In: Halevi, S. (ed.) CRYPTO 2009. LNCS, vol. 5677, pp. 1–17. Springer, Heidelberg (2009)

15. Herrmann, M., May, A.: Solving Linear Equations Modulo Divisors: On Factoring Given Any Bits. In: Pieprzyk, J. (ed.) ASIACRYPT 2008. LNCS, vol. 5350, pp. 406–424. Springer, Heidelberg (2008)
16. Howgrave-Graham, N.: Finding small roots of univariate modular equations revisited. In: Darnell, M.J. (ed.) Cryptography and Coding 1997. LNCS, vol. 1355, pp. 131–142. Springer, Heidelberg (1997)
17. Jochemsz, E., May, A.: A Polynomial Time Attack on RSA with Private CRT-Exponents Smaller Than $N^{0.073}$. In: Menezes, A. (ed.) CRYPTO 2007. LNCS, vol. 4622, pp. 395–411. Springer, Heidelberg (2007)
18. Kocher, P.C.: Timing attacks on implementations of Diffie-Hellman, RSA, DSS, and other systems. In: Koblitz, N. (ed.) CRYPTO 1996. LNCS, vol. 1109, pp. 104–113. Springer, Heidelberg (1996)
19. Lenstra, A.K., Lenstra Jr, H.W., Lovász, L.: Factoring polynomials with rational coefficients. Mathematische Annalen 261(4), 513–534 (1982)
20. May, A.: Cryptanalysis of Unbalanced RSA with Small CRT-Exponent. In: Yung, M. (ed.) CRYPTO 2002. LNCS, vol. 2442, pp. 242–256. Springer, Heidelberg (2002)
21. Menezes, A.J., van Oorschot, P.C., Vanstone, S.A.: Handbook of Applied Cryptography. CRC Press (2001)
22. PKCS #1 standard for RSA. http://www.rsa.com/rsalabs/node.asp?id=2125
23. Quisquater, J.-J., Couvreur, C.: Fast decipherment algorithm for RSA public-key cryptosystem. Electronic Letters 18, 905–907 (1982)
24. Rivest, R.L., Shamir, A., Adleman, L.: A method for obtaining digital signatures and public key cryptosystems. Communications of ACM 21(2), 158–164 (1978)
25. Sarkar, S., Maitra, S.: Partial Key Exposure Attack on CRT-RSA. In: Abdalla, M., Pointcheval, D., Fouque, P.-A., Vergnaud, D. (eds.) ACNS 2009. LNCS, vol. 5536, pp. 473–484. Springer, Heidelberg (2009)
26. Sarkar, S.: Partial Key Exposure: Generalized Framework to Attack RSA. In: Bernstein, D.J., Chatterjee, S. (eds.) INDOCRYPT 2011. LNCS, vol. 7107, pp. 76–92. Springer, Heidelberg (2011)
27. Takayasu, A., Kunihiro, N.: Better Lattice Constructions for Solving Multivariate Linear Equations Modulo Unknown Divisors. In: Boyd, C., Simpson, L. (eds.) ACISP. LNCS, vol. 7959, pp. 118–135. Springer, Heidelberg (2013)
28. van Dijk, M., Gentry, C., Halevi, S., Vaikuntanathan, V.: Fully homomorphic encryption over the integers. In: Gilbert, H. (ed.) EUROCRYPT 2010. LNCS, vol. 6110, pp. 24–43. Springer, Heidelberg (2010)
29. Wiener, M.: Cryptanalysis of Short RSA Secret Exponents. IEEE Transactions on Information Theory 36(3), 553–558 (1990)
30. Yilek, S., Rescorla, E., Shacham, H., Enright, B., Savage, S.: When private keys are public. Results from the 2008 Debian OpenSSL debacle (May 2009)

On the Leakage of Information
in Biometric Authentication

Elena Pagnin, Christos Dimitrakakis,
Aysajan Abidin, and Aikaterini Mitrokotsa$^{(\boxtimes)}$

Chalmers University of Technology, Gothenburg, Sweden
{elenap,chrdimi,aysajan.abidin,aikmitr}@chalmers.se

Abstract. In biometric authentication protocols, a user is authenticated or granted access to a service if her fresh biometric trait *matches* the reference biometric template stored on the service provider. This matching process is usually based on a suitable *distance* which measures the similarities between the two biometric templates. In this paper, we prove that, when the matching process is performed using a specific family of distances (which includes distances such as the Hamming and the Euclidean distance), then information about the reference template is leaked. This leakage of information enables a *hill-climbing* attack that, given a sample that matches the template, could lead to the full recovery of the biometric template (*i.e.* centre search attack) even if it is stored encrypted. We formalise this "leakage of information" in a mathematical framework and we prove that centre search attacks are feasible for any biometric template defined in $\mathbb{Z}_q^n, (q \geq 2)$ after a number of authentication attempts linear in n. Furthermore, we investigate brute force attacks to find a biometric template that matches a reference template, and hence can be used to run a *centre search attack*. We do this in the binary case and identify connections with the *set-covering* problem and *sampling without replacement*.

Keywords: Biometric authentication · Privacy-preservation · Centre search attack · Hill-climbing · Brute force attacks

1 Introduction

While biometric authentication is becoming increasingly popular, the privacy and security risks related to their usage are raising severe concerns. The main threats associated to biometric authentication include profiling and tracking of individuals and identity theft. If successfully performed, any attack that recovers a biometric template may have serious impact since users cannot change their biometric features and biometric data may reveal very sensitive information (*e.g.* genetic [1] information and medical diseases [2]).

Biometric authentication protocols involve comparing fresh biometric data with a stored biometric template. The process is essentially performed by computing some distance or divergence between the fresh and the stored template.

© Springer International Publishing Switzerland 2014
W. Meier and D. Mukhopadhyay (Eds.): INDOCRYPT 2014, LNCS 8885, pp. 265–280, 2014.
DOI: 10.1007/978-3-319-13039-2_16

If the measured distance is less than a predefined threshold, then the user is authenticated; otherwise she is rejected. Many biometric authentication protocols use straightforward choices for the distance, such as the Hamming distance [3,4], the normalised Hamming distance ([5] for iris recognition) and the Euclidean distance [6–9]. In these cases the matching process leaks information that could be exploited by an adversary to recover the stored template. More precisely, the adversary could run an iterative process where he progressively changes the components of an arbitrary biometric template until acceptance. This strategy is known as *hill-climbing* attack [10], due to similarity with the synonymous optimisation technique. When the initial template is an acceptable biometric trait (*e.g.* a fresh sample) this process is called *centre search* attack [10]. Recovering stored biometric templates has more severe impact than just finding an acceptable biometric template. Indeed, the same stored template might be used in multiple biometric authentication systems which may even employ different matching processes. Furthermore, a recovered stored template could be used to find a match in criminal biometric template databases or even compromise health records [11].

Hill climbing attacks involve making incremental changes to a potential solution, until one or more acceptable solutions are found. In our case, the adversary observes how the matcher responds to forged biometric templates. His goal is to recover the stored template from one matching template. Bringer *et al.* [12] presented a hill-climbing strategy that is successful even when a dedicated secure access module (*e.g.* smartcard) is used to perform the biometric authentication process. The matching process considered in [12] involves an adapted Hamming distance with erasures, nevertheless, the adversary is able to recover multiple encrypted biometric templates. Later on, Simoens *et al.* [10] describe multiple attacks (including the centre search attack) that can be mounted by each of the internal entities in a distributed biometric authentication systems.

In the past years privacy-preserving distance computation has been investigated [13–15]. Although these protocols have direct applications to biometric identification and authentication they all suffer from leakage of information when a centre search attack is employed.

The problem of leakage of information due to the employment of distances has also been investigated in other areas not relevant to biometric authentication. For example, the Hamming weight model has been employed in order to successfully perform side channel attacks [16,17] (*e.g.* differential power analysis). It has been shown [16,17] that the power consumption of a device (*e.g.* a smart card) directly depends on the Hamming weight and on the number of changes $0 \leftrightarrow 1$ in the binary vector that is considered during the execution of the attack.

Our Contribution: In this paper, we point out that all biometric authentication protocols that rely on certain distances (including the Hamming and the Euclidean distance) are susceptible to leakage of information and we provide a formal mathematical framework to analyse this. In particular, we generalise the centre search attack and prove that it is efficient and feasible in the binary case as well as when the biometric templates are defined in $\mathbb{Z}_q^n$. In both cases we show that the maximal number of authentication attempts in order to fully recover the stored biometrics corresponding to the given data is linear in n (the size of the biometric string). Our proofs hold also when the Euclidean distance is employed. Thus, we go beyond the Hamming distance case that was described in [10]. We furthermore investigate the preliminary step to the centre search attack: finding a biometric template that matches a reference one. For the binary case, we propose a new algorithm that exploits a tree structure and we compare its performance to standard brute force attacks and to the *optimal* but infeasible attack. Finally, we highlight how the *optimal* solution of finding a matching biometric template connects to the NP-complete *set-covering* problem and *sampling without replacement*. Our proofs are valid for standard as well as for privacy-preserving biometric authentication protocols since the output of the matching process is not affected by the employed protection mechanism (*e.g.* homomorphic encryption). This means that encryption alone cannot mitigate the leakage of information of the matching process. More precisely, this leakage of information leads to full recovery of the stored template for the centre search attack and to a matching template for the brute-force attack. An implication of our work is that achieving security and privacy of biometric templates using the known techniques is challenging.

Outline: The notations and the background material are introduced in Section 2 while Section 3 describes the adversarial model. We generalise the centre search attack in Section 4 in two ways: first to any leaking distance on $\mathbb{Z}_2^n$ and then to any leaking distance on $\mathbb{Z}_q^n$. In addition, we investigate the success probability of finding an acceptable fresh biometric template and compare the bounds for the success probability in different cases in Section 5. Finally, Section 6 summarizes our results.

2 Preliminaries

Notations: Let $q \in \mathbb{Z}$ be a positive integer, $q \geq 2$. The set of n-dimensional vectors with components in $\mathbb{Z}_q = \{0, 1, \cdots, q - 1\}$ is denoted by $\mathbb{Z}_q^n$. The i-th component of a vector $x \in \mathbb{Z}_q^n$ is referred to as $x_i \in \mathbb{Z}_q$. Given a distance $d : \mathbb{Z}_q^n \times \mathbb{Z}_q^n \to \mathbb{R}_{\geq 0}$, a point $x \in \mathbb{Z}_q^n$ and a positive number $\tau \in \mathbb{R}_{>0}$, the d-ball of center x and radius τ is defined as $B_x(\tau) = \{z \in \mathbb{Z}_q^n : d(x, z) \leq \tau\}$. In the following, the binary case ($q = 2$) will always be explicitly written as $\mathbb{Z}_2^n$. If not otherwise specified, $\mathbb{Z}_q^n$ implies $q > 2$. We denote the bit-flip operation as $\bar{\ } : \mathbb{Z}_2 \to \mathbb{Z}_2$, namely $\bar{1} = 0, \bar{0} = 1$. The integer part of a real number τ, is denoted by $\lfloor \tau \rfloor$ (rounding to the closest integer $\leq \tau$).

2.1 Biometric Authentication

A biometric authentication system consists of two main phases: the *enrolment phase* and the *authentication phase*.

The *enrolment phase* is a one-time step: a user (client) $\mathcal{C}$ registers to a trusted party her biometric templates (digital strings b) along with her identity ID. These two pieces of information are then stored in the database of the authentication server $\mathcal{AS}$. Once enrolled in the system, the client can authenticate herself an unlimited number of times.

In the *authentication phase*, the client is required to provide a fresh biometric trait b' as well as her identity ID. These two data are then communicated to the authentication server, which checks if matching templates (fresh b' and stored b) match. If the distance between the user's fresh biometric trait b' and the reference biometric template b is less or equal to a predefined threshold τ, then the client gets authenticated. Otherwise, the system rejects the user.

Without loss of generality we will consider only the two party setting (*i.e.* one client $\mathcal{C}$ and one authentication server $\mathcal{AS}$, as depicted in Figure 1). However, our analysis naturally applies when more than two parties are involved in the biometric authentication process [4,18,19]. Due to privacy concerns, the biometric templates should be protected and not sent in the clear over the network. This implies that often the matching procedure is performed in the encrypted domain. For instance, in multiple privacy-preserving biometric authentication protocols, secure multi-party computation techniques are employed to preserve the privacy of the users. In those protocols usually the biometric data are protected using homomorphic encryption [20], garbled circuits [21] or oblivious transfer [22].

Figure 1 depicts the authentication phase of a biometric authentication system in a two party setting, between a client $\mathcal{C}$ and an authentication server $\mathcal{AS}$. The client presents her fresh biometric and her ID to the authentication system. The sensor $\mathcal{S}$ gets the user's biometric vector b' and her identity. In the privacy-preserving case, $\mathcal{S}$ encrypts b' ($E(b')$) and ID ($\widetilde{\mathsf{ID}}$), otherwise this data is sent in the clear. Subsequently, the two data ($E(b')$, $\widetilde{\mathsf{ID}}$) are sent to the authentication server $\mathcal{AS}$, who retrieves the (possibly encrypted) stored template that corresponds to the user with identity ID. The matching process is then preformed by checking if the distance between the fresh and stored biometric templates is less than a predefined threshold τ (*i.e.* $d(b, b') \leq \tau$). Finally, depending on the outcome of the matching ($\mathsf{Out}_{\mathcal{AS}}$), the authentication server either accepts or rejects the client. Note that even in the privacy-preserving case, where the biometric data is encrypted, the output of the authentication server depends only on the value of $d(b, b')$, *i.e.* the distance between the fresh and the stored biometric vectors. Hence, encryption alone does not mitigate our attacks.

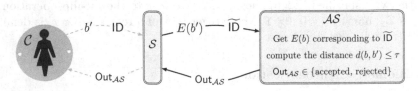

Fig. 1. Authentication phase in a two-party biometric authentication system

The main enablers of the attacks described in this paper are:

(a) A return channel of the biometric authentication process, denoted as Out_{AS} (*e.g.* access granted or not) that is sent by the authentication server to the user after each authentication attempt. In a real-life biometric authentication scenario this could be a door that opens denoting "access granted" when biometric authentication is used for access control in a building.

(b) The fact that the matching process (and so the value of Out_{AS}) is based on a distance that is sensitive to single component variations (see leaking distance Definition 1).

In this paper, we demonstrate that even when secure-multi party computation techniques are employed, it is still possible to disclose the biometric templates as long as a certain family of distances is used to compare the raw (plaintext) biometric data. That is, an attacker can learn information about the value of b (plaintext of stored biometric template) by observing the authentication server's response Out_{AS} to the client's authentication requests, if the response depends on the value of $d(b, b')$. More precisely, if d is a distance that detects component-variation (see Definition 1), and if there exists a function f that enables to retrieve information about the distance of the raw templates, given their possibly encrypted versions, *i.e.* $\exists f$ *s.t.* $f(E(b), E(b')) = d(b, b')$, then the biometric authentication system leaks information (in the non privacy-preserving case $E = \text{id}$, is the identity map and $f = d$ is the given distance). In particular, it is always possible to disclose the original b given a matching b'. For instance, consider the case [4] where $b, b' \in \mathbb{Z}_2^n$, $d = d_H$ is the Hamming distance and E and D are the Goldwasser-Micali [23] encryption and decryption functions, respectively. Then, $d_H(b, b') = \text{HW}(b \oplus b)' = \text{HW}(D(E(b \oplus b'))) = \text{HW}(D(E(b) \times E(b')))$, where HW denotes the Hamming weight of a vector, *i.e.* $\text{HW}(x) = \sum_{i=1}^{n} x_i$. In this case, we have $f = \text{HW} \circ D \circ \times$.

3   Adversarial Model

The main threats in a privacy-preserving biometric authentication protocol are classified as follows [10]:

- *Biometric reference recovery:* the adversary tries to recover the reference (stored) biometric template b.
- *Biometric sample recovery:* the adversary tries to recover (or generate) a fresh biometric template b' that will be acceptable by the biometric authentication system.
- *Identity privacy:* the adversary tries to link a biometric template $b(i)$ of a user i to the user's identity $\text{ID}(i)$.
- *Traceability and distinguishability of users:* the adversary's objective is to distinguish different users and/or trace one user in different authentication attempts.

In this paper, we focus on the two first threats only, as they apply to any biometric authentication system, privacy-preserving or not. We also consider that the adversary $\mathcal{A}$ has access to the output of the authentication process ($\mathsf{Out}_{\mathcal{AS}}$) as well as to the predefined threshold τ used in matching process. The settings for the two attacks are:

- *Biometric reference recovery:* the adversary $\mathcal{A}$ has an acceptable fresh biometric template b' at his disposal and tries to recover the stored template b (*centre search* attack).
- *Biometric sample recovery:* the adversary $\mathcal{A}$ does not have access to an acceptable fresh biometric b' but tries to find an accepted template anyway (brute force attack).

4   Generalisations of the Centre Search Attack

Let $b' \in \mathbb{Z}_q^n$ denote a fresh biometric template and $b \in \mathbb{Z}_q^n$ the reference (stored) template, for $q \geq 2$. The standard *centre search* attack aims at finding the point b in the centre of the *acceptance ball* $B_b(\tau) = \{\, z \in \mathbb{Z}_2^n \; : \; d(b,z) \leq \tau \,\}$. Simoens *et al.* [10] gave an informal description of this attack in the case d is the Hamming distance. Here, we extend this attack to a larger family of distances over $\mathbb{Z}_2^n$ (Theorem 2). In order to do so, we prove in Theorem 1 that any *leaking distance* (cf. Definition 1) over $\mathbb{Z}_2^n$ is *equivalent* to the Hamming distance. In addition, Theorem 3 proves that a centre search attack is feasible also for $b \in \mathbb{Z}_q^n$ when $q > 2$ if a *leaking distance* (e.g. the Euclidean distance) is employed in the matching process.

The family of distances we consider in this paper is defined as follows:

Definition 1 (Leaking distances).   *Let $q \geq 2$, a distance $d : \mathbb{Z}_q^n \times \mathbb{Z}_q^n \to \mathbb{R}_{\geq 0}$, is said to be a leaking distance (to detect component variations) if it can be written as $d(x,y) = h\left(\sum_{i=1}^n |x_i - y_i|^k\right)$, for all $x, y \in \mathbb{Z}_2^n$, $k \in \mathbb{Q}_{>0}$ and $h : \mathbb{R} \to \mathbb{R}_{\geq 0}$ a monotonically strictly increasing positive function.*

The Hamming distance is an example of a leaking distance over $\mathbb{Z}_2$ (take h to be the identity map and $k = 1$). For a general $q \geq 2$, the Euclidean distance detects component variation (h is the square-root function and $k = 2$). Note that *leaking* distances are *reasonable* distances to be used for biometric authentication, as they enable to compare vectors (biometric data) component wise.

In order to simulate the query/access to an oracle, we introduce the following decision function.

Definition 2.   *Let $q \geq 2$, $\tau \in \mathbb{R}_{>0}$ and let $d : \mathbb{Z}_q^n \times \mathbb{Z}_q^n \to \mathbb{R}_{\geq 0}$ be a distance metric. Then, for each $x \in \mathbb{Z}_q^n$, we define a decision function $\delta_x : \mathbb{Z}_q^n \to \{0,1\}$ as $\delta_x(z) = \begin{cases} 0 & \text{if } d(x,z) > \tau \\ 1 & \text{if } d(x,z) \leq \tau \end{cases}$.*

It is easy to see that the decision function δ_x corresponds to the output of the authentication process denoted as $\mathsf{Out}_{\mathcal{AS}}$ in Sections 2 and 3. Firstly, we consider biometric templates as binary vectors. This is for instance the case for iris recognition based biometric authentication [5,24]. We begin by proving that any binary leaking distance can be written in terms of the Hamming distance.

Theorem 1. *Let $d : \mathbb{Z}_2^n \times \mathbb{Z}_2^n \to \mathbb{R}_{\geq 0}$ be a leaking distance on $\mathbb{Z}_2^n$. Then every d-ball corresponds to a d_H-ball, with d_H being the Hamming distance.*

We provide the proof of Theorem 1 in the appendix. Observe that Theorem 1 provides a *boardwalk* among all binary leaking distances. In particular, it enables us to extend all the results concerning Hamming distance to any other leaking distance (on $\mathbb{Z}_2^n$). For example, the *correction* factor for the Euclidean distance on $\mathbb{Z}_2^n$ is $\tau = \tilde{\tau}^2$.

Theorem 2. *Let $d_H : \mathbb{Z}_2^n \times \mathbb{Z}_2^n \to \mathbb{R}_{\geq 0}$ be the Hamming distance and $\tau \in \mathbb{R}_{>0}$. Then, it is possible to determine the bit-values of a string x having access only to a vector $y \in B_x(\tau)$ and in at most $n + 2\tau$ calls to the decision function δ_x (cf. Definition 2).*

The proof of Theorem 2 is provided in the appendix. In light of Theorem 1, we have the natural extension of Theorem 2 to the case of any leaking distance on $\mathbb{Z}_2^n$.

Corollary 1. *For any leaking distance d on $\mathbb{Z}_2^n$, Theorem 2 holds, with $\tau = h^{-1}(\tilde{\tau})$ being the corresponding threshold when $\tilde{\tau}$ is the given radius of the ball for the distance d.*

As a side result, we have:

Corollary 2. *If x is the stored biometric template b, and y is a matching fresh measurement b' satisfying $d(b, b') \leq \tau$, then Theorem 2 provides an algorithm to retrieve b being given b' in a number of authentication attempts linear in bit-length of the biometric templates.*

In the protocol for iris recognition by Daugman [5], the matching process relies on a normalised Hamming distance, which is defined as $\mathsf{NHD}(b, b', X, Y) = \sum_{i=1}^n (b_i \oplus b_i') X_i Y_i / \sum_{i=1}^n X_i Y_i$, for $b, b', X, Y \in \mathbb{Z}_2^n$. In the previous formula the vector X is the mask for the stored biometric template b, while Y masks the fresh trait b'. It is immediate to see that the normalised Hamming distance does not comply with Definition 1, nevertheless it is still possible, given b' and Y, to mount a centre search attack and recover the bits of b that are not blinded by the mask X, *i.e.* b_i such that $X_i = 1$.

Theorem 2 holds only for leaking distances on $\mathbb{Z}_2^n$ as in the proof we exploit the fact that $|x_i - y_i|$ can only assume two values 0 and 1, when $x_i = y_i$ and $x_i \neq y_i$ respectively. However, Theorem 3 generalises the reasoning in Theorem 2 to the non-binary case when any leaking distance is used (such as the Euclidean distance, often used in non-binary biometric authentication protocols).

Theorem 3. *Let $d : \mathbb{Z}_q^n \times \mathbb{Z}_q^n \to \mathbb{R}_{\geq 0}$ be any leaking distance on $\mathbb{Z}_q^n$ (cf. Definition 1) and $\tau \in \mathbb{R}_{>0}$, be a threshold such that $\tau < h(\lfloor \frac{q}{2} \rfloor^k)$, then it is possible to determine the value of the vector $x \in \mathbb{Z}_q^n$ having access only to a vector $y \in B_x(\tau)$ in at most mn calls to the decision function δ_x (as in Definition 2), where $m = \min\{\lfloor 2\tau \rfloor, 2\log q\}$.*

The proof of Theorem 3 is provided in the appendix. Also in this case, if we consider the vectors as biometric templates it holds:

Corollary 3. *Considering x as the stored biometric template b, and y as the fresh matching trait b', then the proof of Theorem 3 provides an algorithm to mount centre search attacks against biometric authentication systems with templates in $\mathbb{Z}_q^n$. And the maximal number of authentication attempts is linear in length (dimension as vectors) of the biometric templates.*

It is important to highlight that the results of this section imply that all biometric authentication protocols that employ a leaking distance in the matching process are vulnerable to the *centre search* attack, and this attack can be performed in an efficient way.

5   Biometric Sample Recovery Attacks in the Binary Case

One of the most severe threats to biometric authentication systems is recovering a stored raw biometric template b (maybe linked to the identity of the user). The knowledge of b provides more information than the knowledge of a fresh trait b', as the same b could be used in multiple biometric authentication systems possibly employing different matching processes (while b' might be rejected). In Section 4 we already presented efficient ways to recover the centre b of a ball, given a point b' *close* to it, namely $b' \in B_b(\tau)$. The question we address now is: *Is there a way to find a matching template b' given access only to δ_b?* The next subsections present four different answers to this question. We discuss the connection between this problem and the *set-covering* problem in Section 5.2.

In the following, we consider only the case in which the biometric traits are binary vectors, *i.e.* $b \in \mathbb{Z}_2^n$, and the employed distance is a leaking distance (cf. Definition 1).

5.1   Blind Brute Force

In the *blind brute force attack*, the attacker randomly chooses a point $b' \xleftarrow{R} \mathbb{Z}_2^n$, and checks the output of the function $\delta_b(b')$. If $\delta_b(b') = 1$, it means that $p \in B_b(\tau)$, so the attacker can easily recover b using this point b' (cf. Theorem 2). Otherwise (*i.e.*, if $\delta_b(b') = 0$), the attacker picks another point at random from $\mathbb{Z}_2^n$ as before. We call this attack *blind brute force* because in each attempt the adversary tries a random point until a point in $B_b(\tau)$ is found.

Let us compute the success probability of this attack after $t \in \mathbb{Z}_{>0}$ attempts. Suppose first that we pick $b' \in \mathbb{Z}_2^n$ uniformly at random. Then the probability of having b' accepted is $\omega := |B_b(\tau)|/|\mathbb{Z}_2^n| = \sum_{k=0}^{\tau} \binom{n}{k}/2^n$. In each attempt, if the trial point is chosen uniformly at random and independently from the previous attempts, then with probability ω this new trial point will be accepted. Let us now introduce binary random variables $X_i = 0$ or 1, for $i = 1, 2, \cdots, t$, and let $\mathbb{P}(X_i = 1) = \omega$ and $\mathbb{P}(X_i = 0) = 1 - \omega$. Obviously, X_i, $i = 1, 2, \cdots, t$, are i.i.d. Bernoulli random variables $X_i \sim \mathcal{B}ern(\omega)$. We are interested in computing $\mathbb{P}\left(\sum_{i=1}^{t} X_i = 1\right)$, the total probability of succeeding once in t attempts. It is not hard to see that $\mathbb{P}\left(\sum_{i=1}^{t} X_i = 1\right) = t\omega(1 - \omega)^{t-1}$, as the random variable $\sum_{i=1}^{t} X_i \sim \mathcal{B}inom(t, \omega)$ has a binomial distribution.

5.2 Sampling without Replacement

Brute Force without Point Replacement. In order to perform a brute force attack *without point replacement* the attacker has to define a set of potential candidates $C \subseteq \mathbb{Z}_2^n$. For the first trial, $C = \mathbb{Z}_2^n$ and the attacker chooses a point $b' \xleftarrow{R} C$ at random. If $\delta_b(b') = 1$, the selected point is inside the acceptance ball, $b' \in B_b(\tau)$, and so the attack is successful. Otherwise, the attacker updates the set of potential candidates $C = C \setminus \{b'\}$, deleting the one point that is not in the acceptance ball. The attack proceeds by randomly picking a point from the updated set C.

Let the random variables X_i, $i = 1, 2, \cdots, t$, be as in the case of the blind brute force attack. Note, however, that now $\mathbb{P}(X_i = 1)$ is different in each attempt. In this case, $\sum_{i=1}^{t} X_i$ follows the Hypergeometric distribution. Therefore, $\mathbb{P}\left(\sum_{i=1}^{t} X_i = 1\right) = B\binom{2^n - B}{t-1}/\binom{2^n}{t}$, where $B = |B_x(\tau)| = \sum_{k=0}^{\tau} \binom{n}{k}$. This attack is intuitively *better* than the blind brute force, but of course the larger the n is, the less efficient it is.

The Tree Algorithm. We propose here a method (Algorithm 1) to find a point $b' \in \mathbb{Z}_2^n$ within distance τ from the unknown biometric template b, given access to the decision function δ_b (as in Definition 2). The central idea of Algorithm 1 is to consider the points of $\mathbb{Z}_2^n$ as leaves of a binary tree of depth n. The tree structure is then exploited to define relatives-relations among the points of $\mathbb{Z}_2^n$ and to ensure that at each unsuccessful trial one can delete non-overlapping portions of the space $\mathbb{Z}_2^n$. More precisely, if a point $p \in \mathbb{Z}_2^n$ is such that $\delta_b(p) = 0$, the algorithm removes from the set of potential centres not only the tried point p, but also its siblings-relatives generated by the τ common ancestor (see Figure 2).

Algorithm 1. The Tree algorithm

Input: $(n, \tau, \delta_b,)$

Output: $b' = b'_1, \cdots, b'_n$ (a matching template)

$C = \mathbb{Z}_2^{n-\tau}$

for $i = 1$ to $2^{n-\tau}$: **do**
  $a \xleftarrow{R} \{C\}$
  $p = \text{generate}(a, \tau)$

  if $\delta_b(b') = 1$ (accepted) **then**
      Return b'
  else
      $C = C \setminus \{a\}$
  end if
end for

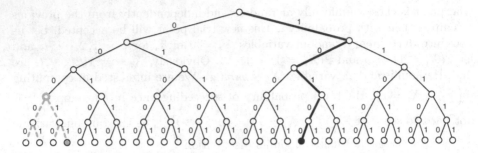

Fig. 2. The fundamental step of the Tree algorithm. Suppose the target biometric template is the vector $b = (10100) \in \mathbb{Z}_2^5$, the black bullet in the tree, and suppose the threshold is set to be $\tau = 2$. Let $a = (000)$ be the selected ancestor, highlighted as a grey circle in the picture. Let $b' = (00011)$ be the leaf randomly generated from a, then $d_H(b', b) > \tau$ and so $\delta_b((00011)) = 0$. In this case the points generated by a (*i.e.* that have a as common ancestor) will be deleted from the set of potential solutions.

The main function called by the algorithm is generate. Its input is the threshold τ and a $(n - \tau)$-dimensional binary vector a. The output is a random leaf $b' \in \mathbb{Z}_2^n$ generated by a (the τ ancestor). That is, generate$(a, \tau) = (a_1, \ldots, a_{n-\tau}, r_1, \ldots, r_\tau) = b'$, where $r_i \in \mathbb{Z}_2, i = 1, \ldots, \tau$ are τ random bits. The set of potential ancestors C is updated at every unsuccessful round, by deleting the chosen ancestor. The tree algorithm uses the Hamming distance.

For a practical implementation, we can store the paths of the tree that lead to the already rejected ancestors, and pick the new node a among the non-already-traversed paths. The running time of the attack is (of course) exponential, as it progressively constructs a binary tree of order $n - \tau$. Nevertheless, the probability to display the whole tree before finding a point that matches the reference template is very low (precisely: $2^{-n+\tau}$).

The *Optimal* Solution. The goal of the attacks described in this section is to find the ball $B_b(\tau) \subset \mathbb{Z}_2^n$ on which δ_b takes the value 1, without any additional information at hand. We have already investigated blind brute force (random tries), brute force without point replacement (remove one point at each unsuccessful trial), and the Tree algorithm (remove 2^τ points at each unsuccessful trial). The *optimal* brute force approach exploits the following idea: if a point $p \in \mathbb{Z}_2^n$ is rejected, *i.e.* $\delta_b(p) = 0$, it means that $b \notin B_p(\tau)$. Hence, the whole ball $B_p(\tau)$ can be removed by the set of potential centres. Intuitively, the *best* one can do to rapidly reduce the size of potential centres, is to use as trial points, points that lie at distance 2τ from each other. This corresponds to covering the space $\mathbb{Z}_2^n$ with the *smallest number* of balls of radius τ. This corresponds to an instance of the well-known *set-covering* problem in a space [25, 26].

More precisely, the optimal biometric sample recovery attack would involve the adversary covering $\mathbb{Z}_2^n$ with a family $\mathfrak{F}$ of balls of radius τ. At this point, the adversary needs to query the oracle (*i.e.* to use the decision function δ_b)

at most $|\mathfrak{F}|$ times, one for each (centre of a) ball in $\mathfrak{F}$. Hence the *best* solution is for $\mathfrak{F}$ a *minimal* covering, *i.e.* $|\mathfrak{F}| = min_{\mathfrak{G} \in \mathscr{C}} |\mathfrak{G}|$, where $\mathscr{C}$ is the set of all possible covering of $\mathbb{Z}_2^n$ with balls of radius τ. This is exactly the set covering problem: to find the minimal number of balls needed to cover a space. It is proven that the set covering problem is NP-complete[26]. This result implies that also providing an *optimal* algorithm for the biometric sample recovery attack is an NP-complete problem. However, there exist some *greedy approximations* that are relatively efficient. In particular, for our case, Theorem 1 in [26] applies directly and hence the number of points that the adversary needs to query is only a factor of $O(\tau \ln(n+1))$ more than the optimal cover.

5.3 Comparisons and Bounds

In order to compare the performance of the four described methods we need to bound the probability that an attacker succeeds in finding a *matching* point, in each case. At the t-th trial, the attacker attempts point $x_t \in Z_2^n$ and observes $y_t \in 0, 1$, with $y_t \triangleq \mathbb{1}_{B_b(\tau)}(x_t) = \delta_b(x_t)$. Let $z_t \in \{0, 1\}$ denote whether or not the attacker has found an acceptable point after t trials and $s_t = \sum_{i=1}^{t} y_t$ be the number of points the attacker has found by time t.

To begin the analysis, we define $\mu_b(\tau) \triangleq |B_b(\tau)|/|\mathbb{Z}_2^n| \in [0, 1]$ to be the relative measure of the acceptance ball around b. In the binary case, dropping the dependence on b, τ, we have $\mu \in [2^{\tau-n}, (n+1)^\tau 2^{-n}]$. Of course, μ is also the probability of acceptance if sampling uniformly.

Blind brute force. In this case the points are selected uniformly without replacement, *i.e.* $x_t \sim \mathcal{U}(Z_2^n)$. It trivially follows that $\mathbb{E}(s_t) = \mu t$. It is also clear that the attack is successful whenever $s_t \geq 1$. For that reason, we shall attempt to bound the probability that this occurs while $\mu t < 1$. As a matter of fact, we can write:

$$\mathbb{P}(s_t \geq 1) = \mathbb{P}(\bigvee_{i=1}^{t} z_i = 1) \leq \sum_{i=1}^{t} \mathbb{P}(z_t = 1) = \mu t \leq (n+1)^\tau 2^{-n} t.$$

where the first inequality becomes an equality whenever $\mu t < 1$.

Sampling without replacement. All the other described approaches correspond to sampling without replacement. In either case, let $\alpha \in [0, 1]$ denote the proportion of points removed at each step. Then, we obtain the following bound:

$$\mathbb{P}(s_t \geq 1) \leq \sum_{i=1}^{t} \mathbb{P}(z_t = 1) \leq \sum_{i=1}^{t} \frac{\mu}{1 - \alpha i} \leq \int_0^t \frac{\mu}{1 - \alpha x} \, dx = \frac{\mu}{\alpha} \log \frac{1}{1 - \alpha t}.$$

For the point-wise replacement algorithm, $\alpha = q^{-n}$, hence there is little effect. For the binary case, we can employ the tree algorithm, $\alpha = 2^{\tau-n}$, which can be a substantial improvement. An unbounded adversary may use an optimal cover, in order to exclude as many points as possible whenever a point is rejected. In fact,

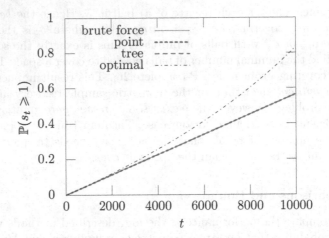

Fig. 3. Visualisation of the bounds for $q = 2, n = 32, \tau = 5$. In this case $\mu \approx 5.6 \times 10^{-5}$.

in the best case, the adversary will be able to remove B points every time a point is rejected, giving a value of $\alpha = B2^{-n}$. To visualise the bounds, we choose some parameters such that there is a clear difference after a small number of iterations (depicted in Figure 3). More precisely, Figure 3 shows the performance of all four methods in terms of an upper bound on their success probability after a number of iterations. The four curves show sampling with replacement (*i.e.* brute force), and three different cases for sampling without replacement. Firstly, removing a single point. Secondly, removing 2^τ points using the tree construction. Finally, removing the maximum number of points B, which is computationally infeasible. There is a significant gain for the last choice, but only after a large portion of the space has already been covered. As when $\alpha \to 0$, $\ln \frac{1}{1-\alpha t} \to \alpha t$, the success probabilities of the first three methods are approximately linear in the size of the space, and hence exponential in the dimension.

The naive no replacement algorithm naturally does not improve significantly over brute force without replacement, since the volume that is excluded at every step is infinitesimal. Obviously, if we are able to remove a significant part of the volume, then we obtain a clear improvement in performance. Only an optimal adversary can do significantly better. However, this would assume either that *set-covering* is in P or that the adversary is computationally unbounded. Consequently, as there is no polynomial algorithm that is significantly better than brute force, biometric authentication schemes based on matching templates are secure against biometric sample recovery attacks.

6   Conclusions

In this paper, we prove that all biometric authentication protocols that employ distances between a template and an fresh biometric in the matching process suffer from leakage of information that could be exploited by an adversary to launch *centre search* attacks. In order to analyse this leakage of information, we provide a mathematical framework and prove that centre search attacks are feasible for any biometric template defined in $\mathbb{Z}_q^n$, $q \geq 2$, after a number of authentication attempts that is linear in n. Our results imply that it is possible to mount this attack on most existing biometric authentication protocols (including privacy-preserving ones) that rely on a Hamming, Euclidean, normalised Hamming distance or any distance that complies with Definition 1.

Furthermore, we investigate whether brute force attacks can be used to recover a matching biometric. We describe four strategies: blind brute force, brute force without replacement, a new algorithm based on a tree structure and the optimal case. Our results demonstrate that improving the success rate in these brute force attacks would imply finding a solution to the NP-complete *set-covering* problem. Thus, this provides some security guarantees of existing biometric authentication protocols as long as the attacker has not access to a matching biometric trait.

A possible countermeasure that could be employed in order to strengthen existing biometric authentication protocols against *centre search* attacks would be the employment of more sophisticated authentication methods. For example, simply using weighted distances in which the weights are secret and different for each user may provide sufficient security. Something similar is already employed in the normalised Hamming distance for which indeed the centre search attack is feasible but only for a subset of the components of the stored biometric template. An alternative and promising direction would be to rely on a mechanism that randomly selects a distance from a pool of distances at each authentication attempt. However, such measures should be incorporated carefully in order not to affect the accuracy of the biometric authentication system.

A   Collected Proofs

Proof (Theorem 1). By hypothesis d is a leaking distance, hence it is of the form $d(x, y) = h(\sum_{i=1}^{n} |x_i - y_i|^k)$, for all $x, y \in \mathbb{Z}_2^n$. Since $h : \mathbb{R} \to \mathbb{R}_{\geq 0}$ is monotonic, it is bijective on its image, in other words it has an inverse $h^{-1} : I \to \mathbb{R}$, where $I = Im(h) = \{w \in \mathbb{R}_{\geq 0} : w = h(z), \exists z \in \mathbb{R}\}$.

Consider the d ball of radius $\tilde{\tau}$ around a point $x \in \mathbb{Z}_2^n$, namely the set $\{y \in \mathbb{Z}_2^n : d(x, y) \leq \tilde{\tau}\}$. We want to prove this d-ball equals a *Hamming distance*-ball centred in x and of radius τ.

Indeed, $d(x, y) \leq \tilde{\tau} \iff h(\sum_{i=1}^{n} |x_i - y_i|^k) \leq \tilde{\tau}$. Noticing that h is increasing implies that h^{-1} is also increasing, one obtains: $\sum_{i=1}^{n} |x_i - y_i|^k \leq h^{-1}(\tilde{\tau})$. In addition, since $|x_i - y_i| \in \{0, 1\}$ we can *ignore* the exponent k in the expression (this is because $0^k = 0$ and $1^k = 1, \forall k \in \mathbb{Q}_{>0}$). Hence, $\sum_{i=1}^{n} |x_i - y_i| \leq h^{-1}(\tilde{\tau})$, but the left hand side of the inequality is exactly the Hamming distance between the points x and y.

To summarise, we have $d(x, y) \leq \tilde{\tau} \iff d_H(x, y) \leq h^{-1}(\tilde{\tau})$. Let us put $\tau = h^{-1}(\tilde{\tau})$, then $\{y \in \mathbb{Z}_2^n : d(x, y) \leq \tilde{\tau}\} = \{y \in \mathbb{Z}_2^n : d_H(x, y) \leq \tau\}$. That is, any d-ball can be described as a d_H-ball (*Hamming distance*-ball) and vice versa. $\square$

Proof (Theorem 2). STEP 1. Find a point w that lies just outside the boundary of $B_x(\tau)$.

By hypothesis $\delta_x(y) = 1$. Let w be the vector obtained from y by flipping the first bit, *i.e.* $w_1 = \bar{y}_1$ and $w_i = y_i$, $\forall i \in \{2, \ldots, n\}$. If w is rejected, that is, if $\delta_x(w) = 0$, it means that y is already on the boundary of $B_x(\tau)$ and we are done by putting $v = y$. Otherwise, proceed by flipping one more bit of y until it exits $B_x(\tau)$. The general step after $k - 1$ trials (flipping bits of y and being accepted) is: set $w = (\bar{y}_1, \ldots, \bar{y}_k, y_{k+1}, \ldots, y_n)$, if $\delta_x(w) = 0$ put $v = (\bar{y}_1, \ldots, \bar{y}_{k-1}, y_k, \ldots, y_n)$. If $\delta_x(w) = 1$, go on and flip the next component. It is quite intuitive that this procedure ends after at most $2\tau + 1$ steps (the worst case is when y is already on the boundary but we move it in the *wrong* direction and cross the ball along its diameter).

STEP 2. Determine the central point x of $B_x(\tau)$.

Note that by STEP 1, we already know the value of the k-th component of x, namely $x_k = v_k$. For $j \in \{1, 2, \ldots, n\} \setminus \{k\}$, consider the vector $v(j)$ defined as $v(j)_i = w_i$, $\forall i \in \{1, \ldots, n\} \setminus \{j\}$. If $\delta_x(v(j)) = 1$, it means that $v(j)$ *compensates* the error (in the k-th component) introduced by w with a *new* correct component (the j-th component). Hence $x_j = v(j)_j$. On the other hand, $\delta_x(v(j)) = 0$ implies that the j-th component of w was correct. Hence, in this case, $x_j = 1 - v(j)_j$. STEP 2 ends after $n - 1$ queries. $\qquad\square$

Proof (Theorem 3). Let $e(i) \in \mathbb{Z}_q^n$ denote the i-th vector of the canonical basis, *i.e.* for each $i = 1, \ldots, n$, $e(i)_i = 1$ and $e(i)_j = 0$, $\forall j \in \{1, \ldots, n\} \setminus \{i\}$. For each of the n components of a biometric template, determine two vectors $v(i), w(i) \in \mathbb{Z}_q^n, i = 1, \ldots, n$ such that: $v(i) = b' + \lambda_1 e(i)$ and $w(i) = b' + \lambda_2 e(i)$, with $\lambda_1 \in \{y_i, q-1-y_i\}$ and $\lambda_2 \in \{0, y_i-1\}$. Moreover, $\delta_x(v(i)) = 1$ but $\delta_x(v(i)+e(i)) = 0$, and $\delta_x(w(i)) = 1$ but $\delta_x(w(i) - e(i)) = 0$. Such pair of vectors exists for each component, as $B_x(\tau)$ is a bounded subset of $\mathbb{Z}_q^n$ and $\tau < h(\lfloor \frac{q}{2} \rfloor^k)$. There are two possible situations:

- $v(i)$ and $w(i)$ are on the boundary of the ball $B_x(\tau)$. In this case the centre of the ball $x \in \mathbb{Z}_q^n$ will have the i-th component equal to the *middle point* $x_i = (v(i)_i + w(i)_i)/2$, $\forall i \in \{1, \ldots, n\}$.
- $v(i)$ and $w(i)$ are not exactly *on* the boundary of the ball $B_b(\tau)$. Since it is $v(i), w(i), x \in \mathbb{Z}_q^n$ the respective distances from the boundary $\epsilon_{v(i)}$ and $\epsilon_{w(i)}$ must be equal (by symmetry). Thus, also in this case $b_i = (v(i)_i + w(i)_i)/2$, $\forall i \in \{1, \ldots, n\}$.

There are two efficient strategies to determine the vectors $v(i), w(i)$:

- *Linear search:* in this case the worst case scenario is when $y = x$, and the adversary needs to try all the points (with components in $\mathbb{Z}_q$) that lie in the diameter of the ball $B_x(\tau)$, that is at most $\lfloor 2\tau \rfloor$ trials.
- *Binary search:* the adversary performs at most $2 \log q$ trials to determine each *external* point, $v(i), w(i)$.

Thus, the maximum number of queries (access to the δ_x function) necessary in order to recover the centre x of a ball in $\mathbb{Z}_q^n$ is bounded by nm, with $m = \min\{\lfloor 2\tau \rfloor, 2\log q\}$.   □

Acknowledgments. We would like to thank the anonymous reviewers for their comments. This work was supported by the FP7-STREP project "BEAT: Biometric Evaluation and Testing", grant number: 284989.

References

1. Penrose, L.: Dermatoglyphic topology. Nature **205**, 544–546 (1965)
2. Bolling, J.: A window to your health. Jacksonville Medicine, Special Issue: Retinal Diseases 51 (2000)
3. Osadchy, M., Pinkas, B., Jarrous, A., Moskovich, B.: SCiFI - A System for Secure Face Identification. In: 2010 IEEE Symposium on Security and Privacy, pp. 239–254 (2010)
4. Bringer, J., Chabanne, H., Izabachène, M., Pointcheval, D., Tang, Q., Zimmer, S.: An application of the goldwasser-micali cryptosystem to biometric authentication. In: Pieprzyk, J., Ghodosi, H., Dawson, E. (eds.) ACISP 2007. LNCS, vol. 4586, pp. 96–106. Springer, Heidelberg (2007)
5. Daugman, J.: How iris recognition works. IEEE Transactions on Circuits and Systems for Video Technology **14**, 21–30 (2004)
6. Erkin, Z., Franz, M., Guajardo, J., Katzenbeisser, S., Lagendijk, I., Toft, T.: Privacy-preserving face recognition. In: Goldberg, I., Atallah, M.J. (eds.) PETS 2009. LNCS, vol. 5672, pp. 235–253. Springer, Heidelberg (2009)
7. Sadeghi, A.-R., Schneider, T., Wehrenberg, I.: Efficient Privacy-preserving face recognition. In: Lee, D., Hong, S. (eds.) ICISC 2009. LNCS, vol. 5984, pp. 229–244. Springer, Heidelberg (2010)
8. Huang, Y., Malka, L., Evans, D., Katz, J.: Efficient privacy-preserving biometric identification. In: NDSS 2011 (2011)
9. Barni, M., Bianchi, T., Catalano, D., Di Raimondo, M., Donida Labati, R., Failla, P., Fiore, D., Lazzeretti, R., Piuri, V., Scotti, F., Piva, A.: Privacy-preserving fingercode authentication. In: Proceedings of the 12th ACM Workshop on Multimedia and Security, pp. 231–240 (2010)
10. Simoens, K., Bringer, J., Chabanne, H., Seys, S.: A framework for analyzing template security and privacy in biometric authentication systems. IEEE Transactions on Information Forensics and Security **7**(2), 833–841 (2012)
11. Jain, A.K., Nandakumar, K., Nagar, A.: Biometric template security. EURASIP J. Adv. Signal Process **2008**, 113:1–113:17 (2008)
12. Bringer, J., Chabanne, H., Simoens, K.: Blackbox security of biometrics. In: Proceedings of the 6th International Conference on Intelligent Information Hiding and Multimenida Signal Processing, pp. 337–340 (2010)
13. Jarrous, A., Pinkas, B.: Secure hamming distance based computation and its applications. In: Abdalla, M., Pointcheval, D., Fouque, P.-A., Vergnaud, D. (eds.) ACNS 2009. LNCS, vol. 5536, pp. 107–124. Springer, Heidelberg (2009)
14. Bringer, J., Chabanne, H., Patey, A.: SHADE: Secure hamming distance computation from oblivious transfer. In: Adams, A.A., Brenner, M., Smith, M. (eds.) FC 2013. LNCS, vol. 7862, pp. 164–176. Springer, Heidelberg (2013)

15. Bringer, J., Chabanne, H., Favre, M., Patey, A., Schneider, T., Zohner, M.: GSHADE: Faster Privacy-preserving Distance Computation and Biometric Identification. In: Proceedings of the 2nd ACM Workshop on Information Hiding and Multimedia Security, pp. 187–198. ACM (2014)
16. Biham, E., Shamir, A.: Power analysis of the key scheduling of the aes candidates. In: Proceedings of the 2nd AES Candidate Conference (1999)
17. Brier, E., Clavier, C., Olivier, F.: Correlation power analysis with a leakage model. In: Joye, M., Quisquater, J.-J. (eds.) CHES 2004. LNCS, vol. 3156, pp. 16–29. Springer, Heidelberg (2004)
18. Barbosa, M., Brouard, T., Cauchie, S., de Sousa, S.M.: Secure biometric authentication with improved accuracy. In: Mu, Y., Susilo, W., Seberry, J. (eds.) ACISP 2008. LNCS, vol. 5107, pp. 21–36. Springer, Heidelberg (2008)
19. Stoianov, A.: Security issues of biometric encryption. In: Proceedings of the 2009 IEEE Toronto International Conference on Science and Technology for Humanity (TIC- STH), pp. 34–39 (2009)
20. Paillier, P.: Public-key cryptosystems based on composite degree residuosity classes. In: Stern, J. (ed.) EUROCRYPT 1999. LNCS, vol. 1592, pp. 223–238. Springer, Heidelberg (1999)
21. Yao, A.C.C.: How to generate and exchange secrets. In: 27th Annual Symposium on Foundations of Computer Science, pp. 162–167. IEEE (1986)
22. Rabin, M.O.: How to exchange secrets with oblivious transfer. IACR Cryptology ePrint Archive 2005, 187 (2005)
23. Goldwasser, S., Micali, S.: Probabilistic encryption and how to play mental poker keeping secret all partial information. In: Proceedings of the 14th Annual ACM Symposium on Theory of Computing, STOC 1982, pp. 365–377. ACM (1982)
24. Bringer, J., Chabanne, H., Cohen, G., Kindarji, B., Zémor, G.: Optimal iris fuzzy sketches. In: Proceedings of the 1st IEEE International Conference on Biometrics: Theory, Applications, and Systems (2007)
25. Chen, L.: New analysis of the sphere covering problems and optimal polytope approximation of convex bodies. Journal of Approximation Theory $133(1)$, 134–145 (2005)
26. Chvatal, V.: A greedy heuristic for the set-covering problem. Mathematics of Operations Research $4(3)$, 233–235 (1979)

Efficient Hardware Design

One Word/Cycle HC-128 Accelerator via State-Splitting Optimization

Ayesha Khalid[1], Prasanna Ravi[2],
Anupam Chattopadhyay[3], and Goutam Paul[4]($\boxtimes$)

[1] Institute for Communication Technologies and Embedded Systems (ICE),
RWTH Aachen University, 52074 Aachen, Germany
`ayesha.khalid@ice.rwth-aachen.de`
[2] National Institute of Technology, Tiruchirappalli (NITT),
Tiruchirappalli 620 015, Tamil Nadu, India
`ravikant64@hotmail.com`
[3] School of Computer Engineering, Nanyang Technological University (NTU),
Singapore, Singapore
`anupam@ntu.edu.sg`
[4] Cryptology and Security Research Unit (CSRU), R.C. Bose Centre for Cryptology
and Security, Indian Statistical Institute, Kolkata 700 108, India
`goutam.paul@isical.ac.in`

Abstract. As today's high performance embedded systems are heterogeneous platforms, a crisp boundary between the software and the hardware ciphers is fast getting murky. This work takes up the design of a dedicated hardware accelerator for HC-128, one of the stream ciphers in the software portfolio of eSTREAM finalists. We discuss a novel idea of splitting states kept in SRAMs into multiple smaller SRAMs and exploit the increased parallel accesses to achieve higher throughput. We optimize the accelerator design with state splitting by different factors. A detailed throughput-area-power analysis of these design points follow along with a benchmarking with the state-of-the-art for HC-128. Our implementation marks an HC-128 ASIC with the highest throughput per area performance reported in the literature till date.

Keywords: eSTREAM · HC-128 · ASIC · Hardware accelerator · Stream cipher implementation

1 Introduction

The eSTREAM [1] competition was a multi-year effort targeted to attract new and efficient stream cipher proposals. Its the only competition of cryptographic proposals that profiled the finalists in separate software and hardware categories. After multiple filtering, the surviving finalists, out of a total of 34 initial proposals, were HC-128, Rabbit, Salsa20/12 and SOSEMANUK in the software profile

Prasanna Ravi: This work was done in part while the second author was visiting
RWTH Aachen as a DAAD summer intern.

© Springer International Publishing Switzerland 2014
W. Meier and D. Mukhopadhyay (Eds.): INDOCRYPT 2014, LNCS 8885, pp. 283–303, 2014.
DOI: 10.1007/978-3-319-13039-2_17

and Grain v1, MICKEY 2.0 and Trivium in the hardware profile. Regarding this categorization, other cryptographic competitions did not follow the same trend. The call for AES [8] announced that the computational efficiency of both hardware and software implementations would be taken up as a decisive factor for the selection of the winner. Similarly, the choice of Keccak as SHA-3 finalist was attributed by NIST to both its good software performance and excellent hardware performance [9]. NESSIE and CRYPTREC projects aiming to identify better cryptographic primitives also did not classify the finalist proposals in the basis of their target platforms.

The eSTREAM categorizes the proposals more suitable for software applications with high throughput requirements in the software profile [1]. However, the high performance requirements coupled with limited energy budgets leave the development of dedicated accelerators in today's heterogeneous systems as an obvious choice. Consequently, the ASIC designs of stream ciphers proposed for software oriented applications have extensively been taken up, e.g., RC4 [11,12,14], Salsa20/12 [15,16], Rabbit [17], SOSEMANUK [18]. Other than these reported efforts for individual ciphers, two major projects took up the ASIC evaluations of the eSTREAM proposals that were most likely to be taken up in the hardware profile [19,20], however, neither of these efforts took up HC-128.

HC-128 is criticized for its large internal state for its unsuitability to be taken up for hardware implementation. Today's heterogeneous systems, however, employ extensively, large memory banks that are time-shared between multiple arbiters. Consequently, its only fair to consider the core of a crypto-processor alone for resource budgeting, as the memory bank is not exclusively used by it. Having external memory banks for programmable cryptographic processors is advantageous, since the memory could also be reused for various versions of an algorithm, offering the user a varying performance/security trade-offs [10,13]. For most of the coarse-grained hardware platforms like FPGAs and CGRAs, block RAM modules are available as macros, that may be configured to desired sized memories and will go wasted when unused.

In the light of this discussion its surprising to have [7] as the sole VLSI implementation effort for HC-128. It discusses several design points for parallelizing various steps in HC-128 keystream generation by having replicas of internal state. Since HC-128 is already notorious for its high memory requirements, having multiple copies of the state in multiple memories only worsens it. We take up an orthogonal approach for efficient parallelization in HC-128 by splitting the memory resources instead of replicating them. Other than completing the jigsaw puzzle for a missing implementation of eSTREAM cipher, this paper also presents parallelization strategies viable for the implementation of other stream ciphers with large internal states (e.g., RC4, WAKE, Py, HC-256, CryptMT etc.).

Section 2 and 3 of this paper discuss a reduced memory HC-128 implementation and its hardware processor design, respectively. We propose a novel state splitting strategy to increase the performance of stream ciphers with large states in Section 4. We study the feasibility of n-way state splitting on HC-128, identify incremental design points and benchmark each against area, throughput, time and latency for a fair evaluation in Section 5. Our HC-128 processor significantly outperforms all reported hardware and software designs in throughput per area. Section 6 provides conclusion and an outlook to this work.

2 Optimized Memory Utilization

The description of HC-128 is provided in Appendix A. Other than two 512-word arrays P and Q, the cipher requires a 1280-word array W used only during initialization. A close observation reveals a separate W array unnecessary, a clever use of the P and the Q arrays could just suffice. We modify the initialization phase, without disturbing the algorithm functionality, to initialization steps given in Table 1. Instead of using W, Key and IV are expanded into the lower half of P memory in *Step 2*. Consequently an offset of 256 is added to the original calculations. *Step 3* updates the entire P array, after which the last 16 locations of P are copied to Q in *Step 4*. *Step 5* updates the Q memory, *Step 6* is the same as the one in original proposal.

Table 1. Modified HC-128 Initialization phase steps

Step 1: Let $K[i+4] = K[i]$ and $IV[i+4] = IV[i]$ for $0 \leq i \leq 3$.

Step 2: The key and IV are expanded to lower half of P memory as follows:
   $P[i+256] = K[i]$, for $0 \leq i \leq 7$;
     $= IV[i-8]$, for $8 \leq i \leq 15$;
     $= f_2(P[i+256-2]) + P[i+256-7]$
       $+f_1(P[i+256-15]) + P[i+256-16] + i$, for $16 \leq i \leq 255$.

Step 3: Update the entire array P as follows: $P[i] = f_2(P[i \boxminus 2]) + P[i \boxminus 7] + f_1(P[i \boxminus 15]) + P[i \boxminus 16] + i$, for $0 \leq i \leq 511$,
Step 4: Last 16 elements of P and copied into Q: $P[i] = Q[i]$, for $496 \leq i \leq 511$,
Step 5: Update the entire array Q as follows: $Q[i] = f_2(Q[i \boxminus 2]) + Q[i \boxminus 7] + f_1(Q[i \boxminus 15]) + Q[i \boxminus 16] + i$, for $0 \leq i \leq 511$,

Step 6: Update the P and Q arrays as follows:
   For $i = 0$ to 511, do
     $P[i] = (P[i] + g_1(P[i \boxminus 3], P[i \boxminus 10], P[i \boxminus 511])) \oplus h_1(P[i \boxminus 12])$;
   For $i = 0$ to 511, do
     $Q[i] = (Q[i] + g_2(Q[i \boxminus 3], Q[i \boxminus 10], Q[i \boxminus 511])) \oplus h_2(Q[i \boxminus 12])$;

Removal of the W memory reduces the memory budget of the cipher from 9KB to 4KB, at the expense of having some extra steps. For weakly programmable processors with very compact instruction set, the increase in the program memory due to these extra steps is trivial. Also the overhead in terms of extra cycles is negligible compared to the total initialization cycles. For the rest of the discussion we refer to the modified initialization in Table 1 for all the processor optimization design points we refer to.

3   Design Space Exploration of HC-128 Accelerator

We present the architecture for an HC-128 stream cipher processor core along with an interface with an external memory bank, that includes P and Q memories. As shown in Fig. 3, the instructions are kept in a program memory, while a simple incrementing PC serves as the address to the memory. By loading the program memory with assembly instructions and setting up the I/Os as shown in the figure, the design can be easily plugged into a System on Chip (SoC) environment. For the rest of the discussion, we call this implementation as *design*1. The I/Os of the processor are discussed below.

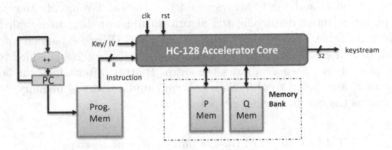

Fig. 1. Block Diagram of HC-128 Accelerator

- *Instruction* is input to the processor core and is 8-bits in width.
- *Key/IV* is taken in during the initialization phase of the algorithm.
- *Keystream* is the 32-bit keystream word generated by the processor.
- *Memory interface* comprises of standard memory control signals for the two SRAMs, data ports and address ports.

3.1   Port Utilization and Limitations

From Table 9 we can see that one keystream word requires 8 memory accesses in total: 5 P memory reads, 2 Q memory reads and one write to the P memory (considering only the *if* block of the algorithm). So hypothetically speaking, with maximum port utilization if possible, the total of 6 accesses to the P memory could be handled in 3 clock cycles for the dual ported P memory, resulting in a possible throughput of 3 cycles/keyword. The maximum utilization of memory ports is however not possible for current SRAMs that require one or more clock cycles for *turnaround* to change access from write to read and vice versa. Consequently, consecutive read-write and write-read accesses from one memory port would require one *turnaround* clock cycle between the two accesses, if an SRAM requiring a single turnaround cycle is considered. For our case, one port is left unused for one cycle after the final writeback to update the P memory.

For a pipelined processor, the limited number of access ports of SRAMs poses a structural hazard. As a workaround, n *nop* (no operation) instructions are inserted between two consecutive instructions. To increase the efficiency, the pipeline stages are designed to target maximum port utilization of memories and thus keeping n as small as possible.

3.2   Processor Resources, Pipeline Architecture and ISA

The processor has a 4 stage pipeline, referred as $EX1$, $EX2$, $EX3$ and $EX4$. These pipeline stages are designed with few critical goals. The first one is to ensure maximum utilization of the memory access ports for fast execution of instructions. Secondly, a judicious division of computationally expensive tasks over various pipeline stages is done so that the critical path of processor does not hamper the throughput. The third goal is ensuring economization of the resources (registers, pipeline registers, combinational blocks), whenever possible. Since the initialization and keystream generation are carried out exclusively, time based resource sharing is possible and exploited.

A total of 9 different instructions are used in the processor as shown in the Table 2. We take up one-hot encoding for the 8-bit instruction word for simpler instruction decoding. Each instruction (other than *nop*) executes a step in the initialization phase or the keystream generation phase as indicated. The instruction *init_kiv* fills the P memory with the Key and the IV values. The dual ported memory requires 8 instructions to fill these 16 locations. Consecutive instructions require no *nops* inserted between two instructions and consequently 8 instructions execute in 8 cycles. Similarly, the instruction *init_p_for_q* copies P memory contents to Q memory. It reads two locations from P memory and writes them to Q memory, requiring 8 instructions in total for the 16 word transfers. No *nops* are required in between. The task division in initialization phase and the keystream generation phase into various pipeline stages is discussed below.

Table 2. Instruction set for HC-128 accelerator

Instruction syntax	Instruction opcode	Related operation	*nops* b/w consecutive instructions	Total cycles for initialization
nop	0x00	-	-	-
init_kiv	0x01	Step1 of Table 1	0	8
init_p_w	0x02	Step2 of Table 1	3	240x4
init_p	0x04	Step3 of Table 1	3	512x4
init_p_for_q	0x08	Step4 of Table 1	0	8
init_q	0x10	Step5 of Table 1	3	512x4
update_p	0x20	Step6 of Table 1	3	512x4
update_q	0x40		3	512x4
keystream	0x80	Table 9	3	-

Initialization Phase: The equation to perform Step 2 of the initialization in Table 1 is rearranged as follows.

$$P[i + 256] = ((f_2(P[i + 256 - 2]) + f_1(P[i + 256 - 15]) + i) + P[i + 256 - 7])$$
$$+P[i + 256 - 16] \quad \text{for } 16 \leq i \leq 1279.$$

This step requires the addition of 5 values, 4 of which are 32-bit number. The pipeline design breaks the critical path of these additions which are slow operations in hardware. Consequently, 3 values are added in $EX2$ pipeline stage (resolving the innermost bracket), followed by the addition of 1 value in each of the next two pipeline stages (resolving the outer brackets). Step 3 and Step 5 of Table 1 are similar to Step 2, as they have the same execution path. The calculation is divided in 4 pipeline stages as follows.

1. In $EX1$, two read accesses to P memory are requested.
2. In $EX2$, functions f_1 and f_2 are carried out on the values read and added together with a 9-bit iteration count i. The result is saved in a 32-bit pipeline register d_reg. In the same pipeline stage, the next memory access is requested, i.e., $P[i + 256 - 7]$.
3. In $EX3$, d_reg is added to the value read from the memory and forwarded to the next pipeline. A memory access is requested for the location $P[i + 256 - 16]$.
4. In $EX4$ the read value from the memory is added to the previously calculated result in pipeline register d_reg and stored back at the memory location $P[i + 256]$ using a write access to the P memory.

Since $EX2$ adds 3 values along with the functionality of f_1 and f_2, this becomes the critical path of the $design1$. As given in the last column of Table 2, the number of $nops$ that must be inserted between two consecutive $init\_p\_w$ instructions is 3. This is to ensure that the $EX4$ of the current instruction and the $EX1$ stage of the following instruction do not overlap, posing a memory access port contention. The instruction $init\_p$ and $init\_q$ also have similar operations and have same number of $nops$ inserted between consecutive instructions.

Keystream Generation Phase: Step 6 of the initialization phase (Table 1) is similar to the keystream generation phase (Table 9), the only difference is that the keystream generated is xored with the updated value of P/Q arrays. Hence the pipeline architecture of $update\_p$, $update\_q$ and $keystream$ is same and given in Fig. 2 It shows only the $update\_p$ and the if part of the algorithm in Table 9 since the $update\_q$ and the $else$ part is trivially similar. The port multiplexing based on the instruction is not shown here. Memory bank access arbitration is also not shown but is straight forward to interpret. The two ports of the dual ported memory are referred as P0 and P1. All accesses going down are read requests except the final writeback in $EX4$ stage. The requested values read are shown by the arrows going up. We discuss the pipeline stage by stage.

1. $EX1$: A 9-bit register i maintains the iteration count of the loop. Two read requests are initiated from the two ports of the P memory with addresses having 9 and 511 as the offsets to register i.
2. $EX2$: The two values read are rotated and xored together to carry out a part of the g_1 function (called g_{11}). We split the g_1 functionality into g_{11} and the g_{12} functions (similarly g_2 is split as g_{21} and g_{22}) as follows.

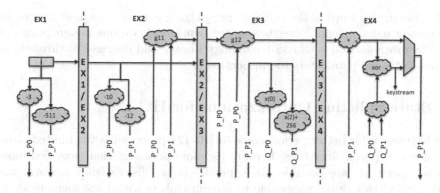

Fig. 2. Pipeline architecture of *design*1 of HC-128 accelerator

- $g_{11}(x, z) = (x \ggg 10) \oplus (z \ggg 23)$.
- $g_{12}(x, y, z) = g_{11}(x, z) + (y \ggg 8)$.

The g_{11} result is passed over to the next pipeline stage using the pipeline register d_reg used also in the initialization phase. The i register value is passed to this stage using a pipeline register i_p and with known offsets of 10 and 12, two more read requests are launched from the two ports of the P memory.

3. *EX3*: The g_{11} result is added to the value read from P0 of the P memory to complete the g_{12} function and the result is stored to d_reg pipeline register. The value read from $P[i-12]$ is used to initialize two read requests according to the h_1 function. $P[i]$ read request is also initialized.

4. *EX4*: The values read from the Q and P memories are taken up to calculate the keystream word (S_k) and the $P[i]$ values as referred in Table 1. The updated word is either $P[i]$ or the one obtained after its xor with the S_k, depending upon the instruction running. A single bit pipeline register holds the select line for this multiplexer.

The 6 accesses to P memory are so distributed that stage *EX2* and *EX3* each perform two reads while *EX4* performs one read and one write. The subsequent instructions should follow no earlier than after 3 *nops*, otherwise the port availability is violated. Consequently, the keystream generation speed is 4 cycles/keyword and 9168 cycles in total are required for the initialization.

The performance of a cipher processor is benchmarked generally by the initialization duration time and the throughput (byte/word/bits per second) of keystream generation phase. We, however, use a more generic term for judging performance, i.e., *cycles per keystream word*. The rationale for this alternate benchmark stems from the fact that mapping a processor design on different CMOS technology libraries will have different critical paths. Consequently, the maximum operating frequency (and the throughput performance) of the designs will differ. Similarly, if the critical path of the processor is dictated by the SRAM access time, the design has a performance dependence on memory modules, since a faster SRAM macro will improve the throughput and vice versa. Hence, having cycles/keystream word as the performance judging parameter is more generic or technology independent. Nevertheless, we benchmark various design points of HC-128 processors on a CMOS technology library and discuss the throughput results in typical parameters too in Section 5.

4   State Splitting Optimizations for HC-128

The bottleneck for better performance in HC-128 processor is the limited number of access ports, since there is room for more parallelization, provided more accesses per cycle are possible. In Appendix B, we discuss some performance enhancement techniques applicable to stream ciphers with large memories and also the scope of mapping these optimizations to HC-128.

This section discusses state splitting for throughput enhancement of cipher implementations. By splitting the state array of stream ciphers, into smaller parts with known address distribution and keeping each smaller part in a separate memory can enable more parallel accesses and consequently enable faster keystream generation. Consider a cipher with one memory M such that $Size(M) = N$ (words or bits or bytes). We divide the memory in S parts with the condition $\sum_{i=1}^{S} Size(m_i) = N$, where m_i are the smaller memories. We now have the possibility to have $S \times 2$ parallel accesses for dual ported memories.

Memory splitting results in a memory bank requiring more power and area. On the contrary, the access times of the smaller memories is smaller, hinting higher operating frequencies for the processors. We plot the area and maximum operating frequency for split SRAMs configurations using *Faraday Memory compiler* [25], for 65nm technology library in Fig. 3. We take up various organizations of a 1024-word memory. Assuming the simplistic case of equal sized split memories, we take up S as a range from 1 till 32 in powers of 2 shown from left to write in the graph. As the memory is split into two banks (512×2), there is a corresponding 20% increase in the area. The design is considered viable only if the corresponding increase in throughput is more than that. This target may be achievable, since there are two factors aiding the throughput performance - the availability of double ports and smaller access time of the memory bank. Hence a design exploiting the available parallelism can achieve much higher performance figures.

Various parameters for splitting should stem from the algorithm in question. The memory contents should be split so that the required accesses for the stream cipher have little, or if possible, no conflict between different memories. A careful study of the HC-128 stream cipher reveals the fact that the access to memories are split into even and odd memory locations. Considering only the if part of keystream generation and when j is even the accesses to P memory are targeted to

- *Even Addresses*: 3 reads from $P[j]$, $P[j \boxminus 10]$, $P[j \boxminus 12]$ and one write to $P[j]$
- *Odd Addresses*: 2 reads from $P[j \boxminus 3]$ and $P[j \boxminus 511]$

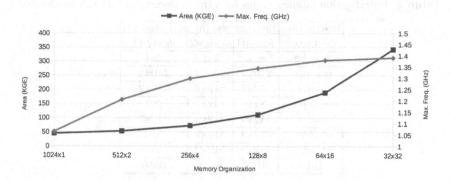

Fig. 3. Area-frequency graph for various configurations of a 1K word memory on 65nm CMOS

Two accesses to the Q memory required in the h_1 function could be targeted to either even or odd memory locations. For odd values of j, the accesses to even and odd locations switch. We redesign the accelerator considering a memory bank consisting of two memories Pe and Po having all even and odd locations of the original P memory, respectively. We refer to this 2-way split design point as *design2*.

We explore the idea of memory splitting further taking up 4-way splitting of the memories (referred as *design3*). P memory is split into 4 equal parts of 128 words each, called $P0$, $P1$, $P2$ and $P3$ such that the last two bits of the address to the original P memory indicate the smaller split memory to be accessed. Hence, if $address$ mod $4 == x$, the contents are in Px memory. Considering the if part of the algorithm in Table 9, we see the following division of the 6 accesses to P memory

- $(j \bmod 4) == 0$: 2 reads each from $P0$, $P1$, 1 read from $P2$, one write to $P0$
- $(j \bmod 4) == 1$: 2 reads each from $P1$, $P2$, 1 read from $P3$, one write to $P1$
- $(j \bmod 4) == 2$: 2 reads each from $P2$, $P3$, 1 read from $P0$, one write to $P2$
- $(j \bmod 4) == 3$: 2 reads each from $P3$, $P0$, 1 read from $P1$, one write to $P3$

4.1   State Splitting by Factor 2: Design2

The external interface of *design2* accelerator interfaces with four memories, namely, Pe, Po, Qe and Qo. We use the same instructions for *design2* as used for *design1*. The pipeline stage design is however altered to maximize parallelism.

Initialization Phase: As more access ports are now available, the structural hazard for subsequent overlapping instructions is less severe. As the initialization phase starts, each *init_kiv* instruction fills up the 4 words of the key input to the accelerator to the first two memory address locations of Pe and Po memory in

Table 3. Initialization latency cycles for various designs of HC-128 accelerator

Instruction syntax	Instructions for initialization		
	$design1$	$design2$	$design3$
init_kiv	8x1	4x1	2x1
init_p_w	240x4	240x2	240x2
init_p	512x4	512x2	512x2
init_p_for_q	8x1	4x1	2x1
init_q	512x4	512x2	512x2
update_p	512x4	512x2	512
update_q	512x4	512x2	512
Total	9168	4584	3556

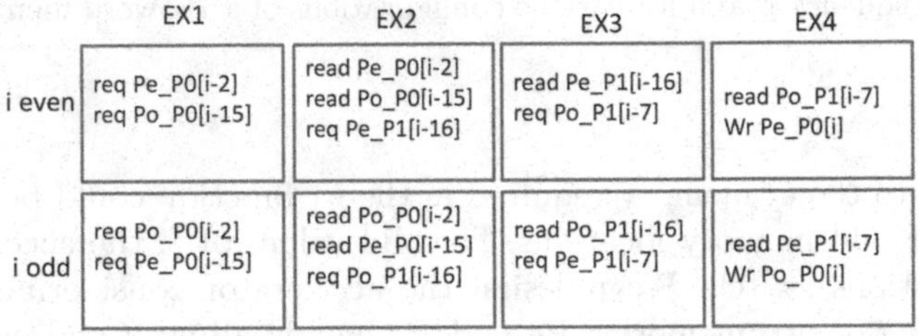

	EX1	EX2	EX3	EX4
i even	req Pe_P0[i-2] req Po_P0[i-15]	read Pe_P0[i-2] read Po_P0[i-15] req Pe_P1[i-16]	read Pe_P1[i-16] req Po_P1[i-7]	read Po_P1[i-7] Wr Pe_P0[i]
i odd	req Po_P0[i-2] req Pe_P0[i-15]	read Po_P0[i-2] read Pe_P0[i-15] req Po_P1[i-16]	read Po_P1[i-16] req Pe_P1[i-7]	read Pe_P1[i-7] Wr Po_P0[i]

Fig. 4. Pipeline architecture for HC-128 initialization instruction

one cycle. Hence in two cycles, the 8 words of key are written to the P memory bank and the next two instructions in two subsequent cycles complete the IV transfer as well.

The pipeline design of instruction init_p_w is shown in Fig. 4. As i increments for each instruction, the target memory for the 5 accesses switches between even and odd. Hence if i is even, $P[i-2]$ resides in memory Pe and otherwise in Po. Both the possibilities are shown in Fig. 4. The pipeline is carefully arranged so that two subsequent instructions require no more than one nop in between. Hence an overlap of $EX1$ and $EX3$ for subsequent even and odd iterations does not pose any hazard. Similarly, the overlap of $EX2$ and $EX4$ causes no port contention. As lesser nops need to be inserted between the consecutive instructions, the initialization phase requires around half as many cycles compared to $design1$, as shown in the Table 3.

Keystream Generation Phase: The pipeline stages operations for update_p, update_q and keystream instruction is shown in Fig. 5. For simplicity, we show only the hardware for the if part of the algorithm in Table 9 with i being even.

1. *EX1*: Three read requests are launched from the two ports of Po and one port of Pe.
2. *EX2*: The g_{11} function is carried out on the values read from the two ports of Pe memory and the result is passed over to the next pipeline stage. Two more read requests are launched with addresses i and $i-10$, while the value

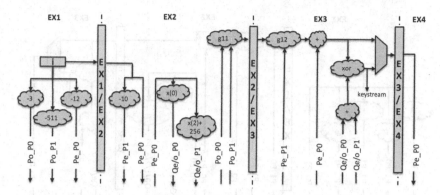

Fig. 5. Pipeline architecture of *design2* of HC-128 accelerator

of i is passed to this stage from the previous one by a pipeline register i_p. Calculation for h_1 is carried out and two read requests are launched to Qo or Qe.

3. *EX3*: During *EX3* stage g_{12} function is performed and addition with the Pe_P1 memory port value is done. If the current instruction is *update_p*, then the result is xored with h_1 function output and stored in the pipeline register. For *keystream* instruction, the xor-ing is skipped.

4. *EX4*: The value from the d_reg register is written to Pe using P0 port.

This pipeline architecture manages a good port utilization in three pipeline stages, i.e., 3 reads in *EX2*, 2 reads in *EX3* and one write in *EX4* out of the 4 ports available per cycle. Subsequent instructions follow after just one *nop*. The overlap of *EX1* with *EX3* due to pipelining causes no port contention, since only 2 ports out of 4 are used. Similarly, the overlap of *EX2* with *EX4* causes no port contention for subsequent even and odd i and uses all 4 ports. Consequently, the throughput doubles, i.e., one keyword is generated in two clock cycles. Port efficiency per keyword generated can be calculated by dividing accesses required per keyword with the accesses possible in two cycles (6/8 for this case).

4.2   State Splitting by Factor 4: Design3

The accelerator core interfaces with 8 smaller memory modules (of 128 words each), i.e., P0, P1, P2, P3, Q0, Q1, Q2, Q3. There are 8 simultaneous ports from each of the P and Q memory. Consequently, the *init_kiv* and the *init_p_for_q* instructions need 2 cycles each for a transfer of 16 values.

In spite of having more access ports, we do not try to speed up the initialization phase instruction *inti_p_w*. This is to ensure that the critical path of the design does not increase accommodating the addition of all the 5 values of in a single pipeline stage. We use the same pipeline stage division as used in *design2* for *init_p_w* instruction; the ports are however tweaked. Since *EX2*, *EX3* and *EX4* utilize only 2, 1 and 2 ports of the memory bank as shown in

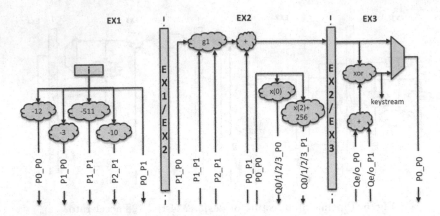

Fig. 6. Pipeline architecture of *design3* of HC-128 accelerator

Fig. 4 respectively, subsequent pipelined instructions use no more than 5 ports simultaneously and hence causes no contention.

The *design3* accelerator uses 3 pipeline stages for *update_p*, *update_q* and *keystream* instructions are shown in Fig. 6. For simplicity, we show only the hardware for the *if* part of the algorithm in Table 9 and the case when (i mod 4) equals 0.

1. *EX1*: We launch all the required read operations simultaneously after adding fixed offsets to iteration count register i.
2. *EX2*: From the values read, the g_1 function is calculated. The result is added to the $P[i]$ value read and is passed over to the next pipeline stage. For h_1 function two requests to Q memory banks are launched.
3. *EX3*: During $EX3$ stage we perform the operations depending on whether we execute *update_p* or *keystream* instruction. For *update_p* instruction, the pipeline register d_reg is xored with the result of the h_1 operation and memory update write is carried out. For *keystream* instruction, the keyword generated is computed by xor-ing the pipeline register with the output of the h_1 function.

From Fig. 6, we can see that in $EX2$ and $EX3$ stages, 5 reads and one write are carried out, respectively. Due to higher number of ports, no port contention happens when consecutive instructions are executed without any *nops*. It results in a throughput of one keyword per clock cycle. When the pipeline is full, an overlap of $EX1$, $EX2$ and $EX3$ will require 6 out of the 8 ports of the P memory bank, indicating a port utilization of 75%.

4.3 Further State Splitting by Higher Factors

Without memory splitting, we managed the 6 P memory accesses in the *if* part of Table 9 in 4 clock cycles. With 2-way and 4-way split, the throughput is doubled, i.e., 2 cycles/word and 1 cycle/word of keystream generation, respectively. A further splitting, i.e., by a factor of 8 should accompany doubling of the throughput (2 words/cycle) to have at least the same port utilization. It calls for unrolling the keystream generation loop twice for generating 2 words per iteration. The accesses double consequently, i.e., 12 accesses to the P memory, 4 accesses to the Q memory for 2 keywords generated.

The hurdle in further improvement of the throughput is the uncertainty of the addresses for Q memory accesses that are required for the calculation of the h_1 function. As shown in the equation for the h_1 function (Appendix A), it requires two reads from the Q memory from the addresses depending on the bytes of $P[j\boxminus12]$. For the case of 4 way splitting ($design3$), no contention occurs, since the dual ported memory can serve two accesses even when both of these Q accesses require the same smaller memory. With an unrolling of 2 and 8 way splitting, we require 4 accesses in one cycle, that can cause a contention of memory ports if more than two of these 4 accesses arbitrate for the same smaller memory. One way of handling that could be to have extra contention detection logic that *freezes* the pipeline for one cycle to complete reads. Consequently, throughput of the processor will not remain deterministic, which is not a desirable feature for many applications. Also the contention detection logic will have a significant overhead in area of the design. Hence, for HC-128, we do not take up the state-splitting any further.

5 Performance Evaluation

The implementation of various designs of HC-128 accelerator is carried out via a high-level synthesis framework, namely, *Synopsys Processor Designer* version 2013.06-SP3 [22]. The high level description of the design is captured by Language of Instruction set Architectures (LISA), that offers rich programming primitives to capture the design of any processor [23]. From LISA, an optimized RTL implementation as well as the entire software toolsuite, can be automatically generated. The algorithm verification is carried out in two abstraction levels, i.e., the cycle accurate instruction set simulation in LISA and the Verilog RTL simulation.

We use *Synopsys Design Compiler* (version G-2012.06) with the Faraday standard cell libraries in topographical mode to carryout synthesis of our design variants. All syntheses are carried out to optimize area. Typical case values of UMC SP/RVT Low-K process 65nm CMOS library are considered. The foundry typical values (1.8 Volt for the core voltage and 25°C for the temperature) are used. The power consumption is estimated by Synopsys Power Compiler based on RTL switching activity annotation. The performance estimates of SRAMs used are obtained using *Faraday Memory compiler* [25], for 65nm technology library.

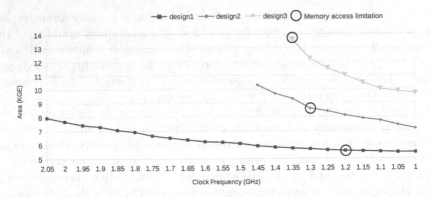

Fig. 7. The Area-throughput trade-off of HC-128 accelerator variants on 65nm CMOS

5.1 Area-Throughput Trade-Off

The three design alternatives are repeatedly synthesized using *compile_ultra* option in an incremental fashion with increasing clock frequency as long as no timing violation is reported. Fig. 7 gives the area-time chart of the design, as the operating frequency is increased from 1GHz in steps of 50MHz. The area figures are given as equivalent NAND gates. The operating frequency of the cores must be chosen based on the access time of the SRAM it interfaces with. For Faraday standard cell library, a 512-word memory needs a minimum access time of 0.83ns, corresponding to the highest operating frequency of *design*1 being 1.2GHz. For small memories, the access time is less, hence the operating frequency of *design*2 and *design*3 can be as high as 1.3 and 1.35GHz, respectively. For the rest of the discussion, we consider only these design points.

Table 4 shows the area estimates of the sequential and the combinational logic, contributing to the core area, along with the memory area for different designs of the accelerator. Quite obviously, the total area of the design is dominated by the memory. The increase in area as the memories are split is noteworthy. As the P and Q memories are divided into 4 parts, the area estimate of the memories with no splitting doubles. Hence, the rationale for this design decision should be justified by a corresponding throughput increase.

Table 4. Area comparison of HC-128 accelerator designs

Design	Clock Freq. (GHz)	Core area (KGE)			Memory			Total Area (KGE)
		Comb.	Sequential	Total	Organization	(Bytes)	(KGE)	
Design1	1.20	5.47	1.64	5.55	2x512	4096	53.22	58.77
Design2	1.30	9.63	1.38	8.60	4x256	4096	72.56	81.16
Design3	1.35	16.24	1.17	13.61	8x128	4096	110.4	124.01

5.2 Power Consumption

The power consumption of an algorithm on a core is a function of the complexity of the design and the clock frequency. From Table 5, we can see that the dynamic power consumptions of the cores contribute majorly to the core power consumption of the design. An increasing trend of power consumption can be seen from *design*1 to *design*3 due to increased complexity as well as higher frequency. From the Faraday memory compiler, the static power of a 512-word memory is reported to be 9.864 μW and the dynamic power consumption is 18.56 pJ/access (or 0.58 pJ/bit access) as specified in Table 5.

Table 5. Power consumption comparison of HC-128 design versions

Design	Clock Freq. (GHz)	Memory Access time (ns)	Memory Power Consumption		Core Power Consumption		
			Static (μW)	Dynamic (pJ/bit access)	Dynamic (μW)	Static (μW)	Total (μW)
*Design*1	1.20	0.83	9.86	0.58	4.05	0.045	4.09
*Design*2	1.30	0.77	6.65	0.53	6.19	0.075	6.26
*Design*3	1.35	0.74	4.99	0.51	8.66	0.136	8.79

5.3 Throughput Efficiency and Initialization Latency

The initialization latency of the three designs is shown in Table 6. As the refinements from *design*1 to *design*3 are taken up, a 3× reduction in the setup time can be seen. Most noteworthy is the throughput improvement from 4 to 1 cycle per word from *design*1 to *design*3; however, there is a corresponding increase in the area resources too. For a fair comparison, we choose area-efficiency or throughput per area (TPA) as our figure of merit. We calculate TPA for area with and without including the memory bank area. *Design*2 and *design*3 report a 1.3 and 1.8× better efficiency compared to *design*1, respectively, when only the core area of designs is considered. With memory area estimates included, the efficiency of the *design*2 and *design*3 is 1.5 and 2.1× better, respectively.

Table 6. Performance comparison of HC-128 design versions

Design	Clock Freq. (GHz)	Initialization latency		Throughput		TPA (Gbps/KGE) with	
		(cycles)	(ms)	(cycles/word)	(Gbps)	Core Area	Total Area
*Design*1	1.20	9168	7.64	4	9.60	1.73	0.16
*Design*2	1.30	4584	3.53	2	20.80	2.42	0.26
*Design*3	1.35	3556	2.63	1	43.20	3.18	0.35

5.4   Comparison with Reported HC-128 Performance

For performance of HC-128 stream cipher on GPPs, eBACS provides a performance benchmarking of stream ciphers on various machines [26]. We take up some of the more recent results available at [27]. The fastest implementation requires 2.86 cycles per byte of keystream generated and 6.19Gbps on an AMD Athlon 64 X2 4200+ processor. For initialization, it requires 23K clock cycles, with 10μs for setup [27]. A study of the implementations of HC-128 on embedded and customizable embedded processors has been done in [7]. The fastest reported implementation is for ARM Cortex-A8 processor with throughput of 0.79Gbps.

The only hardware implementation for HC-128 is in [7] which reports *base implementation, Parallel initialization, Parallel keystream generation* and *Even odd memory splitting* as various design points referred in Table 7 as design 1, 2, 3 and 4, respectively. For a 65nm standard Cell CMOS technology the fasted reported throughput is 22.88Gbps. The SRAM size is being reported in terms of KBytes. We have used memory compiler to give the area results as KGE. For the base implementation, the 1280-word W memory requires 53 KGE area, while each of two 512-word memories (P and Q) need 26.61 KGE each. Due to P and Q memory replication, the area overhead due to memories increases. In terms of throughput, our *design3* is around twice as fast compared to the fastest implementation (even odd memory splitting) in [7]. For a fair comparison, we also tabulate the throughput per area, both with and without the memory bank area. The throughput efficiency of our *design3* (without memory) is around 1.7× better than the best implementation reported (even odd memory splitting) [7], while considering efficiency with area is around 3× better than their best reported results (base implementation) [7].

Table 7. Performance comparison of HC-128 design versions in [7]

Design [7]	Clock Freq. (GHz)	Area (KGE)			Throughput (Gbps)	TPA (Gbps/KGE) with	
		Core	Memory	Total		Core Area	Total Area
1	1.67	8.19	53+(2×26.61)=106.22	114.41	13.36	1.63	0.12
2	1.67	14.21	53+(4×26.61)=159.44	173.65	13.36	0.94	0.08
3	1.67	13.66	53+(8×26.61)=265.88	279.54	17.81	1.30	0.06
4	1.43	12.65	53+(8×26.61)=265.88	278.53	22.88	1.81	0.08

6   Conclusion and Outlook

We propose a novel state splitting idea applicable to any cipher with large internal memories. The consequent area overhead and improvement in terms of lower access times is discussed for real world memory models. The throughput requirement to achieve area efficiency is also presented. We consider the 2-way and the 4-way split accelerator designs for one of the eSTREAM finalist, namely HC-128, along with the limitations of how far this idea could be extended. The designs focus on the maximum port utilization to achieve high throughput. Our 4-way split HC-128 accelerator, with 43.20Gbps outperforms all reported software and hardware implementations by a good margin.

We plan to extend this idea of state splitting to other stream ciphers with throughput constraints tied to memory port availability.

A   Description of HC-128

The eSTREAM [1] Portfolio (revision 1 in September 2008) contains the stream cipher HC-128 [6] in Profile 1 (SW) which is a lighter version of HC-256 [21] stream cipher born as an outcome of 128-bit key limitation imposed in the competition. Several research works exist on the cryptanalysis of HC-128 [2–5,28], however none of them pose a security threat to the cipher. We sketch the HC-128 cipher description in brief from [6].

The following operators are used in HC-128.

$+$: addition modulo 2^{32}, $\boxminus$: subtraction modulo 512.
$\oplus$: bit-wise exclusive OR, $\|$: bit-string concatenation.
$\gg$: right shift, $\ll$: left shift (both defined on 32-bit numbers).
$\ggg$: right rotation operator, $\lll$: left rotation (both defined on 32-bit numbers).

Two internal state arrays P and Q are used in HC-128, each with 512 many 32-bit elements. These arrays are also referred to as S-Boxes. A 128-bit key array $K[0,\dots,3]$ and a 128-bit Initialization Vector $IV[0,\dots,3]$ are used, each entry being a 32-bit element. Let s_t denote the keystream word generated at the t-th step, $t = 0, 1, 2, \dots$.

The following three pairs of functions are used in HC-128.

- $f_1(x) = (x \ggg 7) \oplus (x \ggg 18) \oplus (x \gg 3)$, $f_2(x) = (x \ggg 17) \oplus (x \ggg 19) \oplus (x \gg 10)$.
- $g_1(x, y, z) = ((x \ggg 10) \oplus (z \ggg 23)) + (y \ggg 8)$, $g_2(x, y, z) = ((x \lll 10) \oplus (z \lll 23)) + (y \lll 8)$.
- $h_1(x) = Q[x_{(0)}] + Q[256 + x_{(2)}]$, $h_2(x) = P[x_{(0)}] + P[256 + x_{(2)}]$.

Here $x = x_{(3)} \| x_{(2)} \| x_{(1)} \| x_{(0)}$ is a 32-bit word, with $x_{(0)}, x_{(1)}, x_{(2)}$ and $x_{(3)}$ being the four bytes from right to left.

The key and IV setup of HC-128 recursively loads the P and Q array from expanded key and IV and runs the cipher for 1024 steps to use the outputs to replace the array elements. We refer to this phase as *initialization phase* in the rest of the discussion. It happens in four steps as given in Table 8.

The keystream is generated using the algorithm as given in Table 9. We refer to this phase as *keystream generation* and the output word as *keystream* or *keyword*.

Table 8. HC-128 Initialization phase steps

Step 1: Let $K[i+4] = K[i]$ and $IV[i+4] = IV[i]$ for $0 \leq i \leq 3$.
Step 2: The key and IV are expanded into an array $W[0, \ldots, 1279]$ as follows:
$\quad W[i] = K[i]$,   for $0 \leq i \leq 7$;
$\quad\quad\quad = IV[i-8]$,   for $8 \leq i \leq 15$;
$\quad\quad\quad = f_2(W[i-2]) + W[i-7]$
$\quad\quad\quad\quad + f_1(W[i-15]) + W[i-16] + i$, for $16 \leq i \leq 1279$.
Step 3: Update the arrays P and Q with the array W as follows:
$\quad P[i] = W[i+256]$, for $0 \leq i \leq 511$,
$\quad Q[i] = W[i+768]$, for $0 \leq i \leq 511$.
Step 4: Run the cipher for 1024 steps and use the outputs to replace
the array elements as follows:
$\quad$ For $i = 0$ to 511, do
$\quad\quad P[i] = (P[i] + g_1(P[i \boxminus 3], P[i \boxminus 10], P[i \boxminus 511])) \oplus h_1(P[i \boxminus 12])$;
$\quad$ For $i = 0$ to 511, do
$\quad\quad Q[i] = (Q[i] + g_2(Q[i \boxminus 3], Q[i \boxminus 10], Q[i \boxminus 511])) \oplus h_2(Q[i \boxminus 12])$;

Table 9. HC-128 Keystream generation phase

$i = 0$;
repeat until enough keystream bits are generated
{
$\quad j = i \bmod 512$;
$\quad$ if $(i \bmod 1024) < 512$
$\quad$ {
$\quad\quad P[j] = P[j] + g_1(P[j \boxminus 3], P[j \boxminus 10], P[j \boxminus 511])$;
$\quad\quad s_i = h_1(P[j \boxminus 12]) \oplus P[j]$;
$\quad$ }
$\quad$ else
$\quad$ {
$\quad\quad Q[j] = Q[j] + g_2(Q[j \boxminus 3], Q[j \boxminus 10], Q[j \boxminus 511])$;
$\quad\quad s_i = h_2(Q[j \boxminus 12]) \oplus Q[j]$;
$\quad$ }
$\quad$ end-if
$\quad i = i + 1$;
}
end-repeat

B   Performance Enhancement for HC-128 Accelerator

B.1   Using Multi-ported SRAMs

The number of read/write ports of an SRAM restricts the multiple simultaneous accesses and consequently the throughput performance. The idea of using a multi-ported SRAM for enhancing performance has been analyzed for RC4 stream cipher with single, dual or 5 ported SRAM [14]. In a recent work, a similar idea using tri-ported SRAMs was taken up for FPGA implementations [24]. Extending on these lines, a performance improvement for HC-128 can be done, however, we choose to use dual ported SRAMs since they are the most commonly used configuration of SRAMs and their optimized design are available by many vendors. Most of the latest FPGAs families also have dual ported block SRAMs available that may be configured to different width and depths.

B.2   Loop Unrolling

Unrolling of the keystream generation loop exposes more processing blocks for parallel execution and can boost throughput if this parallelism is effectively exploited. The price to be paid is the extra checks that are needed, catering to all possible data incoherence in the design. Loop unrolling has been taken up for RC4, successfully boosting its throughput [11]. The proposal for HC-128 [6] also presents an optimized software implementation that relies on unrolling the keystream generation loop 16 times. For hardware accelerator designs, unrolling by 2 will require double the number of memory accesses (16 accesses per 2 keywords). Consequently, dealing simultaneously with more access may result in a large area and power hungry design that should be justified with a consequent increase in throughput to be efficient.

B.3   State Replication

The use of copies of state array with multiple instances of SRAMs increases the simultaneous access ports availability and consequently can boost throughput. The idea has been viably extended to RC4 [10] and HC-128 [7] implementations. The penalty is paid in terms of doubling of area resources due to memory duplication as well as the extra writes necessary to keep the memories coherent.

Acknowledgments. We thank the anonymous reviewers whose feedback helped in improvement of the technical as well as the editorial quality of our paper. We are also grateful to the Project CoEC (Centre of Excellence in Cryptology), Indian Statistical Institute, Kolkata, funded by the Government of India, for partial support towards this project.

References

1. eSTREAM: The ECRYPT Stream Cipher Project. www.ecrypt.eu.org/stream
2. Kircanski, A., Youssef, A.M.: Differential Fault Analysis of HC-128. In: Bernstein, D.J., Lange, T. (eds.) AFRICACRYPT 2010. LNCS, vol. 6055, pp. 261–278. Springer, Heidelberg (2010)
3. Maitra, S., Paul, G., Raizada, S., Sen, S., Sengupta, R.: Some observations on HC-128. Designs, Codes and Cryptography 59(1–3), 231–245 (2011)
4. Paul, G., Maitra, S., Raizada, S.: A Theoretical Analysis of the Structure of HC-128. In: Iwata, T., Nishigaki, M. (eds.) IWSEC 2011. LNCS, vol. 7038, pp. 161–177. Springer, Heidelberg (2011)

302   A. Khalid et al.

5. Stankovski, P., Ruj, S., Hell, M., Johansson, T.: Improved distinguishers for HC-128. Designs, Codes and Cryptography **63**(2), 225–240 (2012)
6. Wu, H.: The Stream Cipher HC-128. www.ecrypt.eu.org/stream/hcp3.html
7. Chattopadhyay, A., Khalid, A., Maitra, S., Raizada, S.: Designing High-Throughput Hardware Accelerator for Stream Cipher HC-128. In: IEEE International Symposium on Circuits and Systems (ISCAS), pp. 1448–1451 (2012)
8. Announcing development of a federal information processing standard for advanced encryption standard. National Institute of Standards and Technology, Docket No. 960924272–6272-01, RIN 0693-ZA13 (January 2, 1997). http://csrc.nist.gov/archive/aes/pre-round1/aes_9701.txt
9. Third Round Report of the SHA-3 Cryptographic Hash Algorithm Competition. National Institute of Standards and Technology, NISTIR 7896 (November 2012). http://nvlpubs.nist.gov/nistpubs/ir/2012/NIST.IR.7896.pdf
10. Chattopadhyay, A., Paul, G.: Exploring security-performance trade-offs during hardware accelerator design of stream cipher RC4. In: 20th International Conference on VLSI and System-on-Chip (VLSI-SoC), pp. 251–254 (2012)
11. Sen Gupta, S., Chattopadhyay, A., Sinha, K., Maitra, S., Sinha, B.P.: High Performance Hardware Implementation for RC4 Stream Cipher. IEEE Transactions on Computers **62**(4), 730–743 (2012)
12. Kitsos, P., Kostopoulos, G., Sklavos, N., Koufopavlou, O.: Hardware Implementation of the RC4 stream Cipher. In: Proc. of 46th IEEE Midwest Symposium on Circuits & Systems, Cairo, Egypt, vol. 3, pp. 1363–1366 (2003)
13. Gupta, S.S., Chattopadhyay, A., Khalid, A.: Designing integrated accelerator for stream ciphers with structural similarities. Cryptography and Communications **5**(1), 19–47 (2013)
14. Matthews Jr., D.P.: Methods and apparatus for accelerating ARC4 processing. US Patent Number 7403615. Morgan Hill, CA (July 2008). www.freepatentsonline.com/7403615.html
15. Henzen, L., Carbognani, F., Felber, N., Fichtner, W.: VLSI hardware evaluation of the stream ciphers Salsa20 and ChaCha, and the compression function Rumba. In: 2nd International Conference on Signals, Circuits and Systems, SCS 2008, pp. 1–5 (2008)
16. Yan, J., Heys, H.M.: Hardware implementation of the Salsa20 and Phelix stream ciphers. In: Canadian Conference on Electrical and Computer Engineering, pp. 1125–1128 (2007)
17. Stefan, D.: Hardware framework for the Rabbit stream cipher. In: Bao, F., Yung, M., Lin, D., Jing, J. (eds.) Inscrypt 2009. LNCS, vol. 6151, pp. 230–247. Springer, Heidelberg (2010)
18. Berbain, C., et al.: SOSEMANUK, A fast software-oriented stream cipher. In: Robshaw, M., Billet, O. (eds.) New Stream Cipher Designs. LNCS, vol. 4986, pp. 98–118. Springer, Heidelberg (2008)
19. Good, T., Benaissa, M.: Hardware performance of eStream phase-III stream cipher candidates. In: Proc. of Workshop on the State of the Art of Stream Ciphers, SASC 2008, pp. 163–173 (2008)
20. Grkaynak, F.K., Peter, L., Nico, B., Blattman, R., Goode, V., Marghitola, M., Kaeslin, H., Felber, N., Fichtner, W.: Hardware evaluation of eSTREAM candidates: Achterbahn, Grain, MICKEY, MOSQUITO, SFINKS, TRIVIUM, VEST, ZK-Crypt. eSTREAM, ECRYPT Stream Cipher Project, Report 2006/015 (2006). www.ecrypt.eu.org/stream

21. Wu, H.: A New Stream Cipher HC-256. In: Roy, B., Meier, W. (eds.) FSE 2004. LNCS, vol. 3017, pp. 226–244. Springer, Heidelberg (2004). The full version available at http://eprint.iacr.org/2004/092.pdf
22. Synopsys Processor Designer. www.synopsys.com/Systems/BlockDesign/ processorDev/Pages/default.aspx
23. Chattopadhyay, A., Meyr, H., Leupers, R.: LISA: A Uniform ADL for Embedded Processor Modelling, Implementation and Software Toolsuite Generation. In: Mishra, P., Dutt, N. (eds.) Processor Description Languages, pp. 95–130. Morgan Kaufmann (2008)
24. Tran, T.H., Lanante, L., Nagao, Y., Kurosaki, M., Ochi, H.: Hardware Implementation of High Throughput RC4 Algorithm. In: Proc. IEEE ISCAS 2012, pp. 77–80 (2012)
25. Faraday Memory Compiler. www.faraday-tech.com/html/Product/IPProduct/ LibraryMemoryCompiler/index.htm
26. eBACS: ECRYPT Benchmarking of Cryptographic Systems. http://bench.cr.yp.to/results-stream.html
27. eSTREAM Optimized Code HOWTO. www.ecrypt.eu.org/stream/perf/#results
28. Liu, Y., Qin, T.: The key and IV setup of the stream ciphers HC-256 and HC-128. In: International Conference on Networks Security, Wireless Communications and Trusted Computing, pp. 430–433 (2009)

A Very Compact FPGA Implementation of LED and PHOTON

N. Nalla Anandakumar[1,2]([⊠]), Thomas Peyrin[1], and Axel Poschmann[1,3]

[1] Division of Mathematical Sciences, School of Physical and Mathematical Science,
Nanyang Technological University, Singapore, Singapore
{nallananth,thomas.peyrin,aposchmann}@gmail.com
[2] Hardware Security Research Group, Society for Electronic
Transactions and Security, Chennai, India
[3] NXP Semiconductors, Hamburg, Germany

Abstract. LED and PHOTON are new ultra-lightweight cryptographic algorithms aiming at resource-constrained devices. In this article, we describe three different hardware architectures of the LED and PHOTON family optimized for Field-Programmable Gate Array (FPGA) devices. In the first architecture we propose a round-based implementation while the second is a fully serialized architecture performing operations on a single cell per clock cycle. Then, we propose a novel architecture that is designed with a focus on utilizing commonly available building blocks (SRL16). This new architecture, organized in a complex scheduling of the operations, seems very well suited for recent designs that use serial matrices. We implemented both the lightweight block cipher LED and the lightweight hash function PHOTON on the Xilinx FPGA series Spartan-3 (low-cost) and Artix-7 (high-end) devices and our new proposed architecture provides very competitive area-throughput trade-offs. In comparison with other recent lightweight block ciphers, the implementation results of LED show a significant improvement of hardware efficiency and we obtain the smallest known FPGA implementation (as of today) of any hash function.

Keywords: FPGA · Lightweight cryptography · LED · PHOTON · SRL16

1 Introduction

Lightweight devices such as RFID tags, wireless sensor nodes and smart cards are increasingly common in applications of our daily life. These smart lightweight devices might manipulate sensitive data and thus usually require some security. Classical cryptographic algorithms are not very suitable for this type of applications, especially for very constrained environments, and thus many lightweight cryptographic schemes have been recently proposed (block ciphers [5, 15, 18, 26, 31] or hash functions [2, 6, 17]). The main focus of lightweight cryptography research has been on the trade-offs between cost, security and performance in terms of speed, area and computational power. These primitives can be implemented either in software or in hardware platforms such as

© Springer International Publishing Switzerland 2014
W. Meier and D. Mukhopadhyay (Eds.): INDOCRYPT 2014, LNCS 8885, pp. 304–321, 2014.
DOI: 10.1007/978-3-319-13039-2_18

Field-Programmable Gate Array (FPGA) and Application Specific Integrated Circuit (ASIC). Compared to ASICs, FPGAs offer additional advantages in terms of time-to-market, reconfigurability and cost.

Recently, Guo et al. proposed the lightweight block cipher LED [18] and the lightweight family of hash functions PHOTON [17], for which the hardware performance has only been investigated on ASICs. LED is based on AES-like design principles with a very simple key schedule. The internal unkeyed permutations of PHOTON can also be seen as an AES-like primitive. Up to now, no design space exploration of LED on FPGAs has been published. The proposed architecture is suited for the applications where low-cost FPGAs are deployed such as FPGA-based RFID tags [14] and low-power FPGAs [33] are deployed for battery powered applications such as FPGA-based wireless sensor nodes [13]. Hence, they represents popular platforms (FPGA-based RFID tags, FPGA-based wireless sensor nodes) for lightweight cryptographic applications.

Our Contributions. In this study, we propose three architectures optimized for the implementation of the LED block cipher and the five different flavors of the PHOTON hash functions family on FPGAs. The first architecture computes one round per clock cycle, while the second is based on the architecture presented in LED [18] and PHOTON [17] for ASIC, and adapted in this paper to FPGA with slight modifications. Our most interesting contribution is the third architecture, also serial by nature, which performs the LED and PHOTON computations based on shift registers (SRL16), thanks to a non-trivial scheduling of the successive operations. This structure is actually strictly better than the second one since it achieves lower area and better throughput.

We emphasize that the goal of this paper is to cover a wide variety of new implementation trade-offs offered by crypto primitives using serialized or recursive MDS (Maximum Distance Separable) matrices (for which LED and PHOTON are the main representatives), on a wide variety of different Xilinx FPGA families, ranging from low-cost (Spartan-3) to high-end (Artix-7). Using our novel architecture, based on SRL16, one requires only 77 slices for LED-64 and 112 slices for PHOTON-80 on a Xilinx Spartan 3 (XC3S50) device, and 40 slices for LED-64 and 58 slices for PHOTON-80 on an Artix-7 (XC7A100T) device (while achieving reasonable throughput of 9.93 Mbps and 22.93 Mbps for LED-64, 6.57 Mbps and 18.33 Mbps for PHOTON-80). To the best of our knowledge, it represents the most compact hash function implementations on FPGAs.

The article is structured as follows. First we provide the description of LED and PHOTON in Section 2. Then, we provide in Section 3 and Section 4 our architectures and FPGA implementations of LED and PHOTON respectively. We finally draw conclusions in Section 5.

2   Algorithms Descriptions

In this section, we describe the different versions of LED block cipher [18] and the PHOTON [17] family of hash functions.

2.1 LED

LED is a 64-bit block cipher based on a substitution-permutation network (SPN). It supports any key lengths from 64 to 128 bits. In this article, we will focus on a few main versions: 64-bit key LED (named LED-64) and 128-bit key LED (named LED-128). The number of rounds N depends on the key size, LED-64 has $N = 32$ rounds while LED-128 has $N = 48$ rounds.

One can view the 64-bit internal state as a 4×4 matrix of 4-bit nibbles and the round function as an AES-like permutation composed of the following four operations:

- *AddConstants:* the internal state is bitwise XORed with a round-dependent constant (generated with an LFSR);
- *SubCells:* the PRESENT [7] S-box is applied to each 4-bit nibble of the internal state;
- *ShiftRows:* nibble row i of the internal state is cyclically shifted by i positions to the left;
- *MixColumnsSerial:* each nibble column of the internal state is transformed by multiplying it once with MDS matrix χ^4 (or two times with matrix χ^2, or four times with matrix χ).

$$\chi = \begin{pmatrix} 0 1 0 0 \\ 0 0 1 0 \\ 0 0 0 1 \\ 4 1 2 2 \end{pmatrix} ; \quad (\chi)^2 = \begin{pmatrix} 0 0 1 0 \\ 0 0 0 1 \\ 4 1 2 2 \\ 8 6 5 6 \end{pmatrix} ; \quad (\chi)^4 = \begin{pmatrix} 0 1 0 0 \\ 0 0 1 0 \\ 0 0 0 1 \\ 4 1 2 2 \end{pmatrix}^4 = \begin{pmatrix} 4 1 2 2 \\ 8 6 5 6 \\ B E A 9 \\ 2 2 F B \end{pmatrix}$$

The key schedule of LED is very simple. In the case of LED-64, the key K is repeatedly XORed to the internal state every 4 rounds (with whitening key operation). In the case of LED-128, the key K is divided into two 64-bit subparts $K = K_1 \| K_2$, each XORed alternatively to the internal state every 4 rounds. The 4-round operation between two key addition is called a step.

2.2 PHOTON

In this section we describe the PHOTON family of hash functions, for which five versions exist with digest sizes of 80, 128, 160, 224 and 256 bits. PHOTON is based on the sponge construction. First, after padding, the input message is divided into blocks of r-bit each. At each iteration, the t-bit internal state ($t = r + c$) absorbs the incoming message block by simply XORing it to the r-bit *bitrate* part (the remaining c-bit part is called the *capacity*). Then, after the absorption of the message block, one applies a t-bit permutation P to the internal state. Once all message blocks have been processed the squeezing phase starts. During this phase, for each iteration r' bits are output from the internal state and the permutation P is applied. One continues to squeeze until the proper digest size n is reached.

The PHOTON internal permutation P is also AES-like and consists of 12 rounds. The internal state is represented as a $(d \times d)$ matrix of s-bit cells and each round is defined as the application of 4 operations:

- *AddConstants:* the internal state is bitwise XORed with a round-dependent constant (generated with an LFSR);

- *SubCells:* the S-box is applied to each *s*-bit nibble of the internal state (the PRESENT S-box [7] if $s = 4$, the AES S-box [9] if $s = 8$);
- *ShiftRows:* nibble row i of the internal state is cyclically shifted by i positions to the left;
- *MixColumnsSerial:* each nibble column of the internal state is transformed by multiplying it once with MDS matrix χ^d (or two times with matrix $\chi^{d/2}$, ... , or d times with matrix χ).

The values of t, c, r, r', s and d depend on the hash output size n and we give in Table 1 the 5 versions of PHOTON (we refer to [17] for the various matrices χ depending on the PHOTON versions). Note that one always uses a cell size of 4 bits, except for the PHOTON-256/32/32 version for which one uses 8-bit cells.

Table 1. The 5 versions of PHOTON parameters

	r	r'	c	s	d	t
PHOTON-80/20/16	20	16	80	4	5	100
PHOTON-128/16/16	16	16	128	4	6	144
PHOTON-160/36/36	36	36	160	4	7	196
PHOTON-224/32/32	32	32	224	4	8	256
PHOTON-256/32/32	32	32	256	8	6	288

3 LED Implementations

In this section, we present three different architectures for the FPGA implementation of the lightweight block cipher LED. The first one is a round-based implementation, while the second one is a fully serialized implementation, performing operations on a single cell during each clock cycle. The third one is a novel architecture, also fully serial, but based on the SRL16s and aiming at the smallest area possible. As we are interested in the performance of the plain LED core, we did not include any I/O logic implementation such as a UART interface.

We have also investigated the performance of the LED cipher with different trade-offs. Indeed, the diffusion matrix being serial in LED, one can view the *MixColumnsSerial* diffusion layer as a single application of $(\chi)^4$, or two successive applications of $(\chi)^2$ or four successive applications of (χ).

We have implemented both LED versions (the 64-bit key version LED-64 and the 128-bit key version LED-128) in VerilogHDL and targeted Xilinx FPGAs Spartan-3 XC3S50-5 and Artix-7 XC7A100T-3. We used Mentor Graphics ModelSimPE for simulation purposes and Xilinx ISE v14.4 WebPACK for design synthesis. In Xilinx ISE the design goal is kept balanced and strategy is kept default (unlocked) and the synthesis optimization goal is set to area.

3.1 Round-Based

We give in Figure 1 the block diagram of the round-based implementation of LED. Naturally, the data register (Dreg) is updated after every round operation. The keys are selected according to the key length (K_1 is loaded without modification every four rounds in LED-64, while K_1 and K_2 are loaded alternatively every four rounds for LED-128). Table 2 provides the detailed results of our round-based FPGA implementations of LED with three different approaches concerning the computation of the diffusion matrix: we compute χ^4 by either applying 4 times the matrix χ, or by applying 2 times the matrix χ^2, or by directly applying the entire matrix χ^4. As expected, the last option provides a higher throughput (since we directly compute the entire diffusion matrix), but for the price of higher resource consumption. In contrary, the first option allows to save resources, but at the expense of a lower throughput. The second option offers a trade-off in between.

We also added in Table 2 a comparison with known round-based FPGA implementations of other (lightweight) block ciphers on the same FPGA device. One can see that our LED-64 and LED-128 proposed round-based implementations outperform all the previous works in term of area.

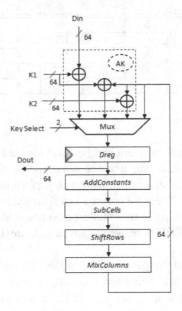

Fig. 1. Architecture of the LED round based encryption module

3.2 Serialized

Our first serialized implementation of LED is derived from the architecture proposed in [18] for ASICs, but with some architectural modifications for the *MCS* state operations in order to improve the performance. This implementation stores the data and key in the registers (FF) and it has a 4-bit wide datapath, i.e. only 4 bits are processed in one clock cycle (see Figure 2). It consists of 4 states: *Init*, *Sbox*, *Srow* and *MCS*: **The Init state** initial data and key values are stored in the data registers and key registers, respectively.

Table 2. FPGA round-based implementation results of LED block cipher

Design	MDS approach	Block Size (bits)	Key Size (bits)	No. of slices	No. of FFs	No. of LUTs	Clock Cycles	Max. freq (MHz)	T/put (Mbps)	Eff. (Mbps/slices)	FPGA Device
LED our paper (Section 3.1)	(χ)	64	64	170	74	326	32	78.78	157.56	0.93	Spartan-3 XC3S50-5
			128	199	76	391	48	78.79	104.8	0.53	
	$(\chi)^2$	64	64	198	74	379	32	87.63	175.3	0.89	
			128	227	76	444	48	87.63	116.54	0.51	
	$(\chi)^4$	64	64	204	74	391	32	98.67	197.35	0.97	
			128	233	76	456	48	98.7	131.2	0.56	
	(χ)	64	64	102	74	206	32	282.77	565.54	5.50	Artix-7 XC7A100T-3
			128	158	76	261	48	282.43	376.57	2.39	
	$(\chi)^2$	64	64	110	74	216	32	290.46	580.97	5.28	
			128	163	76	281	48	291.89	389.18	2.40	
	$(\chi)^4$	64	64	136	74	231	32	334.85	669.7	4.92	
			128	168	76	292	48	333.73	444.97	2.65	
PRESENT [30]		64	128	202	—	—	32	254	508	2.51	Spartan-3 XC3S400-5
AES [16]		128	128	17,425	—	—	—	196.1	25,107	1.44	Spartan-3 XC3S2000-5
AES [8]		128	128	1800	—	—	—	150	1700	0.90	Spartan 3
ICEBERG [32]		64	128	631	—	—	—	254	1016	1.61	Virtex-II
SEA [27]		126	126	424	—	—	—	145	156	0.37	Virtex-II XC2V4000

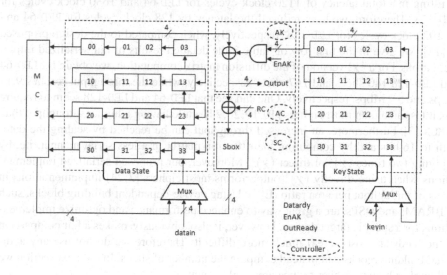

Fig. 2. A serialized architecture of the LED encryption module

The Sbox state is for the simultaneous execution of the *SubCells* (SC) operations, *AddConstants* (AC) operations and XORing the roundkey (AK) every fourth round. It requires 16 clock cycles.

The Srow state is for the execution of the *ShiftRows* operation. It can be performed in 3 clock cycles with no additional hardware cost, because it just shifts the row positions of the state matrix.

The MCS state is for the execution of the *MixColumnsSerial* operation. It calculates the result fully serialized, that is one cell in each clock cycle. It first calculates the topmost cell of the leftmost column (cell 00) by storing the result in the last row of the rightmost column (cell 33) in Figure 2. At the same time, the entire state array is shifted to the left by one position, where the leftmost cells in every row are shifted into the rightmost cells of the row located on top. This way in the subsequent clock cycle the topmost cell of the second column is processed, leading to a serialized *row-by-row* calculation of the *MixColumnsSerial*.

It is to be noted that during the *MixColumnsSerial* operation in the architecture proposed in [18], the result is stored in the last row of the *leftmost* column (cell 30), leading to a serialized *column-by-column* calculation. Our new architecture is strictly better as it saves both area and time: As the leftmost column requires only 1-input FFs instead of 2-input FFs the area requirement is reduced significantly. Our proposed architecture has similarities with the work from [29], regarding the way the storing and rotating of matrices are implemented. Furthermore, it takes only 16 clock cycles to perform the *MixColumnsSerial* instead of the usual 20 clock cycles [18]. This new architecture is applicable to all AES-like permutations that use a serialized *MixColumns* operation and we will also use it for the PHOTON implementations described in Section 4.

This serialized architecture of LED requires 35 clock cycles to perform one round, resulting in a total latency of 1120 clock cycles for LED-64 and 1680 clock cycles for LED-128. Therefore, we have reduced the latency by 128 clock cycles for LED-64 and by 192 clock cycles for LED-128, respectively, when compared to the design proposed in [18]. We give in the first row of Table 3 the detailed results of our serialized implementations. For a (χ) version of the diffusion matrix computation, we obtain for LED-64 and LED-128 140 slices and 167 slices respectively, while the throughput reaches 9.11 Mbps and 5.2 Mbps, respectively. One can see that LED-64 and LED-128 seem to require much less area than most ciphers [10,19,23,24] while having a higher throughput than SIMON [3]. Furthermore, an increased throughput can be reached by scaling the datapath to 16 bits and by computing the diffusion matrix in a less serial manner, i.e. by applying two times $(\chi)^2$ or direct $(\chi)^4$. Moreover, our proposed serialized implementations when using directly $(\chi)^4$ outperforms most ciphers [3,24] implementations in terms of throughput per area ratio (Eff.). Using device-dependent building blocks, such as BRAMs and DSPs, are a great way to enhance performance and optimize implementations for a specific target device. However, it also, obviously, makes a fair comparison of the hardware costs (area) much more difficult. Therefore we do not use any additional building blocks and instead compare the number of slices. In the next section we will explain how to further reduce area and latency.

3.3   Serialized Using SRL16s

Our second serialized implementation of LED is based on the use of a building block of Xilinx Spartan-3 FPGAs called SRL16s [21]. More precisely, SRL16 are look up tables (LUT) that are used as 16-bit shift registers that allow to access (or output) bits of its internal state in two ways (as shown in Figure 3): the last bit of its 16 stages (Q_{15}) is always available, while a multiplexer allows to access one additional bit from any of its internal stages.

The Configurable Logic Blocks (CLBs) are the basic logic units in an FPGA. Each CLB has four slices, but only the two at the left-hand of the CLB can be used as shift registers. Spartan-3 FPGAs can configure some LUTs as a 16-bit shift register without

Table 3. FPGA serialized implementation results of LED block cipher

Design	MDS approach	Data-path (bits)	Block Size (bits)	Key Size (bits)	Area (slices)	No. of FFs	No. of LUTs	Clock Cycles	Max. freq (MHz)	T/put (Mbps)	Eff. (Mbps/slices)	FPGA Device
LED our paper (Section 3.2)	(χ)	4	64	64	140	151	255	1120	159.43	9.11	0.07	Spartan-3 XC3S50-5
				128	167	216	302	1680	137.34	5.2	0.03	
	$(\chi)^2$	8	64	64	169	157	332	608	157.43	16.6	0.10	
				128	203	219	388	912	142.01	9.97	0.05	
	$(\chi)^4$	16	64	64	180	162	342	352	137.5	24.99	0.14	
				128	219	227	414	528	128.73	15.6	0.07	
	(χ)	4	64	64	37	52	78	1120	378	21.6	0.58	Artix-7 XC7A100T-3
				128	40	57	82	1680	368	14.02	0.35	
	$(\chi)^2$	8	64	64	58	95	135	608	380.3	40.03	0.69	
				128	61	104	141	912	356.5	25.02	0.41	
	$(\chi)^4$	16	64	64	78	110	162	352	367.4	66.8	0.86	
				128	82	175	188	528	375.6	45.53	0.56	
LED our paper (Section 3.3)	(χ)	16	64	64	111	80	215	640	119.62	11.96	0.11	Spartan-3 XC3S50-5
				128	122	72	233	960	118.25	7.88	0.06	
	$(\chi)^2$	8	64	64	77	44	148	768	119.19	9.93	0.13	
				128	86	48	167	1152	120.75	6.71	0.08	
	$(\chi)^4$	16	64	64	119	76	228	256	119.27	29.82	0.25	
				128	127	70	248	384	117.87	19.65	0.15	
	(χ)	16	64	64	51	45	113	640	303.9	30.39	0.60	Artix-7 XC7A100T-3
				128	59	55	121	960	308.5	20.57	0.35	
	$(\chi)^2$	8	64	64	40	36	100	768	275.2	22.93	0.57	
				128	50	40	107	1152	302.64	16.81	0.34	
	$(\chi)^4$	16	64	64	63	43	133	256	284.83	71.21	1.13	
				128	69	53	138	384	286.5	47.75	0.70	
PRESENT [34]			64	128	117	—	—	256	114.8	28.46	0.24	Spartan-3 XC3S50-5
HIGHT [34]			64	128	91	—	—	160	163.7	65.48	0.72	Spartan-3 XC3S50-5
xTEA [23]			64	128	254	—	—	112	62.6	35.78	0.14	Spartan-3 XC3S50-5
PRESENT [19]			64	80	271	—	—	—	—	—	—	Spartan-3E XC3S500
SIMON [3]			128	128	36	—	—	—	136	3.60	0.10	Spartan-3E XC3S500
AES [10]			128	128	184	—	—	160	45.6	36.5	0.20	Spartan-3 XC3S50-5
AES [24]			128	128	393	—	—	534	—	16.86	0.04	Spartan-3 XC3S50-5

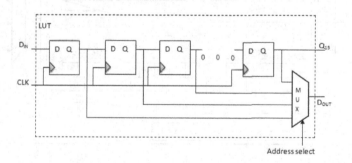

Fig. 3. LUT configured as a shift register

using the flip-flops available in each slice. When a shift register is described in generic HDL code with the global reset signal, it has no impact on shift registers and synthesis tools infer the use of the SRL16s. Moreover, SRL16 is present in almost all XILINX FPGA families and [20] describes a way to use SRL16s on ALTERA devices.

We have investigated possible area reductions by scaling the 64-bit implementation to an 8-bit (when using $(\chi)^2$) and 16-bit datapath (when using χ and $(\chi)^4$) using

SRL16s. As *MixColumnsSerial* requires 16-bit inputs (4 times 4-bit) in every clock cycle, but each SRL16 only allows access to 2 bits, we have to use eight and sixteen SRL16s to store the state, respectively.

Figure 4 shows the block diagram for the SRL16s based implementation of LED with 8-bit datapath when using $(\chi)^2$. It consists of 4 states: Init, SrSc, Re-update and MCS, where the content of each SRL16 is indicated in Table 4 for all the state operations. We also give the SRL16 positions in the long version of the article for 16-bit datapath implementations when using χ and $(\chi)^4$ respectively.

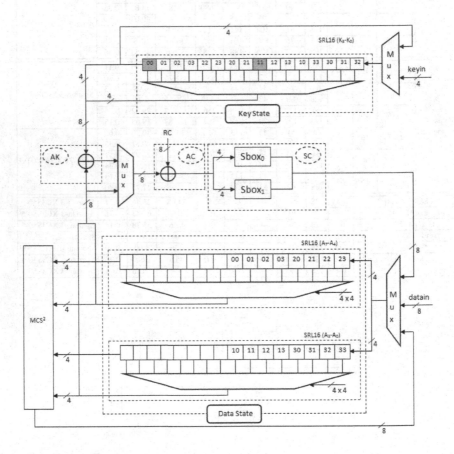

Fig. 4. The block diagram for the SRL16s based implementation of LED with 8-bit datapath when using $(\chi)^2$

The *Init* state: initial data and key values are stored in the data SRL16s and key SRL16s, respectively. A special ordering of the nibbles is required as shown in Table 4 and in Figure 4.

The *SrSc* state: performs *ShiftRows*, *SubCells*, *AddConstants* and *AddRoundKey* simultaneously by clever memory (SRL16) addressing schedule. Table 4 depicts in bold the

Table 4. Content of SRL16s for one round of LED when using $(\chi)^2$ for the 8-bit datapath. Every cell of the content shows the index of a nibble of the state. Printed in bold is the input to the subsequent operation (see also Figure 4). The indices of the next round are indicated with a'.

clk	content of SRL16s (Init)	clk	content of SRL16s (Re-update)
1	00 10	17	01 02 03 20 21 22 23 00 **01** 02 03 22 23 20 21 00 11 12 13 30 31 32 33 11 **12** 13 10 33 30 31 32 11
2	00 01 10 11	18	02 03 20 21 22 23 00 01 **02** 03 22 23 20 21 00 01 12 13 30 31 32 33 11 12 **13** 10 33 30 31 32 11 12
3	00 01 02 10 11 12	19	03 20 21 22 23 00 01 02 **03** 22 23 20 21 00 01 02 13 30 31 32 33 11 12 13 **10** 33 30 31 32 11 12 13
4	00 01 02 03 10 11 12 13	20	20 21 22 23 00 01 02 03 **22** 23 20 21 00 01 02 03 30 31 32 33 11 12 13 10 **33** 30 31 32 11 12 13 10
5	00 01 02 03 20 10 11 12 13 30	21	21 22 23 00 01 02 03 22 **23** 20 21 00 01 02 03 22 31 32 33 11 12 13 10 33 **30** 31 32 11 12 13 10 33
6	00 01 02 03 20 21 10 11 12 13 30 31	22	22 23 00 01 02 03 22 23 **20** 21 00 01 02 03 22 23 32 33 11 12 13 10 33 30 **31** 32 11 12 13 10 33 30
7	00 01 02 03 20 21 22 10 11 12 13 30 31 32	23	23 00 01 02 03 22 23 20 **21** 00 01 02 03 22 23 20 33 11 12 13 10 33 30 31 **32** 11 12 13 10 33 30 31
8	**00** 01 02 03 20 21 22 23 10 **11** 12 13 30 31 32 33	24	**00** 01 02 03 22 23 20 21 00 01 02 03 **22** 23 20 21 **11** 12 13 10 33 30 31 32 11 12 13 10 **33** 30 31 32
	SrSc		*MCS*
9	00 **01** 02 03 20 21 22 23 00 10 11 **12** 13 30 31 32 33 11	25	**01** 02 03 22 23 20 21 00 01 02 03 22 **23** 20 21 00' 12 13 10 33 30 31 32 11 12 13 10 33 **30** 31 32 10'
10	00 01 **02** 03 20 21 22 23 00 01 10 11 12 **13** 30 31 32 33 11 12	26	**02** 03 22 23 20 21 00 01 02 03 22 23 **20** 21 00' 01' 13 10 33 30 31 32 11 12 13 10 33 **31** 32 10' 11'
11	00 01 02 **03** 20 21 22 23 00 01 02 **10** 11 12 13 30 31 32 33 11 12 13	27	**03** 22 23 20 21 00 01 02 03 22 23 20 21 00' 01' 02' 10 33 30 31 32 11 12 13 10 33 20 31 **32** 10' 11' 12'
12	00 01 02 03 20 21 **22** 23 00 01 02 03 10 11 12 13 30 31 32 **33** 11 12 13 10	28	**22** 23 20 21 00 01 02 03 22 23 20 21 00' 01' 02' 03' 33 30 31 32 11 12 13 10 33 30 31 32 **10'** 11' 12' 13'
13	00 01 02 03 20 21 22 **23** 00 01 02 03 22 10 11 12 13 **30** 31 32 33 11 12 13 10 33	29	**23** 20 21 00 01 02 03 22 23 20 21 00' **01'** 02' 03' 20' 30 31 32 11 12 13 10 33 30 31 32 10' **11'** 12' 13' 30'
14	00 01 02 03 **20** 21 22 23 00 01 02 03 22 23 10 11 12 13 30 **31** 32 33 11 12 13 10 33 30	30	**20** 21 00 01 02 03 22 23 20 21 00' 01' **02'** 03' 20' 21' 31 32 11 12 13 10 33 30 31 32 10' 11' **12'** 13' 30' 31'
15	00 01 02 03 20 **21** 22 23 00 01 02 03 22 23 20 10 11 12 13 30 31 **32** 33 11 12 13 10 33 30 31	31	**21** 00 01 02 03 22 23 20 21 00' 01' 02' **03'** 20' 21' 22' 32 11 12 13 10 33 30 31 32 10' 11' 12' **13'** 30' 31' 32'
16	00 01 02 03 20 21 22 23 **00** 01 02 03 22 23 20 21 10 11 12 13 30 31 32 33 **11** 12 13 10 33 30 31 32	32	00 01 02 03 22 23 20 21 **00'** 01' 02' 03' 20' 21' 22' 23' 11 12 13 10 33 30 31 32 10' **11'** 12' 13' 30' 31' 32' 33'

bits that are selected in every clock cycle to achieve this. The round operation starts by bitwise XORing the incoming data with the round key and round constants, then applying this result to two S-boxes (8-bit datapath) or four S-boxes (16-bit datapath), respectively. The first nibbles processed are 00 and 11 (8-bit datapath) and 00, 11, 22, and 33 (16-bit datapath), respectively. In order to perform *ShiftRows, SubCells, Add-Constants* and *AddRoundKey* operations on the whole state, it takes 8 clock cycles (clk 9-16 in Table 4) using an 8-bit datapath, and 4 clock cycles using a 16-bit datapath, respectively.

The *Re-update* state: when using the 8-bit datapath, the 8-bit output from the S-boxes needs to be duplicated within the SRL16s. This is because the *MixColumnsSerial* operation reads four input vectors simultaneously and thus the leftmost bits of the SRL16s must be used. 8 clock cycles (clk 17-24 in Table 4) are required for this step. Note that this state only applies to 8-bit datapath.

The *MCS* state: the 4 x 4-bit input data is read from the bits indicated in bold in Table 4. It starts with the four 4-bit blocks 00, 11, 22 and 33, and using $(\chi)^2$, the resulting 8-bit output is stored in the SRL16s labeled as $00'$, $10'$ (and $20'$, and $30'$, respectively) to indicate the indices of the next round. In the next clock cycle, the input data is 01, 12, 23, and 30, and the corresponding result is labeled as $01'$, $11'$ (and $21'$, and $31'$) and so on. In total 8 clock cycles (clk 25-32) are required to complete the *MixColumnsSerial* layer using $(\chi)^2$, 4 clock cycles when using $(\chi)^4$, and 16 clock cycles when using (χ), respectively. The next round starts with the *SrSc* state (clk 9) and inputs $00'$ and $11'$.

Concerning the key incorporation, the details of SRL16 positions will be provided in the long version of the article for the key when using 8-bit datapath with $(\chi)^2$ (resp. when using $(\chi)^4$ or (χ) for the 16-bit datapath).

For the 8-bit datapath, 24 clock cycles are required in order to complete one round of LED (clk 9-32 in Table 4), resulting in a total latency of 768 clock cycles for LED-64 and 1152 clock cycles for LED-128. Table 3 shows the detailed results of our implementations of LED based on SRL16s for various MDS matrix computation approaches. Our design $(\chi)^2$ only occupies 77 slices for LED-64 and 86 slices for LED-128 respectively, with a corresponding throughput of 9.93Mbps and 6.71Mbps respectively. The throughput can be increased to 29.82Mbps by scaling the 8-bit to a 16-bit datapath and by directly computing the $(\chi)^4$ matrix. It is noteworthy to point out that our SRL16 based implementation on Artix-7 FPGA only occupies 40 slices for LED-64 and 50 slices for LED-128, respectively, with a throughput almost three times increased compared to Spartan-3 devices.

We also give in Table 3 the performance of existing FPGA implementations of some other lightweight block ciphers. As can be seen from the table, our work seems to require much less area than most ciphers [10,19,23,24,34] while having a higher throughput than AES [24] implementations and also yields a better throughput per area ratio (Eff.) compared to most ciphers [23,24]. Compared to FPGA implementations of the lightweight block cipher SIMON [3], we get bigger area requirements but for a higher throughput (and also achieves the better throughput per area ratio (Eff.) when using direct matrix $(\chi)^4$). We remark that HIGHT [34] has a better throughput per area ratio than LED, but in this article our goal with serialised implementations is to reduce area, and not to improve throughput per area ratio. More importantly, one can see in the table that our SRL16 implementation technique both saves area and increases throughput compared to a classical optimized serial implementation. Therefore, we believe this technique is very interesting in order to implement serial-matrix based cryptographic primitives in FPGA technology.

4   PHOTON Implementations

In this section, we present three different architectures for the FPGA implementation of the lightweight hash function PHOTON. As in the previous section, the first architecture is a round-based implementation, the second one a fully serialized implementation, and the third one our new serial architecture based on SRL16s. The diffusion layer in PHOTON is based on a similar serial MDS matrix as in LED, thus we also tested different trade-offs concerning its implementation. We have implemented all PHOTON versions in VerilogHDL and targeted Xilinx FPGAs Spartan-3 XC3S50-5 and Artix-7 XC7A100T-3. Again, we used Mentor Graphics ModelSimPE for simulation purposes and Xilinx ISE v14.4 WebPACK for design synthesis.

4.1 Round-Based

The round-based hardware architecture of the PHOTON hash function implementations is shown in Figure 5. The details of the architecture will be provided in the long version of this article. The architectures were optimized for high throughput and minimal FPGA area resource consumption. The resulting design fits in the smallest Xilinx devices such as Spartan-3 XC3S50 for variants PHOTON-80/20/16, PHOTON-128/16/16 and Spartan-3 XC3S400 for variants PHOTON-160/36/36, PHOTON-224/32/32 and PHOTON-256/32/32 (because Spartan-3 XC3S50 has only 768 Slices). In Table 5, our results are compared to other hardware implementations [1,4]. One can see that our proposed round-based implementations outperform all the previous works in terms of throughput per area ratio (Eff.).

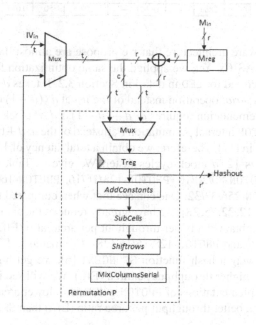

Fig. 5. Architecture of the PHOTON round-based implementations

4.2 Serialized

Similarly to our work on LED in Section 3.2, we have built a serialized implementation of the different PHOTON versions. One can see in Figure 6 that our serialized implementation consists of 6 modules: MCS, IO, AC, SC, ShR, and Controller. These modules

Table 5. FPGA round-based implementation results of PHOTON hash function

Design	MDS approach	Data-path (bits)	Area (slices)	No. of FFs	No. of LUTs	Clock Cycles	Max. freq (MHz)	T/put (Mbps)	Eff. (Mbps/ slices)	FPGA Device
PHOTON-80/20/16	(χ)	100	285	127	565	12	78.53	130.88	0.46	Spartan-3 XC3S50-5
	(χ)	100	142	111	336	12	232.65	387.75	2.73	Artix-7 XC7A100T-3
SPONGENT-88 [1]		88	157	—	—	45	—	17.78	0.11	Spartan-3
PHOTON-128/16/16	(χ)	144	549	172	1022	12	65.39	87.19	0.16	Spartan-3 XC3S50-5
	(χ)	144	204	156	590	12	189.03	252.04	1.24	Artix-7 XC7A100T-3
SPONGENT-128 [1]		136	208	—	—	70	—	11.43	0.06	Spartan-3
PHOTON-160/36/36	(χ)	196	846	243	1534	12	61.03	183.09	0.22	Spartan-3 XC3S400-5
	(χ)	196	429	207	800	12	155.75	467.25	1.10	Artix-7 XC7A100T-3
SPONGENT-160 [1]		176	264	—	—	90	—	8.89	0.03	Spartan-3
PHOTON-224/32/32	(χ)	256	1235	279	2241	12	51.73	137.95	0.11	Spartan-3 XC3S400-5
	(χ)	256	616	267	1292	12	150.79	402.11	0.65	Artix-7 XC7A100T-3
SPONGENT-224 [1]		240	322	—	—	120	—	6.67	0.02	Spartan-3
PHOTON-256/32/32	(χ)	288	2067	300	3673	12	35.34	94.24	0.05	Spartan-3 XC3S400-5
	(χ)	288	865	300	2126	12	112.43	299.81	0.35	Artix-7 XC7A100T-3
SPONGENT-256 [1]		272	357	—	—	140	—	5.71	0.02	Spartan-3
CUBEHASH-256 [4]		—	2883	—	—	—	59	50	0.017	Spartan-3 XC3S5000-5

and the general hardware architecture that we propose are almost the same as the one described in [17] for ASICs. Yet, we applied the same optimization for *MixColumnSerial* that we have described for LED in detail in Section 3.2. It takes $d \cdot d$ clock cycles to perform *MixColumnsSerial* operation instead of the usual $d \cdot (d+1)$ clock cycles [17].

Overall, our implementation requires $d \cdot d + (d-1) + d \cdot d$ clock cycles to perform one round of the PHOTON internal permutation P, instead of the $d \cdot d + (d-1) + d \cdot (d+1)$ clock cycles required in [17]. Therefore, we obtain a total latency of $12 \cdot (2 \cdot d \cdot d + d - 1)$ clock cycles, which is $12 \cdot d$ clock cycles faster. We give in Table 6 our implementation results for PHOTON-80/20/16, PHOTON-128/16/16, PHOTON-160/36/36, PHOTON-224/32/32 and PHOTON-256/32/32. One can see that when compared to previous FPGA implementations [11, 12, 22, 25, 28], we have greatly reduced the area and increased the throughput (and also obtained a better throughput per area ratio (Eff.)) as compared to PHOTON-80/20/16 [12] and PHOTON-128/16/16 [28]. Compared to FPGA implementations [1] of the lightweight hash function SPONGENT [6], we get bigger area requirements but for a much higher throughput per area (Eff.). We will see in the next section that SRL16 based implementations of PHOTON will lead to lower area and much higher throughput and yield a better throughput per area ratio (Eff.) than SPONGENT.

4.3 Serialized using SRL16s

As for LED in Section 3.3, we considered a second serialized implementation of PHOTON hash function based on the use of SRL16s [21]. Our architecture is based on a 20-bit datapath that uses χ. It it is depicted in Figure 7 and consists of 3 states: Init, SrSc and MCS, where the content of each SRL16 positions will be provided in the long version of the article for all the state operations.

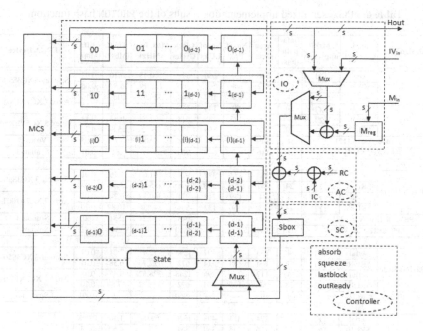

Fig. 6. A serialized architecture of the PHOTON hash function

The *Init* state: after the padding procedure, the IV value is stored into the data SRL16s ($z = s \cdot d$ bits) using a 3×1 multiplexer which drives either the IV input value, updates SrSc state value, or updates MCS state value.

The *SrSc* state: it reads the data values from SRL16s by selecting address taps according to the ShiftRows positions. The round operation starts by bitwise XORing the incoming data with r bits of the message input if applicable, and then adding the constants (round constants and internal constants). Next, the result goes through d S-boxes for a z-bit datapath. Finally, the output of the 4-bit S-boxes is given as input to the blocks of SRL16s. Thus, it takes d clock cycles for a z-bit datapath to perform *AddConstants*, *ShiftRows* and *SubCells* operations on the entire state.

The *MCS* state: it starts with the five 4-bit blocks 00, 11, 22, 33 and 44, and using (χ), the resulting 20-bit output is stored in the SRL16s labeled as 11, 22, 33, 44 and 00′. In the next clock cycle, the input is 01, 12, 23, 34 and 40, and the corresponding result is labeled as 12, 23, 34, 40, 01′ and so on. In total 25 clock cycles are required to complete the *MixColumnsSerial* operation for PHOTON-80/20/16. We have also implemented the remaining 4 versions of PHOTON using same architecture.

$d \cdot d$ clock cycles are required for a z-bit datapath in order to complete the *MixColumnsSerial* operation. Overall, we require $d + d \cdot d$ clock cycles to compute a single round. Since PHOTON has 12 rounds, the total number of cycles required to process one block of message is $12(d + d \cdot d)$. Table 6 describes the performance results of our implementations and compares it with existing FPGA implementations of PHOTON and other lightweight hash functions. Concerning KECCAK-f[200], perhaps we just add

Table 6. FPGA serialized implementation results of the PHOTON hash function

Design	impl. approach	MDS approach	Data-path (bits)	Area (slices)	No. of FFs	No. of LUTs	Clock Cycles	Max. freq (MHz)	T/put (Mbps)	Eff. (Mbps/slices)	FPGA Device
PHOTON-80/20/16	serial	(χ)	4	146	137	256	648	100.43	3.10	0.02	Spartan-3 XC3S50-5
	SRL16	(χ)	20	112	68	203	360	118.19	6.57	0.06	
	serial	(χ)	4	67	134	167	648	329.51	10.17	0.15	Artix-7 XC7A100T-3
	SRL16	(χ)	20	58	89	144	360	329.95	18.33	0.32	
	serial	(χ)	4	82	135	188	648	302.68	9.34	0.11	Virtex-5 XC5VLX50-1
	SRL16	(χ)	20	69	89	159	360	285.2	15.84	0.22	
PHOTON-80/20/16 [12]			4	149	—	—	708	250	7	0.05	Virtex-5
SPONGENT-88 [1]			4	116	—	—	900	—	.81	0.01	Spartan-3
PHOTON-128/16/16	serial	(χ)	4	183	173	317	924	101.60	1.76	0.01	Spartan-3 XC3S50-5
	SRL16	(χ)	24	137	79	250	504	115.67	3.67	0.03	
	serial	(χ)	4	84	179	212	924	360.33	6.24	0.07	Artix-7 XC7A100T-3
	SRL16	(χ)	24	72	99	137	504	342.36	10.87	0.20	
PHOTON-128/16/16 [28]			4	469	—	—	948	30.2	.551	0.001	Spartan-3
SPONGENT-128 [1]			4	144	—	—	2380	—	.34	0.002	Spartan-3
PHOTON-160/36/36	serial	(χ)	4	233	257	407	1248	72.4	2.01	0.01	Spartan-3 XC3S50-5
	SRL16	(χ)	28	164	138	314	672	122.76	6.58	0.04	
	serial	(χ)	4	117	252	296	1248	328.52	9.47	0.08	Artix-7 XC7A100T-3
	SRL16	(χ)	28	89	135	204	672	328.17	17.58	0.20	
SPONGENT-160 [1]			4	193	—	—	3960	—	.2	0.001	Spartan-3
PHOTON-224/32/32	serial	(χ)	4	274	311	493	1620	69.04	1.36	0.005	Spartan-3 XC3S50-5
	SRL16	(χ)	32	176	135	339	864	123.33	4.57	0.03	
	serial	(χ)	4	130	305	328	1620	382.52	7.55	0.06	Artix-7 XC7A100T-3
	SRL16	(χ)	32	96	131	181	864	327.28	12.12	0.13	
SPONGENT-224 [1]			4	225	—	—	7200	—	.11	0.0005	Spartan-3
GRØSTL-224 [22]			64	1276	—	—	—	60	192	—	Spartan-3
PHOTON-256/32/32	serial	(χ)	8	327	335	577	924	42.56	1.47	0.004	Spartan-3 XC3S50-5
	SRL16	(χ)	48	416	160	806	504	58.95	3.74	0.009	
	serial	(χ)	8	157	331	373	924	132.49	4.59	0.03	Artix-7 XC7A100T-3
	SRL16	(χ)	48	159	100	384	504	166.41	10.75	0.07	
SPONGENT-256 [1]			4	241	—	—	9520	—	0.08	.0003	Spartan-3 XC3S200-5
SHABAL-256 [11]			—	499	—	—	—	100	.8	1.60	Spartan-3 XC3S200-5
BLAKE-256 [25]			—	631	—	—	—	—	216.3	0.34	Spartan-3 XC3S50-5
GRØSTL [25]			—	766	—	—	—	—	192.6	0.25	Spartan-3 XC3S50-5
JH [25]			—	558	—	—	—	—	63.7	0.11	Spartan-3 XC3S50-5
KECCAK [25]			—	766	—	—	—	—	46.2	0.06	Spartan-3 XC3S50-5
SKEIN [25]			—	766	—	—	—	—	16.6	0.02	Spartan-3 XC3S50-5
SHA-2 [25]			—	745	—	—	—	—	137.8	0.19	Spartan-3 XC3S50-5

that KECCAK-f[200] is not included in this table as no FPGA implementation of this function has been published so far. As seen from the table, our work provides the smallest area among all known implementations of lightweight hash functions while having a higher throughput and yields a better throughput per area ratio (Eff.) than PHOTON-80/20/16 [12], PHOTON-128/16/16 [28] and the implementation of SPONGENT [1]. We remark that SHABAL [11] has a better throughput per area ratio than PHOTON, but in this article our goal with serialised implementations is to reduce area, and not to improve throughput per area ratio.

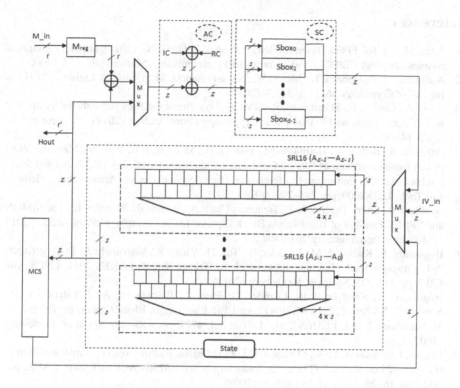

Fig. 7. A serialized architecture of the PHOTON hash function based on SRL16s

5   Conclusion

In this paper, we have analyzed the feasibility of creating a very compact, low cost FPGA implementation of LED and PHOTON. For both primitives, we studied round-based and serial architectures and we implemented several possible tradeoffs when computing the diffusion matrix. In particular, we proposed an SRL16 based architecture, that seems to be very well suited for all cryptographic primitives that use serial matrices. Our results show that LED and PHOTON are very good candidates for lightweight applications, our implementations yield for example the best area of all lightweight hash functions implementations published so far. Future work will include the investigation of side-channel analysis on our implementations and apply countermeasures [29] in order to resist these attacks.

Acknowledgments. Authors would like to thank the anonymous reviewers for their helpful comments. The first author wishes to thank Prof. R. Balsubramanian, Executive Director (SETS) and Sri. S. Thiagarajan, Registrar (SETS) for their support. Thomas Peyrin is supported by the Singapore National Research Foundation Fellowship 2012 (NRF-NRFF2012-06).

References

1. Adas, M.: On The FPGA Based Implementation of SPONGENT (2011). http://ece.gmu.edu/coursewebpages/ECE/ECE646/F11/project/F11_presentations/Marwan.pdf
2. Aumasson, J.-P., Henzen, L., Meier, W., Naya-Plasencia, M.: Quark: A Lightweight Hash. Journal of Cryptology **26**(2), 313–339 (2013)
3. Aysu, A., Gulcan, E., Schaumont, P.: SIMON Says, Break the Area Records for Symmetric Key Block Ciphers on FPGAs. IACR Cryptology ePrint Archive (2014). http://eprint.iacr.org/2014/237
4. Baldwin, B., Byrne, A., Hamilton, M., Hanley, N., McEvoy, R.P., Pan, W., Marnane, W.P.: FPGA Implementations of SHA-3 Candidates: CubeHash, Grøstl, LANE, Shabal and Spectral Hash. In: 12th Euromicro Conference on Digital System Design, Architectures, Methods and Tools, DSD 2009, pp. 783–790. IEEE (2009)
5. Beaulieu, R., Shors, D., Smith, J., Treatman-Clark, S., Weeks, B., Wingers, L.: The SIMON and SPECK Families of Lightweight Block Ciphers. IACR Cryptology ePrint Archive 2013 (2013). http://eprint.iacr.org/2013/404
6. Bogdanov, A., Knežević, M., Leander, G., Toz, D., Varıcı, K., Verbauwhede, I.: SPONGENT: A Lightweight Hash Function. In: Preneel, B., Takagi, T. (eds.) CHES 2011. LNCS, vol. 6917, pp. 312–325. Springer, Heidelberg (2011)
7. Bogdanov, A., Knudsen, L.R., Leander, G., Paar, C., Poschmann, A., Robshaw, M.J.B., Seurin, Y., Vikkelsoe, C.: PRESENT: An Ultra-Lightweight Block Cipher. In: Paillier, P., Verbauwhede, I. (eds.) CHES 2007. LNCS, vol. 4727, pp. 450–466. Springer, Heidelberg (2007)
8. Bulens, P., Standaert, F.-X., Quisquater, J.-J., Pellegrin, P., Rouvroy, G.: Implementation of the AES-128 on Virtex-5 FPGAs. In: Vaudenay, S. (ed.) AFRICACRYPT 2008. LNCS, vol. 5023, pp. 16–26. Springer, Heidelberg (2008)
9. Canright, D.: A Very Compact S-Box for AES. In: Rao, J.R., Sunar, B. (eds.) CHES 2005. LNCS, vol. 3659, pp. 441–455. Springer, Heidelberg (2005)
10. Chu, J., Benaissa, M.: Low area memory-free FPGA implementation of the AES algorithm. In: 2012 22nd International Conference on Field Programmable Logic and Applications (FPL), pp. 623–626. IEEE (2012)
11. Detrey, J., Gaudry, P., Khalfallah, K.: A Low-Area Yet Performant FPGA Implementation of Shabal. In: Biryukov, A., Gong, G., Stinson, D.R. (eds.) SAC 2010. LNCS, vol. 6544, pp. 99–113. Springer, Heidelberg (2011)
12. Eiroa, S., Baturone, I.: FPGA implementation and DPA resistance analysis of a lightweight HMAC construction based on photon hash family. In: FPL, pp. 1–4. IEEE (2013)
13. Engel, A., Liebig, B., Koch, A.: Feasibility Analysis of Reconfigurable Computing in Low-Power Wireless Sensor Applications. In: Koch, A., Krishnamurthy, R., McAllister, J., Woods, R., El-Ghazawi, T. (eds.) ARC 2011. LNCS, vol. 6578, pp. 261–268. Springer, Heidelberg (2011)
14. Feldhofer, M., Aigner, M.J., Baier, T., Hutter, M., Plos, T., Wenger, E.: Semi-passive RFID development platform for implementing and attacking security tags. In: ICITST, pp. 1–6. IEEE (2010)
15. Gong, Z., Nikova, S., Law, Y.W.: KLEIN: A New Family of Lightweight Block Ciphers. In: Juels, A., Paar, C. (eds.) RFIDSec 2011. LNCS, vol. 7055, pp. 1–18. Springer, Heidelberg (2012)
16. Good, T., Benaissa, M.: AES on FPGA from the Fastest to the Smallest. In: Rao, J.R., Sunar, B. (eds.) CHES 2005. LNCS, vol. 3659, pp. 427–440. Springer, Heidelberg (2005)
17. Guo, J., Peyrin, T., Poschmann, A.: The PHOTON Family of Lightweight Hash Functions. In: Rogaway, P. (ed.) CRYPTO 2011. LNCS, vol. 6841, pp. 222–239. Springer, Heidelberg (2011)

18. Guo, J., Peyrin, T., Poschmann, A., Robshaw, M.: The LED Block Cipher. In: Preneel, B., Takagi, T. (eds.) CHES 2011. LNCS, vol. 6917, pp. 326–341. Springer, Heidelberg (2011)
19. Guo, X., Chen, Z., Schaumont, P.: Energy and Performance Evaluation of an FPGA-Based SoC Platform with AES and PRESENT Coprocessors. In: Bereković, M., Dimopoulos, N., Wong, S. (eds.) SAMOS 2008. LNCS, vol. 5114, pp. 106–115. Springer, Heidelberg (2008)
20. Xilinx Inc. AN 307: Altera Design Flow for Xilinx Users (March 2013). http://www.altera.com/literature/an/an307.pdf
21. Xilinx Inc., Using Look-Up Tables as Shift Registers (SRL16) in Spartan-3 Generation FPGAs (May 2005). http://www.xilinx.com/support/documentation/application_notes/xapp465.pdf
22. Jungk, B., Reith, S.: On FPGA-based implementations of Grøstl. IACR Cryptology ePrint Archive (2010). http://eprint.iacr.org/2010/260
23. Kaps, J.-P.: Chai-Tea, Cryptographic Hardware Implementations of xTEA. In: Chowdhury, D.R., Rijmen, V., Das, A. (eds.) INDOCRYPT 2008. LNCS, vol. 5365, pp. 363–375. Springer, Heidelberg (2008)
24. Kaps, J.-P., Sunar, B.: Energy Comparison of AES and SHA-1 for Ubiquitous Computing. In: Zhou, X., Sokolsky, O., Yan, L., Jung, E.-S., Shao, Z., Mu, Y., Lee, D.C., Kim, D.Y., Jeong, Y.-S., Xu, C.-Z. (eds.) EUC Workshops 2006. LNCS, vol. 4097, pp. 372–381. Springer, Heidelberg (2006)
25. Kaps, J.-P., Yalla, P., Surapathi, K.K., Habib, B., Vadlamudi, S., Gurung, S.: Lightweight Implementations of SHA-3 Candidates on FPGAs. In: The Third SHA-3 Candidate Conference (2012)
26. Knudsen, L., Leander, G., Poschmann, A., Robshaw, M.J.B.: PRINTCIPHER: A Block Cipher for IC-Printing. In: Mangard, S., Standaert, F.-X. (eds.) CHES 2010. LNCS, vol. 6225, pp. 16–32. Springer, Heidelberg (2010)
27. Macé, F., Standaert, F.-X., Quisquater, J.-J.: FPGA Implementation(s) of a Scalable Encryption Algorithm. IEEE Transactions on Very Large Scale Integration (VLSI) Systems 16(2), 212–216 (2007)
28. Malka, P.K.: Compact Hardware Implementation of PHOTON Hash Function in FPGA (2011). http://ece.gmu.edu/coursewebpages/ECE/ECE646/F11/project/F11_presentations/Pavan.pdf
29. Moradi, A., Poschmann, A., Ling, S., Paar, C., Wang, H.: Pushing the Limits: A Very Compact and a Threshold Implementation of AES. In: Paterson, K.G. (ed.) EUROCRYPT 2011. LNCS, vol. 6632, pp. 69–88. Springer, Heidelberg (2011)
30. Poschmann, A.Y.: LIGHTWEIGHT CRYPTOGRAPHY: Cryptographic Engineering for a Pervasive World. Phd thesis. Citeseer (2009)
31. Shibutani, K., Isobe, T., Hiwatari, H., Mitsuda, A., Akishita, T., Shirai, T.: *Piccolo*: An Ultra-Lightweight Blockcipher. In: Preneel, B., Takagi, T. (eds.) CHES 2011. LNCS, vol. 6917, pp. 342–357. Springer, Heidelberg (2011)
32. Standaert, F.-X., Piret, G., Rouvroy, G., Quisquater, J.-J.: FPGA implementations of the ICE-BERG block cipher. Integration, the VLSI Journal 40(1), 20–27 (2007)
33. Tuan, T., Rahman, A., Das, S., Trimberger, S., Kao, S.: A 90-nm Low-Power FPGA for Battery-Powered Applications. IEEE Trans. on CAD of Integrated Circuits and Systems 26(2), 296–300 (2007)
34. Yalla, P., Kaps, J.-P.: Lightweight Cryptography for FPGAs. In: International Conference on Reconfigurable Computing and FPGAs, ReConFig 2009, pp. 225–230. IEEE (2009)

S-box Pipelining Using Genetic Algorithms for High-Throughput AES Implementations: How Fast Can We Go?

Lejla Batina [1], Domagoj Jakobovic [2], Nele Mentens [3], Stjepan Picek [1,2],
Antonio de la Piedra [1(✉)], and Dominik Sisejkovic [2]

[1] Digital Security Group, ICIS, Radboud University Nijmegen,
Nijmegen, The Netherlands
{lejla,s.picek,a.delapiedra}@cs.ru.nl
[2] Faculty of Electrical Engineering and Computing, University of Zagreb,
Zagreb, Croatia
{domagoj.jakobovic,dominik.sisejkovic}@fer.hr
[3] KU Leuven ESAT/COSIC and iMinds, Leuven-Heverlee, Belgium
nele.mentens@kuleuven.be

Abstract. In the last few years, several practitioners have proposed a wide range of approaches for reducing the implementation area of the AES in hardware. However, an area-throughput trade-off that undermines high-speed is not realistic for real-time cryptographic applications. In this manuscript, we explore how Genetic Algorithms (GAs) can be used for pipelining the AES substitution box based on composite field arithmetic. We implemented a framework that parses and analyzes a Verilog netlist, abstracts it as a graph of interconnected cells and generates circuit statistics on its elements and paths. With this information, the GA extracts the appropriate arrangement of Flip-Flops (FFs) that maximizes the throughput of the given netlist. In doing so, we show that it is possible to achieve a 50 % improvement in throughput with only an 18 % increase in area in the UMC 0.13 μm low-leakage standard cell library.

Keywords: Real-time cryptography · Genetic Algorithms (GAs) · S-boxes

1 Introduction

Implementations of cryptography are of constant interest for companies making security products. The challenges vary from very compact, low-power/energy to high-speed implementations of both symmetric and asymmetric cryptographic

This work was supported in part by the Technology Foundation STW (project 12624 - SIDES), The Netherlands Organization for Scientific Research NWO (project Pro-FIL 628.001.007) and the ICT COST action IC1204 TRUDEVICE.

algorithms. Ever growing applications require security services, introduce more constraints and real-time crypto has become of paramount importance.

This race for the fastest implementations in both hardware and software is especially difficult for algorithms that feature endless implementation options such as Elliptic Curve Cryptography (ECC) and the AES standard. For example AES can be implemented with table look-ups or via multiple composite field representations, each of which has certain advantages for area, performance and security [1–4]. Moreover, other practitioners have relied on resource sharing and folded architectures for reducing the implementation area [5,6]. However, when considering fast hardware implementations, pipelining is an obvious choice to increase the throughput. Nevertheless, as pipelining implies adding FFs, it also increases the area.

An interesting research challenge is to optimize throughput, while keeping the area under control. More precisely, considering compact options for the AES S-box, namely those relying on composite field arithmetic, which are the best pipelining solutions to maximize the throughput? Strategies for solving this problem typically involve hardware tools and rely mostly on good hardware design practice, but are generally far from straightforward. The goal of this research is to investigate this problem of optimizing the performance via pipelining such that the best throughput is found for a given composite field S-box. For this purpose we use genetic algorithms that have already found their place in other cryptographic applications.

Our contribution is in comprehensively evaluating how to boost the performance of AES implementations based on composite field arithmetic. To this end, we consider a hardware implementation where the S-box is implemented using a polynomial basis in $GF(((2^2)^2)^2)$ as used for example by Satoh et al. [2] and Mentens et al. [3]. We deploy genetic algorithms to find a good solution for the position of the pipelining FFs in order to reduce the critical path as much as possible.

Through our methodology we find a solution that adds one level of pipelining registers in order to increase the throughput of the S-box with 50 % while the extra FFs only increase the area with 18 %. To underline the added value of our approach, we add some statistics on the circuit under investigation, which show that it is far from straightforward to find this solution.

The remainder of this paper is organized as follows. First, in Section 2, we describe how GAs have been coupled with cryptographic applications in the literature as well as different alternatives for exploring the design space of an AES implementation. In Section 3 we illustrate the initial design of the AES S-box based on composite fields that we have selected for this work. Then, in Section 4, we present the framework that we have developed for analyzing Verilog netlists, generating an appropriate input for the GA, evaluating the correctness of the resulting netlist and synthesizing our solution. In Section 5, we describe our results and end in Section 6 with some conclusions.

2 Related Work

Considering previous works on hardware implementations of AES, numerous papers appeared optimizing various implementation properties i.e. throughput, area, power, energy etc. Here we remind the reader to some of those that are using a design choice that is similar to ours. The main focus is on implementations using composite field arithmetic to boost compactness and/or speed.

Satoh *et al.* were the first to introduce a new composite field $GF(((2^2)^2)^2)$-based implementation which resulted in the most compact S-box at the time with a gate complexity of 5.4 kgates. Wolkerstorfer *et al.* [7] used arithmetic in $GF((2^4)^2)$ to achieve an implementation with a gate count comparable to the one presented by Satoh *et al.* (5.7 kgates). Mentens *et al.* [3] and Canright [4] found the best choice of polynomials and representation to optimize the S-box area for polynomial and normal basis respectively. However, Moradi *et al.* have recently published the most compact AES implementation [8] of the size of only 2.4 kgates. This result is obtained by optimizing AES encryption on all layers and pushing the area to this minimum. Hodjat *et al.* [9] have described an ASIC implementation for the same composite field $GF((2^4)^2)$ as Wolkerstorfer. Their approach was to perform an area-throughput trade-off by fully pipelining the architecture and optimizing the key-schedule implementation. One of the more recent works is from Kumar *et al.* [10] considering an FPGA implementation of AES based on inversion in $GF(2^8)$. The critical path they obtain is 4.6 ns, which is substantially slower than our best result (2.6 ns). In general, hardware design tools consist of algorithms that focus on the optimization of area, speed and/or power/energy consumption, but none of the tools handles automatic pipelining to the extend that we do in this paper.

The literature is rich in examples where researchers have been relying on GAs for solving cryptographic problems. For instance, Jhajharia *et al.* and Sokouti *et al.* proposed the utilization of GAs for generating cryptographic keys [11,12]. Further, Zarza *et al.* and Park *et al.* utilized GAs in the context of cryptographic protocol design [13,14]. Carpi *et al.* studied the selection of security parameters for protecting smart cards against fault attacks via GAs [15]. Moreover, there are many successful applications of evolutionary computation when evolving S-boxes. Clark *et al.* used the principles from the evolutionary design of Boolean functions to evolve S-boxes with desired cryptographic properties [16]. They used the Simulated Annealing (SA) heuristic coupled with the hill-climbing algorithm to evolve bijective S-boxes with high non-linearity. Picek *et al.* used genetic algorithms to evolve S-boxes of various sizes that have better resistance to DPA attacks [17,18].

In this work, we use GAs to explore the design space of a standard cell netlist that has many unbalanced and partially overlapping paths from input to output. In order to achieve high throughput, GAs are used to choose the position of the pipelining FFs. We are not aware of prior work addressing pipelining of digital circuits using evolutionary computation.

3   S-box Implementation

In this section we describe the implementation options for the AES S-box that we experiment with for this study. It was shown before in the works of Canright [4] and Mentens *et al.* [3] that the most compact solutions rely on composite field arithmetic.

Considering various arithmetic options in binary extension fields to optimize the inversion operation in the AES S-box, there are basically two implementation options. We can either perform the subfield operations directly in $GF(2^4)$ or we can perform computations in the tower field $GF(((2^2)^2)^2)$, i.e. working in all subfields and using the fact that inversion is linear in $GF(2^2)$. The latter option is the one we choose.

The field $GF(((2^2)^2)^2)$ is considered as an extension of degree 2 over $GF((2^2)^2)$ constructed using the irreducible polynomial $P(x) = x^2 + p_1 x + p_0$, where $p_1, p_0 \in GF((2^2)^2)$. $GF((2^2)^2)$ is a field extension of degree 2 over $GF(2^2)$ using the irreducible polynomial $Q(y) = y^2 + q_1 y + q_0$ with $q_1, q_0 \in GF(2^2)$. $GF(2^2)$ is a field extension of degree 2 over $GF(2)$ using the irreducible polynomial $R(z) = z^2 + z + 1$. The constants are given in Appendix A.

In this case, inversion in $GF(2^2)$ requires only one addition:

$$d = t_1 z + t_0 \in GF(2^2) : d^{-1} = t_1 z + (t_1 + t_0). \tag{1}$$

These operations are implemented as depicted in Fig. 1.

Unlike the inversion in $GF((2^4)^2)$, the building blocks are not implemented as 4-bit look-up tables, but as operations in $GF((2^2)^2)$, which can be computed as follows (with $a_1, a_0, b_1, b_0 \in GF(2^2)$):

- $(a_1 y + a_0) \cdot (b_1 y + b_0) = (a_1 b_1 + a_1 b_0 + a_0 b_1) y + (a_1 b_1 \phi + a_0 b_0);$
- $(a_1 y + a_0)^2 = a_1 y + (a_1 \phi + a_0);$
- $(a_1 y + a_0) \cdot \lambda = (a_1 y + a_0) \omega y = (a_1 + a_0) \omega y + a_1 \omega \phi.$

These equations consist of operations in $GF(2^2)$ that can be computed as follows (with $g_1, g_0, h_1, h_0 \in GF(2)$):

- $(g_1 z + g_0) \cdot (h_1 z + h_0) = (g_1 h_1 + g_0 h_1 + g_1 h_0) z + (g_1 h_1 + g_0 h_0);$
- $(g_1 z + g_0)^2 = a_1 z + (a_1 + a_0);$
- $(g_1 z + g_0) \cdot \phi = (g_1 z + g_0) \cdot z = (g_1 + g_0) z + g_1;$
- $(g_1 z + g_0) \cdot \omega = (g_1 z + g_0) \cdot (z + 1) = g_0 z + (g_1 + g_0).$

4   Methodology

In this section we provide an explanation of the framework that we have developed for interfacing the GA with different modules for analyzing, simulating and synthesizing evolved netlists (Figure 2). First, we parse a Verilog description of a certain circuit, which is, in our case, the S-box design described in Section 3.

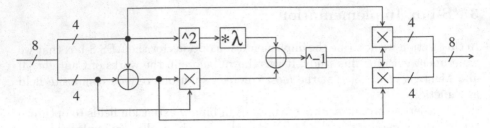

Fig. 1. Schematic of the S-box by Satoh *et al.* [2], where the building blocks are operations in $GF((2^2)^2)$, which are decomposed into operations in $GF(2^2)$

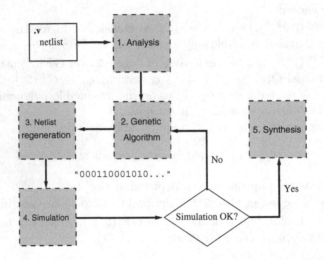

Fig. 2. Workflow of our approach for pipelining the AES S-box

This step provides us with different statistics of the target design such as the number of standard cells (referred to as elements), cell inputs and paths in the circuit. Then, an appropriate input for the GA is generated according to the delay of each element of the netlist. From the GA, we receive a certain arrangement of FFs that maximizes the throughput of the substitution box. The FFs are inserted in a new netlist, simulated and synthesized (Steps 3–5 in Figure 2). In the following sections, we first describe our optimization problem and then we continue on each step and relate our results using the design described in Section 3.

4.1   Pipelining as an Optimization Problem

The task at hand is to insert a combination of FFs in order to increase throughput through a certain number of pipeline stages. A valid solution to this problem requires that all the paths in the circuit contain the same number of FFs, which are placed in a way that minimizes the delay in each pipeline stage. For a given number of stages, the number of FFs at every path is one less than the number of stages (i.e. one FF per path in a two-stage pipeline).

To define this as an optimization problem, we encode each possible solution as a bitstring (a sequence of bits) where each bit represents every input location for all circuit elements. In this encoding, a bit is set to the value "0" if the corresponding input does not have an associated FF and to the value "1" if there is a FF preceding that input (the unmodified circuit is represented with all zeros). The total bitstring length is equal to the sum of all the inputs in the circuit, which in this case amounts to 432. A potential solution is therefore a sequence of bits of length 432, which defines a search space of 2^{432}.

The quality (also called the fitness value) of each potential solution is determined by the delay of the pipeline stage with the greatest delay among all the stages. However, since the optimization algorithm operates with any combination of bits in the search process, a great number of potential solutions are expected to be infeasible, because they will represent a circuit with a different number of FFs in all the paths. To handle that constraint, a penalty factor can be included in the quality estimate to differentiate between feasible and infeasible solutions. The penalty should be great enough such that each feasible solution (a circuit with the same number of FFs in every path), regardless of the delay, is still better than any infeasible solution, to guide the search to valid solutions.

Based on the previous, we define the following fitness function that represents the optimization problem:

$$fitness = max\_delay\_time + (1,000 * number\_invalid\_paths) \qquad (2)$$

Here, max_delay_time presents the longest delay for every pipeline stage and invalid paths are all those that do not have a correct number of FFs. We experimentally set the weight to be 1,000 in the formula above. Intuitively, the weight needs to be large enough such that even in the case that there is only one invalid path, the total fitness should be worse than for the solution without FFs. For that same reason, every larger weight factor would work the same. The optimization objective is the minimization of the fitness function. Note that to calculate the maximum delay, all possible paths in the network need to be traversed, which poses a fairly large computational demand.

In the next section we describe the first step in Figure 2, focused on the analysis of netlists.

4.2 Analysis of Verilog Netlists

Our framework generates statistical information about a circuit represented as a Verilog netlist. In Table 1 we show the statistical details related to our choice of representation. These parameters are extracted using the framework we developed for pipelining the AES substitution box. The number of elements in the

Table 1. Statistics of the preliminary S-box design

Number of elements	Number of inputs	Number of paths	Shortest path	Critical path (ps)
165	432	8,023,409	4	3,884.52

table denotes the number of standard cells. The number of inputs refers to the number of inputs to all standard cells. Finally, the number of paths denotes the number of different possible paths through the circuit from an input to an output.

As depicted, there are too many possible paths to encode all of them into one solution. Since the total number of elements as well as the total number of inputs is relatively small, we decided to encode the possible solutions as bitstrings where each bit represents every input location for all elements.

Given a netlist of the S-box in Verilog, our framework first parses it according to a predefined grammar and then, each element from the UMC 0.13 μm low leakage standard cell library is identified [19]. This is done using a framework developed in python 2.7.5-5 in combination with the pyparsing 2.0.1 library[1].

Relying on that library, we have defined a grammar that deconstructs each entry of a Verilog netlist into a set of cells and their connections. For instance, a NAND gate defined in the standard cell library can appear in a given netlist as "ND2CLD U181 (.I1(\input [0]), .I2(\input [1]), .O(n1));". Hence, the parser must identify that element as a NAND gate (ND2CLD) associated to the U181 identifier. Moreover, it must detect that is connected to the first two inputs of the S-box and that the n1 wire routes its output.

This process is performed by creating a grammar that expects a set of entries consisting of:

- The type of the cell.
- The cell identifier.
- A comma-separated list of inputs and outputs (i.e. I1, I2 and O in the example) connected to the circuit inputs, outputs or internal Verilog wires (i.e. \input [0], \input [1], and n1).

Each element of the netlist is abstracted in a data structure that stores the cell type, the cell identifier, the number of inputs of the cell and all the elements that are connected to their inputs i.e. their adjacent elements. Moreover, the delay associated to each element according to the standard cell library is also stored. This information is later used as an input for the optimization algorithm.

The resulting list contains all the circuit cells together with their number of inputs and their adjacent nodes (that is, the cells that are connected to their inputs) as well as the delay of each element. A small example of the parser output is given below:

---

[1] http://pyparsing.wikispaces.com/

U163	2	U251 U248			146.8
U164	3	U198 U256 U163			86.471
U165	4	U198 U163 U256 U164			98.369
U166	1	U207			59.39
U167	4	U207 U209 U210 U166			114.406

This example describes the number of inputs for the cells U163–U167 (i.e. 2, 3, 4, 1, 4 respectively) together with the cells that are connected to those inputs and the respective delay of the cell according to the standard cell library.

These values are obtained as average values for all possible combinations (transitions from low to high and from high to low) for each element. For each FF element that will be inserted in order to maximize the throughput, we use a D-FF with a single output and no clear, set or enable (QDFFCLD) with an average delay time of 320.35 ps. All delay times are given for a temperature of 25 degrees Celsius, a core voltage of 1.2 V and a load capacitance of 1.5 fF.

Taking the average of low-to-high and high-to-low delays and assuming a low load capacitance of 1.5 fF is an approximation that gives good results. Nevertheless, to improve our methodology, the actual delay information based on the load of each standard cell should be taken into account.

Next, we present the optimization algorithm we used to generate pipelined circuits.

4.3   Genetic Algorithms

In accordance with the given representation, we selected genetic algorithms (GAs) as the optimization method to be used in our experiments.

Prior to going into to the details about genetic algorithms, first we offer a short rationale behind the choice of them. Since there is no previous work that uses any kind of heuristics to evolve the optimal arrangement of FF elements in a combinatorial circuit, we believe we should start with some well-researched algorithm that can be easily adapted.

Genetic algorithms are an evolutionary computation technique that has been successfully applied to various optimization problems. Additionally, bitstring representation is one of several standard representations of GAs [20]. Naturally, there are other heuristic algorithms that also use bitstring representation (e.g. Particle Swarm Optimization [21], Genetic Annealing [22]) that could be used here. In accordance with that, it is not possible to stipulate what algorithm would perform the best. The "No Free Lunch" theorem states that there is no single best algorithm for all the problems, i.e. when averaged over all search problems, all algorithms behave the same [23]. Therefore, only thorough experimental work can give insight into more appropriate algorithms. Further details about genetic algorithms and operators we use are given in Appendix B.

Common Parameters. The parameters used in each run of the algorithm are the following: the number of independent runs for each evolutionary experiment is 30 and the population size is 30. The tournament size k in the tournament selection is equal to 3. Mutation probability is set to 0.45 per individual where we choose it on a basis of a small set of tuning experiments where it showed the best results on average.

Further, our setting has one more important parameter that needs to be set i.e. the number of pipeline stages. With this parameter we control how many levels of FF elements we want in our circuit. From Table 1 we see that the number of elements in the shortest path is 4. Therefore, this path can have only 3 levels of FFs and that is the maximum number of FFs our circuit can have in order to produce a correct output.

Evolutionary Process. After the parameters are set, the GA starts with the generation of the initial population. In this part, the genetic algorithm reads all the elements of the parser output file and for each cell input it reserves one position in the bitstring representation. Notice that our bitstring size is fixed for a given circuit and it can not be dynamically changed during the evolutionary process. This results in the fact that our current setting does not support multiple FF elements one after the other. The initial population is built by creating random bitstrings of the designated length corresponding to randomly setting FFs in the preliminary netlist.

When the initial population is generated, the genetic algorithm starts with the evolution process. In each iteration it randomly chooses k possible solutions (the k tournament size) and eliminates the worst solution among those (this is also to ensure elitism i.e. the best solutions are always propagated to the next generation). The remaining solutions are used as parents which create one offspring via variation operators. The offspring (new solution) then replaces the worst individual in the population and additionally undergoes a mutation with a given probability.

For each offspring, a genetic operator is selected uniformly at random between all operators within an operator class (mutation or crossover). We use simple and mix mutations and uniform [24] and one-point [20, 25] crossover operators. These variation operators are selected among those that are commonly used nowadays.

The evolution process repeats until the stopping criterion is met. In our case, the stopping criterion is based on 50 generations without improvement of the best solution.

4.4   Reconstructing Evolved Individuals to the Netlist

Using a list of structures described in Section 4.2 it is possible to compute all the paths of the circuit based on all the possible combinations for the eight inputs and outputs of the AES substitution box. This is done by transforming the list of cell structures described above into a non-directed graph, where the cells are represented by nodes and their connections by edges. Then, it is possible to extract the connections in the circuit and identify all the paths for all the input-output combinations using a graph exploration algorithm such as the

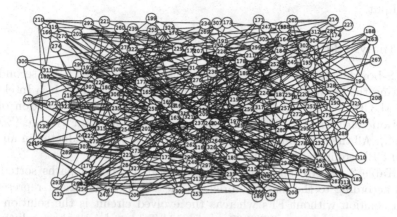

Fig. 3. Graph representation of the S-box connections for each cell identifier

breadth-first search (cf. [26]). We have depicted in Figure 3 how our framework abstracts a Verilog netlist.

From the GA, we obtain the precise arrangement of FFs that will be inserted in the new netlist in order to maximize its throughput as described in Section 4.1. Given the internal structure that we created from a Verilog netlist, now it is possible to reconstruct the circuit according to the output from the GA (Step 3 in Figure 2). Our framework first splits the binary string from the GA in different chunks according to the number of inputs of each element. Then, it associates the respective FFs to each cell input. Moreover, the required Verilog wires that connect each FF to the input/output of the cell are added. For instance, for an XOR gate with two inputs (e.g. XOR2ELD) and an output "11" from the GA, this element would be reconstructed with two FFs attached to their inputs using two wires (e.g. ff_9_q, ff_10_q) as:

```
QDFFCLD FF9 ( .CK(clk), .D(n180), .Q(ff_9_q) );
QDFFCLD FF10 ( .CK(clk), .D(n198), .Q(ff_10_q) );
XOR2ELD U212 ( .I1(ff_9_q), .I2(ff_10_q), .O(n201) );
```

Finally, a test bench with assertions for all the 256 possibilities of the S-box is created for the regenerated netlist. This is used in Mentor Graphics ModelSim 6.5c to guarantee the correctness of the new circuit. The resulting circuit is then synthesized using Synopsys Design Compiler in order to get pre-layout implementation results for the critical path delay and the area.

5   Results

In this section, we present our performance figures for the developed framework that analyzes and reconstructs Verilog netlist. We also show the synthesis results for our best candidates i.e. those substitution boxes that obtained the maximum throughput.

5.1 Analysis Framework

In our S-box design, based on 165 cells (Table 1), we required 0.465 seconds for generating the input for the GA using the described framework in a Intel Core i5-3230M CPU clocked at 2.60GHz (Step 1, Figure 2). The evaluation of one generation consisting of 100 individuals requires 3.5 min. on average (Step 2, Figure 2). All the experiments carried out with GAs were conducted in an Intel i5-3470 CPU equipped with 6 GB of RAM.

In Table 2 we give an overview of the total number of paths sorted into classes according to the maximum delay time. Each initial circuit represents a certain solution without FFs whereas the evolved circuit is the solution with the maximum delay time (2,793.62 as given in Table 3 for the best solution we found). Additionally, in Table 3 we present different statistics for several evolved circuits. Here, column Number of FFs represents the total number of flip-flops in evolved circuit and column Number of generations represents how long was GA running.

Table 2. Number of paths per length class

Class	Initial circuit	Evolved circuit
0 – 500	2	5,570
500 – 1,000	2,164	78,5432
1,000 – 1,500	149,944	3,751.897
1,500 – 2,000	2,026.442	2,639.751
2,000 – 2,500	3,580,150	816,636
2,500 – 3,000	1,899,675	26,411
3,000 – 3,500	361,708	0
3,500 – 4,000	3,324	0

Table 3. Statistics of evolved circuits

Max. delay time (ps)	Number of stages	Number of FFs	Number of generations
2,793.62	2	73	587
2,826.52	2	68	15
2,942.42	2	66	691
3,155.11	2	49	482
3,223.02	2	64	4452
3,247.64	2	42	1434
2,918.92	3	100	618

5.2 Synthesis

As can be seen from Table 3, the best solution the GA finds when dividing the circuit into two stages (i.e. inserting one layer of pipelining FFs) has an estimated critical path of 2,793.62 ns. The best solution with two layers of pipelining FFs to have the shortest critical path is 2,918.92 ns. Intuitively, we would expect the solution with two layers of pipelining to have the shortest critical path. However, because the number of possible solutions is much bigger for a 3-stage circuit than for a 2-stage circuit, the optimal solution found by the GA for the 3-stage circuit is worse than the optimal solution it finds for the 2-stage circuit. Nevertheless, we know that there should exist a better solution for 3 than for 2 when a longer search is performed.

We synthesized both solutions, resulting in Table 4. In order to evaluate the critical path properly, we inserted flip-flops at the inputs and outputs of all the S-boxes. As mentioned before, the implementation corresponds to the state-of-the-art design of Mentens et al. described in [3]. The netlist with 1 stage only contains these input and output flip-flops. The netlists with 2 and 3 stages contains 1 and 2 layers of pipelining flip-flops, respectively. Because of these input and output flip-flops, the netlist with only one stage is larger in area than the composite field S-boxes reported in literature (they do not contain any flip-flops). The table shows that the 2-stage S-box introduces a 50 % improvement in throughput, which is equal to the number of bits at the output (8 in our case) divided by the delay of the critical path. The increase in area is only 18 %. The synthesis results for the critical path are even slightly better than the estimate of the GA. The reason is that the synthesis tool optimizes the generated pipelined netlist again, which leads to further improvements. For the 3-stage S-box, the synthesis results are worse than the estimate of the GA. This is probably due to the fact that there is less room for optimization with two layers of pipelining flip-flops and thus less logic in between the layers.

Table 4. Pre-layout synthesis results of the netlist with 1, 2 and 3 stages.

Number of stages	Critical path (ns)	Throughput (Gbits/s)	Area (μm^2)	Gate count
1	3.9	2.05	2,450	612.50
2	2.6	3.07	2,901	725.25
3	3.2	2.50	3,433	858.25

6 Conclusion

This paper presents a methodology for pipelining composite field AES S-boxes to maximize the throughput using genetic algorithms. The best trade-off between throughput and area results in a throughput of 3.07 Gbits/s and an area of 2,901 μm^2 in a UMC 0.13 μm standard cell library. This comes down to a throughput increase of 50 % with an area overhead of 18 % in comparison to an S-box without pipelining. In order to improve the throughput even more, the design space should be increased with more composite field representations. The GA could also still be optimized, e.g. by making a more intelligent choice of the seed.

Appendix A

Here we give the details of the constants used for our composite field implementations of the AES S-box.

In [2], Satoh *et al.* made the following choices for the coefficients of the irreducible polynomials:

$$p_1 = 1 = \{0001\}_2,$$
$$p_0 = \lambda = \omega y = (z+1)y = \{1100\}_2,$$
$$q_1 = 1 = \{01\}_2,$$
$$q_0 = \phi = z = \{10\}_2.$$

The inverse operation is implemented as

$$\Delta = \delta_1 x + \delta_0 \in GF(((2^2)^2)^2):$$
$$\Delta^{-1} = (\delta_1 x + (\delta_1 + \delta_0)) \cdot (\lambda \delta_1^2 + (\delta_1 + \delta_0)\delta_0)^{-1}, \tag{3}$$
$$\delta = d_1 y + d_0 \in GF((2^2)^2):$$
$$\delta^{-1} = (d_1 y + (d_1 + d_0)) \cdot (\phi d_1^2 + (d_1 + d_0)d_0)^{-1}.$$

Inversion in $GF(2^2)$ requires only one addition:

$$d = t_1 z + t_0 \in GF(2^2): d^{-1} = t_1 z + (t_1 + t_0). \tag{4}$$

Appendix B

Genetic algorithms belong to the evolutionary family of algorithms where the elements of the search space S are arrays of elementary type [27]. We give a short pseudo-code for a genetic algorithm (this is also a pseudo-code for any evolutionary algorithm) in Algorithm 1.

In order to produce new individuals (solutions), the GA uses mutation and crossover operators. Mutation operators use one parent to create one child by applying randomized changes to the parent. The mutation depends on the mutation rate p_m which determines the probability that a change will occur within an individual. Crossover operators modify two or more parents in order to create an offspring via the information contained within parent solutions. Recombination is usually applied probabilistically according to a crossover rate p_c. In this work, we use only operators that work with two parents. Additionally, GAs use selection methods to choose the individuals that will continue to the next generation. We opted here for the steady-state tournament or k-tournament selection method [27]. In this selection from k randomly selected individuals, two with the best fitness values are chosen to evolve and create one offspring, replacing the worst from the tournament [25, 28].

Next, we give a short description of crossover and mutation operators that we use.

Algorithm 1. Genetic algorithm

Input : Parameters of the algorithm
Output : Optimal solution set
$t \leftarrow 0$
$P(0) \leftarrow CreateInitialPopulation$
while $TerminationCriterion$ **do**
  $t \leftarrow t + 1$
  $P'(t) \leftarrow SelectMechanism\ (P(t-1))$
  $P(t) \leftarrow VariationOperators(P'(t))$
end while
Return OptimalSet(P)

One Point Crossover. When performing one point crossover, both parents are split at the same randomly determined crossover point. Subsequently, a new child genotype is created by appending the first part of the first parent with the second part of the second parent [20,25].

Uniform Crossover. Single and multi-point crossover defines cross points as places between positions where an individual can be split. Uniform crossover generalizes this scheme to make every place a potential crossover point. A crossover mask, the same length as the individual structure is created at random and the parity of the bits in the mask indicate which parent will supply the offspring with which bits. The number of effective crossing points in uniform crossover is not fixed, but will average to $l/2$ where l represents string length.

Mix Mutation. Mix (or mixing) mutation randomly chooses one area inside the individual where it will change the bits. First, in that area number of ones and zeros is counted and then random bits are set while preserving the respective number of values [20].

Simple Mutation. In simple mutation every bit is inverted with a predefined mutation probability p_m [20].

References

1. Fischer, V., Drutarovský, M.: Two Methods of Rijndael Implementation in Reconfigurable Hardware. In: Koç, Ç.K., Naccache, D., Paar, C. (eds.) CHES 2001. LNCS, vol. 2162, pp. 77–92. Springer, Heidelberg (2001)
2. Satoh, A., Morioka, S., Takano, K., Munetoh, S.: A compact rijndael hardware architecture with s-box optimization. In: Boyd, C. (ed.) ASIACRYPT 2001. LNCS, vol. 2248, pp. 239–254. Springer, Heidelberg (2001)
3. Mentens, N., Batina, L., Preneel, B., Verbauwhede, I.: A systematic evaluation of compact hardware implementations for the Rijndael S-BOX. In: Menezes, A. (ed.) CT-RSA 2005. LNCS, vol. 3376, pp. 323–333. Springer, Heidelberg (2005)

4. Canright, D.: A very compact s-box for AES. In: Rao, J.R., Sunar, B. (eds.) CHES 2005. LNCS, vol. 3659, pp. 441–455. Springer, Heidelberg (2005)
5. Fischer, V., Drutarovský, M., Chodowiec, P., Gramain, F.: InvMixColumn decomposition and multilevel resource sharing in AES implementations. IEEE Trans. VLSI Syst. **13**(8), 989–992 (2005)
6. Chodowiec, P., Gaj, K.: Very Compact FPGA Implementation of the AES Algorithm. In: Walter, C.D., Koç, Ç.K., Paar, C. (eds.) CHES 2003. LNCS, vol. 2779, pp. 319–333. Springer, Heidelberg (2003)
7. Wolkerstorfer, J., Oswald, E., Lamberger, M.: An ASIC implementation of the AES sboxes. In: Preneel, B. (ed.) CT-RSA 2002. LNCS, vol. 2271, pp. 67–78. Springer, Heidelberg (2002)
8. Moradi, A., Poschmann, A., Ling, S., Paar, C., Wang, H.: Pushing the Limits: A Very Compact and a Threshold Implementation of AES. In: Paterson, K.G. (ed.) EUROCRYPT 2011. LNCS, vol. 6632, pp. 69–88. Springer, Heidelberg (2011)
9. Hodjat, A., Verbauwhede, I.: Area-throughput trade-offs for fully pipelined 30 to 70 gbits/s AES processors. IEEE Trans. Computers **55**(4), 366–372 (2006)
10. Kumar, S., Sharma, V., Mahapatra, K.: Low latency VLSI architecture of S-box for AES encryption. In: 2013 International Conference on Circuits, Power and Computing Technologies (ICCPCT), pp. 694–698 (March 2013)
11. Jhajharia, S., Mishra, S., Bali, S.: Public key cryptography using neural networks and genetic algorithms. In: Parashar, M., Zomaya, A.Y., Chen, J., Cao, J., Bouvry, P., Prasad, S.K. (eds.) IC3, pp. 137–142. IEEE (2013)
12. Sokouti, M., Sokouti, B., Pashazadeh, S., Feizi-Derakhshi, M.R., Haghipour, S.: Genetic-based random key generator (GRKG): a new method for generating more-random keys for one-time pad cryptosystem. Neural Computing and Applications **22**(7–8), 1667–1675 (2013)
13. Zarza, L., Pegueroles, J., Soriano, M., Martínez, R.: Design of cryptographic protocols by means of genetic algorithms techniques. In: Malek, M., Fernández-Medina, E., Hernando, J. (eds.) SECRYPT, pp. 316–319. INSTICC Press (2006)
14. Park, K., Hong, C.: Cryptographic protocol design concept with genetic algorithms. In: Khosla, R., Howlett, R.J., Jain, L.C. (eds.) KES 2005. LNCS (LNAI), vol. 3682, pp. 483–489. Springer, Heidelberg (2005)
15. Carpi, R.B., Picek, S., Batina, L., Menarini, F., Jakobovic, D., Golub, M.: Glitch it if you can: parameter search strategies for successful fault injection. In: Francillon, A., Rohatgi, P. (eds.) CARDIS 2013. LNCS, vol. 8419, pp. 236–252. Springer, Heidelberg (2014)
16. Clark, J.A., Jacob, J.L., Stepney, S.: The design of S-boxes by simulated annealing. New Generation Computing **23**(3), 219–231 (2005)
17. Picek, S., Ege, B., Batina, L., Jakobovic, D., Chmielewski, L., Golub, M.: On Using Genetic Algorithms for Intrinsic Side-channel Resistance: The Case of AES S-box. Proceedings of the First Workshop on Cryptography and Security in Computing Systems, CS2 2014, pp. 13–18. ACM, New York (2014)
18. Picek, S., Ege, B., Papagiannopoulos, K., Batina, L., Jakobovic, D.: Optimality and beyond: The case of 4x4 s-boxes. In: 2014 IEEE International Symposium on Hardware-Oriented Security and Trust, HOST 2014, Arlington, VA, USA, May 6-7, pp. 80–83 (2014)
19. Faraday: Faraday Cell Library 0.13 μm Standard Cell (2004)
20. Holland, J.H.: Adaptation in Natural and Artificial Systems: An Introductory Analysis with Applications to Biology, Control, and Artificial Intelligence. The MIT Press, Cambridge (1992)

21. Kennedy, J., Eberhart, R.: Particle swarm optimization. In: Proceedings of the IEEE International Conference on Neural Networks, vol. 4, pp. 1942–1948 (November 1995)
22. Yao, X.: Optimization by genetic annealing. In: Proc. of 2nd Australian Conf. on Neural Networks, pp. 94–97 (1991)
23. Wolpert, D.H., Macready, W.G.: No Free Lunch Theorems for Optimization. IEEE Transactions on Evolutionary Computation 1(1), 67–82 (1997)
24. Syswerda, G.: Uniform crossover in genetic algorithms. In: Proceedings of the 3rd International Conference on Genetic Algorithms, pp. 2–9. Morgan Kaufmann Publishers Inc., San Francisco (1989)
25. Eiben, A.E., Smith, J.E.: Introduction to Evolutionary Computing. Springer, Heidelberg (2003)
26. Knuth, D.E.: The Art of Computer Programming. Fundamental Algorithms, 3rd edn., vol. 1. Addison Wesley Longman Publishing Co. Inc., Redwood City (1997)
27. Weise, T.: Global Optimization Algorithms Theory and Application (2009)
28. Michalewicz, Z.: Genetic algorithms + data structures = evolution programs, 3rd edn. Springer, London (1996)

Protected Hardware Design

Wire-Tap Codes as Side-Channel Countermeasure
– An FPGA-Based Experiment –

Amir Moradi[✉]

Horst Görtz Institute for IT Security, Ruhr University Bochum,
Bochum, Germany
amir.moradi@rub.de

Abstract. In order to provide security against side-channel attacks a masking scheme which makes use of *wire-tap codes* has recently been proposed. The scheme benefits from the features of binary linear codes, and its application to AES has been presented in the seminal article. In this work – with respect to the underlying scheme – we re-iterate the fundamental operations of the AES cipher in a hopefully more understandable terminology. Considering an FPGA platform we address the challenges each AES operation incurs in terms of implementation complexity. We show different scenarios on how to realize the SubBytes operation as the most critical issue is to deal with the large S-boxes encoded by the underlying scheme. Constructing various designs to actualize a full AES-128 encryption engine of the scheme, we provide practical side-channel evaluations based on traces collected from a Spartan-6 FPGA platform. As a result, we show that – despite nice features of the scheme – with respect to its area and power overhead its advantages are very marginal unless its fault-detection ability is also being employed.

1 Introduction

Nowadays security of embedded devices does rely not only on the underlying modern cipher but also on the way it is implemented. Side-channel analysis (SCA) attacks, which have been brought to the attention of scientific communities since the late 90s [14,15], are amongst the major threats against the security of cryptographic devices. Amongst other countermeasures masking [6,7,17], which by randomizing the secret internals aims at cutting the relation between the side-channel leakages and predictable processes, is the most studied one. It can also be seen as the one that achieving its goals in practice is most challenging. For instance, dealing with glitches in hardware platforms is of major concerns for most of the masking schemes (see [18,20,22,24]). Along the same line different masking approaches like additive [6], multiplicative [1], and affine [9,27] have been introduced. It has been shown in [9] that affine masking, which combines both additive and multiplicative masking, can achieve – in terms of the number of traces – a considerably high level of security.

© Springer International Publishing Switzerland 2014
W. Meier and D. Mukhopadhyay (Eds.): INDOCRYPT 2014, LNCS 8885, pp. 341–359, 2014.
DOI: 10.1007/978-3-319-13039-2_20

Recently a masking scheme which makes use of wire-tap channel concept has been proposed [5]. The scheme which is developed based on the principle of binary linear codes can be seen as an extended version of affine masking. More precisely, in affine masking the bit size of the words, e.g., 8 in case of the AES state bytes, does not change after masking a secret. But in wire-tap approach the length of the masked value is expanded. Another specification of the scheme is regarding the selection of the masks. In affine masking the additive mask has the same length as the unmasked value, and also can take all possible values, e.g., $\{0,1\}^8$ in case of 8-bit AES state bytes. However, they are restricted to the elements of the underlying code (so-called *codewords*) in case of wire-tap code approach. An expanded value is seen as error bits which are added to a codeword, i.e., the mask. This therefore provides two features for the scheme:

- The additive mask can be removed without knowing it; thanks to parity-check matrix of the underlying code which can eliminate the codewords.
- Certain faults which cause the additive mask to be not a codeword anymore can be detected while the main goal of the scheme is to provide security against e.g., power analysis attacks.

The focus of [5] is on an AES encryption engine, and a procedure on how to make a protected version of the cipher under the proposed scheme is introduced. Moreover, by means of simulation results the success rate of first- and second-order CPA [4] attacks as well as MIA [10] for different settings is examined.

Our contribution in this work is in the direction of realizing a hardware implementation of the scheme for a certain setting. We first shortly restate the scheme and its features by concentrating on the AES-128 encryption as the target cipher. With respect to how to implement each operation of the cipher, we try to provide more details that are partially missing in the original work [5]. We also give solution for a couple of issues which we faced during the implementation that hugely affect its performance as well as its security. Based on practical investigations, which are performed on a Spartan-6 FPGA as the implementation platform, we provide a comparison between the underlying scheme and classical Boolean masking from efficiency and security points of view. In short, we show that the additional features that the scheme provides are very marginal considering its high performance overhead when it is implemented correctly.

2   Underlying Scheme

Since the theory behind the wire-tap channel [23,28] and the secrecy of the scheme is given in detail in [5], below we mainly focus on the target cipher, i.e., AES-128 encryption, with respect to side-channel protection. For the sake of consistency we also try to follow the notations given in [5].

2.1   Notations

Suppose a binary linear code C of length n, dimension k, and minimum distance d as $[n, k, d]$. C consists in a generator matrix G of size $(n - k) \times n$ with rows of $(g_1, \ldots, g_{n-k})$, where each row is an n-bit binary vector. A codeword $c \in C$ is generated as $m \cdot G$, where m is an $(n - k)$-bit binary vector. Indeed all 2^{n-k} codewords of C can be generated by m going through the entire space vector $\{0, 1\}^{n-k}$. C also contains a $k \times n$ (binary) parity-check matrix H as $\forall c \in C, \, c \cdot H^T = 0$.

In order to encode a k-bit message x it should be first expanded by means of a $k \times n$ matrix L as $x \cdot L$. Rows of $L = (l_1, \ldots, l_k)$ are n-bit binary vectors which are linearly independent of each other, and none of them is a codeword. The encoding of x is determined by adding (modulo 2) a randomly selected codeword to the result of the expansion as

$$z \; = \; x \cdot L \, \oplus \, m \cdot G.$$

At anytime the encoded message z can be multiplied by H^T to eliminate the added random codeword:

$$z \cdot H^T \; = \; x \cdot L \cdot H^T \oplus m \cdot \underbrace{G \cdot H^T}_{:=0}.$$

The result, i.e., $x \cdot L \cdot H^T$, which is in some cases called syndrome, is related to x. Indeed the decoding (deriving x from z) can be done if $x \to x \cdot L \cdot H^T$ forms a bijection.

2.2   Settings

Considering AES each message x is mapped to a cipher state byte as $k = 8$. Hereafter for the sake of simplicity without losing generality we specify the length of the code $n = 16$ as the evaluations reported in [5] indicate its high level of security. In other words, each state byte of AES is expanded to 16 bits and a $[16, 8, 5]$ binary code [11] is used for masking. This leads to generator matrix $G = \left(- M^T | I_8 \right)$ and parity-check matrix $H = \left(I_8 | M \right)$, where I_8 denotes the identity matrix of size 8×8 and

$$M = \begin{pmatrix} 10001011 \\ 11000101 \\ 11100010 \\ 01110001 \\ 10111000 \\ 01011100 \\ 00101110 \\ 00010111 \end{pmatrix}.$$

For other values of n and parametric definitions of G and H sizes see [5].

By defining the underlying code as above the only missing point is how to define the L matrix. An example of L given in [5] as $L_{\text{simple}} = (l_1, \ldots, l_k)$ is chosen as

A. Moradi

$$l_1 = [1000000000000000],$$
$$l_2 = [0100000000000000],$$
$$l_3 = [0010000000000000],$$
$$l_4 = [0001000000000000],$$
$$l_5 = [0000100000000000],$$
$$l_6 = [0000010000000000],$$
$$l_7 = [0000001000000000],$$
$$l_8 = [0000000100000000].$$

Expanding x by L_{simple} results in x as the left half and 0 as the right half. Indeed, L_{simple} brings a feature as $x \cdot L \cdot H^T = x$. In other words, decoding of $z = x \cdot L \oplus m \cdot G$ is easily done by $z \cdot H^T := x$.

Further, a simple algorithm is given in [5] to randomly generate the L matrix. This resulted in an example (given in [5]) as $L_{\text{specific}} = (l_1, \ldots, l_k)$ as follows:

$$l_1 = [0111110110100101],$$
$$l_2 = [1011000010000101],$$
$$l_3 = [0001011100100111],$$
$$l_4 = [0001110011001101],$$
$$l_5 = [0000110101001100],$$
$$l_6 = [1011110110111111],$$
$$l_7 = [1001111111111111],$$
$$l_8 = [0101100101110101].$$

However, decoding $z = x \cdot L_{\text{specific}}$ by H^T is not possible. It means that $x \rightarrow x \cdot L_{\text{specific}} \cdot H^T$ is not a bijection as

$$\exists x_1, x_2; \ x_1 \cdot L_{\text{specific}} \cdot H^T = x_2 \cdot L_{\text{specific}} \cdot H^T.$$

It in fact prevents L_{specific} to be considered as a valid case of L. In short, examining whether $x \rightarrow x \cdot L \cdot H^T$ is bijective is missing in the algorithm given in [5]. Alternatively it can be checked whether $L \cdot H^T$ has a right inverse Inv as

$$L \cdot H^T \cdot Inv = I_8.$$

Since Inv is an 8×8 binary matrix, its existence can be checked by an exhaustive search for each column separately, in sum in a space of 8×2^8.

As a valid example, replacing l_4 by $[1010011111010111]$ causes L_{specific} to fulfill all the requirements. Hereafter we consider this corrected matrix as $L_{\text{specific}'}$ for further investigations. Also, an example for the corresponding Inv is given in Appendix.

2.3 AES Operations

XOR (AddRoundKey). Suppose x_1 and x_2 are encoded as $z_{i \in \{0,1\}} = x_i \cdot L_{\text{specific}'} \oplus c_i$ by two randomly selected codewords c_1 and c_2. A correct encoding of $x_1 \oplus x_2$ can be made by $z_1 \oplus z_2 = (x_1 \oplus x_2) \cdot L_{\text{specific}'} \oplus (c_1 \oplus c_2)$ since as a property of binary linear codes $(c_1 \oplus c_2)$ is also a codeword.

S-box (SubBytes). If $z = x \cdot L_{\text{specific}'} \oplus c$ is the input of the S-box, $z' = S(x) \cdot L_{\text{specific}'} \oplus c'$ as an encoding of $S(x)$ can be the result of a table lookup $S'(z)$, where $\forall x$ the look-up table S' is precomputed given two independent codewords c and c'.

Permutation (ShiftRows). Since ShiftRows is a byte-wise permutation, the corresponding word-wise permutation of an encoded cipher state results in correct encoding of the cipher state after ShiftRows.

Multiply by 2. MixColumns over a column (x_1, x_2, x_3, x_4) can be realized by an XOR sequence of the input bytes, some multiplied by 2 and 3 in $\text{GF}(2^8)$. Since $x * 3 = (x * 2) \oplus x$, and as explained above XOR of the encoded words is straightforward, the only remaining operation to realize MixColumns is

$$\text{MUL2}(z) = (x * 2) \cdot L_{\text{specific}'} \oplus c''; \ z = x \cdot L_{\text{specific}'} \oplus c; \ c, c'' \in C.$$

A solution that we provide here (not clearly given in [5]) is to find a matrix P such that

$$x \cdot L_{\text{specific}'} \cdot P = (x * 2) \cdot L_{\text{specific}'}.$$

A simple solution is to select $P = L_{\text{specific}'}^{-1} \cdot M2 \cdot L_{\text{specific}'}$, where $L_{\text{specific}'}^{-1}$ denotes the right inverse of $L_{\text{specific}'}$ which as explained before can be found in a space of 8×2^{16}. Also $M2$ stands for binary matrix representation of multiply-by-2 in $\text{GF}(2^8)$. An example for $L_{\text{specific}'}^{-1}$, $M2$, and P are given in Appendix.

Multiplying z by P leads to

$$z \cdot P = x \cdot L_{\text{specific}'} \cdot P \oplus c \cdot P = (x * 2) \cdot L_{\text{specific}'} \oplus c \cdot P.$$

Here the problem is that $c \cdot P$ is not necessarily a codeword. As a solution, also somehow followed by [5], $c'' \oplus c \cdot P$ should be added to the above result to obtain[1]

$$z \cdot P \oplus c'' \oplus c \cdot P = (x * 2) \cdot L_{\text{specific}'} \oplus c'' = \text{MUL2}(z).$$

A question arising here is whether there exists an L such that

$$\overset{?}{\exists} L, \ \forall c \in C; \ c \cdot P \in C.$$

The answer is unfortunately negative as if $c \cdot P$ is a codeword, applying the parity-check matrix should lead to

$$c \cdot P \cdot H^T = c \cdot L^{-1} \cdot M2 \cdot L \cdot H^T = 0.$$

However, $L \cdot H^T$ cannot be 0 following the definition of L (see Section 2.1). So, there is no way to avoid mask correction, i.e., adding $c'' \oplus c \cdot P$.

Algorithm 1 gives an overview of the full AES-128 encryption based on the selected settings and according to the one given in [5]. As explained above and

---

[1] Note that c'' and c must not be necessarily different.

Algorithm 1. AES-128 encryption protected by wire-tap code

Input : Plaintext $\mathcal{X}$, seen as 16 bytes x_i, $i \in \{0, \ldots, 15\}$,
        11 RoundKeys $\mathcal{R}^j$, each seen as 16 bytes r_i^j, $j \in \{0, \ldots, 10\}$
Output: Ciphertext $\mathcal{Y}$, seen as 16 bytes y_i

1 Select 64 random bytes $m_{i \in \{0, \ldots, 63\}}$
2 Generate 4 vectors $\mathcal{C}, \mathcal{C}', \mathcal{C}^1, \mathcal{C}^3$, each seen as 16 codewords as
   $c_i = m_i \cdot G$, $c_i' = m_{i+16} \cdot G$, $c_i^1 = m_{i+32} \cdot G$, $c_i^3 = m_{i+48} \cdot G$, $i \in \{0, \ldots, 15\}$
3 $\mathcal{C}^2 = \mathcal{C} \oplus \mathcal{C}^1$
4 $\mathcal{T} = \mathcal{C}^1 \oplus \texttt{MixColumn}\big(\texttt{ShiftRows}(\mathcal{C}')\big)$ /* Mask Correction */

5 **for** $i \in [\![0, 15]\!]$ **do**
6 $\quad$ $\forall x \in \{0,1\}^8$; $S_i'(x \cdot L_{\text{specific}'} \oplus c_i) = S(x) \cdot L_{\text{specific}'} \oplus c_i'$
7 $\quad$ $z_i = x_i \cdot L_{\text{specific}'} \oplus c_i^1$ /* Plaintext Encoding as $\mathcal{Z}$ */
8 $\quad$ $\forall j \in [\![0, 10]\!]$; $k_i^j = r_i^j \cdot L_{\text{specific}'} \oplus c_i^2$ /* RoundKey Encoding */
9 **end**

10 **for** $j \in [\![0, 9]\!]$ **do**
11 $\quad$ $\forall i \in [\![0, 15]\!]$; $z_i = z_i \oplus k_i^j$ /* $\mathcal{Z} : (\mathcal{X} \oplus \mathcal{R}^j) \cdot L_{\text{specific}'} \oplus \mathcal{C}$ */
12 $\quad$ $\forall i \in [\![0, 15]\!]$; $z_i = S_i'(z_i)$ /* $\mathcal{Z} : \text{SB}(\mathcal{X} \oplus \mathcal{R}^j) \cdot L_{\text{specific}'} \oplus \mathcal{C}'$ */
13 $\quad$ $\mathcal{Z} = \texttt{ShiftRows}(\mathcal{Z})$ /* $\mathcal{Z} : \text{SR}\big(\text{SB}(\mathcal{X} \oplus \mathcal{R}^j)\big) \cdot L_{\text{specific}'} \oplus \text{SR}(\mathcal{C}')$ */
14 $\quad$ **if** $j \neq 9$ **then**
15 $\quad\quad$ $\mathcal{Z} = \texttt{MixColumns}(\mathcal{Z})$ /* $\mathcal{Z} : \text{MC}\big(\text{SR}(\text{SB}(\mathcal{X} \oplus \mathcal{R}^j))\big) \cdot L_{\text{specific}'} \oplus \text{MC}(\text{SR}(\mathcal{C}'))$ */
16 $\quad\quad$ $\mathcal{Z} = \mathcal{Z} \oplus \mathcal{T}$ /* $\mathcal{Z} : \text{MC}\big(\text{SR}(\text{SB}(\mathcal{X} \oplus \mathcal{R}^j))\big) \cdot L_{\text{specific}'} \oplus \mathcal{C}^1$ */
17 $\quad$ **else**
18 $\quad\quad$ $\mathcal{Z} = \mathcal{Z} \oplus \mathcal{C}^3$ /* $\mathcal{Z} : \text{SR}\big(\text{SB}(\mathcal{X} \oplus \mathcal{R}^9)\big) \cdot L_{\text{specific}'} \oplus \text{SR}(\mathcal{C}') \oplus \mathcal{C}^3$ */
19 $\quad$ **end**
20 **end**

21 $\forall i \in [\![0, 15]\!]$; $z_i = z_i \oplus k_i^{10}$ /* $\mathcal{Z} : \mathcal{Y} \cdot L_{\text{specific}'} \oplus \text{SR}(\mathcal{C}') \oplus \mathcal{C}^3 \oplus \mathcal{C}^2$ */
22 $\forall i \in [\![0, 15]\!]$; $w_i = z_i \cdot H^T$ /* $\mathcal{W} : \mathcal{Y} \cdot L_{\text{specific}'} \cdot H^T$ */
23 $\forall i \in [\![0, 15]\!]$; $y_i = w_i \cdot Inv$ /* $\mathcal{Y}$: ciphertext */

---

also as given in line 16 of the algorithm, at every round the masks should be corrected to keep them as valid codewords since MixColumns is not transparent to the underlying code. This in fact limits one of the main features of the scheme as the same mask correction is usually done for a classical Boolean masking scheme (e.g., see DPA contest V4 [26]). Note that the required mask correction is not due to our selection of $L_{\text{specific}'}$ or matrix P. The same scenario is given by the algorithm available in the original work of [5].

The only place in the algorithm which makes use of this feature – as masks can be removed without knowing them – is the last step of the algorithm lines 22 and 23. Indeed these two lines can be replaced by

$$\mathcal{Y} = (\mathcal{Z} \oplus \text{SR}(\mathcal{C}') \oplus \mathcal{C}^3 \oplus \mathcal{C}^2) \cdot L_{\text{specific}'}^{-1},$$

where first the mask is removed then the right inverse of $L_{\text{specific}'}$ is applied to obtain the ciphertext. This way the masks do not need to be valid codewords.

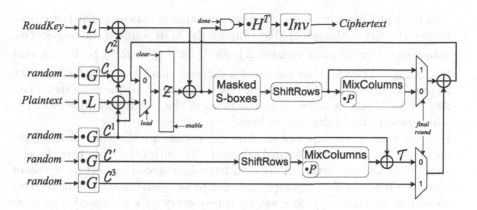

Fig. 1. Overall architecture of the design

So, any $c \in \{0,1\}^{16}$ can be selected as a mask to form $\mathcal{C}$, $\mathcal{C}'$, $\mathcal{C}^1$, and $\mathcal{C}^3$. More precisely, no binary linear code is required to be chosen, and any L – with a right inverse – can be employed to extend the cipher state bytes. It indeed shows that this feature of the scheme has a very marginal application for this perspective.

3   Hardware Design

Regardless of the limiting issue mentioned above, we give the details of the hardware design we developed to realize Algorithm 1. The target platform chosen for our practical experiments is a Spartan-6 LX75 FPGA embedded on a Side-channel AttacK User Reference Architecture (SAKURA-G) [21]. Due to the nature of the algorithm and efficiency as the main feature of hardware platforms, we aim at a round-based architecture, i.e., all operations of a cipher round are performed in parallel. This usually leads to a design with "single clock per round" capability. However, as it is explained later the efficiency that can be reached by the underlying scheme might be much less.

Figure 1 shows an overview of the hardware design which is well matched with Algorithm 1. It should be noted that the RoundKeys $\mathcal{R}$ have been precomputed. As shown in the algorithm and also by the design diagram, the mask of the round input, i.e., $\mathcal{C}^1$, is constant during an encryption. It is essential since for the selected masks the masked S-boxes S' are precomputed once at the start of the encryption. This might be a dangerous situation for a hardware platform as bit flips of the registers – modeled by Hamming distance (HD) – affect the amount of power consumption. Indeed if two values masked by the same mask, $x \oplus m$ and $y \oplus m$, are consecutively saved in a register, $x \oplus m \oplus y \oplus m = x \oplus y$ has an influence on power consumption which might be easily detected by $\mathrm{HD}(x, y)$. Therefore, there are two important facts which should be considered for the design:

- The most problematic case of the above mentioned issue is at the last cipher round. Due to the absence of MixColumns, the XOR result of two consecutive values stored by the state register $\mathcal{Z}$ is $\mathrm{SB}^{-1}\left(\mathrm{SR}^{-1}\left(\mathcal{Y} \oplus \mathcal{R}^{10}\right)\right) \oplus \mathcal{Y}$ expanded by $L_{\mathrm{specific}'}$. Therefore, by means of a different mask at the end of the last round (see lines 16 and 18 of Algorithm 1) the mask of the ciphertext, i.e., $\mathrm{SR}\left(\mathcal{C}'\right) \oplus \mathcal{C}^3 \oplus \mathcal{C}^2$, is different to the mask of the last round input, i.e., $\mathcal{C}^1$. This prevents the addressed problem[2].
- Still the mask of the input of the first two rounds are the same. It may give an opportunity to a side-channel adversary to control the plaintexts thereby simplifying the prediction of consecutive values stored by certain words of the state register $\mathcal{Z}$. Therefore, in order to prevent this the state register should be precharged before saving a new entry at each round (see *clear* signal of state register $\mathcal{Z}$ in Fig. 1).

3.1   Masked S-box

The most challenging part of the design is how to actualize the masked S-boxes S'. The problem is due to the bit length of the S-box input, i.e., 16 bits in our settings, which causes realizing a table with 16×2^{16} bits problematic. In case of our platform there exist 172 instances of 18-kbit Block RAMs (BRAM)[3], 64 of them are required to make this table. However, it consists of only 256 valid entries for the chosen input and output masks.

Direct Mapping. A straightforward solution is to map the encoded S-box input to an 8-bit (masked) value and directly perform the table lookup over a 256-entry table containing the encoded S-box outputs (size of $16 \times 2^8 = 4\,\mathrm{kbits}$ fitting into a 9-kbit BRAM). In order to do so, one can add $m \cdot L_{\mathrm{specific}'}$ to the encoded S-box input as

$$(x \cdot L_{\mathrm{specific}'} \oplus c) \oplus m \cdot L_{\mathrm{specific}'},$$

where $c = m \cdot G$. Now by applying the H^T matrix and Inv as defined in Section 2.2 we obtain

$$((x \oplus m) \cdot L_{\mathrm{specific}'} \oplus m \cdot G) \cdot H^T \cdot Inv = ((x \oplus m) \cdot L_{\mathrm{specific}'} \cdot H^T) \cdot Inv = x \oplus m.$$

Therefore, we can convert the encoded S-box input to its corresponding Boolean masked value. However, this is in contradiction with the concept of the underlying scheme since the operation is performed on classical Boolean masked data not under the defined binary linear code.

---

[2] Indeed, introducing a new mask as $\mathcal{C}^3$ is not required, but to keep the algorithm compatible with that of [5] we kept it in the design.

[3] Each 18-kbit BRAM can be configured as two independent 9-kbit BRAM resulting in total 344 instances. For more information see RAMB16BWER and RAMB8BWER in [29].

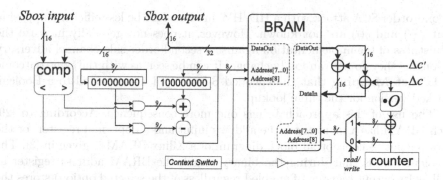

Fig. 2. Block diagram of the masked S-box circuit (left) binary search module, (right) update module

Binary Search. Another solution given in [5] is to make two tables, one to save 16-bit encoded S-box inputs and the other for the corresponding 16-bit encoded S-box outputs. So, a space of $32 \times 2^8 = 8\,\text{kbits}$ is required for these two tables which can easily fit into a half of a 18-kbit BRAM. Therefore, for each masked S-box lookup the input table should be searched.

Since the search through an arbitrary table might be very inefficient, we considered the binary search which requires the tables to be sorted based on the encoded input. Figure 2 shows a block diagram of the binary search module. The BRAM is configured to use a 32-bit data path for input and output and 8 bits for the address[4]. Each 32-bit entry of the BRAM contains a tuple of (input, output) as $(x \cdot L_{\text{specific}'} \oplus c, \; S(x) \cdot L_{\text{specific}'} \oplus c')$. At the start of the search, 10000000 is given to the address of the BRAM. During the next clock cycle comparison result of the given encoded S-box input and the 16-bit part of the BRAM output decides whether 01000000 should be added or subtracted from the previous BRAM address. This procedure is continued till the comparison shows equality. In the worst case it takes 9 clock cycles to obtain the desired encoded S-box output[5].

In contrast to **Direct Mapping**, in this case the address of the BRAM does not necessarily form a classical Boolean masking of the input x as $x \oplus m$. But since it is the index of the table sorted based on the encoded S-box input, it can be modeled as $f(x) \oplus g(m)$, where

- $f(x)$ stands for a function which maps x to the index of the sorted table of $x \cdot L_{\text{specific}'}$, but
- deriving $g(m)$ is more complicated and explained in [19].

Both these two functions are deterministic based on $L_{\text{specific}'}$. They are also linear and can be represented by a matrix multiplication as $g(m) = m \cdot G \cdot O$, where O is the sorting matrix given in [19]. Compared to **Direct Mapping**, a

---

[4] Indeed the address has 9 bits width, we explain later how to control the 9th bit.
[5] It can be done also in 8 clock cycles, but the BRAM content of address 10000000 should be previously saved in a separate register.

second-order SCA attack with a HD/HW model might be less efficient supposing that $f(\cdot)$ and $g(\cdot)$ are not known. However, it does not generally increase the robustness of the implementation against a second-order side-channel adversary as long as they do not randomly change. It can be seen as with the same outcome as **Direct Mapping** that the encoded S-box input is converted to a Boolean masked version for the table lookup.

The use of this approach brings one more consequence. According to [29] each BRAM has a register for the address input and an optional register for the data output. A conceptual block diagram of a Xilinx BRAM is given in [3]. The problem here is that during the binary search the BRAM address register as well as its output register (if enabled regardless of the selected option) stores the consecutive values which are masked by the same mask. As explained before, this leads to a side-channel leakage depending on the XOR result of the corresponding two unmasked values (for a similar observation see [2]). In case of the address register $f(x_1) \oplus f(x_2) = f(x_1 \oplus x_2)$ has a considerable contribution on the amount of power consumption when $f(x_1) \oplus g(m)$ and $f(x_2) \oplus g(m)$ are consecutive addresses given to the BRAM by the binary search module. The same holds for the output register as the difference between two consecutive (input, output) tuples is $\Big((x_1 \oplus x_2) \cdot L_{\text{specific}'},\ \big(S(x_1) \oplus S(x_2)\big) \cdot L_{\text{specific}'} \Big)$. Note that even if the output register is not enabled, the output signals still drive a combinatorial circuit (here MixColumns) and the same leakage can be seen.

In order to avoid such leakage, similar to the state register $\mathcal{Z}$ the BRAM also should be precharged during the binary search process. In our designs we have precharged the BRAM address by 0 before giving it a new address. Therefore, it leads to maximum 18 clock cycles for each masked S-box lookup. Due to the existence of a register for the BRAM input, precharge of the state register $\mathcal{Z}$ can be performed at the same time with the last clock cycle of the binary search. It leads to maximum 19 clock cycles per cipher round, i.e., $190 + 1$ for one complete encryption[6]. This is extremely higher than that of the **Direct Mapping** approach, i.e., $20 + 1$ clock cycles for a complete encryption. At each cipher round, one clock cycle is needed to save in the state register $\mathcal{Z}$ and precharge the BRAM, and one clock cycle for both the mask S-box lookup and precharge of the state register.

In short both above expressed approaches have the same drawback as the S-box lookup needs to be out of the underlying code. Note that it can be prevented only if the (input, output) table is randomly – independent of m – permuted, which causes a search to take in average 128 clock cycles (at most 256).

3.2   Mask Update

As stated before, all the masks – derived from the 64 random bytes m_i – stay unchanged during the whole encryption. Moreover, the same holds for $\mathcal{C}^2$ derived from other masks as well as for $\mathcal{T}$ used for mask corrections at the end of each

---

[6] One last additional clock cycle is required to save the final round output in the state register $\mathcal{Z}$ (see Fig. 1).

round (see Algorithm 1). By giving a new set of m_i new codewords and constants should be computed. Also, the masked S-box tables S' need to be updated. A scheme to use BRAMs as the masked tables and dynamically update them is previously given in [13] called *BRAM Scrambling*. It is mainly based on the dual-port feature of BRAMs and the fact that two tables can fit in one BRAM. Fortunately it is the case for our settings as for both approaches given above two masked tables simply fit into a BRAM. Two 4-kbit tables fit into each 9-kbit BRAM in case of **Direct Mapping**, and one 18-kbit BRAM can hold two 8-kbit tables of **Binary Search** approach.

While the encryption module uses the first half of the BRAM via one of the ports, the other port is used to make the second half updated. After finishing the update process when an encryption is not in progress the context switch is performed thereby using the updated second half for the encryption and updating the first half. Note that the ports used for encryption and update are not swapped. Generally only one bit as the MSB of the 9-bit address changes to switch the halves used by these two modules.

Based on the two approaches given in Section 3.1 the corresponding update modules are slightly different. In case of **Direct Mapping**, it is similar to the one explained in [13]. Suppose m as the random byte from which input mask $c = m \cdot G$ has been derived, and suppose c' as the output mask. The new randoms are denoted by m_{new}, respectively c_{new} and c'_{new}. The XOR of the current and new masks are given to the update module as $\Delta m = m \oplus m_{\text{new}}$ and $\Delta c' = c' \oplus c'_{\text{new}}$.

The update module performs the following operations

- It reads the address $i := x \oplus m$ from one half of the BRAM (the part being used for encryption) as $v := S(x) \cdot L_{\text{specific}'} \oplus c'$.
- It saves $v \oplus \Delta c' := S(x) \cdot L_{\text{specific}'} \oplus c'_{\text{new}}$ at the address $i \oplus \Delta m := x \oplus m_{\text{new}}$ of the other half of the BRAM.

The above procedure is repeated 256 times $\forall i \in [\![0, 255]\!]$ to finish the update process.

For the **Binary Search** approach the update module is a bit more complicated since the updated table should be saved as sorted. Here $\Delta c = c \oplus c_{\text{new}}$ is also given to the update module, and the following operations are performed:

- The address $i := f(x) \oplus g(m)$ is read from one half of the BRAM as $(w, v) := \left(x \cdot L_{\text{specific}'} \oplus c, S(x) \cdot L_{\text{specific}'} \oplus c'\right)$.
- $(w \oplus \Delta c, v \oplus \Delta c') := (x \cdot L_{\text{specific}'} \oplus c_{\text{new}}, S(x) \cdot L_{\text{specific}'} \oplus c'_{\text{new}})$ is stored at the address $i \oplus g(\Delta m) := f(x) \oplus g(m_{\text{new}})$ of another half of the BRAM.

$f(\cdot)$ and $g(\cdot)$ are those functions introduced in Section 3.1. As explained before, $g(m) = m \cdot G \cdot O$, where details of sorting matrix O are given in [19]. Here $g(\Delta m)$ can be computed as

$$\Delta m \cdot G \cdot O = (m \oplus m_{\text{new}}) \cdot G \cdot O = \Delta c \cdot O.$$

A block diagram of the update module for the **Binary Search** approach is shown by Fig. 2. In both cases the update process can be done in 512 clock cycles since each read or write operation needs to be done in a separate clock cycle. Therefore, similar to the scheme presented in [13] – depending on the delay between consecutively given plaintexts – the masks used for a couple of encryptions may not be different.

4   Security Evaluation

4.1   Evaluation

Before providing practical evaluation results, we would like to comment on the security evaluations presented in [5]. If an arbitrary byte x_1 (and x_2) is encoded by all possible 256 codewords we obtain a set $S_1 = \{x_1 \cdot L_{\text{specific'}} \oplus m \cdot G | m \in \{0,1\}^8\}$ (and respectively S_2). As a feature of the scheme – thanks to the parity-check matrix – the masks can be removed without knowing them. However, it means that S_1 and S_2 do not have any overlap, i.e., $\nexists z; z \in S_1, z \in S_2$. Therefore, if the SCA leakages associated to S_1 and S_2 are different, their observation may lead to categorize them based on x_1 and x_2. This means that under certain assumptions an SCA attack will be possible. This assumptions are related to the order of the attack as well as the algebraic degree of the pseudo-Boolean representation of the leakage function. It is indeed the same concept as *low-entropy masking* studied in [16] and [12]. Since the selected code in our settings is an optimal $[16, 8, 5]$ binary code, it can resist against first-order attacks if the algebraic degree of the leakage function $d < 5$. Formally speaking

$$\mathbb{E}\big(\mathscr{L}\left(X \cdot L_{\text{specific'}} \oplus M \cdot G\right) | X = x\big)$$

is constant for any pseudo-Boolean function $\mathscr{L}$ of algebraic degree $d < 5$, where X represents any 8-bit random variable, M is a random variable uniformly distributed on $\{0,1\}^8$, and $\mathbb{E}$ stands for expectation. Similarly it can resist against univariate second-order attacks (without considering the leakage of the mask m or $m \cdot G$) for all leakage functions with algebraic degree $d < 3$. However, a univariate third-order attack works if the leakage function is not linear.

In theory a univariate mutual information analysis (MIA) [10] should be able to distinguish x considering the distribution of $\mathscr{L}\left(x \cdot L_{\text{specific'}} \oplus M \cdot G\right)$. The higher the algebraic degree of $\mathscr{L}$ is, the easier MIA can distinguish x. In fact, these issues have not been considered in the evaluations of [5], and a noisy Hamming weight (HW) model, i.e., a linear $\mathscr{L}$, is considered to show the resistance of the scheme against first-order attacks using simulation results.

4.2   Implementation

For practical experiments we developed three implementations:

- *Profile 1*, with the design of Fig. 1 and $L_{\text{specific'}}$ where the masked S-boxes are realized by **Binary Search** approach,
- *Profile 2*, the same as *Profile 1* except the masked S-boxes which are actualized by **Direct Mapping** technique,

Table 1. Comparison of area and performance of three designed profiles

Profile	Area			Performance (# clock)			
	# Register	# LUT	# BRAM 9k*	Round		Total	
				avg**	max	avg	max
1	3 050	14 499	32	17	19	171	191
2	2 747	13 808	16	2	2	21	21
3	1 595	2 320	16	2	2	21	21

* In case of *Profile 1*, each 18-kbit BRAM is counted as two 9-kbit BRAMs.
** Average performance of a binary search over an N-element array is $O(\log_2(N) - 1)$.

– *Profile 3*, with the same design as the other profiles, but without using the wire-tap approach. In other words, it follows the architecture of Fig. 1, but realizes a classical Boolean masking, i.e., no expansion and no binary linear code is used. The masked S-boxes are also implemented straightforwardly similar to that of *Profile 2* but without any mapping.

As stated before, our implementation platform is a Spartan-6 LX75 FPGA of SAKURA-G. Table 1 represents a comparison of area and performance of our three profiles. It should be emphasized that the area required for KeySchedule and to store the RoundKeys is ignored in the given numbers. Also, the PRNG module which is supposed to provide random numbers is excluded in the comparisons as all three profiles need the same PRNG and the same number of random bytes per encryption. As the last note, *Profiles 1* and *2* are designed to be parametric, i.e., for any selected L and C with length $n = 16$ the area requirements stay unchanged.

As expected, compared to the others *Profile 1* is the largest and slowest design. Also, comparing *Profiles 2* and *3* the overhead of using the wire-tap approach based on our settings becomes clear, i.e., 0.7 times more registers and 4.95 times more logic LUTs.

4.3 Measurements

In order to perform practical evaluations we collected the power traces of the target FPGA by means of a LeCroy digital oscilloscope at the sampling rate of 1 GS/s. The measurements are done by monitoring the voltage drop by a $0.6\,\Omega$ resistor placed at the Vdd path. We also made use of the amplifier embedded on SAKURA-G to increase the level of the signal compared to the electrical noise level. During the measurements the target FPGA is clocked at a frequency of 3 MHz.

As illustrated in Section 3.1, from a security point of view *Profiles 1* and *2* are roughly the same, and evaluation of *Profile 1* faces more challenges due it is much longer traces. Therefore, we provide here only the evaluation result of *Profiles 2* and *3*, a sample trace of each is shown by Fig. 3. As expected, due to its higher area requirements *Profile 2* has much higher power consumption compared to *Profile 3*. Also, the evaluation result of *Profile 1* is given in [19].

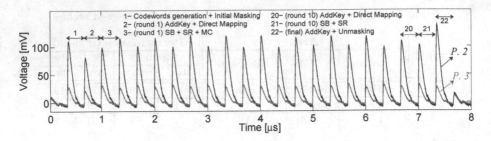

Fig. 3. Sample traces of *Profile 2* and *Profile 3*

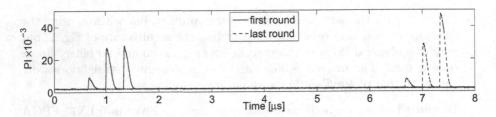

Fig. 4. *Profile 2*, PRNG off, Perceived Information curves based on an S-box output byte of first and last rounds using 1 000 000 traces

PRNG Off. In order to have a reference about the leakage of our platform as well as to verify our measurement setup we first considered *Profile 2* when the PRNG is switched off. Therefore, all 64 bytes $m_{i\in\{0,\dots,63\}}$ (see Algorithm 1) are constant as 0 during the collection of 1 000 000 traces. In other words, in this case only the expansion with $L_{\text{specific}'}$ is performed and all codewords used for encoding are selected as 0.

Our security evaluations are based on the *Information Theoretic* (IT) metric of [25]. It means that we estimate the mutual information between the measured traces and a secret internal which we suppose to know. In order to prevent missing any statistical moments, in all our evaluations the probability distributions are estimated by histograms of 12 bins rather than Gaussian. Following the notations of [25], we estimate the mutual information as

$$I(S; \mathbf{L}) = H[S] + \sum_s \Pr[s] \sum_\mathbf{l} \Pr[\mathbf{l}|s] \cdot \log_2 \Pr[s|\mathbf{l}],$$

where S (a secret internal) is selected as an S-box output in our evaluations[7]. We indeed measure the amount of perceived information [8] as we estimate the distributions by a histogram and consider the leakage model based on an S-box output value.

Figure 4 shows two perceived information curves each of which associated to an S-box output; one for the first round and the other for the last round of

---

[7] Due to the bijective property of AES S-box and the former linear key addition, it indeed leads to the same result if the corresponding plaintext byte is selected.

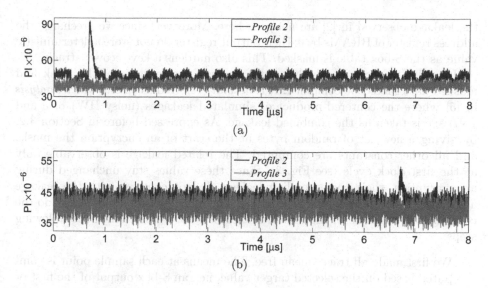

Fig. 5. *Profiles 2* and *3*, PRNG on, Perceived Information curves based on an S-box output byte of (a) first and (b) last round using 50 000 000 traces

encryption. As expected – due to constant masks – information available through the traces either at the first round or at the last round can be easily detected.

PRNG On. By switching the PRNG on we first confirmed the uniform distribution of 64 random bytes given to the design. As stated before, due to a long time required to update the masked S-boxes some encryptions may share the same masks. Therefore, we kept a considerable delay between consecutive encryptions during the measurements in order to make sure that no mask is reused. In other words, one encryption does not start till the last mask update process is finished. In this settings we collected 50 000 000 traces from each of *Profiles 2* and *3*.

Repeating the last experiment as estimating the perceived information between the measured traces and the S-box output led to the curves shown in Fig. 5(a) and Fig. 5(b) for the first and the last round respectively. As shown by the graphics, there is a difference between the estimated perceived information of *Profile 2* and *Profile 3*. It indeed confirms our statement in Section 4.1 about the existence of a univariate leakage in case of *Profile 2* though due to its very low magnitude a practical attack might be very challenging. Note that the same result is observed by evaluation of *Profile 1* as shown in [19].

Attacking an implementation realized by BRAMs is in general harder compared to the corresponding circuit purely implemented by combinatorial elements (see [3]). It becomes more challenging if the implementation is equipped with masking as in the case of our profiles. We should stress that in all of our profiles the BRAMs' ports used for encryption are disabled right after finishing the encryption process; it can be seen by the sample traces of Fig. 3. Therefore, the leakage observed in [2] are not seen here. Moreover, since we precharge the address register of BRAMs by 0, their output register do not store a deterministic value as the S-box table is masked. This also hardens a key-recover attack.

In order to perform a higher-order analysis the leakages of the mask and masked data should be combined. It is actually referred as *second-order analysis* in [5] where the centered-product of simulated leakages (noisy HW) of c and $x \cdot L \oplus c$ is taken as the combined leakage. As expressed before in Section 3.2, by giving a new set of random bytes at the start of an encryption the masks and all other constants are computed. The related leakage is observable only at the first clock cycle (see Fig. 3). Since these values stay unchanged during an encryption, their corresponding leakage do not combine with the leakage associated to the process of masked data. Therefore, we should manually do the combination required for higher-order analyses. So, we performed the following steps:

- We first made all traces mean free. The means at each sample point is computed based on the selected target value, i.e., an S-box output of the first or the last round.
- Next, we obtained the average of a part of a mean-free trace related to the first clock cycle (see Fig. 3).
- At the last step the average value is multiplied to all points of the mean-free trace.

The last two steps are repeated for all mean-free traces independently, and the second step is needed since the leakage associated to the masks may not appear at a specific sample point.

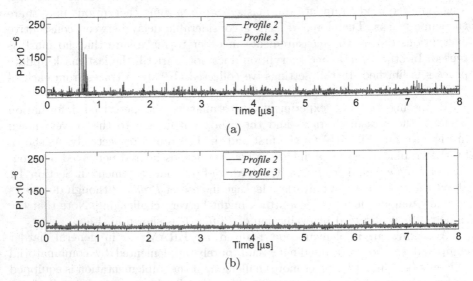

Fig. 6. *Profiles 2* and *3*, PRNG on, Perceived Information curves based on an S-box output byte of (a) first and (b) last round using 50 000 000 center-product preprocessed traces

Finally, we repeated the same experiment as before by estimating the perceived information between the preprocessed traces and an S-box output at both the first and the last round. The corresponding results shown by Fig. 6 indicate the same observation as before, i.e., there exist more exploitable leakages by *Profile 2* compared to *Profile 3*. We should highlight that although the magnitude of perceived information compared to that of Fig. 5 is increased, performing a successful key-recovery attack on both of these two profiles is not an easy task. This is due to the high level of switching noise related to our round-based architectures as well as to the random precharge of BRAM output registers.

5   Conclusions

In this work we have taken an in-depth look at the wire-tap coding approach as a side-channel countermeasure with focus on AES and an FPGA as the target platform. Under certain assumptions and settings we have demonstrated the difficulties a hardware designer may face when implementing the basic modules of the cipher. The most challenging issue is how to realize the masked S-box, that is due to the S-box size and the underlying binary code length. As the encoded S-box cannot easily fit into the memory, we examined a couple of solutions, most of which turns the design into a sort of classical Boolean masking for S-box lookup. The problems we addressed here can be mitigated for other ciphers with a smaller S-box size, where the whole encoded S-box can be straightforwardly implemented as a look-up table.

We have shown that one of the nice features of the scheme as *the possibility to unmask without the knowledge of the mask* is only beneficial at the end of the cipher operations, which can be replaced by a simple XOR. Moreover, due to the intransparency of MixColumns to the underlying binary code the mask correction is unavoidable during the cipher-round computations. These two issues cause the scheme to be not much advantageous compared to classical Boolean masking. Further, our practical implementations on a Spartan-6 FPGA showed a considerable area overhead, i.e., 0.7 times more registers and 5 times more LUTs, compared to a corresponding design of classical Boolean masking which expectedly consumed less energy. Our practical side-channel analyses also indicated that the underlying scheme does not provide a higher level of resistance. Indeed, we have shown that the scheme might be vulnerable to certain attacks while the corresponding Boolean masking design is still robust.

In short, with respect to only power analysis and compared to Boolean masking we do not find a motivating advantage of the scheme – in our settings – as the circuit is more complicated, needs more energy, and is slightly less robust against power analysis attacks. However, we have not considered two advantage of the scheme in our analyses:

- Due to the properties of the binary linear codes, a higher-order version of the scheme can be made with moderate efforts.
- Expansion matrix L can randomly change. This results in requiring another module responsible for generation of a new L following its requirements.

Other matrices like P to multiply by 2, Inv to decode the ciphertext, sorting matrix O, and etc should also be dynamically obtained accordingly. If so, another source of randomness is needed for the system which optimistically complicates a key-recovery attacks.

- The scheme by nature can detect certain faults. Protecting against both DPA and fault attacks is of crucial interest as most of the known countermeasures can deal with only one of them. By the extensive evaluation given here we provided first a roadmap how to implement it, and second an overview about its DPA resistance in practice. As a result, this scheme might be a potential candidate to increase security against both DPA and fault attacks, that the later one should be carefully investigated to determine to which extend it can defeat fault attacks.

Acknowledgments. The author would like to thank Julien Bringer from Morpho (France), Stefan Heyse, Cornel Reuber, and Tobias Schneider from Ruhr University Bochum (Germany) for their helpful discussions and comments.

References

1. Akkar, M.-L., Giraud, C.: An Implementation of DES and AES, Secure against Some Attacks. In: Koç, Ç.K., Naccache, D., Paar, C. (eds.) CHES 2001. LNCS, vol. 2162, pp. 309–318. Springer, Heidelberg (2001)
2. Bhasin, S., Guilley, S., Heuser, A., Danger, J.-L.: From cryptography to hardware: analyzing and protecting embedded Xilinx BRAM for cryptographic applications. J. Cryptographic Engineering **3**(4), 213–225 (2013)
3. Bhasin, S., He, W., Guilley, S., Danger, J.-L.: Exploiting FPGA block memories for protected cryptographic implementations. In: ReCoSoC 2013, pp. 1–8. IEEE (2013)
4. Brier, E., Clavier, C., Olivier, F.: Correlation Power Analysis with a Leakage Model. In: Joye, M., Quisquater, J.-J. (eds.) CHES 2004. LNCS, vol. 3156, pp. 16–29. Springer, Heidelberg (2004)
5. Bringer, J., Chabanne, H., Le, T.H.: Protecting AES against side-channel analysis using wire-tap codes. J. Cryptographic Engineering **2**(2), 129–141 (2012)
6. Chari, S., Jutla, C.S., Rao, J.R., Rohatgi, P.: Towards Sound Approaches to Counteract Power-Analysis Attacks. In: Wiener, M. (ed.) CRYPTO 1999. LNCS, vol. 1666, pp. 398–412. Springer, Heidelberg (1999)
7. Coron, J.-S., Goubin, L.: On Boolean and Arithmetic Masking against Differential Power Analysis. In: Paar, C., Koç, Ç.K. (eds.) CHES 2000. LNCS, vol. 1965, pp. 231–237. Springer, Heidelberg (2000)
8. Durvaux, F., Standaert, F.-X., Veyrat-Charvillon, N.: How to Certify the Leakage of a Chip? In: Nguyen, P.Q., Oswald, E. (eds.) EUROCRYPT 2014. LNCS, vol. 8441, pp. 459–476. Springer, Heidelberg (2014)
9. Fumaroli, G., Martinelli, A., Prouff, E., Rivain, M.: Affine Masking against Higher-Order Side Channel Analysis. In: Biryukov, A., Gong, G., Stinson, D.R. (eds.) SAC 2010. LNCS, vol. 6544, pp. 262–280. Springer, Heidelberg (2011)
10. Gierlichs, B., Batina, L., Tuyls, P., Preneel, B.: Mutual Information Analysis. In: Oswald, E., Rohatgi, P. (eds.) CHES 2008. LNCS, vol. 5154, pp. 426–442. Springer, Heidelberg (2008)

11. Grassl M.: Code Tables: Bounds on the Parameters of Various Types of Codes (June 2008). http://www.codetables.de/
12. Grosso, V., Standaert, F.-X., Prouff, E.: Low Entropy Masking Schemes, Revisited. In: Francillon, A., Rohatgi, P. (eds.) CARDIS 2013. LNCS, vol. 8419, pp. 33–43. Springer, Heidelberg (2014)
13. Güneysu, T., Moradi, A.: Generic Side-Channel Countermeasures for Reconfigurable Devices. In: Preneel, B., Takagi, T. (eds.) CHES 2011. LNCS, vol. 6917, pp. 33–48. Springer, Heidelberg (2011)
14. Kocher, P.C.: Timing Attacks on Implementations of Diffie-Hellman, RSA, DSS, and Other Systems. In: Koblitz, N. (ed.) CRYPTO 1996. LNCS, vol. 1109, pp. 104–113. Springer, Heidelberg (1996)
15. Kocher, P.C., Jaffe, J., Jun, B.: Differential Power Analysis. In: Wiener, M. (ed.) CRYPTO 1999. LNCS, vol. 1666, pp. 388–397. Springer, Heidelberg (1999)
16. Maghrebi, H., Guilley, S., Danger, J.-L.: Leakage Squeezing Countermeasure against High-Order Attacks. In: Ardagna, C.A., Zhou, J. (eds.) WISTP 2011. LNCS, vol. 6633, pp. 208–223. Springer, Heidelberg (2011)
17. Mangard, S., Oswald, E., Popp, T.: Power Analysis Attacks: Revealing the Secrets of Smart Cards. Springer (2007)
18. Mangard, S., Pramstaller, N., Oswald, E.: Successfully Attacking Masked AES Hardware Implementations. In: Rao, J.R., Sunar, B. (eds.) CHES 2005. LNCS, vol. 3659, pp. 157–171. Springer, Heidelberg (2005)
19. Moradi, A.: Wire-Tap Codes as Side-Channel Countermeasure - an FPGA-based experiment. Cryptology ePrint Archive, Report 2014/716 (2014). http://eprint.iacr.org/
20. Moradi, A., Mischke, O., Eisenbarth, T.: Correlation-Enhanced Power Analysis Collision Attack. In: Mangard, S., Standaert, F.-X. (eds.) CHES 2010. LNCS, vol. 6225, pp. 125–139. Springer, Heidelberg (2010)
21. Morita Tech. Side-channel AttacK User Reference Architecture (SAKURA), Further information are available via http://www.morita-tech.co.jp/SAKURA/en/index.html
22. Nikova, S., Rijmen, V., Schläffer, M.: Secure Hardware Implementation of Nonlinear Functions in the Presence of Glitches. J. Cryptology 24(2), 292–321 (2011)
23. Ozarow, L.H., Wyner, A.D.: Wire-Tap Channel II. In: Beth, T., Cot, N., Ingemarsson, I. (eds.) EUROCRYPT 1984. LNCS, vol. 209, pp. 33–50. Springer, Heidelberg (1985)
24. Prouff, E., Roche, T.: Higher-Order Glitches Free Implementation of the AES Using Secure Multi-party Computation Protocols. In: Preneel, B., Takagi, T. (eds.) CHES 2011. LNCS, vol. 6917, pp. 63–78. Springer, Heidelberg (2011)
25. Standaert, F.-X., Malkin, T.G., Yung, M.: A Unified Framework for the Analysis of Side-Channel Key Recovery Attacks. In: Joux, A. (ed.) EUROCRYPT 2009. LNCS, vol. 5479, pp. 443–461. Springer, Heidelberg (2009)
26. TELECOM ParisTech SEN research group. DPA Contest (4th edn.) (2013-2014). http://www.DPAcontest.org/v4/
27. von Willich, M.: A Technique with an Information-Theoretic Basis for Protecting Secret Data from Differential Power Attacks. In: Honary, B. (ed.) Cryptography and Coding 2001. LNCS, vol. 2260, p. 44. Springer, Heidelberg (2001)
28. Wyner, A.D.: The Wire-Tap Channel. Bell System Technical Journal 54(8), 1355–1387 (1975)
29. Xilinx. Spartan-6 Libraries Guide for HDL Designs (April 2012), http://www.xilinx.com/support/documentation/sw_manuals/xilinx14_1/spartan6_hdl.pdf

Differential Power Analysis in Hamming Weight Model: How to Choose among (Extended) Affine Equivalent S-boxes

Sumanta Sarkar, Subhamoy Maitra[✉], and Kaushik Chakraborty

Indian Statistical Institute, 203 B T Road, Kolkata 700 108, India
{sumanta.sarkar,kaushik.chakraborty9}@gmail.com,
subho@isical.ac.in

Abstract. From the first principle, we concentrate on the *Differential Power Analysis* (DPA) in the Hamming weight model. Based on the power related data of an (n, n) permutation S-box, we propose a spectrum (we call it Relative Power Spectrum, RPS in short) at 2^n points each providing a vector containing n coordinates. Each coordinate contains the data related to single-bit DPA, and taking them together we provide relevant results in the domain of multi-bit DPA. For two affine equivalent (n, n) permutation S-boxes F and G, such that $G(x) = F(Ax \oplus b)$, where A is a linear permutation (nonsingular binary matrix) and b is an n-bit vector, the RPSs of F and G are permutations of each other. However, this is not true in general when F and G are affine or extended affine equivalent, i.e., $G(x) = B(F(Ax \oplus b)) \oplus L(x) \oplus c$, where B is a linear permutation, L is a linear mapping, and c is an n-bit vector. In such a case, the RPSs of F and G may not be related by permutation and may contain completely different vectors. We provide the effect of this in terms of DPA both in noise-free and noisy scenarios. Our results guide the designer to choose one S-box among all those in the same (extended) affine equivalence class when DPA in the Hamming weight model is considered. This is an instance where cryptographic advantage is attained by applying (extended) affine equivalence. For example, we provide a family of S-boxes that should replace the $(4, 4)$ S-boxes proposed in relation to the PRINCE block cipher.

Keywords: Cross-correlation · Differential Power Analysis (DPA) · (Extended) Affine Equivalence · Prince · Permutation S-box

1 Introduction

Differential Power Analysis (DPA) is one important area in the field of side-channel attacks in which the information about the secret key is leaked through power traces while the encryption (or decryption) is being executed on the cryptographic platform. The strengths of such attacks are naturally much higher than linear or differential cryptanalysis due to the additional information related to the power traces. It is evident that the S-boxes in block ciphers would be the

© Springer International Publishing Switzerland 2014
W. Meier and D. Mukhopadhyay (Eds.): INDOCRYPT 2014, LNCS 8885, pp. 360–373, 2014.
DOI: 10.1007/978-3-319-13039-2_21

prime target of DPA. From the designers point of view, the S-boxes should be chosen carefully such that they should have high DPA resilience in addition to the resistance to other classical cryptanalytic attacks such as linear and differential cryptanalysis. Unfortunately, recent research in this area [5,6,9,10] provides some results to believe that the S-boxes with good autocorrelation and nonlinearity properties are more vulnerable against DPA. It is also well understood that constant or linear functions, that are good against DPA under certain measures, cannot be used in a cryptosystem.

In [7], the theoretical resistance of AES and DES S-boxes to linear cryptanalysis vis a vis DPA attacks in terms of signal-to-noise ratio (SNR) was investigated. Then, an attempt to quantify the DPA resilience of the S-boxes was made in [10], where the parameter Transparency Order was introduced. The work of [10] has been revisited recently in [6]. Kocher et al [8] presented Differential Power Analysis (DPA) as one form of side channel attack that observes the difference between the power consumed by a single gate when its output flips. Consider that the adversary obtains sufficiently large collection of ciphertexts $E_{\dot{K}}(x)$, where $\dot{K}$ is the round key, E is the encryption function, and x is the corresponding plaintext. The power traces $T_x(t)$ (for example, a series of power related data based on time or may be some summarized data) are collected for a sufficient number of plaintexts x. These samples provide the information about the power consumed by each gate when the output flips. To collect the power traces, the attacker doesn't need any information about the plaintexts, and it is enough to record the power consumption during the computation of E. Based on these power traces, the key recovery can be done off-line. In the single bit DPA scenario, a particular coordinate j is targeted and the attacker tries to build a distinguisher by partitioning the power traces into two bins. This partitioning is done by considering whether the output bit value of the j-th coordinate function of the S-box is zero or one, given the guessed key. Let $\dot{K}$ be the actual round key. The attacker guesses a key K and then the traces are assigned in one of the two bins, say S_0 and S_1, according to the output bit value of the j-th coordinate function. One can compute the Differential Trace [7], denoted by $D_{K,j}$ as

$$D_{K,j} = \frac{1}{|S_1|} \sum_{T_x \in S_1} T_x - \frac{1}{|S_0|} \sum_{T_x \in S_0} T_x, \qquad (1)$$

where T_x is the power trace captured from the output corresponding to the input x. The quantity $D_{K,j}$ works as a distinguisher in this model. According to the theory proposed in [8], the distribution $D_{K,j}$ should show a peak for the correct key $K = \dot{K}$ with a good probability. Before proceeding further, let us provide some background that will help in understanding the scenario better.

1.1 Background

Let $\mathbb{F}_2^n$ be the vector space that contains all the n-bit binary vectors. A (single output) Boolean function of n variables is a mapping from $\mathbb{F}_2^n$ into $\mathbb{F}_2$. We denote the set of n-variable Boolean functions as $\mathcal{B}_n$. The support of a Boolean function

f is defined as $Supp(f) = \{x \in \mathbb{F}_2^n | f(x) = 1\}$. When we use a Boolean function as a cryptographic primitive, we generally consider the functions which output 0 and 1 with equal probability. Thus, we generally consider functions in $\mathcal{B}_n$ for which the cardinality of the support is 2^{n-1}. These are known as balanced functions. By $H(u)$, we mean the number of 1's in the binary string u. Thus for a balanced function $f \in \mathcal{B}_n$, $H(f) = 2^{n-1}$.

An S-box can be seen as a multi-output Boolean function. An (n, m) S-box is a mapping $F = (F_1, \ldots, F_m) : \mathbb{F}_2^n \to \mathbb{F}_2^m$, where each $F_i \in \mathcal{B}_n$, and termed as the coordinate function of F. We may represent the truth table of an (n, n) permutation S-box as a permutation of the set of all n-bit integers. For example, a $(4, 4)$ S-box may be represented as c480ea756fbd1329; here the 4-bits are written in the hexadecimal format and the $2^4 = 16$ outputs are listed in natural order corresponding to the inputs.

As discussed, the DPA provides the attacker a verifier or distinguisher to guess the correct key. The distinguisher, based on the differential traces, works on the hypothesis that for the correct key it takes the maximum value. Now, from the designer's point of view, one should design the S-box in such a manner so that the distribution of the differential traces (for all keys) becomes almost uniform. In such a case, the correct key cannot be distinguished from a wrong key. This gives a security criterion for designing the S-boxes.

The Hamming weight model, introduced in [4], assumes that the power consumption for the output y is proportional to the Hamming weight of y. Therefore, given the input x to the S-box, the power consumption $P_{F(x \oplus \dot{K}) \oplus \beta}$ for the computation $F(x \oplus \dot{K}) \oplus \beta$, where β is the value of the precharge logic, would be $cH(F(x \oplus \dot{K}) \oplus \beta)$, for some constant c. Therefore in the presence of noise, the power observed would be $cH(F(x \oplus \dot{K}) \oplus \beta) + \mathcal{N}$, where by $\mathcal{N}$ we denote the effect of noise. We can normalize the observed power as $H(F(x \oplus \dot{K}) \oplus \beta) + \mathcal{N}/c$ and simply (with a little abuse of notation) we can rewrite the observed power as $H(F(x \oplus \dot{K}) \oplus \beta) + \mathcal{N}$. Let us also briefly point out the relevance of β. The precharge logic is used in case of microcontrollers in which there is a precharge phase where, in the initial part of the clock cycle, the registers are initialized to some fixed value in each round. That is, given the precharge logic assumption for the hardware design, $\beta \in \{0, 1\}^n$ is considered to be a constant.

When the attacker runs the S-box for a guessed key K, and the targeted coordinate j, suppose that the bin S_b contains $H(F(x \oplus \dot{K}) \oplus \beta) + \mathcal{N}$ if the output $F(x \oplus \dot{K}) = b$, for $b \in \{0, 1\}$. Then from (1),

$$D_{K,j}(F) = \frac{1}{|S_1|} \sum_{F_j(x \oplus K)=1} (H(F(x \oplus \dot{K}) \oplus \beta) + \mathcal{N})$$

$$- \frac{1}{|S_0|} \sum_{F_j(x \oplus K)=0} (H(F(x \oplus \dot{K}) \oplus \beta) + \mathcal{N}).$$

In general, a permutation S-box is chosen only (though there are some notable exceptions, for example, DES S-boxes), i.e., $m = n$, and so $|S_0| = |S_1| = 2^{n-1}$.

Therefore, for each $j \in \{1\ldots,n\}$,

$$D_{K,j}(F) = \frac{1}{2^{n-1}}[\sum_{F_j(x \oplus K)=1} (H(F(x \oplus \dot{K}) \oplus \beta) + \mathcal{N})$$

$$- \sum_{F_j(x \oplus K)=0} (H(F(x \oplus \dot{K}) \oplus \beta) + \mathcal{N})]. \qquad (2)$$

Henceforth, we will only be considering permutation S-boxes.

To understand and analyze the properties of Boolean functions, several spectra are exploited. We explain a few of those now. Let $x = (x_1,\ldots,x_n)$ and $\omega = (\omega_1,\ldots,\omega_n)$ both belong to $\mathbb{F}_2^n$, the inner product of x and ω is defined as

$$x \cdot \omega = x_1\omega_1 \oplus \cdots \oplus x_n\omega_n.$$

The Walsh transform of $f(x)$ is an integer valued function over $\mathbb{F}_2^n$ which is defined as

$$W_f(\omega) = \sum_{x \in \{0,1\}^n} (-1)^{f(x) \oplus x \cdot \omega}.$$

The autocorrelation transform of $f(x)$ is again an integer valued function over $\mathbb{F}_2^n$ which is defined as

$$\mathcal{A}_f(\omega) = \sum_{x \in \{0,1\}^n} (-1)^{f(x) \oplus f(x \oplus \omega)}.$$

The Walsh and autocorrelation spectra are important properties in designing Boolean functions that are used as cryptographic primitives. Given $f_1, f_2 \in \mathcal{B}_n$, we define the cross-correlation spectrum between these functions as

$$\mathcal{C}_{f_1,f_2}(\omega) = \sum_{x \in \{0,1\}^n} (-1)^{f_1(x) \oplus f_2(x \oplus \omega)}.$$

In general, it is expected that the maximum absolute value in each of such spectrum should be low for better resistance against cryptanalysis.

Two (n,n) S-boxes F and G are called affine equivalent if for all x, they follow the relation

$$G(x) = B(F(Ax \oplus b)) \oplus c,$$

where A and B are two linear permutations and b and c are two n-bit binary vectors. There is another equivalence criterion for S-boxes that is known as the extended affine equivalence (EA equivalence). Two (n,n) S-boxes F and G are extended affine equivalent if

$$G(x) = B(F(Ax \oplus b)) \oplus L(x) \oplus c,$$

where L is a linear function. It is important to know if two S-boxes are affine or extended affine equivalent, since many cryptographic properties of the S-boxes are invariant under these equivalence relations. It is easy to verify that non-linearity and the maximum absolute value in the autocorrelation spectrum are invariant for the affine or extended affine equivalent S-boxes. Similarly algebraic degree is also invariant under affine or extended affine equivalence of the S-boxes having degree more than 1. The reason is that nonlinearity and absolute autocorrelation values do not alter if linear functions are added to component functions. The same argument holds for the degree if it is more than 1.

Before getting into the technical results, let us provide a brief outline of this paper.

1.2 Organization and Contribution

In this introductory section, we have provided the motivation and necessary background towards this work. In the next section (Section 2), we consider DPA in the noise free scenario. We define the *Relative Power Spectrum* (RPS) of an S-box which comprises of the RPS vectors. The RPS vector corresponding to the correct key is of the form $((-1)^{\beta_1}, \ldots, (-1)^{\beta_n})$ for an (n, n) S-box, where $\beta = (\beta_1, \ldots, \beta_n)$ is the precharge logic. We present some theoretical results in this section that provide better understanding of RPS and how the spectrum is related to DPA.

In Section 3, we consider DPA in the presence of noise. We define the metric *distance* which is the Euclidean distance of an RPS (Relative Power Spectrum) vector from $((-1)^{\beta_1}, \ldots, (-1)^{\beta_n})$. If there had been no noise, the RPS vector corresponding to the correct key would have the least distance, (i.e., 0). However, in the presence of noise, the RPS vector corresponding to the correct key may shift away, and some other RPS vector may become closest to $((-1)^{\beta_1}, \ldots, (-1)^{\beta_n})$ (for example, see Table 4). We also define the rank of a key, which is the position of the RPS vector for the correct key from the vector $((-1)^{\beta_1}, \ldots, (-1)^{\beta_n})$ in terms of distance. Therefore, in the presence of noise, if the correct key has rank higher than 1, then the attacker needs to consider keys that are positioned before. If any vector (other than the vector corresponding to the correct key) in the RPS is very close to $((-1)^{\beta_1}, \ldots, (-1)^{\beta_n})$, then it is highly likely that the wrong key corresponding to that distance may topple the correct key in the noisy scenario. In such a case, it will be difficult for the attacker to locate the correct key.

We also note that the RPS, in general, is not invariant under (extended) affine equivalence. This motivated us to obtain the best possible S-box in the same (extended) affine equivalence class that may resist the DPA. In this direction we study all the $(4, 4)$ permutation S-boxes with nonlinearity 4, degree 3 and absolute autocorrelation value 8. Our result provide a family of S-boxes that may replace the S-boxes proposed in relation to the PRINCE [3] block cipher for better resistance against DPA. We also study the AES [1] S-box and some of its (extended) affine equivalent permutations in this direction. Section 4 concludes the paper.

2   Relative Power Spectrum in the Noise Free Scenario

In this section we analyze the values of $D_{K,j}(F)$ in the noise free scenario. In this case (2) becomes

$$D_{K,j}(F) = \frac{1}{2^{n-1}}\left(\sum_{F_j(x \oplus K)=1} H(F(x \oplus \dot{K}) \oplus \beta) - \sum_{F_j(x \oplus K)=0} H(F(x \oplus \dot{K}) \oplus \beta) \right) \quad (3)$$

First let us provide the exact value of $D_{\dot{K},j}(F)$.

Proposition 1. *For an (n,n) permutation S-box $F : \mathbb{F}_2^n \to \mathbb{F}_2^n$, $D_{\dot{K},j}(F) = (-1)^{\beta_j}$, for every coordinate $j \in \{1,\ldots,n\}$.*

Proof. e have

$$D_{K,j}(F) = \frac{1}{2^{n-1}}[\sum_{F_j(x \oplus K)=1} H(F(x \oplus \dot{K}) \oplus \beta) - \sum_{F_j(x \oplus K)=0} H(F(x \oplus \dot{K}) \oplus \beta)]$$

$$= -\frac{1}{2^{n-1}} \sum_{x \in \mathbb{F}_2^n} (-1)^{F_j(x \oplus K)} H(F(x \oplus \dot{K}) \oplus \beta)$$

$$= -\frac{1}{2^{n-1}} \sum_{x \in \mathbb{F}_2^n} (-1)^{F_j(x \oplus K)} \left(\frac{n}{2} - \frac{1}{2} \sum_{i=1}^{n} (-1)^{F_i(x \oplus \dot{K}) \oplus \beta_i} \right)$$

$$= -\frac{n}{2^n} \sum_{x \in \mathbb{F}_2^n} (-1)^{F_j(x \oplus K)} + \frac{1}{2^n} \sum_{x \in \mathbb{F}_2^n} \sum_{i=1}^{n} (-1)^{F_j(x \oplus K) \oplus F_i(x \oplus \dot{K}) \oplus \beta_i}$$

$$= \frac{1}{2^n} \sum_{i=1}^{n} (-1)^{\beta_i} \mathcal{C}_{F_i,F_j}(K \oplus \dot{K}),$$

since F is permutation, so each F_j is balanced. Suppose $K \oplus \dot{K} = a$, then we can write

$$D_{K,j}(F) = \frac{1}{2^n} \sum_{i=1}^{n} (-1)^{\beta_i} \mathcal{C}_{F_i,F_j}(K \oplus \dot{K}) = \frac{1}{2^n} \sum_{i=1}^{n} (-1)^{\beta_i} \mathcal{C}_{F_i,F_j}(a). \quad (4)$$

If $K = \dot{K}$, i.e., $a = 0$, then $D_{\dot{K},j}(F)$ can be written as,

$$D_{\dot{K},j}(F) = \frac{1}{2^n} \sum_{i=1}^{n} (-1)^{\beta_i} \mathcal{C}_{F_i,F_j}(0)$$

$$= \frac{1}{2^n}[\sum_{i=1, i \neq j}^{n} (-1)^{\beta_i} \mathcal{C}_{F_i,F_j}(0) + (-1)^{\beta_j} \mathcal{A}_{F_j}(0)]$$

$$= \frac{1}{2^n}[\sum_{i=1, i \neq j}^{n} (-1)^{\beta_i} \cdot 0 + (-1)^{\beta_j} 2^n], \text{ since } F \text{ is a permutation}$$

$$= (-1)^{\beta_j}.$$

$\square$

As the value of $D_{K,j}(F)$ is $(-1)^{\beta_j}$ for $K = \dot{K}$, therefore, for the targeted coordinate j, if there is only one key K such that $|D_{K,j}(F)| = 1$, then the attacker simply decides that $\dot{K} = K$. If not, then the attacker chooses another coordinate j and takes the same strategy to single out the correct key. We now analyze the set of all $\{D_{K,j}(F)\}_{K \in \mathbb{F}_2^n, j \in [1,n]}$ in this direction.

Definition 1 (Relative Power Spectrum (RPS)). *For a permutation S-box* $F : \mathbb{F}_2^n \to \mathbb{F}_2^n$, *we call the set of all n-tuples*

$$\mathcal{D}_F = \{(D_{K,1}(F), \dots, D_{K,n}(F))\}_{K \in \mathbb{F}_2^n}$$

as the RPS of F. We also name each n-tuple $(D_{K,1}(F), \dots, D_{K,n}(F))$ *as the RPS Vector corresponding to the key K.*

The RPS vector for the correct key $\dot{K}$ is $((-1)^{\beta_1}, \dots, (-1)^{\beta_n})$, where β is the precharge logic. In Table 1, we present the relative power spectrum of a $(4, 4)$ S-box $F = (F_1, \dots, F_4)$ for $\beta = 0$, where the S-box F can be described as 084c2a6e1953bdf7. For instance, when $K = (1, 0, 0, 0) \oplus \dot{K}$ the value of $D_{K,3}(F)$ is 1.25, and when $K = (1, 1, 1, 0) \oplus \dot{K}$, the value of $D_{K,3}(F)$ is -1.25. When $K \oplus \dot{K} = 0$, that corresponds to the relative power spectrum vector for the correct key $\dot{K}$, which is $(1, 1, 1, 1)$ in this case.

Table 1. RPS of F, where $\beta = 0$

$K \oplus \dot{K}$	$D_{K,1}(F)$	$D_{K,2}(F)$	$D_{K,3}(F)$	$D_{K,4}(F)$
0 0 0 0	1.0	1.0	1.0	1.0
0 0 0 1	−0.5	1.0	1.0	1.0
0 0 1 0	1.0	−0.5	1.0	1.0
0 0 1 1	0.0	0.0	0.5	1.0
0 1 0 0	0.0	0.5	−1.0	1.0
0 1 0 1	−1.0	0.0	−0.5	1.0
0 1 1 0	0.0	−1.0	−1.0	1.0
0 1 1 1	−0.5	−1.0	−1.0	1.0
1 0 0 0	0.0	0.75	1.25	−1.0
1 0 0 1	0.0	0.75	1.25	−1.0
1 0 1 0	0.5	−0.25	0.75	−1.0
1 0 1 1	0.5	−0.25	0.75	−1.0
1 1 0 0	−0.5	0.25	−0.75	−1.0
1 1 0 1	−0.5	0.25	−0.75	−1.0
1 1 1 0	0.0	−0.75	−1.25	−1.0
1 1 1 1	0.0	−0.75	−1.25	−1.0

While the result that the RPS vector for the correct key is $((-1)^{\beta_1}, \dots, (-1)^{\beta_n})$ might have been folklore, we have not seen this in any published work. The value of K for which $D_{K,j}(\cdot)$ attains the maximum [8, Page 5], was assumed to be the correct key. However, our result shows that the correct key corresponds to a unique value, i.e., $(-1)^{\beta_j}$ in the single bit DPA, and $((-1)^{\beta_1}, \dots, (-1)^{\beta_n})$ in the multi-bit DPA. In the noise-free scenario, one may get higher absolute value of $D_{K,j}(\cdot)$ for a wrong key than the correct key. Therefore, in the noisy scenario the attacker should consider the closest distance from $((-1)^{\beta_1}, \dots, (-1)^{\beta_n})$. This motivates us to define the distance profile later in Definition 2, Section 3.

2.1 RPS of S-boxes in the same (Extended) Affine Equivalence Classes

Two (n, n) S-boxes F and G are affine equivalent if there exist two linear permutations A and B, and two n-bit vectors b and c such that $G(x) = B(F(Ax \oplus b)) \oplus c$, for all $x \in \mathbb{F}_2^n$. If F and G are such that $G(x) = B(F(Ax \oplus b)) \oplus L(x) \oplus c$, for all $x \in \mathbb{F}_2^n$, where L is a linear function, then F and G are said to have extended affine equivalence. As discussed in the background, some cryptographic properties are invariant for these two equivalence relations.

We now examine whether RPS too remain invariant under affine and EA equivalence.

Theorem 1. *uppose F and G are two permutation S-boxes that have the following affine equivalence relation $G(x) = F(Ax \oplus b)$. Then the RPS of F and G are permutations of each other.*

Proof. ince $D_{K,j}(F) = \frac{1}{2^n} \sum_{i=1}^n \mathcal{C}_{F_i, F_j}(a)$, the result will depend on the behaviour of $\mathcal{C}_{F_i, F_j}(a)$ under the affine equivalence relation.

Suppose $F = (F_1, \ldots, F_n)$ and $G = (G_1, \ldots, G_n)$. We have

$$\mathcal{C}_{G_i, G_j}(a) = \sum_{x \in \mathbb{F}_2^n} (-1)^{G_i(x) \oplus G_j(x \oplus a)} = \sum_{x \in \mathbb{F}_2^n} (-1)^{F_i(Ax \oplus b) \oplus F_j(A(x \oplus a) \oplus b)}$$

$$= \sum_{A^{-1} x \in \mathbb{F}_2^n} (-1)^{F_i(x \oplus b) \oplus F_j(x \oplus Aa \oplus b)} = \sum_{x \in \mathbb{F}_2^n} (-1)^{F_i(x \oplus b) \oplus F_j(x \oplus Aa \oplus b)}$$

$$= \sum_{x \oplus b \in \mathbb{F}_2^n} (-1)^{F_i(x) \oplus F_j(x \oplus Aa)} = \sum_{x \in \mathbb{F}_2^n} (-1)^{F_i(x) \oplus F_j(x \oplus Aa)} = \mathcal{C}_{F_i, F_j}(Aa).$$

The proof follows from this. ∎

On the other hand, if F and G do not follow the relation as given in Theorem 1, i.e., they are related by the relation $G(x) = B(F(Ax \oplus b)) \oplus L(x) \oplus c$ for some linear functions B, where the following do not hold simultaneously, (i) B is an identity mapping, (ii) L is the zero mapping, (iii) c is zero, then the RPSs are not necessarily permutations of each other. That means for (extended) affine equivalence, the RPS is not invariant in general.

Remark 1. From Theorem 1, we have $D_{K,j}(G) = \frac{1}{2^n} \sum_{i=1}^n \mathcal{C}_{F_i, F_j}(Aa)$. If we consider the relation $G'(x) = F(Ax \oplus b) + c$ for $c = (c_1, \ldots, c_n)$, then $\mathcal{C}_{G'_i, G'_j}(a) = (-1)^{c_i \oplus c_j} \mathcal{C}_{F_i, F_j}(Aa)$. So we get

$$D_{K,j}(G') = \frac{1}{2^n} \sum_{i=1}^n (-1)^{c_i \oplus c_j} \mathcal{C}_{F_i, F_j}(Aa) = \frac{1}{2^n} (-1)^{c_j} \sum_{i=1}^n (-1)^{c_i} \mathcal{C}_{F_i, F_j}(Aa).$$

Thus, even after $G' = G \oplus c$, one may note that in general $D_{K,j}(G)$ and $D_{K,j}(G')$ are not equal due to the involvement of $(-1)^{c_i}$ inside the sum and the value may change. This leads to our understanding that the RPS of two (extended) affine equivalent permutation S-boxes may not be permutation of each other. Similarly, this can be observed in general for extended affine equivalence relation between two permutation S-boxes. For example, take the $(4, 4)$ S-boxes F with the hexadecimal representation $F = 084c2a1563efbd97$ and $G = 0c842e19a367f5db$ that have the relation $G(x) = B(F(x))$, where B is the linear function associated to the matrix $\begin{bmatrix} 1 & 1 & 0 & 0 \\ 1 & 0 & 0 & 0 \\ 0 & 0 & 1 & 0 \\ 0 & 0 & 0 & 1 \end{bmatrix}$. The vector $(0.5, 0.5, 0.75, 0.75)$ is present in the RPS of F but not in the RPS of G.

3   Relative Power Spectrum in the Presence of Noise

As we have seen the RPS vector corresponding to the correct key is $((-1)^{\beta_1}, \ldots, (-1)^{\beta_n})$, we study the Euclidean distances from $((-1)^{\beta_1}, \ldots, (-1)^{\beta_n})$ to all the RPS vectors in the RPS. The set of all distances for all the possible keys is defined as the distance profile.

Definition 2 (Distance profile of an S-box).   *For an (n, n) permutation S-box F, we define the set*

$$\{\frac{1}{n}\sqrt{[(-1)^{\beta_1} - D_{K,1}(F))^2 + \cdots + ((-1)^{\beta_n} - D_{K,n}(F)]^2} | K \in \mathbb{F}_2^n\}$$

as the distance profile of the S-box F.

Obviously, the minimum distance in the set is 0, which occurs for the correct key.

Remark 2. The S-box that has the second minimum distance close to 0 is important in the presence of noise as it is likely that the corresponding vector will get closer to $((-1)^{\beta_1}, \ldots, (-1)^{\beta_n})$. The vector corresponding to the correct key, i.e., $((-1)^{\beta_1}, \ldots, (-1)^{\beta_n})$ itself may shift further away in the noisy scenario, thus garbling the attacker's option to choose the correct key.

   Given an (n, n) permutation S-box F, the second minimum distance is denoted as $\mathcal{D}2_n(F)$.

For explaining our analysis and simulation results, we concentrate on $(4, 4)$ S-boxes. Such an S-box has been used in PRINCE [3] block cipher recently and a family of such permutation S-boxes has been presented in [3]. The number of $(4, 4)$ S-boxes up to affine equivalence is 300 as given in [2]. Among them there are 10 S-boxes that have nonlinearity 4, degree 3 and absolute autocorrelation value 8 which are of the same quality of the 8 S-boxes provided in [3]. These 10 S-boxes from [2] are presented in Table 2.

   We determine the distance profile of all these 10 S-boxes. Below we present the S-box for which the second minimum distance $\mathcal{D}2_4(\cdot)$ is the lowest among others. One such S-box is 084c2a1563db79ef, the third one in Table 2. We present RPS and the corresponding distances in Table 3.

   As we mentioned earlier that the RPS may differ for (extended) affine equivalent S-boxes, therefore we checked all possible permutation S-boxes that are

Table 2. All $(4,4)$ S-box up to affine equivalence with nonlinearity 4, degree 3 and absolute autocorrelation value 8

084c2a1563efbd97, 084c2613a9db75ef, 084c2a1563db79ef, 084c261da937b5ef, 084c261da39b75ef, 084c261da3be9f57, 084c2a1563ef7d9b, 084c261da3be5f97, 084c261da3bef975, 084c261da39b7e5f.

Table 3. Relative power spectrum and the distances of the S-box F which is 084c2a1563db79ef

$K \oplus \tilde{K}$	$D_{K,1}(F)$	$D_{K,2}(F)$	$D_{K,3}(F)$	$D_{K,4}(F)$	Distance of vector
0 0 0 0	1.0	1.0	1.0	1.0	0.000
0 0 0 1	0.5	0.5	0.75	0.75	0.197
0 0 1 0	0.5	0.0	0.75	0.75	0.293
0 0 1 1	−0.5	0.0	1.0	1.0	0.450
0 1 0 0	1.0	0.75	0.5	0.75	**0.153**
0 1 0 1	0.25	0.5	0.25	1.0	0.293
0 1 1 0	0.5	0.25	0.75	0.5	0.265
0 1 1 1	−0.25	0.0	1.0	0.25	0.441
1 0 0 0	−0.25	−0.25	−1.0	0.0	0.712
1 0 0 1	−0.25	0.25	−1.0	−0.5	0.723
1 0 1 0	−0.25	−1.25	−0.25	−0.75	0.838
1 1 0 0	−0.75	−0.25	−0.25	−1.25	0.838
1 1 0 0	−0.25	−0.25	−1.0	−0.5	0.765
1 1 0 1	−0.5	0.5	−1.0	−1.0	0.810
1 1 1 0	−0.25	−1.25	−0.75	−0.75	0.892
1 1 1 1	−0.5	−0.5	−0.75	−1.25	0.888

EA equivalent to each one given in Table 2. We find that for every S-box in Table 2, the minimum value of $\mathcal{D}2_4(\cdot)$ is 0.125 in their respective EA equivalent classes. For example, for the fifth S-box in Table 3, i.e., 084c261da39b75ef (say F), has $\mathcal{D}2_4(F) = 0.293$. There is an EA equivalent S-box G (c480ea756fbd1329) related to F by the relation $G(x) = B(F(x)) \oplus L(x) \oplus c$, where B and L are the linear functions associated to the matrices $\begin{bmatrix} 1 & 0 & 0 & 0 \\ 1 & 1 & 0 & 0 \\ 1 & 1 & 1 & 0 \\ 0 & 1 & 1 & 1 \end{bmatrix}$ and $\begin{bmatrix} 0 & 0 & 1 & 0 \\ 1 & 1 & 0 & 1 \\ 0 & 0 & 1 & 0 \\ 1 & 1 & 0 & 1 \end{bmatrix}$ respectively, and $c = (0,1,0,0)$. We have $\mathcal{D}2_4(G) = 0.125$. The RPS of G is presented in the upper half of Table 4.

In this direction, we have also studied the AES S-box. Since this is an $(8,8)$ S-box, it is not possible to search through all the S-boxes in its (extended) affine equivalence class. However, we made some partial search. For the AES S-box, say F, $\mathcal{D}2_8(F) = 0.291$. We only considered the affine equivalent S-boxes of the form $F \oplus c$, where $c \in \mathbb{F}_2^8$. We obtained a permutation G among these, such that $\mathcal{D}2_8(G) = 0.251$. This shows that reducing the second minimum distance is indeed possible and one may try to obtain further reduction with more search effort.

3.1 Impact of the Noise on the Distance Metric

In this section we will see how the presence of noise can change the distance metric, i.e., we consider $\mathcal{N}$ in Equation 2. First we define a notion of rank of the correct key.

Definition 3 (Key rank). *For a permutation S-box with precharge logic β, the rank of a key is the position of the RPS vector corresponding to that key from the vector $((-1)^{\beta_1}, \ldots, (-1)^{\beta_n})$ in the distance profile of the said S-box.*

We have already shown that when there is no noise, the rank of the correct key is always 1. However, the scenario may change in the presence of noise.

As in [11], we consider the noise to be uniform for the simulation purpose. Because of the noise, the RPS vector corresponding to the correct key may be shifted away from $((-1)^{\beta_1}, \ldots, (-1)^{\beta_n})$, and some other vector may come closer. That means whenever, the correct key does not have rank 1, the attacker fails to recover the correct key uniquely.

Table 4. RPS and the distances of the S-box G=c480ea756fbd1329 for $\beta = (0, 0, 0, 0)$ in noise-free (top, rank of the correct key is 1) and noisy (bottom, rank of the correct key is 3)

$K \oplus \hat{K}$	$D_{K,1}(G)$	$D_{K,2}(G)$	$D_{K,3}(G)$	$D_{K,4}(G)$	Distance of vector
0 0 0 0	1.0	1.0	1.0	1.0	0.000
0 0 0 1	−0.25	0.75	0.75	0.25	0.375
0 0 1 0	0.75	0.75	0.75	0.75	0.125
0 0 1 1	0.0	0.0	1.0	0.5	0.375
0 1 0 0	0.25	−0.75	−0.25	−0.25	0.649
0 1 0 1	−0.5	−1.25	−0.75	−0.5	0.888
0 1 1 0	0.0	−0.5	−1.0	0.5	0.684
0 1 1 1	−1.25	−1.0	−0.5	−0.25	0.897
1 0 0 0	−1.0	−0.75	−0.75	−1.0	0.939
1 0 0 1	0.0	−0.25	−0.75	−0.5	0.701
1 0 1 0	−0.75	−1.0	−1.0	−0.75	0.939
1 1 0 0	0.25	−0.5	−1.0	−0.25	0.723
1 1 0 0	−0.25	0.5	0.75	0.0	0.423
1 1 0 1	0.75	1.25	0.5	0.5	0.197
1 1 1 0	0.0	0.75	0.5	−0.25	0.423
1 1 1 1	1.0	1.0	0.75	0.25	0.197
$K \oplus \hat{K}$	$D_{K,1}(G)$	$D_{K,2}(G)$	$D_{K,3}(G)$	$D_{K,4}(G)$	Distance of vector
0 0 0 0	0.82	0.62	1.44	0.91	0.153
0 0 0 1	−0.55	0.43	1.042	0.18	0.461
0 0 1 0	1.11	0.97	0.60	0.43	0.176
0 0 1 1	−0.14	0.26	1.46	0.78	0.362
0 1 0 0	0.08	−0.37	0.08	−0.38	0.584
0 1 0 1	−0.55	−1.18	−0.52	−0.43	0.848
0 1 1 0	−0.07	−0.37	−1.45	0.15	0.780
0 1 1 1	−1.62	−1.02	−0.40	−0.46	0.969
1 0 0 0	−1.00	−1.13	−1.04	−1.14	1.039
1 0 0 1	−0.28	−0.50	−0.30	−0.44	0.691
1 0 1 0	−1.00	−0.679	−1.35	−0.53	0.957
1 1 0 0	0.06	−0.97	−0.52	−0.57	0.772
1 1 0 0	−0.39	0.072	1.18	−0.39	0.545
1 1 0 1	0.48	0.81	0.77	0.89	0.152
1 1 1 0	−0.09	0.97	0.45	−0.01	0.396
1 1 1 1	1.05	1.05	0.77	0.72	0.092

For example, for G =c480ea756fbd1329, with precharge logic $\beta = 0$ the correct key has rank 1 when there is no noise (upper half of Table 4). Let us now consider the noise, $\mathcal{N}$ as in Equation 2, that takes the values uniformly from $[-0.5, 0.5]$. We observed that in an average 39% of the times the correct key was toppled from the 1st position. Following is an instance where the rank of the correct key drops down to 3 (lower half of Table 4). Note that in this example, the key $K' = \dot{K} \oplus (0,0,1,0)$ was closest to the correct key $\dot{K}$ in terms of the distance. However, in the noisy scenario, the correct key is toppled by the key $K'' = \dot{K} \oplus (1,1,1,1)$, which was further away from $\dot{K}$ than K' in the noise free scenario.

3.2   Impact of Noise in Key Recovery Attack

Given that the correct key may receive lower rank in the presence of noise, the attacker cannot obtain the correct key immediately. Let us now explain some simulation results that identify the rank of the correct key in noisy scenario. As we have discussed in Remark 2, it will be difficult for the attacker to guess the correct key when $\mathcal{D}2_n(\cdot)$ is of small value. In such a case, it is more probable that the rank of the correct key will be more than one. We refer to the S-boxes given in [3, Table 3]. For the S-box $F = 012d47f68c53aeb9$, the value of $\mathcal{D}2_n(F)$ is 0.318, whereas the minimum value of $\mathcal{D}2_n(G)$, where the permutation S-box G (e29cf1654db03a78) is in the extended affine equivalent class of F, is as low as 0.125. Note that G and F are related as: $G(x) = B(F(x)) \oplus L(x) \oplus c$, where the associated matrices of the linear functions B and L are $\begin{bmatrix} 1&0&0&0 \\ 0&1&0&0 \\ 0&0&1&0 \\ 1&1&1&1 \end{bmatrix}$ and $\begin{bmatrix} 1&0&1&0 \\ 1&0&1&0 \\ 1&0&1&0 \\ 1&0&1&1 \end{bmatrix}$, respectively, and $c = (1,1,1,0)$.

Table 5. Average Key rank of F and its extended affine equivalent G in the presence of noise ($\mathcal{N}$ as in Equation 2) to take values uniformly from $[-t, t]$

S-Box	Average key rank			
	$t = 0.25$	$t = 0.5$	$t = 0.75$	$t = 1.0$
F (012d47f68c53aeb9)	1	1.0027	1.1271	1.4976
G (e29cf1654db03a78)	1.0422	1.4689	1.9771	2.5229

Table 6. PRINCE S-boxes and their extended affine equivalent ones having the lowest $\mathcal{D}2_n(\cdot)$

S-box (F) [3]	$\mathcal{D}2_n(F)$	EA Equivalent S-Boxes (H)	$\mathcal{D}2_n(H)$
012d47f68c53aeb9	0.318	e29cf1654db03a78	0.125
012d47f68c9bae53	0.395	456d30f1bca89e27	0.125
012d47f68cb9ae35	0.441	e256f1c9d478a3b0	0.125
012d47f68cB9ae53	0.441	674e30c1af9b8d25	0.125
012d47f68ceba935	0.405	297a0ef5314d68cb	0.125
012d47f68eba59c3	0.330	230c65e49fab78d1	0.125
012d47f68eba93c5	0.306	efc7a95824103d6b	0.125
012d47f68ec95ba3	0.441	230b4796c8ad5fe1	0.125
bf32ac916780e5d4	0.197	8a9325dc1b760ef4	0.125

From Table 5, we see that the S-box G for which $\mathcal{D}2_n(\cdot)$ is smaller, the average rank of the correct key is higher in the presence of noise. Thus the designer would like to choose G instead of F in such a scenario for better resistance against DPA. In this direction, we have studied all the $(4, 4)$ S-boxes presented in relation to PRINCE [3] (the first 8 are from Table 3 and the last one is from the S-layer in Page 5 in that paper). We made exhaustive search for each of such S-boxes, and obtained an (extended) affine equivalent permutation S-box with the minimum possible value of $\mathcal{D}2_n(\cdot)$. The results are presented in Table 6.

4   Conclusions

In this paper, we have studied multi-bit DPA in the Hamming weight model. In this direction, the Relative Power Spectrum (RPS) of an S-box is proposed, and consequently we considered the second minimum distance in the spectrum denoted by $\mathcal{D}2_n(\cdot)$. This measure can be used to determine the resistance of an S-box against DPA; lower the value of $\mathcal{D}2_n(\cdot)$, better the resistance. We have also shown that two permutation S-boxes, which are affine or extended affine equivalent, may not have the same value of $\mathcal{D}2_n(\cdot)$. Given a permutation S-box, we propose to use the one from its extended affine equivalence class having the minimum value of $\mathcal{D}2_n(\cdot)$. We have provided simulation results in relation to the S-boxes used in AES and PRINCE. It is important to study the (extended) affine equivalent permutation S-boxes of different block ciphers in this direction. Our results highlight that the designer can obtain cryptographic advantage against DPA by exploiting (extended) affine equivalence.

Acknowledgments. The authors would like to thank the anonymous reviewers for the detailed comments that improved the technical as well as editorial quality of this paper. This work has been partially supported by Centre of Excellence in Cryptology, Indian Statistical Institute.

References

1. Advanced Encryption Standard, http://en.wikipedia.org/wiki/Rijndael_S-box
2. Bilgin, B., Nikova, S., Nikov, V., Rijmen, V., Stütz, G.: Threshold Implementations of All 3 3 and 4 4 S-Boxes. In: Prouff, E., Schaumont, P. (eds.) CHES 2012. LNCS, vol. 7428, pp. 76–91. Springer, Heidelberg (2012)
3. Borghoff, J., Canteaut, A., Güneysu, T., Kavun, E.B., Knezevic, M., Knudsen, L.R., Leander, G., Nikov, V., Paar, C., Rechberger, C., Rombouts, P., Thomsen, S.S., Yalçın, T.: PRINCE – A Low-Latency Block Cipher for Pervasive Computing Applications. In: Wang, X., Sako, K. (eds.) ASIACRYPT 2012. LNCS, vol. 7658, pp. 208–225. Springer, Heidelberg (2012)
4. Brier, E., Clavier, C., Olivier, F.: Correlation Power Analysis with a Leakage Model. In: Joye, M., Quisquater, J.-J. (eds.) CHES 2004. LNCS, vol. 3156, pp. 16–29. Springer, Heidelberg (2004)
5. Carlet, C.: On Highly Nonlinear S-Boxes and Their Inability to Thwart DPA Attacks. In: Maitra, S., Veni Madhavan, C.E., Venkatesan, R. (eds.) INDOCRYPT 2005. LNCS, vol. 3797, pp. 49–62. Springer, Heidelberg (2005)

6. Chakraborty, K., Maitra, S., Sarkar, S., Mazumdar, B., Mukhopadhyay, D.: Redefining the Transparency Order. http://eprint.iacr.org/2014/367
7. Guilley, S., Hoogvorst, P., Pacalet, R.: Differential Power Analysis Model and Some Results. In: Proceedings of Smart Card Research and Advanced Applications VI, CARDIS 2004, pp. 127–142. Kluwer Academic Publishers (2004)
8. Kocher, P.C., Jaffe, J., Jun, B.: Differential Power Analysis. In: Wiener, M. (ed.) CRYPTO 1999. LNCS, vol. 1666, pp. 388–397. Springer, Heidelberg (1999)
9. Mazumdar, B., Mukhopadhyay, D., Sengupta, I.: Constrained Search for a Class of Good Bijective S-Boxes with Improved DPA Resistivity. IEEE Transactions on Information Forensics and Security 8(12), 2154–2163 (2013)
10. Prouff, E.: DPA Attacks and S-Boxes. In: Gilbert, H., Handschuh, H. (eds.) FSE 2005. LNCS, vol. 3557, pp. 424–441. Springer, Heidelberg (2005)
11. Whitnall, C., Oswald, E.: A Fair Evaluation Framework for Comparing Side-Channel Distinguishers. J. Cryptographic Engineering 1(2), 145–160 (2011)

Confused by Confusion: Systematic Evaluation of DPA Resistance of Various S-boxes

Stjepan Picek[1,3], Kostas Papagiannopoulos[1]([✉]), Barış Ege[1],
Lejla Batina[1,2], and Domagoj Jakobovic[3]

[1] ICIS - Digital Security Group, Radboud University Nijmegen,
Nijmegen, The Netherlands
kostaspap88@gmail.com
[2] ESAT/COSIC, KU Leuven, Leuven, Belgium
[3] Faculty of Electrical Engineering and Computing
University of Zagreb, Zagreb, Croatia

Abstract. When studying the DPA resistance of S-boxes, the research community is divided in their opinions on what properties should be considered. So far, there exist only a few properties that aim at expressing the resilience of S-boxes to side-channel attacks. Recently, the confusion coefficient property was defined with the intention to characterize the resistance of an S-box. However, there exist no experimental results or methods for creating S-boxes with a "good" confusion coefficient property. In this paper, we employ a novel heuristic technique to generate S-boxes with "better" values of the confusion coefficient in terms of improving their side-channel resistance. We conduct extensive side-channel analysis and detect S-boxes that exhibit previously unseen behavior. For the 4×4 size we find S-boxes that belong to optimal classes, but they exhibit linear behavior when running a CPA attack, therefore preventing an attacker from achieving 100% success rate on recovering the key.

1 Introduction

Today, the most practical attacks belong to side-channel analysis (SCA) that target actual implementations of block ciphers in software or hardware. It is well known that the efficiency of side-channel attacks is much greater [1] than linear [2] or differential [3] cryptanalysis. To improve an algorithm's resiliency to SCA, there exist many possible countermeasures such as various hiding and masking schemes [4]. However, all countermeasures come with a substantial increase in cost due to larger memory and area requirements and the decrease in performance of the algorithm implemented.

S-boxes (as the only nonlinear part in many ciphers) have a fundamental role in the security of most modern block ciphers [5] and their "good" cryptographic

---

The work presented in this article has been supported by the European Cooperation in Science and Technology (COST) Action IC1204 (Trustworthy Manufacturing and Utilization of Secure Devices - TRUDEVICE), by the Technology Foundation STW (project 12624 - SIDES), and by the Netherlands Organization for Scientific Research NWO (project ProFIL 628.001.007).

© Springer International Publishing Switzerland 2014
W. Meier and D. Mukhopadhyay (Eds.): INDOCRYPT 2014, LNCS 8885, pp. 374–390, 2014.
DOI: 10.1007/978-3-319-13039-2_22

properties are of utmost importance for the security of encryption schemes in numerous applications.

Although there exist a plethora of cryptographic properties defined for S-boxes in the literature, there are only a handful properties related to the SCA resistance. Currently, the properties related with SCA are SNR (DPA) [6], transparency order [7], criterion for the S-box resilience against CPA attacks [8] and the most recent measure, confusion coefficient [9,10]. Considering the transparency order, which was heavily investigated so far, results from different groups are somewhat conflicting [11–13]. Yet in all previous works the transparency order seems to have a certain influence on DPA resistance. For example, in the 4×4 S-boxes case (as used in e.g. PRESENT [14]), it is shown that one can obtain S-boxes that have better DPA resistance, while retaining properties of optimal S-boxes [15].

Nevertheless, numerous inconclusive results for different ciphers, platforms and leakage models have led to an attempt to redefine the transparency order measure [16]. This new approach also remains to be convincing in practical results.

When considering 8×8 S-boxes, previous results on transparency order suffer from two major drawbacks. The first drawback stems from the fact that an improved S-box (in regards to the transparency order property) may result in deterioration of some properties related with linear and differential resistance of the algorithm (e.g. nonlinearity and δ-uniformity). The second major drawback is the necessity to implement such improved S-boxes as lookup tables. For instance, an improved AES S-box (e.g. derived from heuristic search) loses the algebraic properties that are important for compact implementations [17,18]. Still, there are possible settings where the improvement in DPA resistance makes up for the aforementioned drawbacks. In contrast to this, when considering 4×4 S-boxes, the situation is improved since both implementation options, as a lookup table and as a Boolean function in hardware, are viable.

In this paper, we generate S-boxes with an improved confusion coefficient and we show that this also improves DPA resistance. In order to confirm that, we conduct simulated and practical side-channel analysis on those improved S-boxes.

1.1   Related Work

In 2004, Guilley presents SNR (DPA) measure which is, to our best knowledge, the first property related with DPA resistance [6]. One year later, Prouff presents the transparency order property which is the first DPA-related property for the multi-bit case [7]. The idea proved to be valuable, as in 2012 several papers revisit the topic [11–13,15].

Apart from the transparency-related efforts, a new line of research by Yunsi Fei et al. [9,10,19,20] attempts to actually model the behavior of a cryptographic implementation with respect to side-channel resistance. Starting from DPA-related models [9,19] they expand the concept to CPA attacks [10] and masking [20], while offering a probabilistic model for side-channel analysis.

1.2   Our Contribution

It is evident that using side-channel leakages is a powerful means of cryptanalysis. Therefore, it is important to find effective and efficient countermeasures (or combinations of them) that can prevent this type of cryptanalysis. This paper investigates the option of using heuristically-created S-boxes to increase the resistance to implementation attacks. More specifically, we are the first to use the confusion coefficient as a cipher design parameter. With the assistance of genetic algorithms, we create 4×4 and 8×8 S-boxes that obtain improved confusion coefficient property. For the 4×4 case, we create S-boxes that have increased resistance in the form of "ghost peaks" [4] (defined here as "Phantom" S-boxes), while remaining in optimal classes [5]. For the 8×8 case, we obtain increased resistance, albeit at the cost of classical cryptanalytic properties like nonlinearity and δ-uniformity. We evaluate the newly generated S-boxes in a real world scenario: we implement variants of PRESENT [14] and AES [21] ciphers that employ the new S-boxes and we perform side-channel analysis on them.

The remainder of this paper is organized as follows. We present necessary information about relevant cryptographic properties of S-boxes in Sect. 2. In Sect. 3 we give explanations of the algorithms we use and the analysis of several S-boxes with improved confusion coefficient property. We also compare the properties of our new S-boxes with the ones obtained from random search as well as with the original ones. The side-channel resistance of the newly proposed mappings is presented in Sect. 4, both with simulations and also with experiments on a real target. We conclude the paper in Sect. 5.

2   Preliminaries

In this section we present some background information about side-channel analysis and cryptographic properties of S-boxes that are of interest.

2.1   Side-channel Analysis

Various cryptographic devices can leak the information they process. Those leakages can be exploited by an adversary monitoring side channels such as timing [22], power consumption [22,23] or electromagnetic emanation [24]. Such attacks enable the attacker to obtain otherwise unknown information on the workings of the underlying algorithm, therefore leading to practical attacks on even real-world cryptosystems. As these attacks are the most practical ones among many cryptanalysis efforts, this area attracts quite some interest in the literature. There are recent publications in the literature that focus on modeling the physical leakage of an algorithm with the assumption of a certain leakage model [7,19]. This line of research is aimed to provide a way to evaluate the side-channel resistance of an algorithm at the design phase to help cryptographers improve the overall security of a cryptosystem.

2.2   Cryptographic Properties of S-boxes

As mentioned previously, there exist several properties of S-boxes where each property relates to a certain cryptographic attack. However, here we concentrate only on several basic properties such as bijectivity, linearity and δ-uniformity (as given in [5]) and of course the new measure, confusion coefficient.

The addition modulo 2 is denoted as " ". The inner product of vectors $\bar{a}$ and $\bar{b}$ is denoted as $\bar{a} \cdot \bar{b}$ and equals $\bar{a} \cdot \bar{b} = \oplus_{i=1}^{n} a_i b_i$.

Function F, called S-box or vectorial Boolean function, of size (n, m) is defined as any mapping F from $\mathbb{F}_2^n$ to $\mathbb{F}_2^m$ [7]. When m equals 1 the function is called Boolean function. Boolean functions f_i, where $i \in \{1, ..., m\}$, are coordinate functions of F, where every Boolean function has n variables.

Hamming weight HW of a vector $\bar{a}$, where $\bar{a} \in \mathbb{F}_2^n$, is the number of nonzero positions in the vector.

An (n, m)-function is called **balanced** if it takes every value of $\mathbb{F}_2^m$ the same number 2^{n-m} of times [25].

Linearity L_f can be defined as [26]

$$L_f = max_{\substack{\bar{a} \in \mathbb{F}_2^n \\ \bar{v} \in \mathbb{F}_2^{m*}}} |W_F(\bar{a}, \bar{v})|. \tag{1}$$

where $W_F(\bar{a}, \bar{v})$ is Walsh transform of F [7].

$$W_F(\bar{a}, \bar{v}) = \sum_{\bar{x} \in \mathbb{F}_2^n} (-1)^{\bar{v} \cdot F(\bar{x}) \oplus \bar{a} \cdot \bar{x}}. \tag{2}$$

Nonlinearity N_F of an (n, m)-function F is equal to the minimum nonlinearity of all non-zero linear combinations $\bar{b} \cdot F$, where $\bar{b} \neq 0$, of its coordinate functions f_i [1].

$$N_F = 2^{n-1} - \frac{1}{2} max_{\substack{\bar{a} \in \mathbb{F}_2^n \\ \bar{v} \in \mathbb{F}_2^{m*}}} |W_F(\bar{a}, \bar{v})|, \tag{3}$$

Differential delta uniformity δ represents the largest value in the difference distribution table without counting the value 2^n in the first row and first column position [3, 25, 27].

Recently, Fei et al. introduced a new property that relates with the DPA resistance of S-boxes - **confusion coefficient** [9, 10, 19, 20]. They give a probabilistic model that encompasses the three core parameters of a side-channel attack: the target device, the number of traces and the algorithm under examination. That model manages to separate these three elements and grants us the freedom to explore the cipher design space by solely focusing on the cipher algorithm. The confusion coefficient stems from the probability that an intermediate value ψ is affected by different keys k_c, k_g (Eq. 4). This notion is extended to the confusion matrix $\mathbf{K}$ and $\mathbf{\Sigma_Y}$ (Eq. 6, Eq. 8), which directly influence the attacker's effectiveness SR_{DPA} (Eq. 9).

$$\kappa(k_c, k_g) = Pr[(\psi|k_c) \neq (\psi|k_g)] \tag{4}$$

$$\tilde{\kappa}(k_c, k_{g_i}, k_{g_j}) = Pr[(\psi|k_{g_i}) = (\psi|k_{g_j}), (\psi|k_{g_i}) \neq (\psi|k_c)] \qquad (5)$$

$$\mathbf{K} : (N_k - 1) \times (N_k - 1), \ \mathbf{K_{ij}} = \begin{cases} \kappa(k_c, k_g), & \text{if } i = j \\ \tilde{\kappa}(k_c, k_{g_i}, k_{g_j}), & \text{if } i \neq j \end{cases} \qquad (6)$$

$$\bar{\mu}_Y = 2 \times \epsilon \times \bar{\kappa} \qquad (7)$$

where $\bar{\kappa}$ is the diagonal vector of $\mathbf{K}$ and ϵ is the the theoretic Difference of Means for the correct key.

$$\mathbf{\Sigma_Y} = 16 * \sigma_W/N_{meas} \times \mathbf{K} + 4 \times \epsilon^2/N_{meas} \times (\mathbf{K} - \bar{\kappa} \times \bar{\kappa}^T) \qquad (8)$$

$$SR_{DPA} = \Phi_{N_k-1}(\sqrt{N_{meas}}\mathbf{\Sigma_Y}^{-1/2}\bar{\mu}_Y) \qquad (9)$$

Equation (9) gives the success rate of a DPA attack (SR_{DPA}). It is computed over the cumulative distribution function (Φ_{N_k}) of a multivariate Gaussian distribution, with dimension (N_k) equal to key dimensionality (e.g. 256 for AES if the selection function partitions into 8-bit targets). The number of traces is directly represented in the formulas via N_{meas} (number of measurements). The target device is characterized from the signal to noise ratio ($SNR = \epsilon/\sigma_w$) and the parameters ϵ and σ_w can be computed from side-channel measurements.

Cipher algorithm is isolated by defining and constructing the confusion coefficient κ as given in Eq. (4) and (5). The confusion matrix $\mathbf{K}$ that is subsequently constructed is given in Eq. (6). The matrix elements capture the behavior of both the cipher and the selection function with respect to a specific key (k_c denotes the correct key and k_g the key guesses that stem from the key space).

The confusion coefficient with respect to a specific S-box was initially defined as the probability that 2 different keys will lead to a different S-box output as given in Eq. (4). Intuitively, a high confusion coefficient indicates that the S-box output (or any other intermediate value ψ targeted by a side-channel attack) is very distinctive. Thus, the S-box output is a good candidate for data leakage. Low confusion coefficient values (also referred to as high collision values) make side-channel attacks harder, i.e. they may require an increase in number of traces or SNR to yield the correct key candidate.

Early work from Fei et al. suggests that the confusion coefficient matrix captures the algorithmic behavior of the cipher [9,19]. However, this matrix incorporates all possible confusion coefficients with respect to a given key, making the whole analysis key-dependent. In addition, we consider beneficial to move towards CPA-related models instead of DPA. Thus, we use more recent findings from Fei et al., namely the confusion coefficient for CPA, the confusion coefficient vector and its frequency distribution [10]. We compute the confusion coefficients for a given CPA selection function as shown in Algorithm 1.

Having computed all possible confusion coefficient values κ w.r.t. CPA attack and Hamming weight (HW) power model we compute the confusion coefficient vector. This vector contains all possible coefficient values for every key combination and its frequency distribution is deemed by the Fei et al. to be possible characterizer of side-channel behavior. The natural question that arises is what features of the frequency distribution of the confusion coefficient vector denotes

Algorithm 1. CPA selection function

---
for all key pairs $k_a, k_b, k_a \neq k_b$ do
  for all possible inputs in do
    $\kappa(k_a, k_b) = E[(HW(Sbox(in \oplus k_a)) - HW(Sbox(in \oplus k_b))^2]$
  end for
end for

---

side-channel resistance. We observe that the mean value of the distribution is directly related to the choice of the selection function, i.e. it solely depends on the divide-and-conquer approach that we use in our attack. Moreover, Heuser et al. demonstrate the link between nonlinearity and the distribution of the vector [28]. Specifically, highly nonlinear elements lead to a distribution with low variance. Therefore, we need to find S-boxes that demonstrate a high variance in the confusion coefficient vector distribution. Note that our S-boxes are generated under the Hamming weight leakage assumption – depending on the device this assumption does not always hold.

Two (n, n)-functions S_1 and S_2 are **affine equivalent** only if the following equation holds:

$$S_2(x) = B(S_1(A(x) \oplus a)) \oplus b, \tag{10}$$

where A and B are invertible $n \times n$ matrices and $a, b \in \mathbb{F}_2^n$ are constant values.

Resistance of S-boxes against most of the attacks remains unchanged if affine transformation is applied before and after S-box [5].

2.3 Optimal S-boxes

When considering 4×4 S-boxes, there exist in total 16! bijective 4×4 S-boxes which is approximately 2^{44} options to search from. Leander and Poschmann define optimal S-boxes as those that are bijective, have linearity equal to 8 and δ-uniformity equal to 4 [5]. Since the linearity of 8 is the same as nonlinearity 4, we continue talking about nonlinearity property instead of linearity.

Furthermore, they found that all optimal S-boxes belong to 16 classes, i.e. all optimal S-boxes are affine equivalent to one of those classes [5].

In Appendix A we give values for several relevant properties for 4×4 S-box size.

3 Experimental Setup and Results

When generating S-boxes with good properties, we use genetic algorithms approach as they produced good results in previous works [13, 15]. Additionally, we use random search as a baseline search strategy and affine transformations to check whether confusion coefficient property is affine invariant.

Random Search. In this setting we use Monte Carlo search method to find S-boxes that have good values of confusion coefficient. With this search method we cannot influence the value of any of the S-box properties and we consider it as a baseline search strategy. Here, S-boxes are created uniformly at random.

Genetic Algorithms. In this technique we evolve S-boxes that have good values not only for DPA-related properties, but also for other cryptography relevant properties.

Affine Transformation. As shown in [15] affine transformation affects transparency order values. We employ several different transformations and investigate their influence on confusion coefficient.

Further details about genetic algorithms and affine transformation are given in the following sections.

3.1   Genetic Algorithm

Genetic algorithms (GAs) are a subclass of evolutionary algorithms where the elements of the search space S are arrays of elementary types [29]. Genetic algorithms belong to evolutionary techniques that have been successfully applied to various optimization problems. To be able to evolve new individuals (solutions) GA uses variation operators where the usual ones are mutation and crossover (recombination) operators. Mutation operators are operators that use one parent to create one child by applying randomized changes to parent. Mutation depends on the mutation rate p_m which determines the probability that a change will occur within individual. Recombination operators work on two or more parents to create offspring from the information contained within parent solutions. Recombination is usually applied probabilistically according to a crossover rate p_c. Besides variation operators, it is necessary to decide about selection method. Today, the k-tournament selection method is widely used for this purpose [29]. In this selection k solutions are selected at random and the worst among them is replaced by the offspring of the remaining solutions. Further information about genetic algorithms can be found in [30,31].

Representation and Fitness Functions. There are several possibilities how to represent S-boxes (e.g. truth tables or lookup tables). We decide to use permutation encoding since in this way the bijectivity property is automatically preserved. In this representation, $n \times m$ S-box is defined with an array of 2^m integer numbers with values between 0 and $2^m - 1$. Each of these values occurs exactly once in an array and represents one entry for the S-box lookup table, where inputs are in lexicographical order.

The initial population for GA is built by creating random permutations of the designated length.

Details about recombination operators (crossover and mutation operators) are described in Appendix C.

Maximization of the value of a fitness function is the objective in all evolutionary experiments. A fitness function represents a definition of the problem to solve with genetic algorithm. For fitness function we use a combination of properties as presented in Section 2.

For the 8×8 case, fitness function equals the sum of nonlinearity (N_F) and confusion coefficient variance (κ) properties values as follows.

$$fitness = N_F + \kappa. \tag{11}$$

When investigating 4×4 case, we add to the fitness function δ-uniformity property. In this way we ensure that evolved S-boxes belong to the one of optimal S-boxes classes.

$$fitness = N_F + \kappa + (2^m - \delta). \tag{12}$$

We subtract δ-uniformity value from the maximum obtainable value since we represent the problem as a maximization problem and δ property should be as small as possible. Both fitness function can be easily extended to contain more properties that are of relevance to the evolutionary experiments. As evident from the formulas above, we do not use weight factors in our fitness equations. This is due to the fact that we first want to reach as good as possible nonlinearity (and δ-uniformity for the 4×4 size) values and then for such values find the best possible confusion coefficient values.

Here we emphasize that our approach is not only easily adaptable when adding additional properties, but if we want to change e.g. the leakage model it would only affect one term in the fitness function.

Common Parameters. For every fitness function we run 30 independent runs and population size is equal to 50. Mutation probability is set to 0.3 per individual. The parameters above are the result of a combination of a small number of preliminary experiments and our experience with similar problems; no thorough parameter tuning has been performed. Tournament selection parameter k is set to 3. Evolution process lasts until the stopping criterion is fulfilled, here the criterion is 50 generations without improvement.

In Figures 1(a) and 1(b) we present results for random, evolved and original S-boxes (PRESENT and AES) for sizes 4×4 and 8×8 respectively.

We see that for 4×4 S-box size we obtain maximum confusion coefficient variance of 3.07 while staying in optimal classes. For the 8×8 size, the best confusion coefficient variance we found is 4.057. However, this value comes at a cost of nonlinearity of 98 and δ-uniformity of 12 (AES has nonlinearity 112 and δ-uniformity 4).

3.2 Affine Equivalence

Recall that resistance of an S-box against most of attacks stays the same if affine transformation is applied before and after S-box. Therefore, it is useful to check whether that is true for confusion coefficient property.

As shown before, transparency order property changes under certain affine transformations [15]. Next, we check what happens with confusion coefficient variance property under affine transformation. We apply transformations as

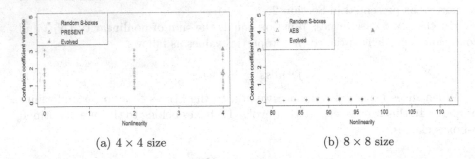

(a) 4×4 size   (b) 8×8 size

Fig. 1. Nonlinearity vs. confusion coefficient variance

listed in Table 1 to AES S-box as well as to representatives to 16 optimal classes for 4×4 size.

Table 1. Affine transformations

Number	Transformation
1	$S(x) \oplus c$
2	$S(B(x) \oplus c)$
3	$A(S(B(x) \oplus c)) \oplus d$
4	$A(S(B(x) \oplus c) \oplus d)$

In this table $c, d \in \mathbb{F}_2^n$ are constants, $\oplus$ represents XOR operation and A and B are invertible matrices.

Following conclusions apply both for 4×4 and 8×8 S-box sizes.

Affine transformation 1 does not change transparency order or confusion coefficient values. To change confusion coefficient property, changes 2, 3 and 4 are applicable. Here we note that our experiments show that the transformations 3 and 4 change confusion coefficient more significantly than the affine transformation 2. The PRESENT S-box has a confusion coefficient variance of 1.709. By applying transformation 2, we succeed in obtaining maximal confusion coefficient variance of 1.803. However, when applying transformations 3 or 4, we obtain maximal confusion coefficient of variance 3.07.

Since affine transformation emerges as a good choice for generating S-boxes with good DPA properties we present result when applying transformation 3 to AES S-box and lexicographical representatives of 16 optimal classes. We opted for transformation 3 since it is one of two transformations that is capable of significantly changing confusion coefficient and we did not observe any significant difference between transformations 3 and 4. For all experiments the procedure consists of applying 25 000 random affine transformations and recording the best obtained results. Our best results are presented in Table 2. Here, captions G0 till G15 represent lexicographical representatives of 16 optimal classes [5].

Table 2. Results for affine transformation 3

S-box	κ variance
PRESENT	1.709
PRESENT transformation	3.07
G3, G4, G5, G6, G7, G11, G12, G13	3.02
G0, G1, G2, G8, G9, G10, G14, G15	3.07
AES	0.111
AES transformation	0.149

We can observe then in the case of 4 × 4 size, affine transformation reaches same maximum values (although different S-boxes) as genetic algorithms for 8 out of 16 optimal classes. Furthermore, division between classes that reach 3.07 and 3.02 is the same as in the case of optimal S-boxes and PRINCE suitable S-boxes [32]. Classes that reach values 3.07 are those that are not suitable for usage in the PRINCE algorithm. For 8 × 8 size affine transformation improves confusion coefficient variance only slightly. The frequency distribution of 4 × 4 S-boxes with κ variance 3.07 can be seen in Figure 2.

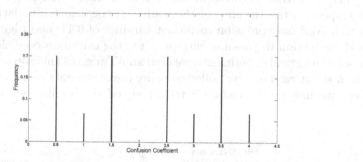

Fig. 2. Frequency distribution of the confusion coefficient for S-boxes with 3.07 variance

In Appendix B we give examples of lookup tables we obtained with GA and affine transformation. When considering 4×4 size we see that both GA and affine transformation reach the same value in both classical properties of interest, as well as for confusion coefficient. Therefore, in this case there is no benefit in applying GA. However, when considering 8×8 case, if interested in as good as possible confusion coefficient value, we observe that GA reaches much better values (at the cost of deterioration of classical properties like nonlinearity and δ-uniformity).

4   Side-Channel Experiments

Although the method explained in the previous sections results in the generation of S-boxes with various values for the confusion coefficient, these S-boxes require a thorough practical analysis. This is required to quantify how much of a change in variance in the confusion coefficient will result in a certain gain in side-channel resistance in terms of the number of measurements required to recover the key.

First, we performed simulations to see how the newly generated S-boxes behave under the Hamming weight model when a certain amount of Gaussian noise is added to the measurements. For the simulated experiments, we used 3 newly generated S-boxes and the PRESENT S-box as the baseline case. One of the 3 newly generated S-boxes is the so called "*Phantom*" S-box that leads to having two key candidates with correlation values equal in magnitude, hence making it more difficult for an attacker who has no knowledge of the exact leakage model of a device to deduce the correct key with 100% accuracy. The "*Phantom*" S-box can be shortly defined as an S-box leading to ghost peaks in the correlation trace after running the attacks. This happens since the S-box outputs have either the same or complementary Hamming weight values for inputs with a particular XOR difference.

Figure 3 presents the logarithm ($\log_2$) of the guessing entropy [33] with respect to the number of traces processed for the attacks. We run the attack on the simulated traces produced with the inclusion of Gaussian noise with mean 0 and standard deviation 16. Important point to note about Figure 3 is that the PRESENT S-box has a confusion coefficient variance of 1.709. Therefore, it can be clearly seen that Figure 3(a) shows a clear distinction in guessing entropy with respect to the variance of the confusion coefficient. Similarly one should note that AES has confusion coefficient variance of 0.11, and Figure 3(b) shows a good distinction in guessing entropy w.r.t. the confusion coefficient variance.

For the practical experiments, we used an ATmega163 microcontroller embedded in a smart card and we collected many measurements using a modified card reader enabling us to produce a trigger signal the oscilloscope and a LeCroy

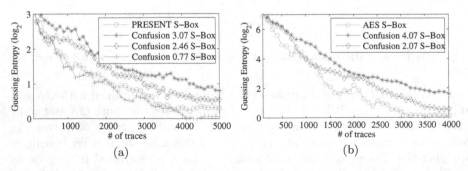

Fig. 3. Guessing entropy of the simulated S-boxes (4×4 in (a), 8×8 in (b)) with respect to the number of traces processed

oscilloscope. To be able to make a fair assessment of the side-channel security of different S-boxes, we collected the information from 50 separate attacks and combined them in terms of guessing entropy in Figure 4. Again it is clear that when the attack is applied using the Hamming weight model, the S-box having the better confusion coefficient value shows better resistance against side-channel attacks. Here an important fact to note here is that the *"Phantom"* S-box exhibits this property only when the Hamming weight model is used. The reason for this behavior is that *"Phantom"* S-boxes lead to having either the same Hamming weight, or the exact opposite Hamming weight in the outputs when the inputs have a certain XOR difference in between. Therefore, when one of the bits is taken into account rather than the Hamming weight of the whole S-box output for mounting the attack, this *"Phantom"* behavior may not necessarily persist. However, if the target leaks the Hamming weight of intermediate values, then the attacker would be forced to use a weaker selection function (bit model) for that particular device, therefore leading to an attack requiring the acquisition of more power traces.

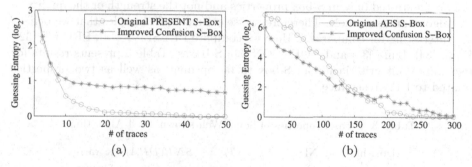

Fig. 4. Guessing entropy of the S-boxes (4×4 in (a), 8×8 in (b)) with respect to the number of traces processed on the AVR microcontroller

It should be noted that the results presented in this paper assume a particular leakage model, namely the Hamming weight model. We have computed the confusion coefficient with this assumption in mind but if the leakage of a device is known, it is straightforward to integrate that leakage model in our genetic algorithms and produce an S-box which will resists to side-channel attacks better in a device leaking in that particular leakage model.

Although we observe that improving the confusion coefficient results in designs which have better side-channel resistance, we do recognize that this cannot be counted as a countermeasure. We believe it is interesting to investigate how an improved S-box interacts with other countermeasures and especially with masking. Since in this work we focus on the practicality of the confusion coefficient metric, it remains as an interesting open question to see whether the S-box improvements are persistent after masking or not.

5   Conclusion

In this work we consider the DPA resistance properties of S-boxes of various sizes. We show it is possible to evolve S-boxes that have better confusion coefficient variance values. Using genetic algorithms we are able to produce both 4×4 and 8×8 size S-boxes that exhibit improved DPA resistance.

Next, we show that an affine transformation changes the confusion coefficient variance property. This fact can be important not only from the theoretical perspective, but also from the practical one. We reiterate that with the genetic algorithms approach change in the leakage model leads only to the change in one fitness function factor. Therefore, we can easily adapt the procedure to other more generic scenarios.

In further work we will concentrate on the interaction between the improved S-boxes and masking countermeasures.

Appendix A

We are interested only in S-box properties and not the strength or the quality of the algorithm as a whole. Therefore, we select a set of examples that we believe are interesting for comparison. Here we compare PRESENT [14], PRINCE [32], Klein [34], Luffa [35] and NOEKEON [36] S-boxes. Table 3 presents results for two important criteria for an S-box to be optimal, as well as two properties related to DPA resistance.

Table 3. S-box Properties of Some Well-known 4×4 Algorithms

Algorithm	NL	δ	T_F	$SNR(DPA)$	κ variance
PRINCE	4	4	3.4	2.129	1.709
PRESENT	4	4	3.533	2.129	1.709
NOEKEON	4	4	3.533	2.187	1.615
Klein	4	4	3.467	1.691	2.742
Luffa	4	4	3.733	2.53	1.191

Appendix B

An example of optimal S-box of size 4×4 and confusion coefficient variance is:

S-box = (0x6, 0x4, 0x7, 0x8, 0x0, 0x5, 0x2, 0xA, 0xE, 0x3, 0xD, 0x1, 0xC, 0xF, 0x9, 0xB)

Evolved 8×8 S-box with variance of 4.057 and nonlinearity 98 is given below.

S-box = (0xb1, 0x23, 0x98, 0x27, 0x4b, 0x14, 0x9, 0x5c, 0x55, 0xa, 0x4a, 0x4c, 0x1b, 0x3a, 0xa2, 0x53, 0xd6, 0xfb, 0x9f, 0x5e, 0xae, 0xde, 0xe7, 0 x9e, 0x4f, 0x97, 0xf7, 0x2d, 0x2e, 0xbe, 0xab, 0x2b, 0x91, 0x87, 0x36 , 0x1c, 0x81, 0x9d, 0xe5, 0x1a, 0xac, 0x1e, 0x5b, 0x86, 0x8c, 0x74, 0x6a, 0x8a, 0x5f, 0x65, 0xd5, 0x3f, 0xfe, 0xd9, 0xf, 0x37, 0xdd, 0x7d , 0xf2, 0xec, 0xf6, 0xe2, 0xb3, 0xaf, 0x77, 0x99, 0xca, 0xb9, 0xbb, 0xd0, 0x6c, 0xa7, 0x3d, 0xcb, 0x17, 0x75, 0x76, 0x4d, 0xad, 0xcf, 0x5 0, 0x68, 0x16, 0x2, 0x12, 0x78, 0x56, 0x1, 0xb0, 0x71, 0x5a, 0x29, 0 x6, 0x69, 0x58, 0x88, 0x8b, 0x6b, 0xe9, 0x8e, 0xc1, 0xc7, 0x6e, 0x63, 0x13, 0xbc, 0x2f, 0x38, 0x96, 0xbd, 0xdc, 0x62, 0xa8, 0x82, 0x24, 0 xa1, 0xb8, 0x0, 0x80, 0x61, 0xcc, 0x83, 0x22, 0x2c, 0xc2, 0xc0, 0xa0, 0x90, 0xf0, 0xdf, 0xdb, 0xba, 0xe8, 0xf9, 0xbf, 0x7c, 0x59, 0x7b, 0 xeb, 0xd8, 0xa3, 0xff, 0xf3, 0xf8, 0xc8, 0x5, 0x64, 0x66, 0xaa, 0xa9, 0xe, 0xb2, 0xd2, 0x19, 0x10, 0x70, 0x45, 0xc, 0x2a, 0x79, 0x3e, 0x5 d, 0x6d, 0xfa, 0xed, 0xda, 0xe1, 0x9a, 0x7f, 0x4e, 0x8d, 0xf5, 0xfc, 0x7a, 0x57, 0xfd, 0xd, 0xe4, 0x95, 0x18, 0xb4, 0xb5, 0x1d, 0x26, 0x4 8, 0x93, 0x67, 0x7, 0x51, 0xd4, 0x34, 0x43, 0x84, 0x9b, 0x92, 0x60, 0x28, 0x49, 0xc6, 0xc4, 0x8, 0x54, 0xa5, 0x41, 0x40, 0xea, 0xa4, 0x44 , 0x35, 0x15, 0x3b, 0xce, 0xf4, 0xd3, 0x33, 0xb6, 0x8f, 0xcd, 0x25, 0xef, 0xb7, 0x3c, 0x46, 0xee, 0x85, 0x32, 0x3, 0xc3, 0x31, 0xb, 0x30, 0x72, 0xd1, 0x20, 0x4, 0xa6, 0xc9, 0x21, 0x89, 0x47, 0x52, 0x7e, 0x 6f, 0x11, 0xc5, 0xf1, 0xd7, 0x39, 0x94, 0x1f, 0xe3, 0x9c, 0xc0, 0x73, 0xe6, 0x42)

Affine transformation of AES S-box with improved confusion coefficient variance of 0.149357 is given next.

S-box = (0x92, 0x21, 0xd1, 0x6c, 0x5c, 0xf2, 0xf5, 0x86, 0xdd, 0x43, 0x8a, 0x2 8, 0xb8, 0xa3, 0x8b, 0xcf, 0x12, 0xca, 0x23, 0x37, 0xa6, 0xb7, 0x3b, 0xc0, 0x20, 0x4b, 0x5b, 0x22, 0xa4, 0x8, 0x96, 0xff, 0xb2, 0x56, 0xe 9, 0xcd, 0x17, 0x13, 0x57, 0x76, 0x19, 0x18, 0x1d, 0x25, 0xa0, 0x70, 0xec, 0x26, 0xef, 0x1e, 0x8e, 0x29, 0x39, 0x78, 0x6b, 0x4d, 0x60, 0x 95, 0x44, 0xfa, 0xab, 0xcb, 0xc8, 0xe, 0xae, 0xf4, 0x79, 0x46, 0xc, 0x85, 0x7c, 0xbf, 0x40, 0x81, 0xd7, 0x3a, 0xf3, 0xbd, 0x2b, 0x27, 0x5 5, 0x90, 0x61, 0x10, 0xde, 0x82, 0xb6, 0xe0, 0x72, 0x4e, 0x35, 0xea, 0x8c, 0xac, 0x77, 0x52, 0xd5, 0x88, 0xdf, 0x64, 0xc1, 0x65, 0x42, 0x 9c, 0x16, 0x15, 0x33, 0x0, 0x7d, 0x4f, 0x98, 0x9f, 0x45, 0xa2, 0x67, 0x69, 0xd6, 0xcc, 0xd0, 0x5e, 0xbb, 0x73, 0x87, 0x6d, 0x74, 0x3f, 0x ad, 0x7, 0x50, 0x7f, 0x4a, 0x1b, 0x68, 0x71, 0xe2, 0x2, 0x3d, 0xc2, 0x38, 0xba, 0xd9, 0xa, 0x32, 0x31, 0x97, 0xda, 0x99, 0x8f, 0xd8, 0x9d , 0xd3, 0x2e, 0x2d, 0x5d, 0xc4, 0x54, 0x58, 0x91, 0x6, 0x6e, 0x51, 0 x3c, 0x6f, 0xfd, 0xf6, 0xf, 0x48, 0x34, 0x4, 0xf7, 0xc6, 0xa7, 0xf1, 0xaa, 0x47, 0x5a, 0x3e, 0x66, 0xdc, 0x6a, 0x3, 0xb3, 0x63, 0xfc, 0x1 a, 0x49, 0x2c, 0xf0, 0xce, 0x36, 0x7e, 0xe7, 0xe5, 0xb5, 0xa1, 0x7a, 0xc9, 0xee, 0x1c, 0xa5, 0x7b, 0xd4, 0x9b, 0x41, 0xd, 0xa8, 0x5, 0x84 , 0xb1, 0x93, 0x2f, 0xbe, 0xc5, 0xb, 0xeb, 0xe1, 0xaf, 0x9a, 0x80, 0 x8d, 0x4c, 0xe4, 0xfb, 0x9e, 0x89, 0x24, 0x2a, 0x83, 0x9, 0x94, 0x53, 0xbc, 0x5f, 0xa9, 0xc7, 0x75, 0xb0, 0x30, 0x1f, 0xb4, 0xdb, 0xf8, 0 xc3, 0xb9, 0xd2, 0xfe, 0x11, 0xed, 0x59, 0xf9, 0xe8, 0x1, 0xe3, 0xe6, 0x62, 0x14)

Appendix C

In all the experiments, the operators are selected uniformly at random from the pool of available operators presented next.

PMX Crossover. First, two crossover positions are chosen randomly, and the segment between them from the first parent is copied to the offspring. Then, starting from the first crossover position check elements in that segment of second parent that have not been copied. For each of those elements i, check the offspring to see what elements j has been copied in its place from first parent.Place those values i into the positions occupied j in parent 2. If the place occupied by j in parent 2 has already been occupied in the offspring by an element k, put i in the position occupied by k in parent 2. After all the elements in crossover segment are finished, the rest of the offspring is filled from parent 2 [30].

PBX Crossover. In this operator first the values in random positions from the first parent are copied to the same positions in the offspring. Next, values from the second parent that are not present in the offspring are copied to it starting from the beginning of the offspring [30].

OX Crossover. Two crossover positions are chosen at random, and the segment between those positions is copied from the first parent to the offspring. Starting from the second crossover point in the second parent, copy unused values to the offspring in the order they appear in the second parent, wrapping around at the end of the list [30].

Inversion Mutation. In this operator, first two positions are chosen at random. Then, the segment between those 2 values are written in reverse order [30].

Insert Mutation. In this operator two positions are selected at random and then the value from one of those position is moved to be next to the other position. Values in the segment between are shuffled to make room for value to be moved [30].

References

1. Carlet, C.: On Highly Nonlinear S-Boxes and Their Inability to Thwart DPA Attacks. In: Maitra, S., Veni Madhavan, C.E., Venkatesan, R. (eds.) INDOCRYPT 2005. LNCS, vol. 3797, pp. 49–62. Springer, Heidelberg (2005)
2. Matsui, M., Yamagishi, A.: A New Method for Known Plaintext Attack of FEAL Cipher. In: Rueppel, R.A. (ed.) EUROCRYPT 1992. LNCS, vol. 658, pp. 81–91. Springer, Heidelberg (1993)
3. Biham, E., Shamir, A.: Differential Cryptanalysis of DES-like Cryptosystems. In: Menezes, A., Vanstone, S.A. (eds.) CRYPTO 1990. LNCS, vol. 537, pp. 2–21. Springer, Heidelberg (1991)
4. Mangard, S., Oswald, E., Popp, T.: Power Analysis Attacks: Revealing the Secrets of Smart Cards (Advances in Information Security). Springer-Verlag New York Inc., Secaucus (2007)
5. Leander, G., Poschmann, A.: On the Classification of 4 Bit S-Boxes. In: Carlet, C., Sunar, B. (eds.) WAIFI 2007. LNCS, vol. 4547, pp. 159–176. Springer, Heidelberg (2007)
6. Guilley, S., Pacalet, R.: Differential Power Analysis Model and Some Results. In: Proceedings of CARDIS. Kluwer Academic Publishers, pp. 127–142 (2004)
7. Prouff, E.: DPA Attacks and S-Boxes. In: Gilbert, H., Handschuh, H. (eds.) FSE 2005. LNCS, vol. 3557, pp. 424–441. Springer, Heidelberg (2005)
8. Guilley, S., Hoogvorst, P., Pacalet, R., Schmidt, J.: Improving Side-Channel Attacks by Exploiting Substitution Boxes Properties. In: International Workshop on Boolean Functions: Cryptography and Applications, ser. BFCA 2014, pp. 1–25 (2007)
9. Fei, Y., Luo, Q., Ding, A.A.: A Statistical Model for DPA with Novel Algorithmic Confusion Analysis. In: Prouff, E., Schaumont, P. (eds.) CHES 2012. LNCS, vol. 7428, pp. 233–250. Springer, Heidelberg (2012)

10. Fei, Y., Ding, A.A., Lao, J., Zhang, L.: A Statistics-based Fundamental Model for Side-channel Attack Analysis. IACR Cryptology ePrint Archive **2014**, 152 (2014)
11. Mazumdar, B., Mukhopadhyay, D., Sengupta, I.: Design and implementation of rotation symmetric S-boxes with high nonlinearity and high DPA resilience. In: 2013 IEEE International Symposium on Hardware-Oriented Security and Trust (HOST), pp. 87–92 (2013)
12. Mazumdar, B., Mukhopadhay, D., Sengupta, I.: Constrained Search for a Class of Good Bijective S-Boxes with Improved DPA Resistivity. IEEE Transactions on Information Forensics and Security (99), 1 (2013)
13. Picek, S., Ege, B., Batina, L., Jakobovic, D., Chmielewski, L., Golub, M.: On Using Genetic Algorithms for Intrinsic Side-channel Resistance: The Case of AES S-box. In: Proceedings of the First Workshop on Cryptography and Security in Computing Systems, ser. CS^2 2014, pp. 13–18. ACM, New York (2014)
14. Bogdanov, A.A., Knudsen, L.R., Leander, G., Paar, C., Poschmann, A., Robshaw, M., Seurin, Y., Vikkelsoe, C.: PRESENT: An Ultra-Lightweight Block Cipher. In: Paillier, P., Verbauwhede, I. (eds.) CHES 2007. LNCS, vol. 4727, pp. 450–466. Springer, Heidelberg (2007)
15. Picek, S., Ege, B., Papagiannopoulos, K., Batina, L., Jakobovic, D.: Optimality and beyond: The case of 4x4 s-boxes. In: 2014 IEEE International Symposium on Hardware-Oriented Security and Trust, HOST 2014, Arlington, VA, USA, May 6-7, pp. 80–83 (2014)
16. Chakraborty, K., Maitra, S., Sarkar, S., Mazumdar, B., Mukhopadhyay, D.: Redefining the Transparency Order, Cryptology ePrint Archive, Report 2014/367 (2014), http://eprint.iacr.org/
17. Canright, D.: A Very Compact S-Box for AES. In: Rao, J.R., Sunar, B. (eds.) CHES 2005. LNCS, vol. 3659, pp. 441–455. Springer, Heidelberg (2005)
18. Canright, D., Batina, L.: A Very Compact "Perfectly Masked" S-Box for AES. In: Bellovin, S.M., Gennaro, R., Keromytis, A.D., Yung, M. (eds.) ACNS 2008. LNCS, vol. 5037, pp. 446–459. Springer, Heidelberg (2008)
19. Luo, Q., Fei, Y.: Algorithmic collision analysis for evaluating cryptographic systems and side-channel attacks. In: HOST 2011, pp. 75–80 (2011)
20. Ding, A.A., Zhang, L., Fei, Y., Luo, P.: A Statistical Model for Higher Order DPA on Masked Devices. IACR Cryptology ePrint Archive **2014**, 433 (2014)
21. Daemen, J., Rijmen, V.: The Design of Rijndael. Springer-Verlag New York Inc., Secaucus (2002)
22. Kocher, P.C.: Timing Attacks on Implementations of Diffie-Hellman, RSA, DSS, and Other Systems. In: Koblitz, N. (ed.) CRYPTO 1996. LNCS, vol. 1109, pp. 104–113. Springer, Heidelberg (1996)
23. Kocher, P.C., Jaffe, J., Jun, B.: Differential Power Analysis. In: Wiener, M. (ed.) CRYPTO 1999. LNCS, vol. 1666, pp. 388–397. Springer, Heidelberg (1999)
24. Quisquater, J.-J., Samyde, D.: ElectroMagnetic Analysis (EMA): Measures and Counter-Measures for Smart Cards. In: Attali, S., Jensen, T. (eds.) E-smart 2001. LNCS, vol. 2140, pp. 200–210. Springer, Heidelberg (2001)
25. Crama, Y., Hammer, P.L.: Boolean Models and Methods in Mathematics, Computer Science, and Engineering, 1st edn. Cambridge University Press, New York (2010)
26. Braeken, A.: Cryptographic Properties of Boolean Functions and S-Boxes, Ph.D. dissertation, Katholieke Universiteit Leuven (2006)
27. Nyberg, K.: Perfect Nonlinear S-Boxes. In: Davies, D.W. (ed.) EUROCRYPT 1991. LNCS, vol. 547, pp. 378–386. Springer, Heidelberg (1991)

28. Heuser, A., Guilley, S., Rioul, O.: A Theoretical Study of Kolmogorov-Smirnov Distinguishers: Side-Channel Analysis vs. Differential Cryptanalysis. IACR Cryptology ePrint Archive **2014**, 8 (2014)
29. Weise, T.: Global Optimization Algorithms - Theory and Application, 2nd ed. Self-Published (January 14, 2009). http://www.it-weise.de/., http://www.it-weise.de/
30. Eiben, A.E., Smith, J.E.: Introduction to Evolutionary Computing. Springer, Heidelberg (2003)
31. Michalewicz, Z.: Genetic algorithms + data structures = evolution programs, 3rd edn. Springer, London (1996)
32. Borghoff, J., Canteaut, A., Güneysu, T., Kavun, E.B., Knezevic, M., Knudsen, L.R., Leander, G., Nikov, V., Paar, C., Rechberger, C., Rombouts, P., Thomsen, S.S., Yalçın, T.: PRINCE – A Low-Latency Block Cipher for Pervasive Computing Applications. In: Wang, X., Sako, K. (eds.) ASIACRYPT 2012. LNCS, vol. 7658, pp. 208–225. Springer, Heidelberg (2012)
33. Standaert, F.-X., Malkin, T.G., Yung, M.: A Unified Framework for the Analysis of Side-Channel Key Recovery Attacks. In: Joux, A. (ed.) EUROCRYPT 2009. LNCS, vol. 5479, pp. 443–461. Springer, Heidelberg (2009)
34. Gong, Z., Nikova, S., Law, Y.W.: KLEIN: A New Family of Lightweight Block Ciphers. In: Juels, A., Paar, C. (eds.) RFIDSec 2011. LNCS, vol. 7055, pp. 1–18. Springer, Heidelberg (2012)
35. Cannière, C., Sato, H., Watanabe, D.: Hash function Luffa: Specification 2.0.1. Submission to NIST (Round 2) (2009). http://www.sdl.hitachi.co.jp/crypto/luffa/
36. Daemen, J., Peeters, M., Assche, G.V., Rijmen, V.: Nessie proposal: the block cipher Noekeon, Nessie submission. (2000), http://gro.noekeon.org/

Elliptic Curves

Binary Edwards Curves Revisited

Kwang Ho Kim[1]([✉]), Chol Ok Lee[1], and Christophe Negre[2,3]

[1] Department of Algebra and Geometry, Institute of Mathematics,
National Academy of Sciences, Pyongyang, D.P.R. of Korea
math.inst@star-co.net.kp
[2] Team DALI, Université de Perpignan, Perpignan, France
[3] LIRMM, UMR 5506, Université Montpellier 2 and CNRS, Perpignan, France

Abstract. We present new formulas for the arithmetic on the binary Edwards curves which are much faster than the-state-of-the-art. The representative speedup are $3M + 2D + S$ for a point addition, $D + S$ for a mixed point addition, S for a point doubling, $M + D$ for a differential addition and doubling. Here M, S and D are the cost of a multiplication, a squaring and a multiplication by a constant respectively. Notably, the new complete differential addition and doubling for complete binary Edwards curves with 4-torsion points need only $5M + D + 4S$ which is just the cost of the fastest (but incomplete) formulas among various forms of elliptic curves over finite fields of characteristic 2 in the literature. As a result the binary Edwards form becomes definitely the best option for elliptic curve cryptosytems over binary fields in view of both efficiency and resistance against side channel attack

Keywords: Elliptic curves · Edwards curves · Binary fields · Montgomery ladder · Countermeasures against Side-Channel Attacks · Complete formulas

1 Introduction

Nowadays, the side channel attacks (SCA) have become the most serious threat for cryptosytems [7,11]. The conditional branches in cryptographic algorithms are the major source from which an adversary may catch up information useful for SCA.

The main operation of elliptic curve cryptosystems (ECC) is the computation of $kP = \underbrace{P + P + \cdots + P}_{k}$ given a scalar k and a curve point P, called the point multiplication or the scalar multiplication (ECSM), where the scalar k is usually served as a secrete key. When a point addition $P+Q$ is performed on Weierstrass-form elliptic curve using the standard formulas, one has to test for and handle separately such cases that any of P, Q and $P + Q$ is the point at infinity, or

---

Date of this document: 2014.09.16. This work has been supported by the National Academy of Sciences, D.P.R. of Korea.

© Springer International Publishing Switzerland 2014
W. Meier and D. Mukhopadhyay (Eds.): INDOCRYPT 2014, LNCS 8885, pp. 393–408, 2014.
DOI: 10.1007/978-3-319-13039-2_23

P is equal to Q. Therefore, in order to avoid such weakness due to conditional branches, the (strongly) unified addition formulas have been presented on various forms of elliptic curves. (See [7] for a survey.) The unified addition formulas can be used for point doublings. However, these formulas do not eliminate the need to check for inputs and outputs at infinity and for such other exceptional cases as points of small order, and so they are vulnerable to the attack (called the exceptional points attack, see [11]) which targets the exceptional cases in the unified formulas.

The complete addition formulas work for all pairs of input points without any exceptional case and hence they are the best ideal way to be resistant against the side channel attacks if they exist [2,3,17]. The first complete addition formulas for elliptic curves over finite fields of odd characteristic has been presented by Edwards in [9].

Since elliptic curves over binary fields have wide applications in contemporary public key cryptosystems, efficient and secure implementation of arithmetic on the curve family have been beeing an active research area [1,13,16,23,30]. In CHES2008, Bernstein, Lange and Farashahi [6] firstly presented Edwards form for elliptic curves over binary fields and deduced the first complete operation formulas for binary ordinary elliptic curves. However, as mentioned in [6], their formulas are not as fast as previous (incomplete) formulas. But, the performance of ECSM is a key factor which determines the practical usefulness of ECC.

Sequentially, there have been some efforts [24,26] to speed up ECSM on the binary Edwards curves. But the works have not decreased the costs of the group operation formulas.

On the other side, very recently, many works [8,10,12,19–21,29,32] have been reported to develop new forms, new coordinates and more efficient group operation formulas for binary elliptic curves which outperform the binary Edwards curves. Remarkably, Oliveira, López, Aranha, and Rodríguez-Henríquez [29] (CHES 2013 best paper award) have improved the point doubling by S, the mixed addition by $3S$, the projective point addition by $2M + 2S$ for Weierstrass-form curves, by devising a novel λ-coordinates. However, their formulas on Weierstrass-form curves have no the completeness.

Throughout this paper, we will denote the cost of an inversion, a multiplication, a squaring and a multiplication by a fixed constant (curve coefficient) in the base field as I, M, S and D respectively. Note that for binary fields of cryptographical interest, $I \geq 8M$, $S \approx 0.1 \sim 0.2M, D < M$ (if the curve coefficient is small enough then $D \ll M$, but generally $D \approx 0.5 \sim 1M$) and the cost of field addition is small enough to neglect, as shown in [14–16,29].

Contributions of this Paper. We speed up the (complete) formulas for the binary Edwards curve arithmetic which have been firstly presented in [6] and were the fastest ones on the curves to the best of our knowledge. Significant improvements are obtained in both affine and (standard) projective coordinates. In detail, we save $M + 2D - 2S$ for an affine addition (AA), $3M + 2D + S$ for a projective point addition (PPA), $D + S$ for a mixed point addition (MPA), S for a projective point doubling (PPD), $M + D$ (or $3D - S$) for a mixed differential addition and doubling (MDAD) in general cases. In the cases of curves with 4-torsion points, our savings are $M + D - 2S$ for AA, $2M + D + S$ for PPA, $D + 2S$ for MPA, S for PPD, D for MDAD. We show that for complete binary Edwards curves, the new formulas keep the property of completeness.

Notably, the new MDAD formulas for curves with 4-torsion points cost $5M + D + 4S$. In comparison, the fastest MDAD formulas in the literature give just the same number of operations, but they are not complete. Therefore binary Edwards curves become definitely the speed leader for protected variable-point-and-variable-scalar-multiplications on elliptic curves over binary fields.

The remainder of this paper is organized as follows: In Section 2, we review Edwards form of elliptic curves over binary fields and the best formulas for arithmetic on it in the literature. New fast explicit formulas are developed in Section 3. The results of this paper are compared to previous works in Section 4. Finally, we conclude in Section 5.

2   Binary Edwards Curves

This section reviews the binary Edwards curves and the best point operation formulas on them. All materials described here can be found in [6].

Definition 1. *Let $d_1, d_2 \in \mathbb{F}_{2^n}$ with $d_1 \neq 0$ and $d_2 \neq d_1^2 + d_1$. The binary Edwards curve with coefficients d_1 and d_2 is the non-singular affine algebraic curve*

$$E_{B,d_1,d_2} : (x + y)(d_1 + d_2(x + y)) = xy(1 + x)(1 + y). \tag{2.1}$$

When $Tr_{\mathbb{F}_{2^n}|\mathbb{F}_2}(d_2) = d_2 + d_2^2 + \cdots + d_2^{2^{n-1}} = 1$ is satisfied, the addition law on E_{B,d_1,d_2} is complete. This means that there are no exceptions to the addition law.

Definition 2. *Let $d_1, d_2 \in \mathbb{F}_{2^n}$ with $d_1 \neq 0$ and $Tr_{\mathbb{F}_{2^n}|\mathbb{F}_2}(d_2) = 1$. The curve E_{B,d_1,d_2} is called complete binary Edwards curve with coefficients d_1 and d_2.*

The projective closure of the binary Edwards curve E_{B,d_1,d_2} has two singular points: a blowup for the singular points is presented in [6]. In terms of the group law on E_{B,d_1,d_2}, the point $(0, 0)$ is the neutral element and (y, x) is the negative of point (x, y).

Via the map $(x, y) \mapsto (u, v)$ defined by

$$u = \frac{d_1(d_1^2 + d_1 + d_2)(x + y)}{xy + d_1(x + y)}, v = d_1(d_1^2 + d_1 + d_2)(\frac{x}{xy + d_1(x + y)} + d_1 + 1),$$
$$\tag{2.2}$$

Edwards model E_{B,d_1,d_2} is birationally equivalent to Weierstrass model of non-supersingular elliptic curve

$$v^2 + uv = u^3 + (d_1^2 + d_2)u^2 + d_1^4(d_1^4 + d_1^2 + d_2^2) \tag{2.3}$$

with $j-$invariant $\frac{1}{d_1^4(d_1^4+d_1^2+d_2^2)}$. An inverse map is given as follows:

$$x = \frac{d_1(u + d_1^2 + d_1 + d_2)}{u + v + (d_1^2 + d_1)(d_1^2 + d_1 + d_2)}, y = \frac{d_1(u + d_1^2 + d_1 + d_2)}{v + (d_1^2 + d_1)(d_1^2 + d_1 + d_2)}. \quad (2.4)$$

The binary Edwards curves have an addition law which corresponds to the usual addition law on an elliptic curve in Weierstrass form by the birational equivalence above: the sum (x_3, y_3) of two points (x_1, y_1) and (x_2, y_2) on E_{B,d_1,d_2} is defined by

$$x_3 = \frac{d_1(x_1 + x_2) + d_2(x_1 + y_1)(x_2 + y_2) + (x_1 + x_1^2)(x_2(y_1 + y_2 + 1) + y_1 y_2)}{d_1 + (x_1 + x_1^2)(x_2 + y_2)},$$

$$y_3 = \frac{d_1(y_1 + y_2) + d_2(x_1 + y_1)(x_2 + y_2) + (y_1 + y_1^2)(y_2(x_1 + x_2 + 1) + x_1 x_2)}{d_1 + (y_1 + y_1^2)(x_2 + y_2)}.$$

$$(2.5)$$

The addition law on E_{B,d_1,d_2} is strongly unified which means it can be used with two identical inputs, i.e., to double.

For $n \geq 3$, each ordinary elliptic curve over $\mathbb{F}_{2^n}$ is birationally equivalent over $\mathbb{F}_{2^n}$ to a complete binary Edwards curve. As aforementioned, the addition law has the very nice feature of completeness for complete binary Edwards curves.

Through §5, §6, §7 in [6], Bernstein, Lange and Farashahi presented explicit formulas in both affine coordinates and standard projective coordinates that we show below. See also the Explicit-Formulas Database [4].

A. Affine point addition $(x_1, y_1) + (x_2, y_2) = (x_3, y_3)$: $I + 11M + 3D + 2S$

$$x_3 = y_1 + \frac{d_2(x_1 + y_1)(x_2 + y_2) + d_1(x_1 + y_1 + x_2) + (x_1 + x_1^2)(x_2 y_2 + x_2)}{d_1 + (x_1 + x_1^2)(x_2 + y_2)},$$

$$y_3 = x_1 + \frac{d_2(x_1 + y_1)(x_2 + y_2) + d_1(x_1 + y_1 + y_2) + (y_1 + y_1^2)(x_2 y_2 + y_2)}{d_1 + (y_1 + y_1^2)(x_2 + y_2)}.$$

$$(2.6)$$

B. Affine ω-coordinates differential addition and doubling $\omega_i = x_i + y_i(0 \leq i \leq 4)$, $2(x_1, y_1) = (x_4, y_4)$, $(x_1, y_1) + (x_2, y_2) = (x_3, y_3)$ with known $(x_0, y_0) = (x_2, y_2) - (x_1, y_1)$: $I + 7M + 2D + 3S$ ($I + 4M + 2D + 3S$ if $d_1 = d_2$) Bernstein-Lange-Farashahi formulas for the affine point doubling are:

$$x_4 = 1 + \frac{d_1 + d_2(x_1^2 + y_1^2) + y_1^2 + y_1^4}{d_1 + x_1^2 + y_1^2 + (d_2/d_1)(x_1^4 + y_1^4)},$$

$$y_4 = 1 + \frac{d_1 + d_2(x_1^2 + y_1^2) + x_1^2 + x_1^4}{d_1 + x_1^2 + y_1^2 + (d_2/d_1)(x_1^4 + y_1^4)}.$$

$$(2.7)$$

In particular, if $d_1 = d_2$,

$$x_4 = \frac{d_1(x_1^2 + y_1^2) + x_1^2 + x_1^4}{d_1 + x_1^2 + y_1^2 + x_1^4 + y_1^4}, y_4 = 1 + x_4 + \frac{d_1}{d_1 + x_1^2 + y_1^2 + x_1^4 + y_1^4}. \quad (2.8)$$

From above affine doubling formulas,

$$\omega_4 = \frac{\omega_1^2 + \omega_1^4}{d_1 + \omega_1^2 + (d_2/d_1)\omega_1^4} \tag{2.9}$$

and if $d_1 = d_2$,

$$\omega_4 = 1 + \frac{d_1}{d_1 + \omega_1^2 + \omega_1^4}. \tag{2.10}$$

Differential addition is also possible by $\omega-$coordinates only:

$$\omega_3 = \omega_0 + \frac{\omega_1\omega_2(1+\omega_1)(1+\omega_2)}{d_1 + \omega_1\omega_2((1+\omega_1)(1+\omega_2) + (\frac{d_2}{d_1}+1)\omega_1\omega_2)}. \tag{2.11}$$

In the case $d_1 = d_2$,

$$\omega_3 = \omega_0 + 1 + \frac{d_1}{d_1 + \omega_1\omega_2(1+\omega_1)(1+\omega_2)}. \tag{2.12}$$

Therefore, in the affine $\omega-$ coordinates, a differential addition and doubling, i.e., one step of a Montgomery ladder costs $I + 7M + 2D + 3S$ considering the computations shared in the differential addition and the doubling ($I + 4M + 2D + 3S$ if $d_1 = d_2$), using Montgomery's simultaneous inversion trick.

A point $(X : Y : Z)$ in the standard projective coordinates corresponds to affine point $(X/Z, Y/Z)$. Using (2.6)-(2.12), Bernstein, Lange and Farashahi presented explicit projective formulas which are given in next subsections.

C. Projective point addition $(X_1 : Y_1 : Z_1) + (X_2 : Y_2 : Z_2) = (X_3 : Y_3 : Z_3)$: $18M + 6D + 3S$ or $18M + 7D + 2S$ ($16M + 3D + 2S$ or $16M + 4D + S$ if $d_1 = d_2$)

$$A = X_1 \cdot X_2, B = Y_1 \cdot Y_2, C = Z_1 \cdot Z_2, D = d_1 \cdot C, E = C^2, F = D^2,$$
$$G = (X_1 + Z_1) \cdot (X_2 + Z_2), H = (Y_1 + Z_1) \cdot (Y_2 + Z_2), K = (X_1 + Y_1) \cdot (X_2 + Y_2),$$
$$I = A + G, J = B + H, U = C \cdot (F + d_1 \cdot K \cdot (K + I + J + C)),$$
$$V = U + D \cdot F + K \cdot (d_2 \cdot (d_1 \cdot E + G \cdot H + A \cdot B) + (d_1 + d_2) \cdot I \cdot J),$$
$$X_3 = V + D \cdot (A + D) \cdot (G + D), Y_3 = V + D \cdot (B + D) \cdot (H + D),$$
$$Z_3 = U + (d_1 + d_2) \cdot C \cdot K^2.$$

D. Mixed point addition $(X_1 : Y_1 : Z_1) + (X_2 : Y_2 : 1) = (X_3 : Y_3 : Z_3)$: $13M + 3D + 3S$ (the same cost when $d_1 = d_2$)

$$W_1 = X_1 + Y_1, W_2 = X_2 + Y_2, A = X_2^2 + X_2, B = Y_2^2 + Y_2,$$
$$D = W_1 \cdot Z_1, E = d_1 \cdot Z_1^2, H = (E + d_2 \cdot D) \cdot W_2,$$
$$I = d_1 \cdot Z_1, U = E + A \cdot D, V = E + B \cdot D, Z_3 = U \cdot V,$$
$$X_3 = Z_3 \cdot Y_2 + (H + X_1 \cdot (I + A \cdot (Y_1 + Z_1))) \cdot V,$$
$$Y_3 = Z_3 \cdot X_2 + (H + Y_1 \cdot (I + B \cdot (X_1 + Z_1))) \cdot U.$$

E. Projective point doubling $2(X_1 : Y_1 : Z_1) = (X_4 : Y_4 : Z_4)$: $2M + 3D + 6S$ ($2M + 2D + 6S$ if $d_1 = d_2$)

$$A = X_1^2, B = A^2, C = Y_1^2, D = C^2, E = Z_1^2, F = d_1 \cdot E^2,$$
$$G = (d_2/d_1) \cdot (B + D), H = A \cdot E, I = C \cdot E, J = H + I, K = G + d_2 \cdot J,$$
$$Z_4 = F + J + G, X_4 = K + H + D, Y_4 = K + I + B.$$

F. Mixed ω-coordinates differential addition and doubling $W_i = X_i + Y_i (1 \leq i \leq 4)$, $2(X_1 : Y_1 : Z_1) = (X_4 : Y_4 : Z_4)$, $(X_1 : Y_1 : Z_1) + (X_2 : Y_2 : Z_2) = (X_3 : Y_3 : Z_3)$ with known ω_0 where $(x_0 : y_0 : 1) = (X_2 : Y_2 : Z_2) - (X_1 : Y_1 : Z_1)$: $6M + 4D + 4S$ ($5M + 2D + 4S$ if $d_1 = d_2$)

$$C = W_1 \cdot (W_1 + Z_1), D = W_2 \cdot (W_2 + Z_2), E = Z_1 \cdot Z_2, F = W_1 \cdot W_2,$$
$$V = C \cdot D, U = V + (\sqrt{d_1} \cdot E + \sqrt{d_2/d_1 + 1} \cdot F)^2,$$
$$W_3 = V + \omega_0 \cdot U, Z_3 = U, W_4 = C^2, Z_4 = W_4 + (\sqrt[4]{d_1} \cdot Z_1 + \sqrt[4]{d_2/d_1 + 1} \cdot W_1)^4.$$

3   New Fast Explicit Formulas

In this section we will show that all of the formulas (A-F) of §2 can be significantly improved. For clear comparison, the subsections that describe our improvements to (A-F) of §2 are labeled as (AA-FF) respectively. Among the binary Edwards curves, the case $d_1 = d_2$ will be of great interest, since the group arithmetic on the curves is significantly simpler than the case $d_1 \neq d_2$ as shown in the previous works.

Here, we would like to point out that the (complete) binary Edwards curves with the coefficients $d_1 = d_2$ are not too special. In fact, as stated in the lemma 1 below which also can be seen as a direct consequence of theorem 4.3 in [6], every binary Edwards curve (complete binary Edwards curve, resp.) with the coefficients $d_1 = d_2$ corresponds one-to-one with an ordinary Weierstrass-form curve $E_{W,0,a_6} : v^2 + uv = u^3 + a_6$ (with $Tr_{\mathbb{F}_{2^n}|\mathbb{F}_2}(a_6) = 1$, resp.) by birational equivalence and hence represent a half (a quarter, resp.) of the ordinary elliptic curves over binary fields. Note that in the context of elliptic curves, the concept of birational equivalence preserving the neutral elements coincides with one of isomorphism, since by definition the curves are nonsingular [21].

Lemma 1. *An ordinary elliptic curve over $\mathbb{F}_{2^n}$ is birationally equivalent over $\mathbb{F}_{2^n}$ to a binary Edwards curve with $d_1 = d_2$ if and only if it has a rational point of order 4. About half (exactly 2^{n-1} classes) of all $2^n - 1$ isomorphism classes of ordinary elliptic curves over $\mathbb{F}_{2^n}$ which have a rational point of order 4 are birationally equivalent over $\mathbb{F}_{2^n}$ to a complete binary Edwards curve with $d_1 = d_2$.*

Proof. very ordinary elliptic curve over binary field $\mathbb{F}_{2^n}$ is isomorphic over $\mathbb{F}_{2^n}$ to a Weierstrass-form curve $E_{W,a_2,a_6} : v^2 + uv = u^3 + a_2 u^2 + a_6$, where $a_6 \in \mathbb{F}_{2^n}^*$

and $a_2 \in \{0, \alpha\}$ with α a fixed element in $\mathbb{F}_{2^n}$ of trace $Tr_{\mathbb{F}_{2^n}|\mathbb{F}_2}(\alpha) = 1$. The curve E_{W,a_2,a_6} has a $\mathbb{F}_{2^n}$−rational point of order 4 if and only if $a_2 = 0$ [15]. The curve $E_{W,0,a_6}$ is isomorphic to the curve $E_{W,d_1+d_1^2,a_6}$ for every $d_1 \in \mathbb{F}_{2^n}$ by the isomorphism $v \rightarrow v + d_1u$. Again, with $d_1 = \sqrt[8]{a_6}$, $E_{W,d_1+d_1^2,a_6}$ is birationally equivalent to the binary Edwards curve E_{B,d_1,d_1} by (2.2) and (2.4). E_{B,d_1,d_1} is a complete binary Edwards curve if and only if $Tr_{\mathbb{F}_{2^n}|\mathbb{F}_2}(d_1) = 1$. These complete the proof of Lemma 1.

AA. New formulas - Affine point addition $(x_1, y_1) + (x_2, y_2) = (x_3, y_3)$: $I + 10M + 1D + 4S$

The following strongly unified formulas, given (x_1, y_1) and (x_2, y_2) on the binary Edwards curve E_{B,d_1,d_2}, compute the sum $(x_3, y_3) = (x_1, y_1) + (x_2, y_2)$ if it is defined:

$$x_3 = \frac{d_2(x_1 + y_1)(x_1 + y_1 + x_2 + y_2) + (x_1 + x_1^2)((x_2 + y_1)^2 + x_2 + y_1)}{d_1 + (x_1 + x_1^2)(x_2 + y_2)} + x_2 + y_1,$$

$$y_3 = \frac{d_2(x_1 + y_1)(x_1 + y_1 + x_2 + y_2) + (y_1 + y_1^2)((y_2 + x_1)^2 + y_2 + x_1)}{d_1 + (y_1 + y_1^2)(x_2 + y_2)} + y_2 + x_1.$$

$$(3.13)$$

Note that we have changed (2.6) using the fact that the point (x_1, y_1) satisfies the curve equation (2.1). For complete binary Edwards curves the denominators $d_1 + (x_1 + x_1^2)(x_2 + y_2)$ and $d_1 + (y_1 + y_1^2)(x_2 + y_2)$ cannot be zero.

Above formulas can be implemented using $I + 10M + 1D + 4S$ by sequence of operations:

$$S = x_1 + x_1^2, T = y_1 + y_1^2, E = x_2 + y_1, F = y_2 + x_1,$$

$$A = d_1 + S \cdot (x_2 + y_2), B = d_1 + T \cdot (x_2 + y_2), Z = A \cdot B,$$

$$D = \frac{1}{Z}, Z1 = D \cdot B, Z2 = D \cdot A, M = d_2 \cdot (x_1 + y_1) \cdot (E + F),$$

$$U = S \cdot (E^2 + E), V = T \cdot (F^2 + F),$$

$$x_3 = Z1 \cdot (M + U) + E, y_3 = Z2 \cdot (M + V) + F.$$

BB. New formulas - Affine ω-coordinates differential addition and doubling $\omega_i = x_i + y_i (0 \le i \le 4)$, $2(x_1, y_1) = (x_4, y_4)$, $(x_1, y_1) + (x_2, y_2) = (x_3, y_3)$ with known $(x_0, y_0) = (x_2, y_2) - (x_1, y_1)$: $I + 7M + 2D + 3S$ ($I + 4M + 1D + 3S$ if $d_1 = d_2$)
From the Lemma 3.1 of [6] (or from the page 261 therein), we know:

$$\omega_3 + \omega_0 = \frac{(\omega_1 + \omega_1^2)(\omega_2 + \omega_2^2)}{d_1 + (\omega_1 + \omega_1^2)(\omega_2 + \omega_2^2) + (\frac{d_2}{d_1} + 1)\omega_1^2\omega_2^2}, \qquad (3.14)$$

$$\omega_3\omega_0 = \frac{d_1(\omega_1^2 + \omega_2^2)}{d_1 + (\omega_1 + \omega_1^2)(\omega_2 + \omega_2^2) + (\frac{d_2}{d_1} + 1)\omega_1^2\omega_2^2} \qquad (3.15)$$

and in the case $d_1 = d_2$

$$\omega_3 = \omega_0 + 1 + \frac{d_1}{d_1 + \omega_1\omega_2(1 + \omega_1)(1 + \omega_2)}. \tag{3.16}$$

Let us assume that $\omega_0 \neq 0, 1$.(We will describe why the assumption is valid for cryptographic applications, in the end of this section.)

From (3.14), it follows

$$\frac{1}{d_1 + (\omega_1 + \omega_1^2)(\omega_2 + \omega_2^2) + (\frac{d_2}{d_1} + 1)\omega_1^2\omega_2^2} = \frac{\omega_3 + \omega_0}{(\omega_1 + \omega_1^2)(\omega_2 + \omega_2^2)}. \tag{3.17}$$

Substituting (3.17) to (3.15), we have $\frac{\omega_3\omega_0}{d_1(\omega_1^2 + \omega_2^2)} = \frac{\omega_3 + \omega_0}{(\omega_1 + \omega_1^2)(\omega_2 + \omega_2^2)}$, i.e.,

$$\omega_3 = \frac{d_1(\omega_1^2 + \omega_2^2)}{(\omega_1 + \omega_1^2)(\omega_2 + \omega_2^2) + \frac{1}{\omega_0} \cdot d_1(\omega_1^2 + \omega_2^2)}. \tag{3.18}$$

In the case $d_1 = d_2$, from (3.16) we have $\frac{1}{d_1 + \omega_1\omega_2(1+\omega_1)(1+\omega_2)} = \frac{\omega_3 + \omega_0 + 1}{d_1}$ and substituting it to (3.15), $\frac{\omega_3\omega_0}{d_1(\omega_1^2 + \omega_2^2)} = \frac{\omega_3 + \omega_0 + 1}{d_1}$.

In the case $d_1 = d_2$, therefore we have

$$\omega_3 = 1 + \frac{\omega_1^2 + \omega_2^2}{(\frac{1}{\omega_0}) \cdot (\omega_1^2 + \omega_2^2) + 1} \tag{3.19}$$

or

$$\omega_3 = \omega_0 + 1 + \frac{1}{(\frac{1}{\omega_0 + \omega_0^2}) \cdot (\omega_1^2 + \omega_2^2 + \omega_0)}.$$

Here it should be noted that $\frac{1}{\omega_0 + \omega_0^2}$ is a base-point-dependent constant fixed during the scalar multiplication. Therefore, the cost of new formulas for the differential addition is $I + 3M + D + 2S(I + M + 2S$ if $d_1 = d_2)$. Combining with Bernstein-Lange-Farashahi doubling formulas((2.9) and (2.10)) and using Montgomery's simultaneous inversion trick, an ω-coordinates differential addition and doubling, i.e., one step of a Montgomery ladder costs $I + 7M + 2D + 3S$ for general d_1, d_2 and $I + 4M + 1D + 3S$ in the case $d_1 = d_2$. Here it should be taken into account that a squaring ω_1^2 is shared in the differential addition and the doubling.

We will show that under the assumption $\omega_0 \neq 0, 1$, the denominators are never zero and hence the formulas are complete.

Lemma 2. *On the complete binary Edwards curves, if $\omega_0 \neq 0$ then $(\omega_1 + \omega_1^2)(\omega_2 + \omega_2^2) + \frac{1}{\omega_0} \cdot d_1(\omega_1^2 + \omega_2^2) \neq 0$.*

Proof. Suppose that $(\omega_1 + \omega_1^2)(\omega_2 + \omega_2^2) + \frac{1}{\omega_0} \cdot d_1(\omega_1^2 + \omega_2^2) = 0$. Two cases are possible: $(\omega_1 + \omega_1^2)(\omega_2 + \omega_2^2) = \omega_1^2 + \omega_2^2 = 0$ or $(\omega_1 + \omega_1^2)(\omega_2 + \omega_2^2) = \frac{1}{\omega_0} \cdot d_1(\omega_1^2 + \omega_2^2) \neq 0$. If we are in the first case, then it follows $\omega_3 = \omega_0$ from (3.14) and $\omega_3\omega_0 = 0$ from (3.15). Namely, $\omega_3 = \omega_0 = 0$ which contradicts the assumption $\omega_0 \neq 0$ of the lemma. If we are in the second case that $(\omega_1 + \omega_1^2)(\omega_2 + \omega_2^2) = \frac{1}{\omega_0} \cdot d_1(\omega_1^2 + \omega_2^2) \neq 0$, then from (3.14) and (3.15) it follows that $(\omega_3 + \omega_0)\omega_0 = \omega_3\omega_0$, i.e., $\omega_0 = 0$, again it is a contradiction.

Lemma 3. *On the complete binary Edwards curves with $d_1 = d_2$, if $\omega_0 \neq 0, 1$ then $\frac{1}{\omega_0} \cdot (\omega_1^2 + \omega_2^2) + 1 \neq 0$.*

Proof. Suppose that $\frac{1}{\omega_0} \cdot (\omega_1^2 + \omega_2^2) + 1 = 0$, or equivalently $\omega_0 = \omega_1^2 + \omega_2^2$. Then by (3.15) we have $\omega_3 = \frac{d_1}{d_1 + (\omega_1 + \omega_1^2)(\omega_2 + \omega_2^2)}$ and then by (3.16) $\omega_3 + \omega_0 + 1 = \omega_3$, namely $\omega_0 = 1$ which contradicts the assumption $\omega_0 \neq 1$ of the lemma.

Above we presented new fast point operation formulas in the affine coordinates, which require at least a field inversion. However, in practice, inversion free formulas using the projective coordinates are generally much faster than affine formulas since the field inversion is much more expensive than a field multiplication. By the projectivization of (3.13)-(3.14), we can present new fast explicit projective formulas which are described in next subsections.

CC. New formulas - Projective point addition $(X_1 : Y_1 : Z_1) + (X_2 : Y_2 : Z_2) = (X_3 : Y_3 : Z_3)$: $15M + 4D + 2S$ $(14M + 2D + S$ if $d_1 = d_2)$
The explicit formulas

$$A = X_1 \cdot X_2, B = Y_1 \cdot Y_2, C = Z_1 \cdot Z_2, D = d_1 \cdot C, F = D^2, L = \frac{1}{d_1} \cdot F,$$

$$G = (X_1 + Z_1) \cdot (X_2 + Z_2), H = (Y_1 + Z_1) \cdot (Y_2 + Z_2), K = (X_1 + Y_1) \cdot (X_2 + Y_2),$$

$$I = A + G, J = B + H, U = K \cdot (K + I + J + C),$$

$$V = K \cdot (\frac{d_2}{d_1} \cdot (L + F + (G + D + B) \cdot (A + D + H) + (A + D) \cdot (G + D) +$$

$$(B + D) \cdot (H + D) + I \cdot J) + IJ),$$

$$X_3 = V + C \cdot ((A + D)(G + D) + L + F + U),$$

$$Y_3 = V + C \cdot ((B + D)(H + D) + L + F + U),$$

$$Z_3 = C \cdot (U + L + (1 + \frac{d_2}{d_1}) \cdot K^2)$$

use just $15M + 4D + 2S$.

Our improvement of $3M + 2D + S$ came from some modifications on Bernstein-Lange-Farashahi addition formulas (C. of §2). First of all we eliminate two $D's$ using the property of projective coordinates by which it holds $(X_3 : Y_3 : Z_3) = (d_1 X_3 : d_1 Y_3 : d_1 Z_3)$. Then we use an equality $GH + AB = (G + D + B)(A + D + H) + (A + D)(G + D) + (B + D)(H + D) + D^2$ to save one M. Remainder of the speedup came from rearranging the computations in Bernstein-Lange-Farashahi addition formulas. For complete binary Edwards curves the denominator Z_3 cannot be zero.

If $d_1 = d_2$ then the explicit formulas

$$A = X_1 \cdot X_2, B = Y_1 \cdot Y_2, C = Z_1 \cdot Z_2, D = d_1 \cdot C, F = D^2, L = \frac{1}{d_1} \cdot F,$$

$$G = (X_1 + Z_1) \cdot (X_2 + Z_2), H = (Y_1 + Z_1) \cdot (Y_2 + Z_2), K = (X_1 + Y_1) \cdot (X_2 + Y_2),$$

$$I = A + G, J = B + H, U = K \cdot (K + I + J + C),$$

$$V = K \cdot (L + F + (G + D + B) \cdot (A + D + H) +$$
$$(A + D) \cdot (G + D) + (B + D) \cdot (H + D)),$$
$$X_3 = V + C \cdot ((A + D)(G + D) + L + F + U),$$
$$Y_3 = V + C \cdot ((B + D)(H + D) + L + F + U),$$
$$Z_3 = C \cdot (U + L)$$

use $14M + 2D + S$. For complete binary Edwards curves the denominator Z_3 cannot be zero.

DD. New formulas - Mixed point addition $(X_1 : Y_1 : Z_1) + (X_2 : Y_2 : 1) = (X_3 : Y_3 : Z_3)$: $13M + 2D + 2S$ ($13M + 2D + S$ if $d_1 = d_2$)
The explicit formulas

$$U = d_1 \cdot Z_1 + (X_2 + X_2^2) \cdot (X_1 + Y_1), V = d_1 \cdot Z_1 + (Y_2 + Y_2^2) \cdot (X_1 + Y_1),$$
$$W = d_2 \cdot (X_2 \cdot Z_1 + Y_2 \cdot Z_1) \cdot (X_2 Z_1 + Y_2 Z_1 + X_1 + Y_1), Z_3 = Z_1 \cdot U \cdot V,$$
$$X_3 = V \cdot (W + (X_1 + Y_2 Z_1) \cdot ((X_2 + X_2^2) \cdot (X_1 + Z_1 + Y_2 Z_1) + U)),$$
$$Y_3 = U \cdot (W + (Y_1 + X_2 Z_1) \cdot ((Y_2 + Y_2^2) \cdot (Y_1 + Z_1 + X_2 Z_1) + V))$$

use just $13M + 2D + 2S$. For complete binary Edwards curves the denominator Z_3 cannot be zero. If $d_1 = d_2$ then the explicit formulas

$$A = X_1 \cdot X_2, B = Y_1 \cdot Y_2, D = d_1 \cdot Z_1, F = D^2, L = \frac{1}{d_1} \cdot F,$$
$$G = (X_1 + Z_1) \cdot (X_2 + Z_2), H = (Y_1 + Z_1) \cdot (Y_2 + Z_2), K = (X_1 + Y_1) \cdot (X_2 + Y_2),$$
$$I = A + G, J = B + H, U = K \cdot (K + I + J + C),$$
$$V = K \cdot (L + F + (G + D + B) \cdot (A + D + H) + (A + D) \cdot (G + D) + (B + D) \cdot (H + D)),$$
$$X_3 = V + Z_1 \cdot ((A + D)(G + D) + L + F + U),$$
$$Y_3 = V + Z_1 \cdot ((B + D)(H + D) + L + F + U),$$
$$Z_3 = Z_1 \cdot (U + L)$$

use $13M + 2D + 1S$. For complete binary Edwards curves the denominator Z_3 cannot be zero.

EE. New formulas - Projective point doubling $2(X_1 : Y_1 : Z_1) = (X_4 : Y_4 : Z_4)$: $2M + 3D + 5S$ ($2M + 2D + 5S$ if $d_1 = d_2$)
The improvement is an easy speedup of the Bernstein-Lange-Farashahi formulas. By simply rewriting the Bernstein-Lange-Farashahi doubling formulas we get following doubling formulas

$$A = X_1 + Y_1, B = A^2, C = A \cdot Z_1, D = X_1 \cdot (X_1 + Z_1),$$
$$E = \sqrt{\frac{d_2}{d_1} + 1} \cdot B, F = \sqrt{d_2} \cdot C, G = \sqrt{d_1} \cdot Z_1^2, H = E + F, I = B + C, \quad (3.20)$$
$$X_4 = (H + D)^2, Y_4 = (H + I + D)^2, Z_4 = (E + I + G)^2$$

which use $2M + 3D + 5S$. Note that our doubling formulas require one less squaring than Bernstein-Lange-Farashahi doubling formulas, keeping the count of other field operations including field additions. For complete binary Edwards curves the denominator Z_4 cannot be zero. In particular, if $d_1 = d_2$ then the explicit formulas

$$A = X_1 + Y_1, B = A^2, C = A \cdot Z_1, D = X_1 \cdot (X_1 + Z_1),$$
$$F = \sqrt{d_1} \cdot C, G = \sqrt{d_1} \cdot Z_1^2, I = B + C, \tag{3.21}$$
$$X_4 = (F + D)^2, Y_4 = (F + I + D)^2, Z_4 = (I + G)^2$$

use $2M + 2D + 5S$. Notice that our doubling formulas require not only one less field squaring, but also less field additions by two than Bernstein-Lange-Farashahi doubling formulas. For complete binary Edwards curves the denominator Z_4 cannot be zero.

FF. New formulas - Mixed ω-Coordinates differential addition and doubling $W_i = X_i + Y_i (1 \leq i \leq 4)$, $2(X_1 : Y_1 : Z_1) = (X_4 : Y_4 : Z_4)$, $(X_1 : Y_1 : Z_1) + (X_2 : Y_2 : Z_2) = (X_3 : Y_3 : Z_3)$ with known ω_0 where $(x_0 : y_0 : 1) = (X_2 : Y_2 : Z_2) - (X_1 : Y_1 : Z_1)$: $5M + 3D + 4S$ or $6M + D + 5S$ ($5M + D + 4S$ if $d_1 = d_2$)

We derive formulas for ω-coordinates projective differential addition and doubling by using ω-coordinates differential addition (3.18) and doubling (2.9). The explicit formulas

$$C = W_1 \cdot (W_1 + Z_1), D = W_2 \cdot (W_2 + Z_2),$$
$$E = (W_1 + W_2) \cdot (W_1 + W_2 + Z_1 + Z_2) + C + D,$$
$$V = C \cdot D, U = d_1 \cdot E^2,$$
$$Z_3 = V + \frac{1}{\omega_0} \cdot U, W_3 = U,$$
$$W_4 = C^2, Z_4 = W_4 + (\sqrt[4]{d_1} \cdot Z_1 + \sqrt[4]{d_2/d_1 + 1} \cdot W_1)^4$$

use $5M + 3D + 4S$. For complete binary Edwards curves, if $\omega_0 \neq 0$ then the denominators are never zero.

If $M \leq 2D + S$ then we will present the explicit formulas

$$C = W_1 \cdot (W_1 + Z_1), D = W_2 \cdot (W_2 + Z_2),$$
$$E = (W_1 + W_2) \cdot (W_1 + W_2 + Z_1 + Z_2) + C + D,$$
$$V = C \cdot D, U = d_1 \cdot E^2,$$
$$W_3 = V + \frac{1}{\omega_0} \cdot U, Z_3 = U,$$
$$W_4 = C^2, Z_4 = W_4 + ((\sqrt[4]{d_1} + W_1) \cdot (\sqrt[4]{d_2/d_1 + 1} + Z_1) + C + W_1^2)^4 + d_1 + d_2$$

which use $6M + D + 5S$. For complete binary Edwards curves if $\omega_0 \neq 0$, then the denominators are never zero.

If $d_1 = d_2$, then the explicit formulas, which have been derived by using ω-coordinates differential addition (3.19) and doubling (2.10),

$$C = (W_1 \cdot Z_2 + W_2 \cdot Z_1)^2, D = (Z_1 \cdot Z_2)^2, E = \frac{1}{\omega_0} \cdot C,$$

$$Z_3 = D + E, W_3 = C + Z_3,$$

$$W_4 = (W_1 \cdot (W_1 + Z_1))^2, Z_4 = W_4 + d_1 \cdot Z_1^4$$

use $5M + D + 5S$. For complete binary Edwards curves if $\omega_0 \neq 0, 1$, then the denominators are never zero.

By using the co-Z trick, we can further save one S. Here, a triple (W_1, W_2, Z) is used to represent a pair of affine points with ω-coordinates $\frac{W_1}{Z}, \frac{W_2}{Z}$ and one step of the Montgomery ladder is represented as formulas which output (W_3, W_4, Z') on the input (W_1, W_2, Z). The explicit formulas

$$C = (W_1 + W_2)^2, D = Z^2, E = \frac{1}{\omega_0} \cdot C, V = D + E,$$

$$U = C + V, S = (W_1 \cdot (W_1 + Z))^2, T = S + d_1 \cdot D^2,$$

$$W_3 = U \cdot T, W_4 = V \cdot S, Z' = V \cdot T$$

use just $5M + D + 4S$. Note that in the derivation we have taken into account that $(W, Z) = (\lambda W, \lambda Z)$ for every nonzero $\lambda \in \mathbb{F}_{2^n}$.

For complete binary Edwards curves, if $\omega_0 \neq 0, 1$ then the denominator $Z' = ((W_1^2 + W_1 Z)^2 + d_1 Z^4 + (d_2/d_1 + 1)W_1^4)(\frac{1}{\omega_0}(W_1 + W_2)^2 + Z^2) \neq 0$. In fact, $(W_1^2 + W_1 Z)^2 + d_1 Z^4 + (d_2/d_1 + 1)W_1^4 \neq 0$ is evident, since $(W_1^2 + W_1 Z)^2 + d_1 Z^4 + (d_2/d_1 + 1)W_1^4 = 0$ is equivalent to $d_2 = \frac{d_1}{W_1^2} + (\frac{d_1}{W_1^2})^2$ which contradicts the condition for the curve to be complete and $\frac{1}{\omega_0}(W_1 + W_2)^2 + Z^2 \neq 0$ was already shown by the lemma 3.

These formulas assume that ω_0 is known, or checked, to be neither zero nor one. If $\omega_0 = 0$ or $\omega_0 = 1$, then one must resort to the previous formulas for ω_3. But $\omega_0 = 0$ or $\omega_0 = 1$ (in the case $d_1 = d_2$) means that the base point of the scalar multiplication is $(0,0)$ or $(1,1)$ which have the order 1 or 2, or, $(0,1)$ or $(1,0)$ (in the case $d_1 = d_2$) which have the order 4. But in cryptographic applications, it is typical to use a base point of odd order in order to thwart Pohlig-Hellman attack. Consequently the practice justifies $\omega_0 \neq 0, 1$ and the differential formulas work for arbitrary $\omega_1, \omega_2, \omega_3, \omega_4$, i.e., complete.

4   Comparison with Previous Works

Table 1 shows the cost comparison between the new formulas and Bernstein-Lange-Farashahi formulas on the general binary Edwards curves.

Table 1. Comparison of operation costs on the binary Edwards curves with $d_1 \neq d_2$

Operation	Bernstein-Lange -Farashahi [6]	This Paper	Saved Costs
Affine addition	$I + 11M + 3D + 2S$	$I + 10M + D + 4S$	$M + 2D - 2S$
Projective doubling	$2M + 3D + 6S$	$2M + 3D + 5S$	S
Projective addition	$18M + 6D + 3S$	$15M + 4D + 2S$	$3M + 2D + S$
Mixed addition	$13M + 3D + 3S$	$13M + 2D + 2S$	$D + S$
Mixed differential addition and doubling	$6M + 4D + 4S$	$5M + 3D + 4S$ $(6M + D + 5S)$	$M + D$ $(3D - S)$

Table 2. Comparison of operation costs on the binary Edwards curves with $d_1 = d_2$

Operation	Bernstein-Lange -Farashahi [6]	This Paper	Saved Costs
Affine addition	$I + 11M + 2D + 2S$	$I + 10M + D + 4S$	$M + D - 2S$
Affine differential addition and doubling	$I + 4M + 2D + 3S$	$I + 4M + D + 3S$	D
Projective doubling	$2M + 2D + 6S$	$2M + 2D + 5S$	S
Projective addition	$16M + 3D + 2S$	$14M + 2D + S$	$2M + D + S$
Mixed addition	$13M + 3D + 3S$	$13M + 2D + S$	$D + 2S$
Mixed differential addition and doubling	$5M + 2D + 4S$	$5M + D + 4S$	D

Binary Edwards curves with 4-torsion points have been found attractive for the cryptographic applications, since they have much faster point operation formulas than ones on the general curves. The table 2 shows the improvements this paper achieves in the case.

Next, we compare the performance of the binary Edwards curves with other curve forms as to protected variable-point-variable-scalar-multiplications.

In order to protect ECSM from simple power analysis attack (SPA), various countermeasures have been proposed using a variety of curve representations with aim at making the sequence of operations regular and keeping the efficiency; e.g., the double-addition-always method (DAA), atomic double and addition formulas based method (ADA), unified addition formula based method (UA), Montgomery ladder [25,27] and Joye's ladder [18] based methods. But DAA and ADA need insertion of dummy instructions or operations into ECSM, which becomes the target of the safe-error attack, a kind of fault attack (see [11]).

Montgomery ladder and Joye's ladder may also be amenable to attacks if they depend on incomplete DAD formulas. It should be noted that when a scalar k is less than the order of base point P, implementors using Montgomery ladder or Joye's ladder based on incomplete formulas for ECSM kP may need not to consider any exceptional case that the formulas exclude as it is warranted that such cases would never appear and so the implementation resists SPA. But there exists another attack class called as differential SCA which is much more intricate

Table 3. Costs per bit and security of Montgomery ladder for various curve forms

Model	Cost	Completeness
Weierstrass [4]	$5M + D + 4S$	no
Binary Edwards [6]	$5M + 2D + 4S$	yes
Gaudry-Lubicz [12]	$5M + D + 5S$	yes
Hessian [10]	$5M + 2D + 4S$	yes
Binary Huff [8]	$5M + D + 5S$	yes
μ_4-normal [21]	$5M + D + 5S$	yes
Wu-Tang-Feng [32]	$5M + D + 5S$	yes
Binary Edwards (This Paper)	$5M + D + 4S$	yes

than SPA and can be applied if a scalar is fixed in many implementations of ECSM [11]. A common countermeasure regarding that attack is to randomize the scalar by adding any multiple of the group order before starting new ECSM. Therefore it should not be assumed for scalars to be less than the group order.

Hence the best way to thwart SPA is to use Montgomery ladder and Joye's ladder relying on differential addition and doubling (DAD) formulas with the completeness. But unfortunately, all of the complete DAD formulas in the literature are slower than the fastest non-complete DAD formulas which is achieved on the Weierstrass form. In the table 3, we compare the costs per bit of Montgomery ladder for various forms of nonsupersingular elliptic curves over binary fields with 4-torsion points. The comparison shows that only the complete binary Edwards curves are highest in both of efficiency and security.

5   Conclusion

We presented new fast point operation formulas for binary Edwards curves which outperform state-of-the-art formulas. Importantly, the faster formulas keep to have property of completeness for complete binary Edwards curves.

If in a polynomial basis representation the definition polynomial of base field are not of special form, the modular reduction can be an expensive part of field multiplication. In such case, it is effective to make use of so called lazy reduction trick that by an example computes $a \cdot b + c \cdot d$ not by two separate field multiplications (which need for reduction two times) followed an addition, but by two multiplications without reduction, addition of the results and then only one reduction. We here omitted the description of point operation algorithms which optimally realize the presented formulas using that trick and regarding number of required registers, since deriving such optimized algorithms from the presented formulas is almost immediate with a little care. (see [22, 28] for example).

To sum up, equipped with the new formulas and algorithms, the binary Edwards form is solidly the best option for ECC over binary fields not only regarding side-channel resistance, but also in efficiency.

Acknowledgments. The authors wish to thank all the anonymous reviewers of Indocrypt 2014 for their valuable comments and suggestions to improve this paper.

References

1. Al-Daoud, E., Mahmod, R., Rushdan, M., Kilicman, A.: A new addition formula for elliptic curves over $GF(2^n)$. IEEE Trans. Comput. **51**(8), 972–975 (2002)
2. Bernstein, D.J., Birkner, P., Joye, M., Lange, T., Peters, C.: Twisted Edwards Curves. In: Vaudenay, S. (ed.) AFRICACRYPT 2008. LNCS, vol. 5023, pp. 389–405. Springer, Heidelberg (2008)
3. Bernstein, D.J., Lange, T.: A complete set of addition laws for incomplete Edwards curves. J. Number Theory **131**, 858–872 (2011)
4. Bernstein, D.J., Lange, T.: Explicit-formulas database (2014). http://www.hyperelliptic.org/EFD/
5. Bernstein, D.J., Lange, T.: Faster Addition and Doubling on Elliptic Curves. In: Kurosawa, K. (ed.) ASIACRYPT 2007. LNCS, vol. 4833, pp. 29–50. Springer, Heidelberg (2007)
6. Bernstein, D.J., Lange, T., Rezaeian Farashahi, R.: Binary Edwards Curves. In: Oswald, E., Rohatgi, P. (eds.) CHES 2008. LNCS, vol. 5154, pp. 244–265. Springer, Heidelberg (2008)
7. Blake, I.F., Seroussi, G., Smart, N.P.: Advances in Elliptic Curve Cryptography. Cambridge University Press, New York (2005)
8. Devigne, J., Joye, M.: Binary Huff Curves. In: Kiayias, A. (ed.) CT-RSA 2011. LNCS, vol. 6558, pp. 340–355. Springer, Heidelberg (2011)
9. Edwards, H.M.: A normal form for elliptic curves. Bulletin of the American Mathematical Society **44**, 393–422 (2007)
10. Farashahi, R.R., Joye, M.: Efficient Arithmetic on Hessian Curves. In: Nguyen, P.Q., Pointcheval, D. (eds.) PKC 2010. LNCS, vol. 6056, pp. 243–260. Springer, Heidelberg (2010)
11. Fan, J., Gao, X., Mulder, E.D., Schaumont, P., Preneel, B., Verbauwhede, I.: State-of-the-art of secure ECC implementation: A survey on known side-channel attacks and countermeasures. In: HOST 2010, pp. 30–41. Springer, Heidelberg (2010)
12. Gaudry, P., Lubicz, D.: The arithmetic of characteristic 2 Kummer surfaces and of elliptic Kummer lines. Finite Fields and Their Applications **15**(2), 246–260 (2009)
13. Ghosh, S., Kumar, A., Das, A., Verbauwhede, I.: On the Implementation of Unified Arithmetic on Binary Huff Curves. In: Bertoni, G., Coron, J.-S. (eds.) CHES 2013. LNCS, vol. 8086, pp. 349–364. Springer, Heidelberg (2013)
14. Hankerson, D., Karabina, K., Menezes, A.: Analyzing the Galbraith-Lin-Scott point multiplication method for elliptic curves over binary fields. IEEE Trans. Comput. **58**(10), 1411–1420 (2009)
15. Hankerson, D., Menezes, A., Vanstone, S.: Guide to Elliptic Curve Cryptography. Springer-Verlag New York Inc., Secaucus (2003)
16. Hankerson, D., Hernandez, J.L., Menezes, A.: Software Implementation of Elliptic Curve Cryptography over Binary Fields. In: Paar, C., Koç, Ç.K. (eds.) CHES 2000. LNCS, vol. 1965, pp. 1–24. Springer, Heidelberg (2000)
17. Hisil, H., Wong, K.K.-H., Carter, G., Dawson, E.: Twisted Edwards Curves Revisited. In: Pieprzyk, J. (ed.) ASIACRYPT 2008. LNCS, vol. 5350, pp. 326–343. Springer, Heidelberg (2008)

18. Joye, M., Yen, S.-M.: The Montgomery Powering Ladder. In: Kaliski Jr., B.S., Koç, Ç.K., Paar, C. (eds.) CHES 2002. LNCS, vol. 2523, pp. 291–302. Springer, Heidelberg (2003)
19. Kohel, D.: Addition law structure of elliptic curves. Journal of Number Theory 131(5), 894–919 (2011)
20. Kohel, D.: A normal form for elliptic curves in characteristic 2. In: Arithmetic, Geometry, Cryptography and Coding Theory (AGCT 2011), Luminy, talk notes (March 15, 2011)
21. Kohel, D.: Efficient Arithmetic on Elliptic Curves in Characteristic 2. In: Galbraith, S., Nandi, M. (eds.) INDOCRYPT 2012. LNCS, vol. 7668, pp. 378–398. Springer, Heidelberg (2012)
22. Kim, K.H., Kim, S.I.: A new method for speeding up arithmetic on elliptic curves over binary fields (2007). http://eprint.iacr.org/2007/181
23. Kim, K.H., Negre, C.: Point multiplication on supersingular elliptic curves defined over fields of characteristic 2 and 3. SECRYPT 2008, pp. 373–376. INSTICC Press (2008)
24. Lin, Q., Zhang, F.: Halving on binary Edwards curves (2010). http://eprint.iacr.org/2010/004
25. López, J., Dahab, R.: Fast Multiplication on Elliptic Curves over $GF(2^m)$ without Precomputation. In: Koç, Ç.K., Paar, C. (eds.) CHES 1999. LNCS, vol. 1717, pp. 316–327. Springer, Heidelberg (1999)
26. Moloney, R., O'Mahony, A., Laurent, P.: Efficient implementation of elliptic curve point operations using binary Edwards curves (2010). http://eprint.iacr.org/2010/208
27. Montgomery, P.L.: Speeding the Pollard and elliptic curve methods of factorization. Mathematics of Computation 48, 243–264 (1987)
28. Negre, C., Robert, J.-M.: Impact of Optimized Field Operations AB, AC and $AB + CD$ in Scalar Multiplication over Binary Elliptic Curve. In: Youssef, A., Nitaj, A., Hassanien, A.E. (eds.) AFRICACRYPT 2013. LNCS, vol. 7918, pp. 279–296. Springer, Heidelberg (2013)
29. Oliveira, T., López, J., Aranha, D.F., Rodríguez-Henríquez, F.: Lambda Coordinates for Binary Elliptic Curves. In: Bertoni, G., Coron, J.-S. (eds.) CHES 2013. LNCS, vol. 8086, pp. 311–330. Springer, Heidelberg (2013)
30. Taverne, J., Faz-Hernández, A., Aranha, D.F., Rodríguez-Henríquez, F., Hankerson, D., López, J.: Speeding scalar multiplication over binary elliptic curves using the new carry-less multiplication instruction. Journal of Cryptographic Engineering 1, 187–199 (2011)
31. Taverne, J., Faz-Hernández, A., Aranha, D.F., Rodríguez-Henríquez, F., Hankerson, D., López, J.: Software Implementation of Binary Elliptic Curves: Impact of the Carry-Less Multiplier on Scalar Multiplication. In: Preneel, B., Takagi, T. (eds.) CHES 2011. LNCS, vol. 6917, pp. 108–123. Springer, Heidelberg (2011)
32. Wu, H., Tang, C., Feng, R.: A New Model of Binary Elliptic Curves. In: Galbraith, S., Nandi, M. (eds.) INDOCRYPT 2012. LNCS, vol. 7668, pp. 399–411. Springer, Heidelberg (2012)

Summation Polynomial Algorithms for Elliptic Curves in Characteristic Two

Steven D. Galbraith$^{(\boxtimes)}$ and Shishay W. Gebregiyorgis

Mathematics Department, University of Auckland, Auckland, New Zealand
S.Galbraith@math.auckland.ac.nz, sgeb522@aucklanduni.ac.nz

Abstract. The paper is about the discrete logarithm problem for elliptic curves over characteristic 2 finite fields $\mathbb{F}_{2^n}$ of prime degree n. We consider practical issues about index calculus attacks using summation polynomials in this setting. The contributions of the paper include: a new choice of variables for binary Edwards curves (invariant under the action of a relatively large group) to lower the degree of the summation polynomials; a choice of factor base that "breaks symmetry" and increases the probability of finding a relation; an experimental investigation of the use of SAT solvers rather than Gröbner basis methods for solving multivariate polynomial equations over $\mathbb{F}_2$.

We show that our new choice of variables gives a significant improvement to previous work in this case. The symmetry-breaking factor base and use of SAT solvers seem to give some benefits in practice, but our experimental results are not conclusive. Our work indicates that Pollard rho is still much faster than index calculus algorithms for the ECDLP over prime extension fields $\mathbb{F}_{2^n}$ of reasonable size.

Keywords: ECDLP · Summation polynomials · Index calculus

1 Introduction

Let E be an elliptic curve over a finite field $\mathbb{F}_{2^n}$ where n is prime. The elliptic curve discrete logarithm problem (ECDLP) is: Given $P, Q \in E(\mathbb{F}_{2^n})$ to compute an integer a, if it exists, such that $Q = aP$. As is standard, we restrict attention to points P of prime order r. This problem is fundamental to elliptic curve cryptography. One approach to solving the ECDLP is to use Semaev's summation polynomials [20] and index calculus ideas of Gaudry, Diem, and others [4–10,13,16,18]. The main idea is to specify a factor base and then to try and "decompose" random points $R = uP + wQ$ as a sum $P_1 + \cdots + P_m$ of points in the factor base. Semaev's summation polynomials allow to express the sum $P_1 + \cdots + P_m - R = \infty$, where ∞ is the identity element, as a polynomial equation over $\mathbb{F}_{2^n}$, and then Weil descent reduces this problem to a system of polynomial equations over $\mathbb{F}_2$.

There is a growing literature on these algorithms. Much of the previous research has been focused on elliptic curves over $\mathbb{F}_{q^n}$ where q is prime or a prime power, and n is small.

© Springer International Publishing Switzerland 2014
W. Meier and D. Mukhopadhyay (Eds.): INDOCRYPT 2014, LNCS 8885, pp. 409–427, 2014.
DOI: 10.1007/978-3-319-13039-2_24

Our Work. This paper is about the case $\mathbb{F}_{2^n}$ where n is prime. Other work (for example [5–7,18]) has focussed on asymptotic results and theoretical considerations. Instead, we focus on very practical issues and ask about what can actually be computed in practice today. In other words, we follow the same approach as Huang, Petit, Shinohara and Takagi [15] and Shantz and Teske [19].

We assume throughout that the ECDLP instance cannot be efficiently solved using the Gaudry-Hess-Smart approach [12] or its extensions, and that the point decomposition step of the algorithm is the bottleneck (so we ignore the cost of the linear algebra).

The goal of our paper is to report on our experiments with three ideas:

(1) We describe a choice of variables for binary Edwards curves that is invariant under the action of a relatively large group (generated by the action of the symmetric group and addition by a point of order 4). This allows the summation polynomials to be re-written with lower degree, which in turn speeds up the computation of relations.

(2) We consider a factor base that "breaks symmetry" and hence significantly increases the probability that relations exist. It may seem counterintuitive to use symmetric variables (in order to reduce the degree) when using a non-symmetric factor base, but if one designs the factor base correctly then this is seen to be possible.

The basic idea is as follows. The traditional approach has relations $R = P_1 + \cdots + P_m$ where $P_i \in \mathcal{F} = \{P \in E(\mathbb{F}_{2^n}) : x(P) \in V\}$ where $V \subseteq \mathbb{F}_{2^n}$ is some $\mathbb{F}_2$-vector subspace of dimension l. Instead, we demand $P_i \in \mathcal{F}_i$ over $1 \leq i \leq m$ for m different factor bases $\mathcal{F}_i = \{P \in E(\mathbb{F}_{2^n}) : x(P) \in V + v_i\}$ where $v_i \in \mathbb{F}_{2^n}$ are elements of a certain form so that the sets $V + v_i$ are all distinct. (Diem [7] has also used different factor bases $\mathcal{F}_i$, but in a different way.) The probability of finding a relation is increased by a factor approximately $m!$, but we need m times as many relations, so the total speedup is approximately by a factor of $(m-1)!$.

(3) We experiment with SAT solvers rather than Gröbner basis methods for solving the polynomial systems. This is possible since we obtain a system of multivariate polynomial equations over $\mathbb{F}_2$, rather than over larger fields. (SAT solvers have been considered in cryptanalysis before, e.g. [3,17].)

Our conclusions are: The new coordinates for binary Edwards curves give a significant improvement over previous work on elliptic curves in characteristic 2. The use of SAT solvers may potentially enable larger factor bases to be considered (however, it seems an "early abort" strategy should be taken, as we will explain). Symmetry breaking seems to give a moderate benefit when n is large compared with lm.

Our overall conclusion is that, despite these new ideas, for parameters of interest for actual computation it is still slower to use summation polynomials to solve an ECDLP instance in characteristic 2 elliptic curves than to use Pollard rho. Hence, summation polynomial algorithms do not seem to be a useful tool for attacking current ECDLP challenge curves.

The paper is organised as follows. Section 2 recalls previous work. Section 3 recalls binary Edwards curves and introduces our new variables. Section 4 shows how to do the index calculus attack in this setting and discusses the symmetry-breaking idea. Section 5 discusses the use of SAT solvers, while Section 6 reports on our experimental results.

2 Index Calculus Algorithms and Summation Polynomials

We briefly recall the basic ideas of these methods and introduce our notation. Let $P \in E(\mathbb{F}_{q^n})$ have prime order r and suppose $Q = aP$. One chooses an appropriate factor base $\mathcal{F} \subseteq E(\mathbb{F}_{q^n})$, computes random points $R = uP + wQ$ and then tries to write $R = P_1 + \cdots + P_m$ for $P_i \in \mathcal{F}$. Each successful decomposition of the point R is called a "relation". Let $\ell = \#\mathcal{F}$. Writing $\mathcal{F} = \{F_1, \ldots, F_\ell\}$ we can write the j-th relation as $u_j P + w_j Q = \sum_{i=1}^{\ell} z_{j,i} F_i$ and store the relation by storing the values (u_j, w_j) and the vector $(z_{j,1}, \ldots, z_{j,\ell})$. When enough relations (more than $\#\mathcal{F}$) are found then one can apply (sparse) linear algebra to find a kernel vector of the matrix $M = (z_{j,i})$ and hence obtain a pair of integers u and w such that $uP + wQ = 0$ from which we can solve for $a \equiv -uw^{-1} \mod r$ as long as $w \not\equiv 0 \mod r$. The details are standard.

The difficult part of the algorithm is decomposing random points over the factor base. This is the main focus of our paper. We will ignore the linear algebra step as, for the parameters considered in the paper, its cost will always be insignificant.

2.1 Summation Polynomials

Semaev [20] introduced summation polynomials for elliptic curves E in Weierstrass form over fields $\mathbb{K}$ of odd characteristic. The mth summation polynomial $f_m(x_1, x_2, \ldots, x_m) \in \mathbb{K}[x_1, x_2, \ldots, x_m]$ has the following defining property. Let $X_1, X_2, \ldots, X_m \in \mathbb{K}$. Then $f_m(X_1, X_2, \ldots, X_m) = 0$ if and only if there exist $Y_1, Y_2, \ldots, Y_m \in \overline{\mathbb{K}}$ such that $(X_i, Y_i) \in E(\overline{\mathbb{K}})$ for all $1 \le i \le m$ and $(X_1, Y_1) + (X_2, Y_2) + \cdots + (X_m, Y_m) = \infty$.

Gaudry and Diem noted that, for elliptic curves $E(\mathbb{F}_{q^n})$ over extension fields, there are choices of factor base for which the problem of finding solutions to summation polynomials can be approached using Weil descent with respect to $\mathbb{F}_{q^n}/\mathbb{F}_q$. In other words, the problem of solving $f_{m+1}(x_1, \ldots, x_m, x(R))$ for $x_i \in \mathbb{F}_q$ can be reduced to a system of multivariate polynomial equations over $\mathbb{F}_q$. The details are standard.

To solve the system of multivariate polynomial equations, the current most effective approach (see [8, 15]) is to perform the F_4 or F_5 algorithm for the graded reverse lex order, followed by the FGLM algorithm [11].

2.2   Degree Reduction via Symmetries

The summation polynomials have high degree, which makes solving them difficult. Since the summation polynomial is invariant under the action of the symmetric group S_m, Gaudry [13] observed that re-writing the polynomial in terms of invariant variables reduces the degree and speeds up the resolution of the system of equations. As well as lowering the degree of the polynomials, this idea also makes the solution set smaller and hence faster to compute using the FGLM algorithm.

Faugère et al. [9,10] have considered action by larger groups (by using points of small order) for elliptic curves over $\mathbb{F}_{q^n}$ where n is small (e.g., $n = 4$ or $n = 5$) and the characteristic is $\neq 2, 3$. Their work gives further reduction in the cost of solving the system.

For a point $P = (x, y)$ on a twisted Edwards curve we have $-P = (-x, y)$ and so it is natural to construct summation polynomials in terms of the y-coordinate (invariant under $P \mapsto -P$). Accordingly Faugère et al. [9] define their factor base as

$$\mathcal{F} = \{P = (x, y) \in \mathbb{F}_{q^n} : y \in \mathbb{F}_q\}.$$

Further, the addition of P with the point $T_2 = (0, -1)$ (which has order 2) satisfies $P + T_2 = (-x, -y)$. Note that $P \in \mathcal{F}$ if and only if $P + T_2 \in \mathcal{F}$. Hence, for each decomposition $R = P_1 + P_2 + \cdots + P_n$, there exist 2^{n-1} further decompositions, such as

$$R = (P_1 + T_2) + (P_2 + T_2) + P_3 + \cdots + P_n.$$

It follows that the dihedral coxeter group $D_n = (\mathbb{Z}/2\mathbb{Z})^{n-1} \rtimes S_n$ of order $2^{n-1}n!$ acts on the set of relations $R = P_1 + \cdots + P_n$ for any given point R (and all these relations correspond to solutions of the summation polynomial). It is therefore natural to write the summation polynomial $f_{n+1}(y_1, y_2, \ldots, y_n, y(R))$ in terms of new variables that are invariant under the group action. For further details see [9].

A recent idea (originating in the work of Joux-Vitse [16] for $E(\mathbb{F}_{q^n})$) is to consider relations with fewer summands $R = P_1 + \cdots + P_m$ with $m < n$. Joux and Vitse take $m = n - 1$ so the probability of a relation is reduced from $1/n!$ to $1/(q(n-1)!)$. The cost of solving the polynomial system is significantly reduced, but the running time increases by the factor q. Shantz and Teske [19] call this the "delta method".

2.3   The Case of $\mathbb{F}_{2^n}$ Where n Is Prime

Following Diem [7] we define the factor base in terms of an $\mathbb{F}_2$-vector space $V \subset \mathbb{F}_{2^n}$ of dimension l. A typical choice for the factor base in the case of Weierstrass curves is $\mathcal{F} = \{P \in E(\mathbb{F}_{2^n}) : x(P) \in V\}$, and one wants to decompose random points as $R = P_1 + \cdots + P_m$ for $P_i \in \mathcal{F}$.

As above, the symmetric group S_m of order $m!$ acts on the set of relations $R = P_1 + \cdots + P_m$ for any given point R (and all these relations correspond to solutions of the summation polynomial). It is therefore natural to write the summation polynomial $f_{m+1}(x_1, x_2, \ldots, x_m, x(R))$ in terms of new variables that are invariant under the group action. In this example, such variables are the elementary symmetric polynomials in the x_i. This approach gives polynomials of lower degree.

Huang et al. [15] observe that it is hard to combine re-writing the summation polynomial in terms of symmetric variables and also using a factor base defined with respect to an arbitrary vector subspace of $\mathbb{F}_{2^n}$. The point is that if $x_1, \ldots, x_m \in V$ then it is not necessarily the case that the value of the symmetric polynomial $e_2 = x_1 x_2 + x_1 x_3 + \cdots + x_{m-1} x_m$ (or higher ones) lies in V. Hence, one might think that one cannot use symmetries in this setting.

Section 3 of [15] considers prime n and the new idea of "both symmetric and non-symmetric variables". It is suggested to use a "special subspace" V that behaves relatively well under multiplication: $x_i, x_j \in V$ implies $x_i x_j \in V'$ for a somewhat larger space V'. The experiments in [15], for n prime in the range $17 \leq n \leq 53$, and $m = 3$ and $l \in \{3, 4, 5, 6\}$, show a significant decrease of the degree of regularity (the highest step degree reached) during Gröbner basis computations. However, the decrease in the degree of regularity is at the expense of an increased number of variables, which in turn increases the complexity of the Gröbner basis computations (which roughly take time N^{3D} and require N^{2D} memory, where N is the number of variables and D is the degree of regularity).

Huang et al. [15] exploit the action of S_m on the summation polynomials but do not exploit points of order 2 or 4. One of our contributions is to give coordinates that allow to exploit larger symmetry groups in the case of elliptic curves over binary fields. We are able to solve larger experiments in this case (e.g., taking decompositions into $m = 4$ points, while [15] could only handle $m = 3$). For more details of our experiments see Section 6.

3   Edwards Elliptic Curves in Characteristic Two

We study binary Edwards curves [1] since the addition by points of order 2 and 4 is nicer than when using the Weierstrass model as was done in [9,10]. Hence we feel this model of curves is ideally suited for the index calculus application. Let $d_1, d_2 \in \mathbb{F}_{2^n}$ be such that $d_1 \neq 0$ and $d_2 \neq d_1^2 + d_1$. A binary Edwards curve has equation $E_{d_1, d_2} : d_1(x + y) + d_2(x^2 + y^2) = xy + xy(x + y) + x^2 y^2$. It is symmetric in the variables x and y with the following group law [1].

1. The identity element is the point $P_0 = (0, 0)$.
2. For a point $P = (x, y) \in E_{d_1, d_2}$, its negation is given by $-P = (y, x)$.
3. Let $P_1 = (x_1, y_1), P_2 = (x_2, y_2) \in E_{d_1, d_2}$, then $P_3 = (x_3, y_3) = P_1 + P_2$ is given by,

$$x_3 = \frac{d_1(x_1 + x_2) + d_2(x_1 + y_1)(x_2 + y_2) + (x_1 + x_1^2)(x_2(y_1 + y_2 + 1) + y_1 y_2)}{d_1 + (x_1 + x_1^2)(x_2 + y_2)}$$

$$y_3 = \frac{d_1(y_1 + y_2) + d_2(x_1 + y_1)(x_2 + y_2) + (y_1 + y_1^2)(y_2(x_1 + x_2 + 1) + x_1 x_2)}{d_1 + (y_1 + y_1^2)(y_2 + y_2)}.$$

4. The point $T_2 = (1, 1) \in E_{d_1, d_2}$ is invariant under negation so it has order 2. For any point $P = (x, y) \in E_{d_1, d_2}$ we have $P + T_2 = (x + 1, y + 1)$.

For summation polynomials with these curves, the best choice of variable is $t = x + y$. This is the natural choice, consistent with previous work [9,13], as this function is invariant under the action of $[-1] : P \mapsto -P$.

The function $t : E \to \mathbb{P}^1$ has degree 4. Given a value $t \in \mathbb{F}_{2^n}$ there are generically four points $P = (x, y) \in E(\overline{\mathbb{F}}_2)$ having the same value for $t(P)$, namely $(x, y), (y, x), (x + 1, y + 1), (y + 1, x + 1)$.

When we come to define the factor base, we will choose a vector subspace V of $\mathbb{F}_{2^n}/\mathbb{F}_2$ of dimension l and will define the factor base to be the set of points corresponding to $t(P) = x(P) + y(P) \in V$.

Lemma 1. *Let E_{d_1, d_2} be a binary Edwards curve over $\mathbb{F}_{2^n}$ such that $d_1 = d_2$ and define $t(P) = x(P) + y(P)$. Define the mth summation polynomials*

$$f_2(t_1, t_2) = t_1 + t_2$$
$$f_3(t_1, t_2, t_3) = (d_1 + t_1 t_2 (t_1 + 1)(t_2 + 1)) t_3^2 + (t_1 t_2 + (t_1 + 1)(t_2 + 1)) t_3 + d_1 (t_1 + t_2)^2$$

and, for all $m \geq 4$ and $1 \leq j \leq m - 3$,

$$f_m(t_1, \ldots, t_m) = Res_t(f_{m-j}(t_1, t_2, \ldots, t_{m-j-1}, t), f_{j+2}(t_{m-j}, t_{m-j+1}, \ldots, t_m, t))$$

where Res_t denotes the resultant. For any points $P_1, \ldots, P_m \in E_{d_1, d_1}(\overline{\mathbb{F}}_2)$ such that $P_1 + \cdots + P_m = P_0$, then $f_m(t(P_1), \ldots, t(P_m)) = 0$. Conversely, given any $t_1, \ldots, t_m \in \overline{\mathbb{F}}_2$ such that $f_m(t_1, \ldots, t_m) = 0$, then there exist points $P_1, \ldots, P_m \in E_{d_1, d_1}(\overline{\mathbb{F}}_2)$ such that $t(P_i) = t_i$ for all $1 \leq i \leq m$ and $P_1 + \cdots + P_m = P_0$. For $m \geq 2$, the polynomials have degree 2^{m-2} in each variable.

The proof of this result, and a more general statement for the case $d_2 \neq d_1$, are given in the full version of this paper. Note that the degree bound 2^{m-2} is consistent with the arguments on page 44 (Sections 2 and 3.1) of [10]: Since $\deg(t) = 4$ we would expect polynomials of degree 4^{m-1}, but t is invariant and so factors through a 2-isogeny, so we get degree 2^{m-1}. The further saving of a factor 2 follows since $t(-P) = t(P)$.

3.1 Action of Symmetric Group

Since the equation $P_1 + \cdots + P_m$ is symmetric it follows that the summation polynomials for binary Edwards curves are symmetric. Hence

$$f_{m+1}(t_1, t_2, \ldots, t_m, t(R)) \in \mathbb{F}_{2^n}[t_1, t_2, \ldots, t_m]^{S_m}$$

where S_m is the symmetric group and the right hand side denotes the ring of polynomials invariant under the group S_m. Hence, it is possible to express the summation polynomials in terms of the elementary symmetric polynomials $(e_1, e_2, \ldots, e_m)$ in the variables t_i, as they are generators of the ring $\mathbb{F}_{2^n}[t_1, \ldots, t_m]^{S_m}$.

Since the elementary symmetric polynomial e_i has degree i, it is natural to expect the polynomial to have lower degree after this change of variables. Another way to explain this degree reduction is to note that each relation $R = P_1 + \cdots + P_m$ comes in an orbit of size (at least, when the points P_i are all distinct) $m!$. This implies that the number of solutions to the polynomial when expressed in terms of the e_i is smaller than the original polynomial, and this is compatible with a lowering of the degree.

3.2   Action of Points of Order 2

It was proposed in [9,10] to consider the action of small torsion points to further lower the degree of the summation polynomials. This idea also allows to effectively reduce the size of the factor base when performing the linear algebra. Hence, it is important to exploit torsion points as much as possible. Of the previous papers, [9] only considers odd characteristic, while [10] considers even characteristic (and even goes as far as summation polynomials of 8 points!) but only for curves in Weierstrass form and using a point of order 2. In this section we consider these ideas for binary Edwards curves, and in the next section extend to using a point of order 4.

Fix a vector space $V \subset \mathbb{F}_{2^n}$ of dimension l. The factor base will be

$$\mathcal{F} = \{P \in E_{d_1,d_2}(\mathbb{F}_{2^n}) : t(P) \in V\}.$$

We expect $\#\mathcal{F} \approx \#V$, and our experiments confirm this.

As mentioned in Section 3, if $P = (x,y) \in E_{d_1,d_2}$ then $P + T_2 = (x+1, y+1)$. Note that $t(P + T_2) = (x+1) + (y+1) = x + y = t(P)$ and so the function t is already invariant under addition by T_2. Since the factor base is defined in terms of $t(P)$ we have that $P \in \mathcal{F}$ implies $P + T_2 \in \mathcal{F}$. In other words, our choice of variables is already invariant under the action of adding a 2-torsion point.

Let $R = P_1 + \cdots + P_m$ and let $u = (u_1, \ldots, u_{m-1}) \in \{0,1\}^{m-1}$. Then

$$R = (P_1 + u_1 T_2) + (P_2 + u_2 T_2) + \cdots + (P_{m-1} + u_{m-1} T_2) + \left(P_m + \left(\sum_{i=1}^{m-1} u_i \right) T_2 \right).$$

This gives an action of the group $(\mathbb{Z}/2\mathbb{Z})^{m-1}$ on the set of relations $R = P_1 + \cdots + P_m$. Combining with the action of the symmetric group, we have that the Dihedral Coxeter group $D_m = (\mathbb{Z}/2\mathbb{Z})^{m-1} \rtimes S_m$ acts on the set of relations, and hence on the summation polynomial. Each relation $R = P_1 + \cdots + P_m$ generically comes in an orbit of size $2^{m-1} m!$.

Since the variables t_i are already invariant under addition by T_2, it follows that $f_{m+1}(t_1, t_2, \ldots, t_m, t(R)) \in \mathbb{F}_{2^n}[t_1, t_2, \ldots, t_m]^{D_m}$. Hence it can be written in terms of the elementary symmetric polynomials e_i, as they are the generators of the ring $\mathbb{F}_{2^n}[t_1, t_2, \ldots, t_m]^{D_m}$. This reduces its degree and we experience a speed-up in the FGLM algorithm due to the reduction in the size of the set of solutions.

To speed-up the linear algebra, the factor base can be reduced in size. Recall that each solution $(t_1, \ldots, t_m)$ corresponds to many relations. Let us fix, for each t, one of the four points $\{P, -P, P+T_2, -P+T_2\}$, and put only that point into our factor base. Hence the size of $\mathcal{F}$ is exactly the same as the number of $t \in V$ that correspond to elliptic curve points, which is roughly $\frac{1}{4}\#V$.

Then, for a point R, given a solution $f_{m+1}(t_1, \ldots, t_m, t(R)) = 0$ there is a unique value $z_0 \in \{0, 1\}$, unique points $P_1, \ldots, P_m \in \mathcal{F}$, and unique choices of sign $z_1, \ldots, z_m \in \{-1, 1\}$ such that $R + z_0 T_2 = \sum_{i=1}^{m} z_i P_i$. It follows that the matrix size is reduced by a factor of $1/4$ (with one extra column added to store the coefficient of T_2) which means we need to find fewer relations and the complexity of the linear algebra, which has a complexity of $\tilde{O}(m\#\mathcal{F}^2)$ using the Lanczos or Wiedemann algorithm, is reduced by a factor of $(1/4)^2$.

3.3 Action of Points of Order 4

We now consider binary Edwards curves in the case $d_1 = d_2$. Then $T_4 = (1, 0) \in E_{d_1, d_1}$ and one can verify that $T_4 + T_4 = (1, 1) = T_2$ and so T_4 has order four. The group generated by T_4 is therefore $\{P_0, T_4, T_2, -T_4 = (0, 1)\}$.

For a point $P = (x, y) \in E_{d_1, d_1}$ we have $P + T_4 = (y, x + 1)$. Hence $t(P + T_4) = t(P) + 1$. We construct our factor base $\mathcal{F}$ such that $(x, y) \in \mathcal{F}$ implies $(y, x+1) \in \mathcal{F}$. For example, we can choose a vector subspace $V \subseteq \mathbb{F}_{2^n}$ such that $v \in V$ if and only if $v + 1 \in V$, and set $\mathcal{F} = \{P \in E_{d_1, d_1}(\mathbb{F}_{2^n}) : t(P) \in V\}$.

If $R = P_1 + P_2 + \cdots + P_m$ is a relation and $(u_1, \ldots, u_{m-1}) \in \{0, 1, 2, 3\}^{m-1}$ then we also have

$$R = (P_1 + [u_1]T_4) + (P_2 + [u_2]T_4) + \cdots + (P_{m-1} + [u_{m-1}]T_4) + (P_m + [u_m]T_4) \quad (1)$$

for $u_m = -\sum_{i=1}^{m-1} u_i$. Hence, one can consider the group $G_m = (\mathbb{Z}/4\mathbb{Z})^{m-1} \rtimes S_m$ acting on the summation polynomial. To express the summation polynomial in terms of invariant variables it suffices to note that the invariants under the action $t \mapsto t + 1$ in characteristic 2 are $t(t + 1) = t^2 + t$ (this is mentioned in Section 4.3 of [10]). Hence,

$$s_2 = (t_1^2 + t_1)(t_2^2 + t_2) + \cdots + (t_{m-1}^2 + t_{m-1})(t_m^2 + t_m),$$
$$\vdots \qquad\qquad\qquad\qquad\qquad\qquad (2)$$
$$s_m = (t_1^2 + t_1)(t_2^2 + t_2) \cdots (t_m^2 + t_m).$$

are invariant variables. One might also expect to use $e_1 + e_1^2 = t_1 + t_1^2 + \cdots + t_m + t_m^2$, but since the addition by T_4 cancels out in equation (1) we actually have that $e_1 = t_1 + \cdots + t_m$ remains invariant. Hence, we can use the invariant variables $e_1, s_2, \ldots, s_m$, which are the generators of the ring $\mathbb{F}_{2^n}[t_1, t_2, \ldots, t_m]^{G_m}$.

It is clear that we further halve the size of the factor base by choosing a unique representative of the orbit under the action. Overall, the factor base is reduced in total by a factor of $1/8$ over the basic method.

4   The Index Calculus Algorithm

We now present the full index calculus algorithm combined with the new variables introduced in Section 3. We work in $E(\mathbb{F}_{2^n}) := E_{d_1,d_1}(\mathbb{F}_{2^n})$ where n is prime and E_{d_1,d_1} is a binary Edwards curve with parameters $d_2 = d_1$. We choose an integer m (for the number of points in a relation) and an integer l. Considering $\mathbb{F}_{2^n}$ as a vector space over $\mathbb{F}_2$ we let V be a vector subspace of dimension l. More precisely, we will suppose $\mathbb{F}_{2^n}$ is represented using a polynomial basis $\{1, \theta, \ldots, \theta^{n-1}\}$ where $F(\theta) = 0$ for some irreducible polynomial $F(x) \in \mathbb{F}_2[x]$ of degree n. We will take V to be the vector subspace of $\mathbb{F}_{2^n}$ over $\mathbb{F}_2$ with basis $\{1, \theta, \ldots, \theta^{l-1}\}$.

We start with the standard approach, leaving the symmetry-breaking to Section 4.2. We define the factor base $\mathcal{F} = \{P \in E(\mathbb{F}_{2^n}) : t(P) \in V\}$, where $t(x, y) = x + y$. Relations will be sums of the form $R = P_1 + P_2 + \cdots + P_m$ where $P_i \in \mathcal{F}$. We heuristically assume that $\#\mathcal{F} \approx 2^l$. Under this heuristic assumption we expect the number of points in $\{P_1 + \cdots + P_m : P_i \in \mathcal{F}\}$ to be roughly $2^{lm}/m!$. Hence, the probability that a uniformly chosen point $R \in E(\mathbb{F}_{2^n})$ can be decomposed in this way is heuristically $(2^{lm}/m!)/2^n = 1/(m!2^{n-lm})$. Hence we would like to choose m and l so that lm is not too much smaller than n.

To compute relations we evaluate the summation polynomial at the point R to get $f_{m+1}(t_1, t_2, \ldots, t_m, t(R)) \in \mathbb{F}_{2^n}[t_1, t_2, \ldots, t_m]$. If we can find a solution $(t_1, t_2, \ldots, t_m) \in V^m$ satisfying $f_{m+1}(t_1, t_2, \ldots, t_m, t(R)) = 0$ then we need to determine the corresponding points, if they exist, $(x_i, y_i) \in E(\mathbb{F}_{2^n})$ such that $t_i = x_i + y_i$ and $(x_1, y_1) + \cdots + (x_m, y_m) = R$. Finding (x_i, y_i) given t_i is just taking roots of a univariate quartic polynomial. Once we have m points in $E(\mathbb{F}_{2^n})$, we may need to check up to 2^{m-1} choices of sign (and also determine an additive term $z_{j,0}T_4$, since our factor base only includes one of the eight points for each value of $t_i(t_i + 1)$) to be able to record the relation as a vector. The cost of computing the points (x_i, y_i) is almost negligible, but checking the signs may incur some cost for large m.

When a relation exists (i.e., the random point R can be written as a sum of m points in the factor base) then there exists a solution $(t_1, \ldots, t_m) \in V^m$ to the polynomial system that can be lifted to points in $E(\mathbb{F}_{2^n})$. When no relation exists there are two possible scenarios: Either there is no solution $(t_1, \ldots, t_m) \in V^m$ to the polynomial system, or there are solutions but they don't lift to points in $E(\mathbb{F}_{2^n})$. In both cases, the running time of detecting that a relation does not exist is dominated by the Gröbner basis computation and so is roughly the same.

In total we will need $\#\mathcal{F} + 1 \approx \#V = 2^l$ relations. Finally, these relations are represented as the system of equations $u_j P + w_j Q = z_{j,0}T_4 + \sum_{P_i \in \mathcal{F}} z_{j,i}P_i$ where $M = (z_{j,i})$ is a sparse matrix with at most $m+1$ non-zero entries per row. Let r be the order of P (assumed to be odd). If S is any vector in the kernel of the matrix (meaning $SM \equiv 0 \pmod{r}$), then writing $u = S(u_1, \ldots, u_{\ell+1})^T$ and $w = S(w_1, \ldots, w_{\ell+1})^T$. We have $uP + wQ = 0$ (the T_4 term must disappear if r is odd) and so $u + wa \equiv 0 \mod r$ and we can solve for the discrete logarithm a. The details are given in Algorithm 1.

4.1   The Choice of Variables

Recall that our summation polynomials $f_{m+1}(t_1, t_2, \ldots, t_m, t(R))$ can be written in terms of the invariant variables $(e_1, s_2, \ldots, s_m)$. Here we are exploiting the full group $(\mathbb{Z}/4\mathbb{Z})^{m-1} \rtimes S_m$. Note that $t(R) \in \mathbb{F}_{2^n}$ is a known value and can be written as $t(R) = r_0 + r_1\theta + r_2\theta^2 + \cdots + r_{n-1}\theta^{n-1}$ with $r_i \in \mathbb{F}_2$.

As noted by Huang et al. [15], and using their notation, let us write t_j, e_1, and s_j in terms of binary variables with respect to the basis for $\mathbb{F}_{2^n}$. We have

$$t_j = \sum_{i=0}^{l-1} c_{j,i}\theta^i \tag{3}$$

for $1 \leq j \leq m$, which is a total of lm binary variables $c_{j,i}$. Set $k = \min(\lfloor n/(2(l-1))\rfloor, m)$. The invariant variables $e_1, s_2, \ldots, s_m$ can be written as,

$$e_1 = \sum_{i=0}^{l-1} d_{1,i}\theta^i, \quad s_j = \sum_{i=0}^{2j(l-1)} d_{j,i}\theta^i \text{ for } 2 \leq j \leq k, \text{ and for } k < j \leq m, s_j = \sum_{i=0}^{n-1} d_{j,i}\theta^i.$$

Suppose that $n \approx lm$. Then $k = n/(2(l-1)) \approx m/2$ and so we suppose it takes the value $\acute{m} = \lceil \frac{m}{2} \rceil$. Then the number of binary variables $d_{j,i}$ is $N = l + (4(l-1)+1) + (6(l-1)+1) + \cdots + (2\acute{m}(l-1)+1) + \acute{m}n \approx (m^2l + mn)/2$.

Writing the evaluated summation polynomial as $G(e_1, s_2, \ldots, s_m)$ we now substitute the above formulae to obtain a polynomial in the variables $d_{j,i}$. Apply Weil descent to the polynomial to get $\phi_1 + \phi_2\theta + \cdots + \phi_n\theta^{n-1} = 0$. where the ϕ_i are polynomials over $\mathbb{F}_2$ in the $d_{j,i}$. This forms a system of n equations in the N binary variables $d_{j,i}$. We add the field equations $d_{j,i}^2 - d_{j,i}$ and then denote this system of equations by sys_1.

One could attempt to solve this system using Gröbner basis methods. For each candidate solution $(d_{j,i})$ one would compute the corresponding solution $(e_1, s_2, \ldots, s_m)$ and then solve a univariate polynomial equation (i.e., take roots) to determine the corresponding solution $(t_1, \ldots, t_m)$. From this one determines whether each value t_j corresponds to an elliptic curve point $(x_j, y_j) \in E(\mathbb{F}_{2^n})$ such that $x_j + y_j = t_j$. If everything works ok then one forms the relation.

However, the approach just mentioned is not practical as the number N of binary variables is too large compared with the number of equations. Hence, we include the $lm < n$ variables $c_{j,\tilde{i}}$ (for $1 \leq j \leq m$, $0 \leq \tilde{i} \leq l-1$) to the problem, and add a large number of new equations relating the $c_{j,\tilde{i}}$ to the $d_{j,i}$ via the t_j and equations (2) and (3). This gives N additional equations in the $N + lm$ binary variables. After adding the field equations $c_{j,\tilde{i}}^2 - c_{j,\tilde{i}}$ we denote this system of equations by sys_2. Finally we solve $sys_1 \cup sys_2$ using Gröbner basis algorithms $F4$ or $F5$ using the Degree Reverse lexicographic ordering. From a solution, the corresponding points P_j are easily computed.

4.2   Breaking Symmetry

We now explain how to break symmetry in the factor base while using the new variables as above. Again, suppose $\mathbb{F}_{2^n}$ is represented using a polynomial basis

Algorithm 1. Index Calculus Algorithm on Binary Edwards Curve

1: Set $N_r \leftarrow 0$
2: **while** $N_r \leq \#\mathcal{F}$ **do**
3:     Compute $R \leftarrow uP + wQ$ for random integer values u and w
4:     Compute summation polynomial $G(e_1, s_2, \ldots, s_m) := f_{m+1}(e_1, s_2, \ldots, s_m, t(R))$ in the variables $(e_1, s_2, \ldots, s_m)$
5:     Use Weil descent to write $G(e_1, s_2, \ldots, s_m)$ as n polynomials in binary variables $d_{j,i}$
6:     Add field equations $d_{j,i}^2 - d_{j,i}$ to get system of equations sys_1
7:     Buld new polynomial equations relating the variables $d_{j,i}$ and $c_{j,\tilde{\imath}}$
8:     Add field equations $c_{j,\tilde{\imath}}^2 - c_{j,\tilde{\imath}}$ to get system of equations sys_2
9:     Solve system of equations $sys_1 \cup sys_2$ to get $(c_{j,\tilde{\imath}}, d_{j,i})$
10:    Compute corresponding solution(s) $(t_1, \ldots, t_m)$
11:    For each t_j compute, if it exists, a corresponding point $P_j = (x_j, y_j) \in \mathcal{F}$
12:    **if** $z_1 P_1 + z_2 P_2 + \cdots + z_m P_m + z_0 T_4 = R$ for suitable $z_0 \in \{0, 1, 2, 3\}, z_i \in \{1, -1\}$
    then
13:        $N_r \leftarrow N_r + 1$
14:        Record z_i, u, w in a matrix M for the linear algebra
15: Use linear algebra to find non-trivial kernel element and hence solve ECDLP

---

and take V to be the subspace with basis $\{1, \theta, \ldots, \theta^{l-1}\}$. We choose m elements $v_i \in \mathbb{F}_{2^n}$ (which can be interpreted as vectors in the n-dimensional $\mathbb{F}_2$-vector space corresponding to $\mathbb{F}_{2^n}$) as follows: $v_1 = 0$, $v_2 = \theta^l = (0, 0, \ldots, 0, 1, 0, \ldots, 0)$ where the 1 is in position l. Similarly, $v_3 = \theta^{l+1}$, $v_4 = \theta^{l+1} + \theta^l$, $v_5 = \theta^{l+2}$ etc. In other words, v_i is represented as a vector of the form $w_0 \theta^l + w_1 \theta^{l+1} + w_2 \theta^{l+2} \cdots = (0, \ldots, 0, w_0, w_1, w_2, \ldots)$ where $\cdots w_2 w_1 w_0$ is the binary expansion of $i - 1$. Note that the subsets $V + v_i$ in $\mathbb{F}_{2^n}$ are pair-wise disjoint.

Accordingly, we define the factor bases to be $\mathcal{F}_i = \{P \in E(\mathbb{F}_{2^n}) : t(P) \in V + v_i\}$ for $1 \leq i \leq m$, where $t(x, y) = x + y$. The decomposition over the factor base of a point R will be a sum of the form $R = P_1 + P_2 + \cdots + P_m$ where $P_i \in \mathcal{F}_i$ for $1 \leq i \leq m$. Since we heuristically assume that $\#\mathcal{F}_i \approx 2^l$, we expect the number of points in $\{P_1 + \cdots + P_m : P_i \in \mathcal{F}_i\}$ to be roughly 2^{lm}. Note that there is no $1/m!$ term here. The entire purpose of this definition is to break the symmetry and hence increase the probability of relations. Hence, the probability that a uniformly chosen point $R \in E(\mathbb{F}_{2^n})$ can be decomposed in this way is heuristically $2^{lm}/2^n = 1/2^{n-lm}$.

There is almost a paradox here: Of course if $R = P_1 + \cdots + P_m$ then the points on the right hand side can be permuted and the point T_2 can be added an even number of times, and hence the summation polynomial evaluated at $t(R)$ is invariant under D_m. On the other hand, if the points P_i are chosen from distinct factor bases $\mathcal{F}_i$ then one does not have the action by S_m, so why can one still work with the invariant variables $(e_1, s_2, \ldots, s_m)$?

To resolve this "paradox" we must distinguish the computation of the polynomial from the construction of the system of equations via Weil descent. The summation polynomial does have an action by D_m (and G_m), and so that action should be exploited. When we do the Weil descent and include the definitions of the factor bases $\mathcal{F}_i$, we then introduce some new variables. As noted by Huang et al. [15], expressing the invariant variables with respect to the variables from the construction of the factor bases is non-trivial. But it is this stage where we introduce symmetry-breaking.

It follows that (in the case $m = 4$) $e_1 = t_1 + t_2 + t_3 + t_4 = d_{1,0} + d_{1,1}\theta + \cdots + d_{1,l-1}\theta^{l-1}$ can be represented exactly as before. But the other polynomials are less simple. For example, $s_2 = (t_1^2 + t_1)(t_2^2 + t_2) + \cdots + (t_3^2 + t_3)(t_4^2 + t_4)$ previously had highest term $d_{2,4l-4}\theta^{4l-4}$ but now has highest terms $d_{2,4l-4}\theta^{4l-4} + d_{2,4l-2}\theta^{4l-2} + \theta^{4l+2}$. Hence, we require one more variable than the previous case, and things get worse for higher degree terms. So the symmetry breaking increases the probability of a relation but produces a harder system of polynomial equations to solve.

An additional consequence of this idea is that the factor base is now roughly m times larger than in the symmetric case. So the number of relations required is increased by a factor m, and so the speedup over previous methods is actually by a factor approximately $m!/m = (m-1)!$. Also, the cost of the linear algebra is increased by a factor m^2 (though the system of linear equations is structured in blocks and so some optimisations may be possible). When using a point of order 4 with binary Edwards curves, the linear algebra cost is reduced (in comparison with the naive method) by a factor $(m/8)^2$.

For large q and small n, it seems that symmetry-breaking is not a useful idea, as the increase in number of variables becomes a huge problem that is not compensated by the $(m-1)!$ factor. However, for small q and large n the situation is less clear. To determine whether the idea is a good one, it is necessary to perform some experiments (see Section 6).

5   SAT Solvers

Shantz and Teske [19] discuss a standard idea [2, 22, 23] that they call the "hybrid method", which is to partially evaluate the system at some random points before applying Gröbner basis algorithms. They argue (Section 5.2) that it is better to just use the "delta method" $(n - ml > 0)$, where m is the number points in a relation and 2^l is the size of the factor base. The main observation of Shantz and Teske [19] is that using smaller l speeds up the Gröbner basis computation at the cost of decreasing the probability of getting a relation. So, they try to find such an optimal l value.

Our choice of coordinates for binary Edwards curves helps us lower the degree of our systems. As a result we were able to make successful experiments for $m = 4$ and $l \in \{3, 4\}$ using Gröbner basis algorithms, as reported in Table 3. For $l > 4$, values such that $n - ml > 0$ suffered high running times as the result of increased number of variables coming from our invariant variables.

To increase the range for these methods, we investigated other approaches to solving systems of multivariate polynomial equations over a binary field. In particular, we experimented with SAT solvers. We used Minisat 2.1 [21], coupled with the Magma system for converting the polynomial system into conjunctive normal form (CNF). On the positive side, our experiments show that SAT solvers can be faster and, more importantly, handle larger range of values for l. As is shown in Table 1, we can hope to work with l up to 7 (only shown for $n = 53$), whereas Gröbner basis methods are limited to $l \in \{3, 4\}$ in our experiments.

However, on the negative side, the running time of SAT solvers varies a lot depending on many factors. They are randomised algorithms. Further, they seem to be faster when there is a solution of low hamming weight. They can be slow on some systems having solutions, and are usually slow when no solution exists. This behavior is very different to the case of Gröbner basis methods, which perform rather reliably and are slightly better when the system of equations has no solution.

Hence, we suggest using SAT solvers with an "early abort" strategy: One can generate a lot of instances and run SAT solvers in parallel and then kill all instances that are still running after some time threshold has been passed.

Table 1. Comparison of solving polynomial systems, when there exists a solution to the system, in experiment 2 using SAT solver (Minisat) versus Gröbner basis methods for $m = 4$. #Var and #P_{equ} are the number of variables and the number of polynomial equations respectively. Mem is average memory used in megabytes by the SAT solver or Gröbner basis algorithm. P_{succ} is the percentage of times Minisat halts with solutions within 200 seconds.

Experiment 2 with SAT solver Minisat						
n	l	#Var	#P_{equ}	T_{Inter}	T_{SAT}	Mem	P_{succ}
17	3	54	59	0.35	7.90	5.98	94%
	4	67	68	0.91	27.78	9.38	90%
19	3	54	61	0.37	3.95	6.07	93%
	4	71	74	1.29	18.38	18.05	86%
23	3	54	65	0.39	1.53	7.60	87%
	4	75	82	2.15	5.59	14.48	83%
	5	88	91	4.57	55.69	20.28	64%
29	4	77	90	3.01	7.23	19.05	87%
	5	96	105	9.95	39.41	32.87	67%
	6	109	114	21.23	15.87	43.07	23%
	7	118	119	36.97	26.34	133.13	14%
31	4	77	92	3.14	17.12	20.52	62%
	5	98	109	11.80	33.48	45.71	57%
	6	113	120	26.23	16.45	118.95	12%
	7	122	125	44.77	21.98	148.95	8%
37	4	77	98	3.41	26.12	29.97	59%
	5	100	117	13.58	48.19	50.97	40%
	6	119	132	41.81	42.85	108.41	11%
	7	134	143	94.28	40.15	169.54	6%
41	4	77	102	3.08	19.28	27.59	68%
	5	100	121	15.71	27.14	49.34	65%
	6	123	140	65.25	31.69	89.71	13%
43	4	77	104	2.97	17.77	28.52	68%
	5	100	123	13.85	29.60	54.83	52%
47	4	77	108	3.18	11.40	29.93	59%
	5	100	127	14.25	27.56	61.55	43%
53	4	77	114	11.02	27.88	32.35	75%
	5	100	131	14.68	34.22	64.09	62%
	6	123	152	49.59	41.55	123.38	11%
	7	146	171	192.20	67.27	181.20	4%

Experiment 2 with Gröbner basis: F_4						
n	l	#Var	#P_{equ}	T_{Inter}	T_{GB}	Mem
17	3	54	59	0.29	0.29	67.24
	4	67	68	0.92	51.79	335.94
19	3	54	61	0.33	0.39	67.24
	4	71	74	1.53	33.96	400.17
23	3	54	65	0.26	0.31	67.24
	4	75	82	2.52	27.97	403.11
29	3	54	71	0.44	0.50	67.24
	4	77	90	3.19	35.04	503.87
31	3	54	73	0.44	0.58	67.24
	4	77	92	3.24	9.03	302.35
37	3	54	79	0.36	0.43	67.24
	4	77	98	3.34	9.07	335.94
41	3	54	83	0.40	0.54	67.24
	4	77	102	3.39	17.19	382.33
43	3	54	85	0.43	0.53	67.24
	4	77	104	3.44	9.09	383.65
47	3	54	89	0.50	0.65	67.24
	4	77	108	3.47	9.59	431.35
53	3	54	95	0.33	0.40	67.24
	4	77	114	11.43	11.64	453.77

A similar idea is mentioned in Section 7.1 of [17]. This is also related to the use of "restarting strategies" [14] in SAT solvers. This could allow the index calculus algorithm to be run for a larger set of parameters. The probability of finding a relation is now decreased. The probability that a relation exists must be multiplied by the probability that the SAT solver terminates in less than the time threshold, in the case when a solution exists. We denote this latter probability by P_{succ} in Table 1. A good design of factorbase with low hamming weight could also favour the running times of SAT solvers.

6 Experimental Results

We conducted several experiments using elliptic curves E over $\mathbb{F}_{2^n}$. We always use the $m + 1$-summation polynomial to find relations as a sum of m points in the factor base. The factor base is defined using a vector space of dimension

Table 2. Comparison of solving our systems of equations, having a solution, using Gröbner basis methods in experiment 1 and experiment 2 for $m = 3$. Notation is as above. '*' indicates that the time to complete the experiment exceeded our patience threshold.

Experiment 1

n	l	D_{reg}	#Var	#P$_{equ}$	T_{Inter}	T_{GB}
17	5	4	42	44	0.08	13.86
19	5	4	42	46	0.08	18.18
	6	4	51	52	0.18	788.91
23	5	4	42	50	0.10	35.35
	6	4	51	56	0.21	461.11
	7	*	*	*	*	*
29	5	4	42	56	0.11	31.64
	6	4	51	62	0.25	229.51
	7	4	60	68	0.60	5196.18
	8	*	*	*	*	*
31	5	4	42	58	0.12	5.10
	6	5	51	64	0.27	167.29
	7	5	60	70	0.48	3259.80
	8	*	*	*	*	*
37	5	4	42	64	0.18	0.36
	6	4	51	70	0.34	155.84
	7	4	60	76	0.75	1164.25
	8	*	*	*	*	*
41	5	4	42	68	0.16	0.24
	6	4	51	74	0.36	251.37
	7	4	60	80	0.77	1401.18
	8	*	*	*	*	*
43	5	4	42	70	0.19	0.13
	6	4	51	76	0.38	176.67
	7	3	60	82	0.78	1311.23
	8	*	*	*	*	*
47	5	4	42	74	0.19	0.14
	6	4	51	80	0.54	78.43
	7	*	*	*	*	*
	8	*	*	*	*	*
53	5	4	51	80	0.22	0.19
	6	5	51	86	0.45	1.11
	7	4	60	92	1.20	880.59
	8	*	*	*	*	*

Experiment 2

n	l	D_{reg}	#Var	#P$_{eq}$	T_{Inter}	T_{GB}
17	5	4	54	56	0.02	0.41
19	5	3	56	60	0.02	0.48
	6	4	62	63	0.03	5.58
23	5	4	60	68	0.02	0.58
	6	4	68	73	0.04	2.25
	7	*	*	*	*	*
29	5	4	62	76	0.03	0.12
	6	4	74	85	0.04	2.46
	7	4	82	90	0.07	3511.14
	8	*	*	*	*	*
31	5	4	62	78	0.03	0.36
	6	4	76	89	0.05	2.94
	7	4	84	94	0.07	2976.97
	8	*	*	*	*	*
37	5	4	62	84	0.04	0.04
	6	4	76	95	0.06	4.23
	7	4	90	106	0.09	27.87
	8	*	*	*	*	*
41	5	4	62	88	0.03	0.04
	6	4	76	99	0.06	0.49
	7	4	90	110	0.09	11.45
	8	*	*	*	*	*
43	5	3	62	90	0.04	0.05
	6	4	76	101	0.06	5.35
	7	4	90	112	0.10	15.360
	8	*	*	*	*	*
47	5	4	62	94	0.04	0.06
	6	4	76	105	0.06	1.28
	7	4	90	116	0.13	8.04
	8	4	104	127	0.16	152.90
53	5	3	62	100	0.04	0.02
	6	4	76	111	0.06	0.19
	7	4	90	122	0.14	68.23
	8	4	104	133	0.19	51.62

l. In our experiments we follow the approach of Huang et al. [15] and examine the effect of different choices of variables on the computation of intermediate results and degree of regularity D_{reg} (as it is the main complexity indicator of $F4$ or $F5$ Gröbner basis algorithms: the time and memory complexities are roughly estimated to be $N^{3D_{\text{reg}}}$ and $N^{2D_{\text{reg}}}$ respectively where N is the number of variables). Our hope is to get better experimental results from exploiting the symmetries of binary Edwards curves.

Experiment 1: For the summation polynomials we use the variables $e_1, e_2, \ldots, e_m$, which are invariants under the group $D_m = (\mathbb{Z}/2\mathbb{Z})^{m-1} \rtimes S_m$. The factor base is defined with respect to a fixed vector space of dimension l.

Experiment 2: For the summation polynomials we use the variables $e_1, s_2, \ldots, s_m$ from equation (2), which are invariants under the group $G_m = (\mathbb{Z}/4\mathbb{Z})^{m-1} \rtimes S_m$. The factor base is defined with respect to a fixed vector space V of dimension l such that $v \in V$ if and only if $v + 1 \in V$.

Experiment 3: For the summation polynomials we use the variables $e_1, s_2, \ldots, s_m$, which are invariants under the group $(\mathbb{Z}/4\mathbb{Z})^{m-1} \times S_m$. We use symmetry-breaking to define the factor base by taking affine spaces (translations of a vector space of dimension l).

Table 1 compares Minisat with Gröbner basis methods (experiment 2) for $m = 4$. We denoted the set-up operations (lines 4 to 8 of Algorithm 1) by T_{Inter}, while T_{SAT} and T_{GB} denote the time for line 9. In all our experiments, timings are averages of 100 trials except for values of $T_{\text{GB}} + T_{\text{Inter}} > 200$ seconds (our patience threshold), in this case they are single instances.

Table 3. Comparison of solving our systems of equations, having a solution, using Gröbner basis methods in experiment 1 and experiment 2 for $m = 4$. Notation is as above. The second column already appeared in Table 1.

			Experiment 1								Experiment 2		
n	l	D_{reg}	#Var	#P_{equ}	T_{Inter}	T_{GB}	n	l	D_{reg}	#Var	#P_{equ}	T_{Inter}	T_{GB}
17	3	5	36	41	590.11	216.07	17	3	4	54	59	0.29	0.29
	4	*	*	*	*	*		4	4	67	68	0.92	51.79
19	3	5	36	43	564.92	211.58	19	3	4	54	61	0.33	0.39
	4	*	*	*	*	*		4	4	71	74	1.53	33.96
23	3	5	36	47	1080.14	146.65	23	3	4	54	65	0.26	0.31
	4	*	*	*	*	*		4	4	75	82	2.52	27.97
29	3	5	36	53	1069.49	232.49	29	3	4	54	71	0.44	0.50
	4	*	*	*	*	*		4	4	77	90	3.19	35.04
31	3	5	36	55	837.77	118.11	31	3	4	54	73	0.44	0.58
	4	*	*	*	*	*		4	4	77	92	3.24	9.03
37	3	5	36	61	929.82	178.04	37	3	4	54	79	0.36	0.43
	4	*	*	*	*	*		4	4	77	98	3.34	9.07
41	3	4	36	65	1261.72	217.22	41	3	4	54	83	0.40	0.54
	4	*	*	*	*	*		4	4	77	102	3.39	17.19
43	3	4	36	67	1193.13	220.25	43	3	4	54	85	0.43	0.53
	4	*	*	*	*	*		4	4	77	104	3.44	9.09
47	3	4	36	71	1163.94	247.78	47	3	4	54	89	0.50	0.65
	4	*	*	*	*	*		4	4	77	108	3.47	9.59
53	3	4	36	77	1031.93	232.110	53	3	4	54	95	0.33	0.40
	4	*	*	*	*	*		4	4	77	114	11.43	11.64

The main observation of this experiment is we can handle larger values of l with Minisat in a reasonable amount of time than Gröbner basis methods. But the process has to be repeated $1/P_{succ}$ times on average, as the probability of finding a relation is decreased by P_{succ}. We also observe that the memory used by Minisat is much lower than that of the Gröbner basis algorithm. We do not report experiments using Gröbner basis method for values of $l > 4$ as they are too slow and have huge memory requirements.

Table 2 compares experiment 1 and experiment 2 in the case $m = 3$. Gröbner basis methods are used in both cases. Experiments in [15] are limited to the case

Table 4. Comparison of solving our systems of equations using Gröbner basis methods having a solution in experiment 3 and experiment 2 when $m = 3$. Notation is as in Table 1. For a fair comparison, the timings in the right hand column should be doubled.

Experiment 3							Experiment 2						
n	l	D_{reg}	#Var	#P_{equ}	T_{Inter}	T_{GB}	n	l	D_{reg}	#Var	#P_{equ}	T_{Inter}	T_{GB}
37	5	3	68	90	0.04	0.25	37	5	4	62	84	0.04	0.04
	6	4	80	99	0.07	5.67		6	4	76	95	0.06	4.23
	7	*	*	*	*	*		7	4	90	106	0.09	27.87
41	5	4	68	94	0.05	0.39	41	5	4	62	88	0.03	0.04
	6	3	80	103	0.07	4.55		6	4	76	99	0.06	0.49
	7	4	93	113	0.11	1905.21		7	4	90	110	0.09	11.45
43	5	4	68	96	0.05	0.21	43	5	3	62	90	0.04	0.05
	6	4	80	105	0.08	4.83		6	4	76	101	0.06	5.35
	7	3	94	116	0.12	100.75		7	4	90	112	0.10	15.360
47	5	4	68	100	0.05	0.17	47	5	4	62	94	0.04	0.06
	6	3	80	109	0.08	3.88		6	4	76	105	0.06	1.28
	7	3	94	120	0.11	57.61		7	4	90	116	0.13	8.04
53	5	3	68	106	0.06	0.08	53	5	3	62	100	0.04	0.02
	6	4	80	115	0.09	12.75		6	4	76	111	0.06	0.19
	7	3	94	126	0.14	11.38		7	4	90	122	0.14	68.23
59	5	4	68	112	0.06	0.05	59	5	4	62	106	0.04	0.02
	6	4	80	121	0.10	0.59		6	3	76	117	0.07	0.11
	7	4	94	132	0.16	13.60		7	4	90	128	0.11	4.34
61	5	4	68	114	0.06	0.04	61	5	4	62	108	0.04	0.02
	6	4	80	123	0.11	0.46		6	3	76	119	0.07	0.09
	7	4	94	134	0.16	8.61		7	4	90	130	0.11	5.58
67	5	3	68	120	0.07	0.02	67	5	4	62	114	0.04	0.02
	6	3	80	129	0.11	0.17		6	4	76	125	0.07	0.07
	7	4	94	140	0.16	121.33		7	4	90	136	0.11	0.94
71	5	3	68	124	0.07	0.02	71	5	4	62	118	0.04	0.02
	6	3	80	133	0.12	0.12		6	4	76	129	0.07	0.04
	7	4	94	144	0.18	2.06		7	3	90	140	0.12	0.25
73	5	3	68	126	0.08	0.02	73	5	4	62	120	0.05	0.02
	6	3	80	135	0.12	0.11		6	4	76	131	0.07	0.03
	7	4	94	146	0.18	1.47		7	3	90	142	0.13	0.22
79	5	3	68	132	0.08	0.02	79	5	4	62	126	0.05	0.02
	6	4	80	141	0.12	0.07		6	4	76	137	0.08	0.03
	7	4	94	152	0.19	0.62		7	4	90	148	0.12	0.33
83	5	3	68	136	0.08	0.02	83	5	4	62	130	0.05	0.02
	6	4	80	145	0.13	0.04		6	4	76	141	0.09	0.03
	7	3	94	156	0.21	0.29		7	4	90	152	0.13	0.13
89	5	3	68	142	0.09	0.02	89	5	4	62	136	0.05	0.02
	6	3	80	151	0.14	0.03		6	4	76	147	0.09	0.03
	7	3	94	162	0.21	0.17		7	4	90	158	0.13	0.05
97	5	3	68	150	0.09	0.02	97	5	4	62	144	0.05	0.02
	6	3	80	159	0.14	0.03		6	4	76	155	0.09	0.03
	7	4	94	170	0.22	0.10		7	4	90	166	0.13	0.04

$m = 3$ and $l \in \{3, 4, 5, 6\}$ for prime degree extensions $n \in \{17, 19, 23, 29, 31, 37, 41, 43, 47, 53\}$. Exploiting greater symmetry (in this case experiment 2) is seen to reduce the computational costs. Indeed, we can go up to $l = 8$ with reasonable running time for some n, which is further than [15]. The degree of regularity stays ≤ 4 in both cases.

Table 3 considers $m = 4$, which was not done in [15]. For the sake of comparison, we gather some data for experiment 1 and experiment 2 (the data for experiment 2 already appeared in Table 1). Again, exploiting greater symmetry (experiment 2) gives a significant decrease in the running times, and the degree of regularity D_{reg} is slightly decreased. The expected degree of regularity for $m = 4$, stated in [18], is $m^2 + 1 = 17$. The table shows that our choice of coordinates makes the case $m = 4$ much more feasible.

Our idea of symmetry breaking (experiment 3) is investigated in Table 4 for $m = 3$. Some of the numbers in the second column already appeared in Table 2. Recall that the relation probability is increased by a factor $3! = 6$ in this case, so one should multiply the timings in the right hand column by $(m - 1)! = 2$ to compare overall algorithm speeds. The experiments are not fully conclusive (and there are a few "outlier" values that should be ignored), but they suggest that symmetry-breaking can give a speedup in many cases when n is large.

For larger values of n, the degree of regularity D_{reg} is often 3 when using symmetry-breaking while it is 4 for most values in experiment 2. The reason for this is unclear, but we believe that the performance we observe is partially explained by the fact that the degree of regularity stayed at 3 as n grows. More discussion is given in the full version of the paper.

7 Conclusions

We have suggested that binary Edwards curves are most suitable for obtaining coordinates invariant under the action of a relatively large group. Faugère et al. [9] studied Edwards curves in the non-binary case and showed how the symmetries can be used to speed-up point decomposition. We show that these ideas are equally applicable in the binary case.

The idea of a factor base that breaks symmetry allows to maximize the probability of finding a relation. For large enough n (keeping m and l fixed) this choice can give a small speed-up compared with previous methods.

SAT solvers often work better than Gröbner methods, especially in the case when the system of equations has a solution with low hamming weight. They are non-deterministic and the running time varies widely depending on inputs. Unfortunately, most of the time SAT solvers are slow (for example, because the system of equations does not have any solutions). We suggest an early abort strategy that may still make SAT solvers a useful approach.

We conclude by analysing whether these algorithms are likely to be effective for ECDLP instances in $E(\mathbb{F}_{2^n})$ when $n > 100$. The best we can seem to hope for in practice is $m = 4$ and $l \leq 10$. Since the probability of a relation is roughly $2^{lm}/2^n$, so the number of trials needed to find a relation is at least

$2^n/2^{ml} \geq 2^{n-40} \geq \sqrt{2^n}$. Since solving a system of equations is much slower than a group operation, we conclude that our methods are worse than Pollard rho. This is true even in the case of static-Diffie-Hellman, when only one relation is required to be found. Hence, we conclude that elliptic curves in characteristic 2 are safe against these sorts of attacks for the moment, though one of course has to be careful of other "Weil descent" attacks, such as the Gaudry-Hess-Smart approach [12]. atively

Acknowledgments. We thank Claus Diem, Christophe Petit and the anonymous referees for their helpful comments.

References

1. Bernstein, D., Lange, T., Farashahi, R.R.: Binary Edwards Curves. In: Oswald, E., Rohatgi, P. (eds.) CHES 2008. LNCS, vol. 5154, pp. 244–265. Springer, Heidelberg (2008)
2. Bettale, L., Faugère, J.-C., Perret, L.: Hybrid approach for solving multivariate systems over finite fields. J. Math. Crypt. 3, 177–197 (2009)
3. Courtois, N.T., Bard, G.V.: Algebraic Cryptanalysis of the Data Encryption Standard. In: Galbraith, S.D. (ed.) Cryptography and Coding 2007. LNCS, vol. 4887, pp. 152–169. Springer, Heidelberg (2007)
4. Diem, C.: On the discrete logarithm problem in elliptic curves over non-prime finite fields. In: Lecture at ECC 2004 (2004)
5. Diem, C.: On the discrete logarithm problem in class groups of curves. Mathematics of Computation 80, 443–475 (2011)
6. Diem, C.: On the discrete logarithm problem in elliptic curves. Compositio Math. 147(1), 75–104 (2011)
7. Diem, C.: On the discrete logarithm problem in elliptic curves II. Algebra and Number Theory 7(6), 1281–1323 (2013)
8. Faugère, J.-C., Perret, L., Petit, C., Renault, G.: Improving the Complexity of Index Calculus Algorithms in Elliptic Curves over Binary Fields. In: Pointcheval, D., Johansson, T. (eds.) EUROCRYPT 2012. LNCS, vol. 7237, pp. 27–44. Springer, Heidelberg (2012)
9. Faugère, J.-C., Gaudry, P., Huot, L., Renault, G.: Using Symmetries in the Index Calculus for Elliptic Curves Discrete Logarithm. Journal of Cryptology (to appear, 2014)
10. Faugère, J.-C., Huot, L., Joux, A., Renault, G., Vitse, V.: Symmetrized summation polynomials: Using small order torsion points to speed up elliptic curve index calculus. In: Nguyen, P.Q., Oswald, E. (eds.) EUROCRYPT 2014. LNCS, vol. 8441, pp. 40–57. Springer, Heidelberg (2014)
11. Faugère, J.-C., Gianni, P., Lazard, D., Mora, T.: Efficient Computation of zero-dimensional Gröbner bases by change of ordering. Journal of Symbolic Computation 16(4), 329–344 (1993)
12. Gaudry, P., Hess, F., Smart, N.P.: Constructive and destructive facets of Weil descent on elliptic curves. J. Crypt. 15(1), 19–46 (2002)
13. Gaudry, P.: Index calculus for abelian varieties of small dimension and the elliptic curve discrete logarithm problem. Journal of Symbolic Computation 44(12), 1690–1702 (2009)

14. Gomes, C.P., Selman, B., Kautz, H.: Boosting combinatorial search through randomization. In: Mostow, J., Rich, C. (eds.) Proceedings AAAI 1998, pp. 431–437. AAAI (1998)
15. Huang, Y.-J., Petit, C., Shinohara, N., Takagi, T.: Improvement of Faugère et al.'s Method to Solve ECDLP. In: Sakiyama, K., Terada, M. (eds.) IWSEC 2013. LNCS, vol. 8231, pp. 115–132. Springer, Heidelberg (2013)
16. Joux, A., Vitse, V.: Cover and Decomposition Index Calculus on Elliptic Curves Made Practical - Application to a Previously Unreachable Curve over $\mathbb{F}_{p^6}$. In: Pointcheval, D., Johansson, T. (eds.) EUROCRYPT 2012. LNCS, vol. 7237, pp. 9–26. Springer, Heidelberg (2012)
17. McDonald, C., Charnes, C., Pieprzyk, J.: Attacking Bivium with MiniSat, ECRYPT Stream Cipher Project, Report 2007/040 (2007)
18. Petit, C., Quisquater, J.-J.: On Polynomial Systems Arising from a Weil Descent. In: Wang, X., Sako, K. (eds.) ASIACRYPT 2012. LNCS, vol. 7658, pp. 451–466. Springer, Heidelberg (2012)
19. Shantz, M., Teske, E.: Solving the Elliptic Curve Discrete Logarithm Problem Using Semaev Polynomials, Weil Descent and Gröbner Basis Methods - An Experimental Study. In: Fischlin, M., Katzenbeisser S. (eds.) Buchmann Festschrift. LNCS, vol. 8260, pp. 94–107. Springer, Heidelberg (2013)
20. Semaev, I.: Summation polynomials and the discrete logarithm problem on elliptic curves, Cryptology ePrint Archive, Report 2004/031 (2004)
21. Sörensson, N., Eén, N.: Minisat 2.1 and Minisat++ 1.0 SAT race 2008 editions, SAT, pp. 31–32 (2008)
22. Yang, B.-Y., Chen, J.-M.: Theoretical analysis of XL over small fields. In: Wang, H., Pieprzyk, J., Varadharajan, V. (eds.) ACISP 2004. LNCS, vol. 3108, pp. 277–288. Springer, Heidelberg (2004)
23. Yang, B.-Y., Chen, J.-M., Courtois, N.: On asymptotic security estimates in XL and Gröbner bases-related algebraic cryptanalysis. In: Lopez, J., Qing, S., Okamoto, E. (eds.) ICICS 2004. LNCS, vol. 3269, pp. 401–413. Springer, Heidelberg (2004)

A Quantum Algorithm for Computing Isogenies between Supersingular Elliptic Curves

Jean-François Biasse[1], David Jao[2($\boxtimes$)], and Anirudh Sankar[2]

[1] Department of Combinatorics and Optimization, Institute for Quantum Computing, University of Waterloo,
Waterloo, ON N2L 3G1, Canada
[2] Department of Combinatorics and Optimization, University of Waterloo,
Waterloo, ON N2L 3G1, Canada
{jbiasse,djao,asankara}@uwaterloo.ca

Abstract. In this paper, we describe a quantum algorithm for computing an isogeny between any two supersingular elliptic curves defined over a given finite field. The complexity of our method is in $\tilde{O}(p^{1/4})$ where p is the characteristic of the base field. Our method is an asymptotic improvement over the previous fastest known method which had complexity $\tilde{O}(p^{1/2})$ (on both classical and quantum computers). We also discuss the cryptographic relevance of our algorithm.

Keywords: Elliptic curve cryptography · Quantum safe cryptography · Isogenies · Supersingular curves

1 Introduction

The computation of an isogeny between two elliptic curves in an important problem in public key cryptography. It occurs in particular in Schoof's algorithm for calculating the number of points of an elliptic curve [23], and in the analysis of the security of cryptosystems relying on the hardness of the discrete logarithm in the group of points of an elliptic curve [16,17]. In addition, cryptosystems relying on the hardness of computing an isogeny between elliptic curves have been proposed in the context of quantum-safe cryptography [5,7,15,22,26]. For the time being, they perform significantly worse than other quantum-safe cryptosystems such as those based on the hardness of lattice problems. However, the schemes are worth studying since they provide an alternative to the few quantum-resistant cryptosystems available today.

In the context of classical computing, the problem of finding an isogeny between two elliptic curves defined over a finite field $\mathbb{F}_q$ of characteristic p has exponential complexity in p. For ordinary curves, the complexity is $\tilde{O}(q^{1/4})$ (here $\tilde{O}$ denotes the complexity with the logarithmic factors omitted) using the algorithm of Galbraith and Stolbunov [12]. In the supersingular case, the method of Delfs and Galbraith [9] is the fastest known technique, having complexity $\tilde{O}(p^{1/2})$.

© Springer International Publishing Switzerland 2014
W. Meier and D. Mukhopadhyay (Eds.): INDOCRYPT 2014, LNCS 8885, pp. 428–442, 2014.
DOI: 10.1007/978-3-319-13039-2_25

With quantum computers, the algorithm of Childs, Jao and Soukharev [6] allows the computation of an isogeny between two ordinary elliptic curves defined over a finite field $\mathbb{F}_q$ and having the same endomorphism ring in subexponential time $L_q(1/2, \sqrt{3}/2)$. This result is valid under the Generalized Riemann Hypothesis, and relies on computations in the class group of the common endomorphism ring of the curves. The fact that this class group in an abelian group is crucial since it allows one to reduce this task to a hidden abelian shift problem. In the supersingular case, the class group of the endomorphism ring is no longer abelian, thus preventing a direct adaptation of this method. The fastest known method for finding an isogeny between two isogenous supersingular elliptic curve is a (quantum) search amongst all isogenous curves, running in $\tilde{O}(p^{1/2})$. The algorithm of Childs, Jao and Soukharev [6] leads directly to attacks against cryptosystems relying on the difficulty of finding an isogeny between ordinary curves [7, 22, 26], but those relying on the hardness of computing isogenies between supersingular curves [5, 15] remain unaffected to this date.

Contribution. Our main contribution is the description of a quantum algorithm for computing an isogeny between two given supersingular curves defined over a finite field of characteristic p that runs in time $\tilde{O}(p^{1/4})$. Moreover, our algorithm runs in subexponential time $L_p(1/2, \sqrt{3}/2)$ when both curves are defined over $\mathbb{F}_p$. Our method is a direct adaptation of the algorithm of Delfs and Galbraith [9] within the context of quantum computing, using the techniques of Childs, Jao, and Soukharev [6] to achieve subexponential time in the $\mathbb{F}_p$ case. We address the cryptographic relevance of our method in Section 6.

2   Mathematical Background

An elliptic curve over a finite field $\mathbb{F}_q$ of characteristic $p \neq 2, 3$ is an algebraic variety given by an equation of the form

$$E : \quad y^2 = x^3 + ax + b,$$

where $\Delta := 4a^3 + 27b^2 \neq 0$. A more general form gives an affine model in the case $p = 2, 3$ but it is not useful in the scope of this paper since we derive an asymptotic result. The set of points of an elliptic curve can be equipped with an additive group law. Details about the arithmetic of elliptic curves can be found in many references, such as [25, Chap. 3].

Let E_1, E_2 be two elliptic curves defined over $\mathbb{F}_q$. An isogeny $\phi \colon E_1 \to E_2$ is a non-constant rational map defined over $\mathbb{F}_q$ which is also a group homomorphism from E_1 to E_2. Two curves are isogenous over $\mathbb{F}_q$ if and only if they have the same number of points over $\mathbb{F}_q$ (see [28]). Two curves over $\mathbb{F}_q$ are said to be isomorphic over $\overline{\mathbb{F}}_q$ if there is an $\overline{\mathbb{F}}_q$-isomorphism between their group of points. Two such curves have the same j-invariant given by $j := 1728 \frac{4a^3}{4a^3 + 27b^2}$. In this paper, we treat isogenies as mapping between (representatives of) $\overline{\mathbb{F}}_q$-isomorphism classes of elliptic curves. In other words, given two j-invariants $j_1, j_2 \in \mathbb{F}_q$, we wish to construct an isogeny between (any) two elliptic curves E_1, E_2 over $\mathbb{F}_q$ having j-invariant j_1 (respectively j_2). Such an isogeny exists if and only if $\Phi_\ell(j_1, j_2) = 0$ for some ℓ, where $\Phi_\ell(X, Y)$ is the ℓ-th modular polynomial.

Let E be an elliptic curve defined over $\mathbb{F}_q$. An isogeny between E and itself defined over $\mathbb{F}_{q^n}$ for some $n > 0$ is called an endomorphism of E. The set of endomorphisms of E is a ring that we denote by $\mathrm{End}(E)$. For each integer m, the multiplication by m map on E is an endomorphism. Therefore, we always have $\mathbb{Z} \subseteq \mathrm{End}(E)$. Moreover, to each isogeny $\phi\colon E_1 \to E_2$ corresponds an isogeny $\hat{\phi}\colon E_2 \to E_1$ called its dual isogeny. It satisfies $\phi \circ \hat{\phi} = [m]$ where $m = \deg(\phi)$. For elliptic curves over a finite field, we know that $\mathbb{Z} \subsetneq \mathrm{End}(E)$. In this particular case, $\mathrm{End}(E)$ is either an order in an imaginary quadratic field (and has $\mathbb{Z}$-rank 2) or an order in a quaternion algebra ramified at p and ∞ (and has $\mathbb{Z}$-rank 4). In the former case, E is said to be ordinary while in the latter it is called supersingular.

An order $\mathcal{O}$ in a field K such that $[K : \mathbb{Q}] = n$ is a subring of K which is a $\mathbb{Z}$-module of rank n. The notion of ideal of $\mathcal{O}$ can be generalized to fractional ideals, which are sets of the form $\mathfrak{a} = \frac{1}{d}I$ where I is an ideal of $\mathcal{O}$ and $d \in \mathbb{Z}_{>0}$. The invertible fractional ideals form a multiplicative group $\mathcal{I}$, having a subgroup consisting of the invertible principal ideals $\mathcal{P}$. The ideal class group $\mathrm{Cl}(\mathcal{O})$ is by definition $\mathrm{Cl}(\mathcal{O}) := \mathcal{I}/\mathcal{P}$. In $\mathrm{Cl}(\mathcal{O})$, we identity two fractional ideals $\mathfrak{a}, \mathfrak{b}$ if there is $\alpha \in K$ such that $\mathfrak{b} = (\alpha)\mathfrak{a}$. The ideal class group is finite and its cardinality is called the class number $h_\mathcal{O}$ of $\mathcal{O}$. For a quadratic order $\mathcal{O}$, the class number satisfies $h_\mathcal{O} \leq |\Delta| \log |\Delta|$, where Δ is the discriminant of $\mathcal{O}$.

The endomorphism ring of an elliptic curve plays a crucial role in most algorithms for computing isogenies between curves. The class group of $\mathrm{End}(E)$ acts transitively on isomorphism classes of elliptic curves (that is, on j-invariants of curves) having the same endomorphism ring. More precisely, the class of an ideal $\mathfrak{a} \subseteq \mathcal{O}$ acts on the isomorphism class of curve E with $\mathrm{End}(E) \simeq \mathcal{O}$ via an isogeny of degree $\mathcal{N}(\mathfrak{a})$ (the algebraic norm of $\mathfrak{a}$). Likewise, each isogeny $\varphi\colon E \to E'$ where $\mathrm{End}(E) = \mathrm{End}(E') \simeq \mathcal{O}$ corresponds (up to isomorphism) to the class of an ideal in $\mathcal{O}$. From an ideal $\mathfrak{a}$ and the ℓ-torsion (where $\ell = \mathcal{N}(\mathfrak{a})$), one can recover the kernel of φ, and then using Vélu's formulae [29], one can derive the corresponding isogeny.

Given $\ell > 0$ prime, the ℓ-isogeny graph between (isomorphism classes of) elliptic curves defined over $\mathbb{F}_q$ is a graph whose vertices are the j-invariants of curves defined over $\mathbb{F}_q$ having an edge between j_1 and j_2 if and only if there exists an ℓ-isogeny ϕ between some two curves E_1, E_2 defined over $\mathbb{F}_q$ having j-invariant j_1 (respectively j_2). Note that while the curves E_1 and E_2 are required to be defined over $\mathbb{F}_q$, the isogeny ϕ is not. When $\ell \nmid q$, the ℓ-isogeny graph is connected. In this case, finding an isogeny between E_1 and E_2 amounts to finding a path between the j-invariant j_1 of E_1 and the j-invariant j_2 of E_2 in the ℓ-isogeny graph. Most algorithms for finding an isogeny between two curves perform a random walk in the ℓ-isogeny graph for some small ℓ. Our method is based on this strategy.

3   High Level Description of the Algorithm

Our algorithm to find an isogeny between supersingular curves E, E' defined over $\mathbb{F}_q$ of characteristic p is based on the approach of Galbraith and Delfs [9], which exploits the fact that it is easier to find an isogeny between supersingular curves when they are defined over $\mathbb{F}_p$. The first step consists of finding an isogeny between E and E_1 (respectively between E' and E_2) where E_1, E_2 are defined over $\mathbb{F}_p$. On a quantum computer, we achieve a quadratic speedup for this first step using Grover's algorithm [13]. We then present a novel subexponential time quantum algorithm to find an isogeny between E_1 and E_2.

All isomorphism classes of supersingular curves over $\overline{\mathbb{F}}_q$ admit a representative defined over $\mathbb{F}_{p^2}$. As pointed out in [9], it is a well-known result that the number of supersingular j-invariants (that is, of isomorphism classes of supersingular curves defined over $\mathbb{F}_{p^2}$) is

$$\#S_{p^2} = \left\lfloor \frac{p}{12} \right\rfloor + \begin{cases} 0 & \text{if } p \equiv 1 \bmod 12, \\ 1 & \text{if } p \equiv 5, 7 \bmod 12, \\ 2 & \text{if } p \equiv 11 \bmod 12, \end{cases}$$

where S_{p^2} is the set of supersingular j-invariants in $\mathbb{F}_{p^2}$. A certain proportion of these j-invariants in fact lie in $\mathbb{F}_p$; we denote this set by S_p. The number of such j-invariants satisfies

$$\#S_p = \begin{cases} \frac{h(-4p)}{2} & \text{if } p \equiv 1 \bmod 4, \\ h(-p) & \text{if } p \equiv 7 \bmod 8, \\ 2h(-p) & \text{if } p \equiv 3 \bmod 8, \end{cases}$$

where $h(d)$ is the class number of the maximal order of $\mathbb{Q}(\sqrt{d})$ (See [8, Thm. 14.18]). As $h(d) \in \tilde{O}(\sqrt{d})$, we have $\#S_p \in \tilde{O}(\sqrt{p})$ (while $\#S_{p^2} \in O(p)$). The method used in [9] to find an isogeny path to a curve defined over $\mathbb{F}_p$ has complexity $\tilde{O}(\sqrt{p})$ (mostly governed by the proportion of such curves), while the complexity of finding an isogeny between curves defined over $\mathbb{F}_p$ is $\tilde{O}(p^{1/4})$.

Following this approach, we obtain a quantum algorithm for computing an isogeny between two given supersingular curves defined over a finite field of characteristic p that has (quantum) complexity in $\tilde{O}(p^{1/4})$. As illustrated in Figure 3, the search for a curve defined over $\mathbb{F}_p$, which is detailed in Section 4, has complexity $\tilde{O}(p^{1/4})$. Then, the computation of an isogeny between curves defined over $\mathbb{F}_p$, which we describe in Section 5, has subexponential complexity.

Theorem 1 (Main result). *Algorithm 1 is correct and runs under the Generalized Riemann Hypothesis in quantum complexity*

- *$\tilde{O}(p^{1/4})$ in the general case.*
- *$L_q(1/2, \sqrt{3}/2)$ when both curves are defined over $\mathbb{F}_p$,*

where $L_p(a, b) := e^{b \log(p)^a \log\log(p)^{1-a}}$.

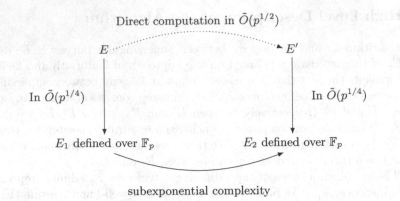

Fig. 1. $\tilde{O}(p^{1/4})$ method for supersingular curves

Algorithm 1. Isogeny computation between supersingular curves defined over a finite field

---

Input: Supersingular curves E, E' defined over $\mathbb{F}_q$ of characteristic p.
Output: An isogeny between E and E'
1: Find $\phi\colon E \to E_1$ where E_1 is defined over $\mathbb{F}_p$ by using Algorithm 2
2: Find $\psi\colon E' \to E_2$ where E_2 is defined over $\mathbb{F}_p$ by using Algorithm 2
3: Find $\alpha\colon E_1 \to E_2$ by using Algorithm 3
4: **return** $\widehat{\psi} \circ \alpha \circ \phi$

Proof. Steps 1 and 2 run in complexity $\tilde{O}(p^{1/4})$ as shown in Section 4 while Step 3 runs in subexponential complexity as shown in Section 5. Moreover, Steps 1 and 2 can be skipped if both curves are defined over $\mathbb{F}_p$.

4   The Quantum Search for a Curve Defined over $\mathbb{F}_p$

Given a supersingular elliptic curve E defined over $\mathbb{F}_{p^2}$, we describe in this section how to find an isogeny path to a curve E' defined over $\mathbb{F}_p$. Our method has complexity $\tilde{O}(p^{1/4})$ and is based on a quantum search amongst a set of short isogeny paths initiating from E.

With classical algorithms, searching an unsorted database of N elements cannot be done in time faster than $O(N)$. With a quantum computer, Grover's algorithm [13] allows us to find an element x in the database such that $C(x) = 1$ (assuming all other elements y satisfy $C(y) = 0$) in complexity $O(\sqrt{N})$ with success probability greater than $1/2$. A rigorous analysis of the run time appears in Boyer et al. [2], which also contains a generalization to a multiple target search. The elements of the database are states that are encoded on n bits where $N = 2^n$, and condition $C(x)$ is assumed to be evaluated in unit time on these states.

The ℓ-isogeny graph for a prime $\ell \nmid p$ is a Ramanujan graph [10, Sec. 2]. This property allows us to evaluate the probability that an ℓ-isogeny path reaches a certain subset of the vertices. The following proposition applies this to the problem of finding a path leading to the subset S_p of the set S_{p^2} of all the vertices of the graph.

Proposition 1. *Under the Generalized Riemann Hypothesis, there is a probability at least $\frac{\pi}{2^7} \frac{1}{p^{1/2}}$ that a random 3-isogeny path of length*

$$\lambda \geq \frac{\log\left(\frac{2}{\sqrt{6}e^{\gamma}}p^{3/4}\right)}{\log\left(\frac{2}{\sqrt{3}}\right)}$$

passes through a supersingular j-invariant defined over $\mathbb{F}_p$, where γ is the Euler constant.

Proof. This is a direct application of [10, Prop. 2.1] which states that for $c \geq 2\sqrt{\ell}$ and $k = \ell+1$, a random ℓ-isogeny walk (for $\ell \nmid p$) of length at least $\frac{\log((2|G|/|S|^{1/2})}{\log(k/c)}$ starting from a given curve will hit a subset S of the vertices G with probability at least $\frac{|S|}{2|G|}$. We apply this to $G = S_{p^2}$ and $S = S_p$, knowing that $|G| \geq p/12$, and that under the Generalized Riemann Hypothesis [14], the class number of the maximal order of $\mathbb{Q}(\sqrt{-d})$ satisfies

$$h(d) \geq (1 + o(1)) \cdot \frac{\pi}{12e^{\gamma}} \frac{\sqrt{d}}{\log\log(d)}.$$

A direct substitution of these values allows us to obtain the desired result.

When p is large enough, the probability that all of $\log(2)\frac{e^{\gamma}}{\pi}p^{1/2}$ random 3-isogeny-paths of length λ defined in Proposition 1 initiating from a given curve do not hit any supersingular j-invariant defined over $\mathbb{F}_p$ is

$$\left(1 - \frac{\pi}{e^{\gamma}} \cdot \frac{1}{p^{1/2}}\right)^{\log(2)\frac{e^{\gamma}}{\pi}p^{1/2}} = e^{\log(2)\frac{e^{\gamma}}{\pi}p^{1/2}\cdot\log\left(1-\frac{\pi}{e^{\gamma}}\cdot\frac{1}{p^{1/2}}\right)}$$

$$\sim e^{\log(2)\frac{e^{\gamma}}{\pi}p^{1/2}\cdot\left(-\frac{\pi}{e^{\gamma}}\cdot\frac{1}{p^{1/2}}\right)} = \frac{1}{2}.$$

Therefore, a set of $N := \log(2)\frac{e^{\gamma}}{\pi}p^{1/2}$ random 3-isogeny-paths of length λ contains at least one that passes through S_p with probability at least $1/2$. A quantum search with Grover's algorithm yields our target isogeny path (which exists with probability $1/2$) in complexity $O(\sqrt{N}) = O(p^{1/4})$. Let us formalize this search. At each node corresponding to the j-invariant j_0, the polynomial $\Phi_3(j_0, X)$ has four roots, one corresponding to the father of the node, and j_1, j_2, j_3 corresponding to its children. Therefore, each path can be encoded in $\{0, 1, 2\}^{\lambda}$. As $\lambda \in O(\log(p))$, the number of bits needed to encode such a path is also in

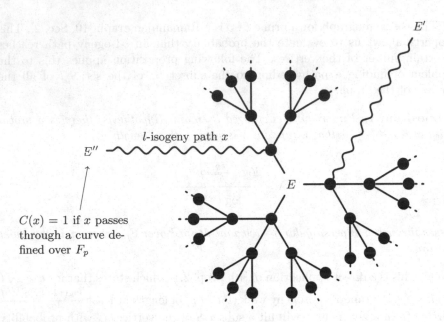

E'

l-isogeny path x

E''

E

$C(x) = 1$ if x passes
through a curve de-
fined over F_p

Fig. 2. Quantum walk to a curve defined over $\mathbb{F}_p$

$O(\log(p))$. Note that the actual computation of a 3-isogeny between representatives of the isomorphism classes of two given j-invariants is classical and used in Section 5. It can be done in polynomial time. At the beginning of the algorithm, we compute a random injection

$$f\colon [1,\cdots,N] \to \{\text{3-isogeny paths of length } \lambda \text{ starting from } E\}.$$

For our search, we use the function C defined on $x \in [1,\cdots,N]$ by

$$C_f(x) := \begin{cases} 1 & \text{if } f(x) \text{ passes through } S_p, \\ 0 & \text{otherwise.} \end{cases}$$

Proposition 2. *Algorithm 2 has success probability $1/4$ and expected run time in $\tilde{O}(p^{1/4})$.*

Proof. The complexity derives from the analysis of Grover's algorithm. The only difference is that the evaluation of C_f is done in polynomial time, thus inducing terms in $\log(p)$ in the complexity.

Remark 1. We can find an isogeny between two given supersingular curves E, E' directly by using a quantum search method. It suffices to apply the above method to the trivial subset $S = \{j(E')\}$ of size 1. The corresponding complexity is in $\tilde{O}(p^{1/2})$.

Algorithm 2. Quantum walk to a curve defined over $\mathbb{F}_p$

Input: Supersingular curve E defined over $\mathbb{F}_q$ of characteristic p.
Output: E' defined over $\mathbb{F}_p$ and $\phi\colon E \to E'$

1: $\lambda := \left\lceil \dfrac{\log\left(\frac{2}{\sqrt{6e7}}p^{3/4}\right)}{\log\left(\frac{2}{\sqrt{3}}\right)} \right\rceil$.

2: Choose $f\colon [1, \cdots, N] \to \{\text{3-isogeny paths of length } \lambda \text{ starting from } E\}$ randomly.
3: Use Grover's algorithm to find $x \in [1, \cdots, N]$ such that $C_f(x) = 1$.
4: Compute the isogeny path $\phi_1, \cdots, \phi_\lambda$ corresponding to x.
5: **return** $\phi_1 \circ \cdots \circ \phi_\lambda, \phi_1 \circ \cdots \circ \phi_\lambda(E)$.

---

5   Computing an Isogeny between Curves Defined over $\mathbb{F}_p$

We now present a quantum algorithm for computing an isogeny between supersingular curves defined over $\mathbb{F}_p$ in subexponential time $L_p(1/2, \sqrt{3}/2)$. Our approach relies on the correspondence between these curves and elliptic curves with complex multiplication by a quadratic order described by Delfs and Galbraith [9], and on the quantum subexponential algorithm for ordinary curves of Childs, Jao and Soukharev [6].

General strategy. The endomorphism ring $\mathrm{End}(E)$ of a supersingular curve is an order in a quaternion algebra, but as shown in [9, Th. 2.1], the ring $\mathrm{End}_{\mathbb{F}_p}(E)$ of endomorphisms defined over $\mathbb{F}_p$ is isomorphic to an order $\mathcal{O}$ in the quadratic number field $K := \mathbb{Q}(\sqrt{-p})$. More specifically, it is equal to either $\mathbb{Z}[\sqrt{-p}]$ or the maximal order $\mathcal{O}_K$. There is a transitive action of $\mathrm{Cl}(\mathcal{O})$ on the $\mathbb{F}_p$-isomorphism classes of supersingular elliptic curves defined over $\mathbb{F}_p$. As in the ordinary case, the class of an ideal $\mathfrak{a}$ acts via an isogeny of degree $\mathcal{N}(\mathfrak{a})$. Therefore, for each supersingular curve E defined over $\mathbb{F}_p$ with endomorphism ring isomorphic to $\mathcal{O}$, we have an injective function

$$f_E\colon \mathrm{Cl}(\mathcal{O}) \longrightarrow \mathbb{F}_p - \text{isomorphism classes of curves over } \mathbb{F}_p$$
$$[\mathfrak{b}] \longmapsto \text{action of } [\mathfrak{b}] \text{ on the class of } E$$

Given two supersingular curves E_1 and E_2 defined over $\mathbb{F}_p$, the problem of finding the ideal class $[\mathfrak{a}]$ such that $f_{E_2}(x) = f_{E_1}([\mathfrak{a}] \cdot x)$ for all x is an instance of the *hidden abelian shift problem*. We solve it to find the ideal class $[\mathfrak{a}]$ such that the class of E_2 is the image of the action of $[\mathfrak{a}]$ on the class of E_1. Then we find the corresponding isogeny $\phi\colon E_1 \to E_2'$ where E_2' lies in the same $\mathbb{F}_p$-isomorphism class as E_2. Finally, we use the method described in [11, Appendix A.2] to calculate the $\mathbb{F}_p$-isomorphism between E_2' and E_2. The composition of both maps is an isogeny between E_1 and E_2. The procedure is summarized in Algorithm 3.

The action of $\mathrm{Cl}(\mathcal{O})$. Let K be the quadratic number field $\mathbb{Q}(\sqrt{-p})$ having maximal order $\mathcal{O}_K$. By [9, Prop. 2.5], there is a one-to-one correspondence

$$\left\{ \begin{array}{l} \text{Supersingular elliptic curves} \\ \text{defined over } \mathbb{F}_p \end{array} \right\} \leftrightarrow \left\{ \begin{array}{l} \text{Elliptic curves } E \text{ over } \mathbb{C} \text{ with} \\ \mathrm{End}(E) \in \{\mathbb{Z}[\sqrt{-p}], \mathcal{O}_K\} \end{array} \right\}.$$

Algorithm 3. Computation of an isogeny between supersingular curves over $\mathbb{F}_p$

Input: Supersingular curves E_1, E_2 defined over $\mathbb{F}_p$.
Output: An isogeny $\phi \colon E_1 \to E_2$.
1: Compute an isogeny $\phi_1 \colon E_1 \to E_1'$ with $\mathrm{End}(E_1') = \mathcal{O}_K$.
2: Compute an isogeny $\phi_2 \colon E_2 \to E_2'$ with $\mathrm{End}(E_2') = \mathcal{O}_K$.
3: Solve the hidden abelian shift problem to find $[\mathfrak{a}] \in \mathrm{Cl}(\mathcal{O}_K)$ such that the action of $[\mathfrak{a}]$ on the isomorphism class of E_1' is the class of E_2'.
4: Deduce an isogeny $\phi_3 \colon E_1' \to E_2''$ where E_2'' is $\mathbb{F}_p$-isomorphic to E_2'.
5: Find the $\mathbb{F}_p$ isomorphism $\alpha \colon E_2'' \to E_2'$.
6: **return** $\widehat{\phi_2} \circ \alpha \circ \phi_3 \circ \phi_1$.

In one direction, this correspondence is given by the Deuring lift, while in the other direction, it is given by the reduction at a place $\mathfrak{P}$ above p. Moreover, we have a bijective map

$$
\begin{array}{ccc}
\text{Classes of curves with } \mathrm{End}(E) = \mathcal{O} & \longrightarrow & \text{Classes of curves with } \mathrm{End}(E)_{\mathbb{F}_p} = \mathcal{O} \\
\text{Isomorphism class of } E & \longmapsto & \mathbb{F}_p\text{-isomorphism class of } \bar{E}
\end{array}
$$

where $\bar{E}$ is the reduction of E modulo $\mathfrak{P}$. Therefore, the $\mathbb{F}_p$-isomorphism classes of curves over $\mathbb{F}_p$ with $\mathbb{F}_p$-endomorphism ring $\mathcal{O}$ are in one-to-one correspondence with isomorphism classes of complex curves with endomorphism ring $\mathcal{O}$. The class group of $\mathcal{O}$ acts on these complex curves, therefore inducing by modular reduction an action on the curves over $\mathbb{F}_p$. Indeed, the class $[\mathfrak{a}]$ of an ideal $\mathfrak{a} \subseteq \mathcal{O}$ acts on the class of a complex curve E via an isogeny $\phi \colon E \to E'$ with $\deg(\phi) = \mathcal{N}(\mathfrak{a})$. By [9, Prop.2.6], this gives us by reduction an isogeny $\bar{\phi} \colon \bar{E} \to \bar{E}'$. From the correspondence between isomorphism classes over $\mathbb{C}$ and $\mathbb{F}_p$-isomorphism classes over $\mathbb{F}_p$, we get a group action of $\mathcal{O}$ on $\mathbb{F}_p$-isomorphism classes of supersingular curves defined over $\mathbb{F}_p$.

Computing the action of $\mathrm{Cl}(\mathcal{O})$. Our method to solve the hidden abelian shift problem is based on the algorithm described by Childs, Jao and Soukharev [6] which relies on a (classical) subexponential algorithm to compute the action of $[\mathfrak{a}] \in \mathrm{Cl}(\mathcal{O})$ on isomorphism classes of ordinary curves. In this paragraph, we show how to compute (classically) the action of $\mathrm{Cl}(\mathcal{O})$ on $\mathbb{F}_p$-isomorphism classes of supersingular curves $E : Y^2 = X^3 + aX + b = 0$ with $\mathrm{End}_{\mathbb{F}_p}(E) \simeq \mathcal{O}$. In a nutshell, it is similar to the approach of Childs, Jao and Soukharev [6], except that the role of $\mathrm{End}(E)$ is replaced by $\mathrm{End}_{\mathbb{F}_p}(E)$.

The first step consists of finding split prime ideals $\mathfrak{p}_1, \cdots, \mathfrak{p}_k$ having norm $\mathcal{N}(\mathfrak{p}_i) \leq L_p(1/2, \sqrt{3}/2)$ such that $[\mathfrak{a}] = [\mathfrak{p}_1]^{e_1} \cdots [\mathfrak{p}_k]^{e_k}$. This way, the action of $[\mathfrak{a}]$ can be calculated as the composition of the action of the $[\mathfrak{p}_i]$ for $i \leq k$. The subexponential classical strategy for performing this decomposition is standard in class group computation and discrete logarithm resolution. In this paper, we use the particular version described in [6, Alg. 1].

Once $[\mathfrak{a}]$ has been successfully rewritten, evaluating its action reduces to evaluating that of $[\mathfrak{p}]$ where $\mathfrak{p}$ is a split prime ideal with $\mathcal{N}(\mathfrak{p}) = \ell$. Let us denote by $\bar{E}$ a representative of the $\mathbb{F}_p$-isomorphism class on which we want to evaluate the action of $[\mathfrak{p}]$ and by E its Deuring lift (which we do not actually compute). Amongst the $\ell + 1$ complex roots of $\Phi_\ell(j(E), X)$ (where $\ell \nmid p$, $\ell \neq 2$), only two reduce to j-invariants defined over $\mathbb{F}_p$. One of them corresponds to the action of $[\mathfrak{p}]$ on the isomorphism class of E while the other one is the result of the action of $[\bar{\mathfrak{p}}]$ (where $\bar{\mathfrak{p}}$ is the complex conjugate of $\mathfrak{p}$). The other roots correspond to ascending or descending isogenies.

Let j be one of the roots mentioned above. As described in Bröker, Charles and Lauter [4, Section 3], there are two methods for computing the equation of a curve E' in the isomorphism class identified by j. One method is to use the Atkin-Elkies formulas given by Schoof in [23, Sec. 7] to compute $E' : Y^2 = X^3 + a'X + b'$ where

$$
a' = -\frac{1}{48} \frac{j'^{\,2}}{j(j - 1728)} \,, \quad b' = -\frac{1}{864} \frac{j'^{\,3}}{j^2(j - 1728)} \,, \quad j' = -\frac{18\,b}{\ell\,a} \frac{\Phi_{\ell,X}(j(E), j)}{\Phi_{\ell,Y}(j(E), j)} j(E),
$$

with $\Phi_{\ell,X}(X,Y) = \frac{\partial \Phi_\ell}{\partial X}(X,Y)$ and $\Phi_{\ell,Y}(X,Y) = \frac{\partial \Phi_\ell}{\partial Y}(X,Y)$. Reduction modulo $\mathfrak{P}$ of the above formulas yield an equation of a supersingular curve defined over $\mathbb{F}_p$ in the $\mathbb{F}_p$-isomorphism class corresponding to the class of complex curves having j-invariant j. This method can fail in the event that one of the terms appearing in a denominator (namely, j, $j - 1728$, or $\Phi_{\ell,Y}(j(E), j)$) equals zero. The second method is to use division polynomials to construct $E[\ell]$ explicitly over a field extension. One then checks each of the possible $\ell + 1$ cyclic ℓ-subgroups of $E[\ell]$ until the correct kernel is found.

In the case of ordinary elliptic curves, the j-invariants $j = 0$ and $j = 1728$ that induce failure in the first method can often be avoided (for example, if they do not belong to the isogeny class in question), and the term $\Phi_{\ell,Y}(j(E), j)$ never vanishes as long as $\ell < 4 \cdot |\operatorname{disc}(\operatorname{End}(E))|$. In the supersingular case, we found experimentally that the $\Phi_{\ell,Y}(j(E), j)$ term does often vanish even when $\ell < 4 \cdot |\operatorname{disc}(\operatorname{End}(E))|$, necessitating the second approach, which works unconditionally.

To determine if j was the j-invariant of the isomorphism class resulting from the action of the class of $\mathfrak{p}$ or its conjugate, we first compute the kernel $C \subset \bar{E}[\ell]$ of the isogeny between $\bar{E}$ and $\bar{E}'$ by the approach described by Schoof [23, Sec. 8] and used by Bröker Charles and Lauter [4]. The ideal $\mathfrak{p}$ is of the form $\mathfrak{p} = \ell\mathcal{O} + (c + d\sqrt{-p})\mathcal{O}$, and it induces an action on the points P of $\bar{E}$ given by $\mathfrak{p} \cdot P = [\ell]P + [c]P + [d]\pi_p(P)$ where π_p is the p-th power Frobenius endomorphism. If $\mathfrak{p} \cdot P = 0$ for all $P \in C$, our choice was correct; otherwise, we redo the computation with the other root of $\Phi_\ell(j(E), X)$.

Proposition 3. *The running time of Algorithm 4 is* $L_p\left(\frac{1}{2}, \frac{\sqrt{3}}{2}\right)$

Proof. The proof of complexity follows from the considerations of [6, Sec. 4.1]. $\square$

Solving the abelian shift problem. As we have an action of $\mathrm{Cl}(\mathcal{O})$ on the $\mathbb{F}_p$-isomorphism classes of supersingular curves defined over $\mathbb{F}_p$ that we can compute in subexponential time, we can readily apply the same method as in [6, Sec. 5] to

---

Algorithm 4. Action of $[\mathfrak{a}] \in \mathrm{Cl}(\mathcal{O})$

Input: A supersingular curve E defined over $\mathbb{F}_p$, a quadratic order $\mathcal{O} \simeq \mathrm{End}_{\mathbb{F}_p}(E)$ and an ideal $\mathfrak{a} \subseteq \mathcal{O}$.

Output: A supersingular curve E' defined over $\mathbb{F}_p$ in the $\mathbb{F}_p$-isomorphism class resulting from the action of $[\mathfrak{a}]$ on the class of E.

1: Find $(\mathfrak{p}_i)_{i \leq k}$ with $\mathfrak{p}_i \nmid (2) \cdot (\#E(\mathbb{F}_p))$ and $\mathcal{N}(\mathfrak{p}_i) \leq L_p(1/2, \sqrt{3}/2)$ such that $[\mathfrak{a}] = \prod_i [\mathfrak{p}_i]$

2: **for** $i \leq k$ **do**

3:    Compute $\Phi_i(X,Y)$ where $l = \mathcal{N}(\mathfrak{p}_i)$.

4:    Find the two roots j_1, j_2 of $\Phi_l(j(E), X)$ defined over $\mathbb{F}_p$.

5:    Compute E' of j-invariant j_1 using the method of [23, Sec. 7].

6:    Compute the kernel C of the isogeny $E \to E'$ using the method of [23, Sec. 8].

7:    If there exists $P \in C$ such that $[c]P + [d]\pi_p(P) \neq 0$, where c and d are integers such that $\mathfrak{p} = (\ell, c + d\pi_p)$, go back to Step 5 and use j_2 instead of j_1.

8:    $E \leftarrow E'$.

9: **end for**

10: **return** E'.

---

solve the hidden abelian shift problem. Childs, Jao and Soukharev considered two quantum algorithms. The first one is Kuperberg's approach based on a Clebsch-Gordan sieve on coset states [19]. The other one relies on Regev's algorithm [21]. In this way we obtain the following result.

Proposition 4 (Theorem 5.4 of [6]). *On a quantum computer, the hidden abelian shift of Step 3 in Algorithm 3 can be solved in time $L_p(1/2, \sqrt{3}/2)$ under the Generalized Riemann Hypothesis.*

Climbing the volcano. Steps 1 and 2 of Algorithm 3 ensure that the curves between which we are trying to compute an isogeny have the same endomorphism ring. As mentioned in [9], a supersingular elliptic curve defined over $\mathbb{F}_p$ has $\mathbb{F}_p$-endomorphism ring satisfying $\mathrm{End}_{\mathbb{F}_p}(E) \simeq \mathcal{O}$ for $\mathcal{O} \in \{\mathbb{Z}[\sqrt{-p}], \mathcal{O}_K\}$. This means that the isogeny volcano has at most two levels, namely the crater and the ground level. In Steps 1 and 2 of Algorithm 3 we climb to the crater. This step can be done by computing a single 2-isogeny. As shown in [9, Sec. 2], if $\Phi_2(j(E), X)$ has three roots, then $\mathrm{End}_{\mathbb{F}_p}(E) = \mathcal{O}_K$ and we do nothing. Otherwise, $\Phi_2(j(E), X)$ has one root, which is the j-invariant of an isogenous curve E' on the crater (that is, with $\mathrm{End}_{\mathbb{F}_p} = \mathcal{O}_K$). In this case, we know that $\mathrm{End}_{\mathbb{F}_p}(E) = \mathbb{Z}[\sqrt{-p}] \neq \mathcal{O}_K$ and we compute $\phi \colon E \to E'$.

6   Cryptographic Relevance

The main motivation for our result is its impact on existing cryptosystems relying on the hardness of finding an isogeny between two given curves. Those that use ordinary elliptic curves [7,22,26] are not affected by our method. The subexponential algorithm of Childs, Jao and Soukharev [6] already provides a quantum subexponential attack against these.

De Feo-Jao-Plût cryptographic schemes. In [10] (which is an extended version of [15]), De Feo, Jao and Plût presented a key exchange protocol, an encryption protocol and a zero knowledge proof of identity all relying on the difficulty of computing an isogeny between supersingular curves. More specifically, given a secret point S of a curve E over $\mathbb{F}_{p^2}$, and a public point R, they exploit the commutative diagram

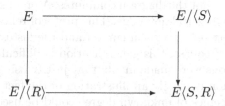

The unified treatment of these three cryptographic schemes around the above commutative diagram yields situations where the degree of the secret isogenies is known and in $O(\sqrt{p})$. Therefore, there is a classical attack in $O(p^{1/4})$ and a quantum attack relying on a claw algorithm [24] with complexity $O(p^{1/6})$. Given these results, our $p^{1/4}$ quantum algorithm does not yield the fastest attack. Moreover, it is not even clear that finding an arbitrary isogeny between two given curves yields an attack at all since the cryptosystems described in [10] rely on the difficulty of finding an isogeny of given degree, while our method returns an isogeny of arbitrary degree. Note that a recent contribution of Jao and Soukharev [18] uses similar methods to describe a quantum-resistant undeniable signature scheme.

Even though our work does not directly yield a faster attack against the existing schemes of [10], it does introduce the possibility that choosing a base curve E defined over $\mathbb{F}_p$ may be insecure. The base curve E is a public parameter, chosen once and for all at the time the system is initialized, and never changed during the life of the system. If this base curve is defined over $\mathbb{F}_p$, then Step 1 of Algorithm 1 becomes trivial. While this situation is not fatal, it does seem to be cause for some concern, provided that the arbitrary degree obstacle mentioned above can be overcome; at the very least, it decreases by half the amount of work the attacker must perform. De Feo et al. [10, Section 4.1] propose two methods for choosing the base curve. One of these methods uses random walks, and would not normally produce base curves defined over $\mathbb{F}_p$. The other method uses Bröker's algorithm [3] to produce a supersingular curve which is then used directly as the base curve. This method does sometimes produce curves defined over $\mathbb{F}_p$, and in light of our results, we recommend avoiding this method out of an abundance of caution.

Generalizations of De Feo-Jao-Plût. It is possible to conceive of potential generalizations of the cryptosystems presented in [10] to a situation where the degree of the isogenies is unknown, in which case our algorithm would yield the fastest (quantum) attack. For example, let us sketch how this could be done for the zero knowledge proof of identity. Assume Peggy knows a secret kernel C_1 and

$\psi: E \rightarrow E/C_1$. At each round, Peggy draws a kernel C_2 coprime with C_1 and publishes $E, E/C_1, E/C_2, E_{\langle C_1, C_2 \rangle}$. Then Vic flips a coin b.

- If $b = 0$, she asks to know $E \rightarrow E/C_2$ and $E/C_1 \rightarrow E/\langle C_1, C_2 \rangle$.
- If $b = 1$, she asks to know $E/C_2/C_1 \rightarrow E/\langle C_1, C_2 \rangle$.

The kernels can be drawn from coprime power-products of ideals of small norm in $\text{End}(E)$, thus ensuring that the diagram commutes. A proof similar to that of [10, Th. 6.3] shows that it is zero knowledge. This protocol relies on the difficulty of finding an isogeny between two given curves, and the fastest quantum attack is our $p^{1/4}$ algorithm. Of course, this generalization is difficult to make practical, and many optimizations were made in [10] that justify using isogenies of known degrees. Our suggestion is merely an illustration of the fact that adaptations to the case of secret isogenies of unknown degree could be used.

The Charles-Goren-Lauter hash function. In [5], Charles, Gore and Lauter described a cryptographic hash function using supersingular elliptic curves. More specifically, its preimage resistance relies on the difficulty of finding an isogeny between two given (supersingular) elliptic curves over $\mathbb{F}_{p^2}$. In this context, our algorithm directly yields the fastest known quantum attack in $\tilde{O}(p^{1/4})$.

7 Example

For the purposes of validating our algorithm, we implemented Algorithm 4 in the MAGMA Computational Algebra System [1, 20] (Algorithm 4 being the only algorithm in this work which can be implemented on a classical computer). We present an example calculation here.

Let E be the curve $y^2 = x^3 + 77x + 42$ over $\mathbb{F}_p$. $p = 101$, $\ell = 83$, and $\mathfrak{a} = (\ell, 27 + \pi_p)$ (corresponding to the quadratic form $83x^2 - 54xy + 10y^2$). This example was chosen to be small enough to allow direct calculation of the answer, in order to check the correctness of our work. Using the division polynomial-based method of [4, Section 3.2], we find that the 83-isogenous curve corresponding to $\mathfrak{a}$ is $E' : y^2 = x^3 + 40x + 65$, having j-invariant $j(E') = 66$. In order to redo the calculation according to the method specified in Algorithm 4, we used Sutherland's `smoothrelation` program [27] to find the relation

$$\mathfrak{p}_{83} = \bar{\mathfrak{p}}_5^{25} \bar{\mathfrak{p}}_7^{16} \bar{\mathfrak{p}}_{11}^3.$$

We then calculated the chain of isogenies corresponding to the right side of the equation. This calculation results in the curve $E'' : y^2 = x^3 + 44x + 24$ which has the same j-invariant as E'.

8 Conclusion

We described the fastest known quantum method for computing isogenies between supersingular curves, both for the general case and when the curves are defined over $\mathbb{F}_p$. In the general case, the quantum complexity of our attack is $\tilde{O}(p^{1/4})$. Some cryptographic applications of our work include a faster quantum preimage attack against the Charles-Goren-Lauter hash function, and a recommendation to avoid using base curves defined over $\mathbb{F}_p$ in De Feo-Jao-Plût type schemes.

Acknowledgments. The first author thanks Luca De Feo for helpful discussions on the quantum safe protocols based on isogenies between supersingular curves described in [10]. This work was supported by an NSERC Discovery Grant.

References

1. Bosma, W., Cannon, J., Playoust, C.: The Magma algebra system. I. the user language. J. Symbolic Comput. **24**(3–4), 235–265 (1997)
2. Boyer, M., Brassard, G.: P. Høyer, and A. Tapp. Tight bounds on quantum searching. Fortschritte Der Physik **46**, 493–505 (1998)
3. Bröker, R.: Constructing supersingular elliptic curves. J. Comb. Number Theory **1**(3), 269–273 (2009)
4. Bröker, R., Charles, D., Lauter, K.: Evaluating Large Degree Isogenies and Applications to Pairing Based Cryptography. In: Galbraith, S.D., Paterson, K.G. (eds.) Pairing 2008. LNCS, vol. 5209, pp. 100–112. Springer, Heidelberg (2008)
5. Charles, D., Lauter, K., Goren, E.: Cryptographic hash functions from expander graphs. Jornal of Cryptology **22**, 93–113 (2009)
6. Childs, A., Jao, D., Soukharev, V.: Constructing elliptic curve isogenies in quantum subexponential time. Journal of Mathematical Cryptology **8**(1), 1–29 (2013)
7. Couveignes, J.-M.: Hard homgeneous spaces. http://eprint.iacr.org/2006/291
8. Cox, D. A.: Primes of the form $x^2 + ny^2$. John Wiley & Sons (1989)
9. Delfs, C., Galbraith, S.: Computing isogenies between supersingular elliptic curves over $\mathbb{F}_p$. The Proceedings of the 11th Algorithmic Nnumber Theory Symposium (ANTS XI) (to appear)
10. De Feo, L., Jao, D., Plût, J.: Towards quantum-resistant cryptosystems from supersingular elliptic curve isogenies. Journal of Mathematical Cryptology (to appear, 2014). http://eprint.iacr.org/2011/506
11. Galbraith, S.: Constructing isogenies between elliptic curves over finite fields. LMS Journal of Computation and Mathematics 2, 118–138 (1999)
12. Galbraith, S., Stolbunov, A.: Improved algorithm for the isogeny problem for ordinary elliptic curves. Applicable Algebra in Engineering, Communication and Computing **24**(2), 107–131 (2013)
13. Grover, L.: A fast quantum mechanical algorithm for database search. In: Proceedings of the Twenty-eighth Annual ACM Symposium on Theory of Computing, STOC 1996, pp. 212–219. ACM, New York (1996)
14. Littlewood, J.: On the class number of the corpus $p(\sqrt{k})$. Proc. London Math. Soc. 27, 358–372 (1928)
15. Jao, D., De Feo, L.: Towards Quantum-Resistant Cryptosystems from Supersingular Elliptic Curve Isogenies. In: Yang, B.-Y. (ed.) PQCrypto 2011. LNCS, vol. 7071, pp. 19–34. Springer, Heidelberg (2011)
16. Jao, D., Miller, S.D., Venkatesan, R.: Expander graphs based on GRH with an application to elliptic curve cryptography. J. Number Theory **129**(6), 1491–1504 (2009)

17. Jao, D., Miller, S.D., Venkatesan, R.: Do All Elliptic Curves of the Same Order Have the Same Difficulty of Discrete Log? In: Roy, B. (ed.) ASIACRYPT 2005. LNCS, vol. 3788, pp. 21–40. Springer, Heidelberg (2005)
18. Jao, D., Soukharev, V.: Isogeny-Based Quantum-Resistant Undeniable Signatures. In: Mosca, M. (ed.) PQCrypto 2014. LNCS, vol. 8772, pp. 160–179. Springer, Heidelberg (2014)
19. Kuperberg, G.: A subexponential-time quantum algorithm for the dihedral hidden subgroup problem. SIAM J. Comput. **35**(1), 170–188 (2005)
20. MAGMA Computational Algebra System. http://magma.maths.usyd.edu.au/
21. Regev, O.: A subexponential time algorithm for the dihedral hidden subgroup problem with polynomial space. arXiv:quant-ph/0406151
22. Rostovtsev, A., Stolbunov, A.: Public-key cryptosystem based on isogenies. IACR Cryptology ePrint Archive **2006**, 145 (2006)
23. Schoof, R.: Counting points on elliptic curves over finite fields. Journal de théorie des nombres de Bordeaux **7**, 219–254 (1995)
24. Seiichiro, T.: Claw finding algorithms using quantum walk. Theoretical Computer Science 410(50), 5285–5297 (2009), Mathematical Foundations of Computer Science (MFCS 2007)
25. Silverman, J.: The arithmetic of elliptic curves, vol. 106. Graduate texts in Mathematics. Springer (1992)
26. Stolbunov, A.: Constructing public-key cryptographic schemes based on class group action on a set of isogenous elliptic curves. Adv. in Math. of Comm. **4**(2), 215–235 (2010)
27. Sutherland, A.: `smoothrelation`. http://math.mit.edu/~drew/smooth_relation_v1.2.tar.
28. Tate, J.: Endomoprhisms of abelian varieties over finite fields. Inventiones Mathematica **2**, 134–144 (1966)
29. Vélu, J.: Isogénies entre courbes elliptiques. C. R. Acad. Sci. Paris Sér. A-B, 273, A238–A241 (1971)

Author Index